Classical Mathematical Logic

Classical Mathematical Logic

The Semantic Foundations of Logic

Richard L. Epstein

with contributions by
Lesław W. Szczerba

Princeton University Press
Princeton and Oxford

Published by Princeton University Press, 41 William Street, Princeton,
New Jersey 08540.

In the United Kingdom: Princeton University Press, 3 Market Place,
Woodstock, Oxfordshire OX20 1SY.

Library of Congress Cataloging-in-Publication Data
Epstein, Richard L., 1947–
Classical mathematical logic: the semantic foundations of logic / Richard L. Epstein ; with contributions by Lesław W. Szczerba.
p. cm.
Includes bibliographical references and indexes.
ISBN-13: 978-0-691-12300-4 (cloth: acid-free paper)
ISBN-10: 0-691-12300-4 (cloth: acid-free paper)
1. Logic, Symbolic and mathematical. 2. Semantics (Philosophy) I. Title.

QA9.E67 2006
511.3--dc22

2005055239

British Library Cataloging-in-Publication Data is available.

This book has been composed in Times and the font QLogicPS created by the author.

The publisher would like to acknowledge the author of this volume for providing camera-ready copy from which this book was printed.

The moral rights of the author have been asserted.

Printed on acid-free paper. ∞

pup.princeton.edu

Printed in the United States of America

10 9 8 7 6 5 4 3 2 1

Dedicated to my love, my wife

Carolyn Kernberger

who rescued an old pound dog.

Contents

VIII Functions

IX The Abstraction of Models

X Axiomatizing Classical Predicate Logic

XI The Number of Objects in the Universe of a Model

XII Formalizing Group Theory

XIII Linear Orderings

XIV Second-Order Classical Predicate Logic

XV The Natural Numbers

XVI The Integers and Rationals

XVII The Real Numbers

The Semantic Foundations of Logic

Contents of Other Volumes

Propositional Logics

The Basic Assumptions of Propositional Logic
Classical Propositional Logic
Relatedness Logic: The Subject Matter of a Proposition
A General Framework for Semantics for Propositional Logics
Dependence Logics
Modal Logics
Intuitionism
Many-Valued Logics
A Paraconsistent Logic: J_3
Translations Between Logics
The Semantic Foundations of Logic

Predicate Logic

Propositions and Propositional Logic
The Internal Structure of Propositions
A Formal Language for Predicate Logic
Semantics
The Logical Form of a Proposition
Identity
Quantifiers
Descriptive Names
Functions
Quantifying over Predicates: Second-Order Logic
Language, the World, and Predicate Logic

Logics of Sense and Reference (projected)

Preface

Mathematics has always been taken as an exemplar of careful reasoning. Yet in the 19th century serious problems arose about how to reason in mathematics, principally in trying to resolve the nature of the infinitesimal calculus, as I describe in the Introduction below. The reasoning appropriate to mathematics became itself a subject of study for mathematicians.

Formal languages were devised to replace the speech of mathematicians. Those languages were based on the modern mathematicians' view of the world as made up of things and propositions as being assertions about relationships among things. Formal semantics—meanings for the languages—were devised to investigate inferences, where only valid inferences were deemed acceptable. In keeping with mathematics as the principal area of application of the logic, the greatest possible abstraction was made from ordinary reasoning in establishing the semantics. Being the simplest possible formalization of reasoning consonant with the assumption that the world is made up of things is what makes the subject classical.

In this volume I carefully trace the path of abstraction that leads to classical mathematical logic in order to clarify the scope and limitations of the subject. In the Introduction I outline that. In Chapters I–XI I set out the language and semantics of the logic and give an axiomatization of the semantics.

Chapters XII–XIX then present applications of the resulting logic to mathematics, beginning with the most abstract part of mathematics, group theory, and progressing through the study of linear orderings, the natural numbers, the integers, the rationals, and finally the real numbers. The real numbers can be understood axiomatically (Chapter XVII), as constructed from the rationals (Appendix to Chapter XVII), or geometrically (Chapters XVIII and XIX). The formalization of axiomatic geometry and its relationship to the formalization of the real numbers is important because it requires a careful analysis of how theories of different subjects in different languages are related, which is the subject of Chapter XX on translations between theories.

By paying attention to differences that were previously ignored along the path of abstraction, we can devise models of reasoning that take into account more of the world and our reasoning, for, above all, logic is the art of reasoning well. Though much of that project is for a subsequent volume, in this volume I present two examples. In Chapter XXI I show how we can relax the assumption that every name must refer to an object. In Chapter XXII I show how by no longer considering propositions to be types we can lift the restriction that we cannot reason about our logic within our logic.

This volume is meant to be a self-contained textbook, requiring as background only some facility in mathematical reasoning but mastery of no particular area of mathematics or logic. To that end, I have repeated parts of previous volumes of the series *The Semantic Foundations of Logic*, particularly in Chapters I–VIII and XIV. What is lacking from this text, and which is covered in *Predicate Logic*, is the extent to which we can use this logic to formalize reasoning from ordinary language.

Exercises to master the material are included, with underlined exercises containing material that is either inessential to mastery of the text or is unusually difficult. Sections printed in smaller type are subordinate to the main story of the text and may be skipped.

Socorro, New Mexico
2005

Acknowledgments

Parts of Chapters I–VIII and XIV appeared in different form in earlier volumes of *The Semantic Foundations of Logic* and are reprinted here by permission of Wadsworth Publishing Company.

I am grateful to Fernando Ferreira and John Burgess for help on my analysis of second-order logic in Chapter XIV.

I am indebted to Douglas Burke, George Weaver, and Peter Eggenberger for comments on a draft of the material on the logic of quantifying over names.

I am grateful to Fred Kroon for offering many suggestions for improving Chapter XXI on free logic. Parts of that material appear in *Loqique and Analyse* and are reprinted here with permission.

Much of the material in Chapter XXII first appeared as "A theory of truth based on a medieval solution to the liar paradox" in *History and Philosophy of Logic*, vol. 13, and is reprinted here with permission. I am grateful to Benson Mates, George Hughes, and Peter Eggenberger who helped me to formulate that theory, and to Walter Carnielli, Fred Kroon, Newton da Costa, John Corcoran, and Howard Blair for suggestions for improving it.

The photo of Ralph on p. 522 is by Serafina Kernberger, and I am grateful for her assistance.

I am also grateful to Lesław W. Szczerba for allowing his lectures and research on axiomatic geometry to be presented in Chapters XVIII and XIX. The mathematics is based on his book with Marek Kordos, *Geometria dla Nauczycieli*, as supplemented and presented by Szczerba in lectures at the University of California in 1973, Victoria University of Wellington in 1977, and at Iowa State University in 1981. His paper with Grzegorz M. Lewandowski, "An axiom system for the one-dimensional theory of the betweeness relation", forms the basis of Chapter XVIII.B. Due to illness, Professor Szczerba has not been able to proofread the version of his work here, so any mistakes should be ascribed solely to me.

Julianna Bueno, Walter Carnielli, Rodrigo Freire, Steven Givant, Mircea Tobosaru, and Bogdan Udrea read portions of an earlier draft and I am glad for the suggestions they made. Phil Watson read the entire penultimate draft and made so many good suggestions that I will be long indebted to him.

To all these people, and any others I have inadvertently forgotten, I am most grateful. Much that is good in this book is due to them; the mistakes and confusions are mine alone. It is with great pleasure I thank them here.

Introduction

Classical mathematical logic is an outgrowth of several trends in the 19th century.

In the early part of the 19th century there was a renewed interest in formal logic. Since at least the publication of *Logic or the Art of Thinking* by Antoine Arnauld and Pierre Nicole in 1662, formal logic had meant merely the study of the Aristotelian syllogisms. Richard Whately, *1827*, and others began a more serious study of that subject, though still far below the level of clarity and subtlety of the medieval logicians (compare Chapter XXII).

Shortly thereafter, George Boole, Augustus De Morgan and others began a mathematization of reasoning. This was part of a general trend of abstraction in mathematics. Using algebraic techniques, *Boole, 1854*, treated the study of Aristotelian syllogisms mathematically; *De Morgan, 1847*, showed the limitations of the Aristotelian tradition in dealing with inferences that depended on relations.

At roughly the same time, problems working with the calculus, particularly justifications and theorems about approximation procedures, prompted many mathematicians to re-examine their conception of the real numbers. Richard Dedekind, *1872*, and Karl Weierstrass offered new analyses, which were elaborated in the work of Georg Cantor, *1895–1897*, on theories of collections—now called 'sets' (Chapter IX.B). Defining the real numbers in terms of infinite collections or limits of infinite sequences of rationals (Chapter XVII) led to a view of mathematics as the study of relations among objects. Functions were understood entirely as relations, with the idea of functions as processes entirely abandoned. Two kinds of paradoxes resulted from this work. The set-theoretic paradoxes were those that led to a contradiction from postulating that there is a set of all sets, or that there is a set all of whose elements are not elements of themselves. These showed that the intuition that any objects whatsoever could be brought together as a collection was faulty. Another kind of paradox came from the definitions of the real numbers and the view of functions as relations. Constructions were given of a continuous function that is nowhere differentiable and of a space-filling curve.

The discovery of non-Euclidean geometries led to a greater interest in the procedure of axiomatizing. Starting with a basic system of axioms, one could obtain different geometries depending on whether one took Euclid's parallel postulate or one of two postulates that contradicted it. By the end of the 19th century Giuseppe Peano, *1889*, had set out his axiomatization of the natural numbers and David Hilbert, *1899*, had perfected an axiomatization of plane geometry.

In the early years of the 20th century Alfred North Whitehead and Bertrand Russell, *1910–1913*, brought together many of these trends. They showed how to axiomatize logic, and within that axiomatization, apparently avoiding paradoxes related to the set-theoretic ones, they were able to develop a very substantial part of mathematics.

Leopold Löwenheim and others, on the other hand, began to develop the idea of a model of an axiom system, with models for the different kinds of plane geometry as archetype.

It was only in the 1930s with the work of Alfred Tarski, Kurt Gödel, Hilbert, and many others that something recognizably like the work in this text was developed. (See *van Heijenoort, 1967* for seminal works by these and others, along with introductory essays on those works.)

The word 'mathematical' in 'classical mathematical logic', then, had two meanings: the mathematization of models of reasoning, and the use of formal logic to formalize reasoning in mathematics.

This development of formal logic stood aside from considerations of other parts of Aristotle's *Organon.* The study of reasoning with non-valid inferences, the analysis of cause-and-effect reasoning, and the use of Aristotelian logic for reasoning about masses and qualities were studied by John Stuart Mill, *1874*, and others, but were not of interest to mathematical logicians. Even the use of formal logic for the analysis of reasoning in ordinary language was almost completely ignored. (Those subjects and their relation to formal logic are presented in *Epstein, 1994* and *Epstein, 2001*.)

The orientation of formal logic, following the mathematicians' lead, was in terms of reasoning about things. Processes were considered fully analyzed by casting them in terms of relations (Chapter VIII), and within the mathematical community the paradoxes of the calculus came to be seen as natural consequences of fully satisfactory definitions. The notions of qualities and predicates came to be highly restricted, being identified in formal logic with their extensions (Chapter IX). The assumption that reasoning was to be restricted to reasoning about things (Chapter III) came to dominate to such an extent that it was no longer explicitly mentioned. The 'classical' in 'classical mathematical logic' in the 20th century came to mean the extreme restriction of the notion of predicate to just its extension: those objects of which it is true. This was in opposition to other views of reasoning in which epistemological, psychological, modal, or other aspects of propositions and predicates were taken into account (see *Epstein, 1990*).

In this text we we will see the basic outlines of the limitations as well as the scope of modeling reasoning within classical mathematical logic.

I Classical Propositional Logic

To begin our analysis of reasoning we need to be clear about what kind of thing is true or false. Then we will look at ways to reason with combinations of those things.

A. Propositions

When we argue, when we prove, we do so in a language. And we seem to be able to confine ourselves to declarative sentences in our reasoning.

I will assume that what a sentence is and what a declarative sentence is are well enough understood by us to be taken as *primitive*, that is, undefined in terms of any other fundamental notions or concepts. Disagreements about some particular examples may arise and need to be resolved, but our common understanding of what a declarative sentence is will generally suffice.

So we begin with sentences, written (or uttered) concatenations of inscriptions (or sounds). To study these we may ignore certain aspects, such as what color ink they are written in, leaving ourselves only certain features of sentences to consider in reasoning. The most fundamental is whether they are true or false.

In general we understand well enough what it means for a simple sentence such as 'Ralph is a dog' to be true or to be false. For such sentences we can regard truth as a primitive notion, one we understand how to use in most applications, while falsity we can understand as the opposite of truth, the not-true. Our goal is to formalize truth and falsity for more complex and controversial sentences.

Which declarative sentences are true or false, that is, have a *truth-value*? It is sufficient for our purposes in logic to ask whether we can agree that a particular sentence, or class of sentences as in a formal language, is declarative and whether it is appropriate for us to assume it has a truth-value. If we cannot agree that a certain sentence such as 'The King of France is bald' has a truth-value, then we cannot reason together using it. That does not mean that we adopt different logics or that logic is psychological; it only means that we differ on certain cases.

Propositions A *proposition* is a written or uttered declarative sentence used in such a way that it is true or false, but not both.

Other views of propositions

There are other views of what propositions are. Some say that what is true or false is not the sentence, but the meaning or thought expressed by the sentence. Thus 'Ralph is a dog' is not a proposition; it expresses one, the very same one expressed by 'Ralph is a domestic canine'.

Platonists take this one step further. A *platonist*, as I use the term, is someone who believes that there are abstract objects not perceptible to our senses that exist independently of us. Such objects can be perceived by us only through our intellect. The independence and timeless existence of such objects account for objectivity in logic and mathematics. In particular, propositions are abstract objects, and a proposition is true or is false, though not both, independently of our even knowing of its existence.

But a platonist, as a much as a person who thinks a proposition is the meaning of a sentence or a thought, reasons in language, using declarative sentences that, they say, represent, or express, or point to propositions. To reason with a platonist it is not necessary that I believe in abstract propositions or thoughts or meanings. It is enough that we can agree that certain sentences are, or from the platonist's viewpoint represent, propositions. Whether for the platonist such a sentence expresses a true proposition or a false proposition is much the same question as whether, from my point of view, it is true or is false.[1]

---

[1] See Example 2 of Chapter XXII for a fuller comparison of the platonist view of propositions to the one adopted here. In my *Propositional Logics* I define a proposition to be a written or uttered declarative sentence that we agree to view as being true or false, but not both. The discussion in that book shows that definition to be equivalent to the one here.

B. Types

When we reason together, we assume that words will continue to be used in the same way. That assumption is so embedded in our use of language that it's hard to think of a word except as a *type*, that is, as a representative of inscriptions that look the same and utterances that sound the same. I don't know how to make precise what we mean by 'look the same' or 'sound the same'. But we know well enough in writing and conversation what it means for two inscriptions or utterances to be *equiform*.

Words are types We will assume that throughout any particular discussion equiform words will have the same properties of interest to logic. We therefore identify them and treat them as the same word. Briefly, *a word is a type*.

This assumption, while useful, rules out many sentences we can and do reason with quite well. Consider 'Rose rose and picked a rose.' If words are types, we have to distinguish the three equiform inscriptions in this sentence, perhaps as 'Rose$_{\text{name}}$ rose$_{\text{verb}}$ and picked a rose$_{\text{noun}}$'.

Further, if we accept this agreement, we must avoid words such as 'I', 'my', 'now', or 'this', whose meaning or reference depends on the circumstances of their use. Such words, called *indexicals*, play an important role in reasoning, yet our demand that words be types requires that they be replaced by words that we can treat as uniform in meaning or reference throughout a discussion.

Suppose now that I write down a sentence that we take to be a proposition:

All dogs bark.

Later I want to use that sentence in an argument, say:

If all dogs bark, then Ralph barks. All dogs bark.
Therefore, Ralph barks.

But we haven't used the same sentence twice, for sentences are inscriptions.

Since words are types, we can argue that these two equiform sentences should both be true or both false. It doesn't matter to us where they're placed on the paper, or who said them, or when they were uttered. Their properties for logic depend only on what words (and punctuation) appear in them in what order. Perhaps it is a simplification, but let's agree that any property that differentiates them won't be of concern to our reasoning.

Propositions are types In any discussion in which we use logic we'll consider a sentence to be a proposition only if any other sentence or phrase that is composed of the same words in the same order can be assumed to have the same properties of concern to logic during that discussion. We therefore identify equiform sentences or phrases and treat them as the same sentence. Briefly, *a proposition is a type.*

It is important to identify both sentences and phrases, for in the inference above we want to identify the phrase 'all dogs bark' in 'If all dogs bark, then Ralph barks.' with 'All dogs bark.'.

The device I just used of putting *single quotation marks* around a word or phrase is a way of naming that word or phrase or any linguistic unit. We need some such convention because confusion can arise if it's not clear whether a word or phrase is being used or referred to as a word or phrase. For example, when I write:

The Taj Mahal has eleven letters

I don't mean that the building has eleven letters, but that the phrase does, and I should write:

'The Taj Mahal' has eleven letters

When we do this we say that we *mention* the word or phrase in quotation marks and that the entire inscription including the quotes is a *quotation name* of the word or phrase. Otherwise, we simply *use* the word or phrase, as we normally do. Note that the assumption that words are types is essential in using quotation names. We can also indicate we are mentioning a linguistic unit by italicizing it or putting it in display format. And single quotation marks are used for quoting direct speech, too.

The device of enclosing a word or phrase in double quotation marks is equivalent to a wink or a nod in conversation, indicating that we're not to be taken literally or that we don't really subscribe to what we're saying. Double quotes are called *scare quotes*, and they allow us to get away with "murder".

Exercises for Sections A and B

1. Of the following, (i) Which are declarative sentences? (ii) Which contain indexicals? (iii) Which are propositions?
   a. Ralph is a dog.
   b. I am 2 meters tall.
   c. Is any politician not corrupt?
   d. Feed Ralph.
   e. Ralph didn't see George.
   f. Whenever Juney barks, Ralph gets mad.
   g. If anyone should say that cats are nice, then he is confused.
   h. If Ralph should say that cats are nice, then he is confused.
   i. If Ralph should say that cats are nice, then Ralph is confused.
   j. Dogs can contract myxomatosis.
   k. $2 + 2 = 4$.
   l. $\dfrac{d\,e^x}{dx} = e^x$
2. Explain why we cannot take sentence types as propositions if we allow the use of indexicals in our reasoning.

C. The Connectives of Propositional Logic

We begin with propositions that are sentences. There are two features of such sentences that contribute to our reasoning: their *syntax*, by which we mean the analysis of their form or grammar, and their *semantics*, by which we mean an analysis of their truth-values and meaning or content. These are inextricably linked. The choice of what forms of propositions we'll study leads to what and how we can mean, and the meaning of the forms leads to which of those forms are acceptable.

There are so many properties of propositions we could take into account in our reasoning that we must begin by restricting our attention to only a few. Our starting point is *propositional logic* where we ignore the internal structure of propositions except as they are built from other propositions in specified ways.

There are many ways to connect propositions to form a new proposition. Some are easy to recognize and use. For example, 'Ralph is a dog and dogs bark' can be viewed as two sentences joined by the connective 'and'.

Some other common connectives are: 'but', 'or', 'although', 'while', 'only if' 'if . . . then . . .', 'neither . . . nor . . .'. We want to strike a balance between choosing as few as possible to concentrate on in order to simplify our semantic analyses, and as many as possible so that our analyses will be broadly applicable.

Our starting point will be the four traditional basic connectives of logic: 'and', 'or', 'if . . . then . . . ', 'not'. In English 'not' is used in many different ways. Here we take it as the connective that precedes a sentence as in *It's not the case that . . .* , or other uses that can be assimilated to that understanding. These four connectives will give us a rich enough grammatical basis to begin our logical investigation.

These English connectives have many connotations and properties, some of which may be of no concern to us in logic. For example, in American English 'not' is usually said more loudly than the surrounding words in the sentence. We'll replace these connectives with formal symbols to which we will give fairly explicit and precise meanings based on our understanding of the English words.

symbol	*what it will be an abstraction of*
$\wedge$	and
$\vee$	or
$\neg$	it's not the case that
$\rightarrow$	if . . . then . . .

Thus, a complex sentence we might study is 'Ralph is a dog $\wedge$ dogs bark', corresponding to the earlier example.

Parentheses are a further formal device important for clarity. In ordinary speech we might say, 'If George is a duck then Ralph is a dog and Dusty is a horse'. But it's not clear which of the following is meant:

If George is a duck, then: Ralph is a dog and Dusty is a horse.

If George is a duck, then Ralph is a dog; and Dusty is a horse.

Such ambiguity should have no place in our reasoning. We can require the use of parentheses to enforce one of those two readings:

George is a duck $\rightarrow$ (Ralph is a dog $\wedge$ Dusty is a horse)
(George is a duck $\rightarrow$ Ralph is a dog) $\wedge$ Dusty is a horse

To lessen the proliferation of quotation marks in naming linguistic items, we can assume that *a formal symbol names itself* when confusion seems unlikely. Thus, we can say that $\wedge$ is a formal connective. Here is some more terminology.

The sentence formed by joining two sentences by $\wedge$ is a *conjunction* of them; each of the original propositions is a *conjunct* we *conjoin* with the other. When $\vee$ joins two sentences it is a *disjunction* formed by *disjoining* the *disjuncts*. The proposition 'Ralph is a dog or Ralph does not bark' is not a disjunction, because the formal symbol '$\vee$' does not appear in it.

The sentence formed by putting $\neg$ in front of another sentence is the *negation* of it; it is *a negation*. For example, the negation of 'Ralph is a dog' is '$\neg$(Ralph is a dog)', where the parentheses are used for clarity.

The symbol $\rightarrow$ is called the *arrow* or the *conditional*. The result of joining two sentences with $\rightarrow$ is a *conditional*; the proposition on the left is called the *antecedent*, and the one on the right is the *consequent*.

Exercises for Section C

1. Classify each of the following as a conjunction, disjunction, negation, conditional, or none of those. If a conditional, identify the antecedent and consequent.
   a. Ralph is a dog $\wedge$ dogs bark.
   b. Ralph is a dog $\rightarrow$ dogs bark.
   c. $\neg$ (cats bark).
   d. Cats bark $\vee$ dogs bark.
   e. Cats are mammals and dogs are mammals.
   f. Either Ralph is a dog or Ralph isn't a dog.
   g. $\neg$ cats bark $\rightarrow$ ($\neg$ (cats are dogs)).
   h. Cats aren't nice.
   i. Dogs bark $\vee$ ($\neg$ dogs bark).
   j. Ralph is a dog.
   k. It is possible that Ralph is a dog.
   l. Some dogs are not white.
2. a. Write a sentence that is a negation of a conditional, the antecedent of which is a conjunction.
   b. Write a sentence that is a conjunction of disjunctions, each of whose disjuncts is either a negation or has no formal symbols in it.
3. Write a sentence that might occur in daily speech that is ambiguous but which can be made precise by the use of parentheses, indicating at least two ways to parse it.
4. List at least three words or phrases in English not discussed in the text that are used to form a proposition from one or more propositions and that you believe are important in a study of reasoning.

D. A Formal Language for Propositional Logic

1. Defining the formal language

We have no dictionary, no list of all propositions in English, nor do we have a method for generating all propositions, for English is not a fixed, formal, static language. But by using variables we can introduce a rigid formal language to make precise the syntax of the propositions we will study.

Let $p_0, p_1, \ldots$ be *propositional variables.* These can stand for any propositions, but the intention is that they'll be ones whose internal structure won't be under consideration. Our formal language is built from these using the connectives $\neg$, $\rightarrow$, $\wedge$, $\vee$ and parentheses. This will be the *formal language.* To be able to talk about that language, indeed even to give a precise definition of it, we need further variables, the *metavariables* A, B, C, A_0, A_1, $A_2, \ldots$ to stand for any of $p_0, p_1, p_2, \ldots$ or complex expressions formed from those, and p, q to stand for any of $p_0, p_1, p_2, \ldots$. The analogue of a sentence in English is a *well-formed formula*, or *wff.* We could define that by:

a. (p_i) is a wff for each $i = 0, 1, 2, \ldots$.

b. If A, B are wffs, then so are $(\neg A)$, $(A \rightarrow B)$, $(A \wedge B)$, and $(A \vee B)$.

c. Only such concatenations of symbols as arise from repeated applications of (a) and (b) are wffs.

But clause (c) is too vague. So instead we use an *inductive definition*, that is, one that correlates the structure of what is being defined to the natural numbers.

Wffs and the formal language $L(\neg, \rightarrow, \wedge, \vee, p_0, p_1, \ldots)$

i. For each $i = 0, 1, 2, \ldots$, (p_i) is an *atomic* wff, to which we assign the number 1.

ii. If A and B are wffs and the maximum of the numbers assigned to A and to B is *n*, then each of $(\neg A)$, $(A \rightarrow B)$, $(A \wedge B)$, $(A \vee B)$ is a *compound* wff to which we assign the number $n + 1$. These are *compound wffs.*

iii. A concatenation of symbols is a *wff* if and only if it is assigned some number $n \geq 1$ according to (i) or (ii) above.

We usually refer to this language as $L(\neg, \rightarrow, \wedge, \vee)$ since we usually use the same propositional variables.

Propositions are written or uttered sentences. Since we previously agreed to view words and propositions as types, let's now agree that the *symbols and formulas of the formal language are types*, too.

Now we need to know that there is only one way to parse each wff.

A platonist definition of the formal language

Platonists conceive of the formal language as a complete infinite collection of formulas. Infinite collections, they say, are real, though abstract, entities. So they define the formal language as:

> $L(\neg, \rightarrow, \wedge, \vee, p_0, p_1, \ldots)$ is the smallest collection containing (p_i) for each $i = 0, 1, 2, \ldots$ and closed under the formation of wffs, that is, if A, B are in the collection, so are $(\neg A)$, $(A \rightarrow B)$, $(A \wedge B)$, and $(A \vee B)$.

One collection is said to be smaller than another if it is contained in or equal to the other collection.

Platonists argue that their definition is clearer and more precise because it replaces condition (c) with strict criteria on collections. Whatever clarity is gained, however, depends on our accepting that there are completed infinite totalities.

2. The unique readability of wffs

We need to show that our definitions do not allow for a wff to be ambiguous. That is, no wff can be read as, for example, both $A \rightarrow B$ and $C \wedge D$.

Theorem 1 Unique readability of wffs
There is one and only one way to parse each wff.

Proof To prove this we will use a proof by induction; in an appendix to this chapter I present a short review of that method.

To each primitive symbol α of the formal language assign an integer $\lambda(\alpha)$ according to the following chart:

$\neg$	$\rightarrow$	$\wedge$	$\vee$	p_i	(	)
0	0	0	0	0	−1	1

To the concatenation of symbols $\alpha_1 \alpha_2 \cdots \alpha_n$ assign the number:

$$\lambda(\alpha_1 \alpha_2 \cdots \alpha_n) = \lambda(\alpha_1) + \lambda(\alpha_2) + \cdots + \lambda(\alpha_n)$$

First I'll show that for any wff A, $\lambda(A) = 0$, using induction on the number of occurrences of $\neg$, $\rightarrow$, $\wedge$, $\vee$ in A.

If there are no occurrences, then A is atomic, that is, for some $i \geq 0$, A is (p_i). Then $\lambda(\text{`('}) = -1$, $\lambda(p_i) = 0$, and $\lambda(\text{`)'}) = 1$. Adding, we have $\lambda(A) = 0$.

Suppose the lemma is true for every wff that has fewer occurrences of these symbols than A does. Then there are 4 cases, which we can't yet assume are distinct: A arises as $(\neg B)$, $(B \rightarrow C)$, $(B \wedge C)$, or $(B \vee C)$. By induction $\lambda(B) = \lambda(C) = 0$, so in each case by adding, we have $\lambda(A) = 0$.

Now I'll show that reading from the left, if α is an initial segment of a wff (reading from the left) other than the entire wff itself, then $\lambda(\alpha) < 0$; and if α is a final segment other than the entire wff itself, then $\lambda(\alpha) > 0$. So no proper initial or final segment of a wff is a wff. To establish this I will again use induction on the number of occurrences of connectives in the wff. I'll let you establish the base case for atomic wffs, where there are no (that is, zero) connectives.

Now suppose the lemma is true for any wff that contains $\leq n$ occurrences of the connectives. If A contains $n + 1$ occurrences, then it must have (at least) one of the forms given in the definition of wffs. If A has the form $(B \wedge C)$, then an initial segment of A must have one of the following forms:

i. (

ii. $(\beta$ where β is an initial segment of B

iii. $(B \wedge$

iv. $(B \wedge \gamma$ where γ is an initial segment of C

For (ii), $\lambda(\text{`('}) = -1$ and by induction $\lambda(\beta) < 0$, so $\lambda(\text{`(}\beta\text{'}) < 0$. I'll leave (i), (iii), and (iv) to you. The other cases (for $\neg$, $\rightarrow$, and $\vee$) follow similarly, and I'll leave those and the proof for final segments to you.

Now to establish the theorem we proceed through a number of cases by way of contradiction. Suppose we have a wff that could be read as both $(A \wedge B)$ and $(C \rightarrow D)$. Then $A \wedge B)$ must be the same as $C \rightarrow D)$. In that case either A is an initial part of C or C is an initial part of A. But then $\lambda(A) < 0$ or $\lambda(C) < 0$, which is a contradiction, as we proved above that $\lambda(A) = \lambda(C) = 0$. Hence, A is C. But then we have that $\wedge B)$ is the same as $\rightarrow D)$, which is a contradiction.

Suppose $(\neg A)$ could be parsed as $(C \rightarrow D)$. Then $(\neg A$ and $(C \rightarrow D$ must be the same. So D would be a final segment of A other than A itself, but then $\lambda(D) > 0$, which is a contradiction. The other cases are similar, and I'll leave them to you. ■

Theorem 1 shows that every wff has exactly one number assigned to it in the inductive definition of wffs. Hence we can make the following definition.

The length of a wff The length of a wff is the number assigned to it in the inductive definition of wffs.

For example:

The length of (p_0) is 1.

The length of $((p_1) \wedge (p_{32}))$ is 2.

The length of $(((p_0) \rightarrow (p_1)) \rightarrow ((p_1) \rightarrow (p_0)))$ is 3.

Now when we want to show that all wffs have some property we can use the following method of proof.

Induction on the length of wffs First show that the atomic wffs have the property and then show that if all wffs of length n have it, so do all wffs of length $n + 1$.

Excessive parentheses can make it difficult to read a formal wff, for example:

$$(((p_0) \rightarrow ((p_1) \wedge (p_{32}))) \rightarrow ((((p_{13}) \wedge (p_6)) \vee (p_{317})) \rightarrow (p_{26})))$$

So we use an informal convention to eliminate parentheses: $\neg$ binds more strongly than $\wedge$ and $\vee$, which bind more strongly than $\rightarrow$. Thus $\neg A \wedge B \rightarrow C$ is to be read as $(((\neg A) \wedge B) \rightarrow C)$. We can also dispense with the outermost parentheses and those surrounding a variable, and we can use [ ] in place of the usual parentheses in informal discussions. So the example above can be written informally as:

$$(p_0 \rightarrow p_1 \wedge p_{32}) \rightarrow ([(p_{13} \wedge p_6) \vee p_{317}] \rightarrow p_{26})$$

Now it is easier to see that the length of this wff is 5.

We can write extended conjunctions or disjunctions without parentheses if we adopt the convention that conjuncts or disjuncts are *associated to the left*. For example, $p_1 \wedge p_2 \wedge p_3 \wedge p_4$ abbreviates $((p_1 \wedge p_2) \wedge p_3) \wedge p_4$.

Formal wffs exemplify the forms of propositions. To talk about the forms of wffs of the formal language, we use *schema*, that is, formal wffs with the variables for propositions replaced by metavariables. These are the skeletons of wffs. For example, $p_{136} \vee \neg p_{136}$ is an *instance* of $A \vee \neg A$.

We can now give an inductive definition of *subformula*:

If C has length 1, then it has just one subformula, C itself.

If C has length $n + 1$ then:

- If C has the form $\neg A$, its subformulas are C, A, and the subformulas of A.
- If C has the form $A \wedge B$, $A \vee B$, or $A \rightarrow B$, its subformulas are C, A, B, and the subformulas of A and of B.

We say p_i *appears in* A if (p_i) is a subformula of A, and write 'A(p)' to mean that p appears in A, and 'C(A)' to mean that A is a subformula of C.

In some proofs about the language it is useful to have a numbering of all wffs that assigns distinct numbers to distinct wffs. There are many ways to number all wffs, and it won't matter for what follows which one we use. I include a particular one here for completeness.

Numbering of wffs

This numbering relies on the Fundamental Theorem of Arithmetic: for each natural number there is, except for the order of multiplication, one and only one way to factor it into primes. We first define a function from wffs to natural numbers:

$$f(p_n) = 2^n$$

If f(A) and f(B) are defined, then: $f(\neg A) = 3^{f(A)}$; $f(A \vee B) = 11^{f(A)} \cdot 13^{f(B)}$;
$f(A \wedge B) = 5^{f(A)} \cdot 7^{f(B)}$; $f(A \rightarrow B) = 17^{f(A)} \cdot 19^{f(B)}$.

Then we number all wffs by taking the nth wff to be the wff that is assigned the nth largest number that is an output of the function f.

3. Realizations

The formal language is meant to give us the forms of the propositions we'll analyze. No wff such as $p_0 \wedge \neg p_1$ is true or false; that is the form of a proposition, not a proposition. Only when we first fix on a particular interpretation of the formal connectives and then assign propositions to the variables, such as 'p_0' stands for 'Ralph is a dog' and 'p_1' stands for 'Four cats are sitting in a tree', do we have a semi-formal proposition, 'Ralph is a dog $\wedge$ $\neg$(four cats are sitting in a tree)', one that can have a truth-value. We may read this as 'Ralph is a dog and it's not the case that four cats are sitting in a tree' so long as we remember that we've agreed that all that 'and' and 'it's not the case that' mean will be captured by the interpretations we'll give for $\wedge$ and $\neg$.

Realizations A *realization* is an assignment of propositions to some or all of the propositional variables. The *realization of a formal wff* is the formula we get when we replace the propositional variables appearing in the formal wff with the propositions assigned to them; it is a *semi-formal wff*. The *semi-formal language* for that realization is the collection of realizations of formal wffs all of whose propositional variables are realized. We call the semi-formal language of a realization *semi-formal English* or *formalized English*. I will use the same metavariables A, B, C, A_0, A_1, A_2, . . . for semi-formal wffs.

For example, we could take as a realization the following assignment of propositions to variables:

(1)

p_0	'Ralph is a dog'
p_1	'Four cats are sitting in a tree'
p_2	'Four is a lucky number'
p_3	'Dogs bark'
p_4	'Juney is barking loudly'
p_5	'Juney is barking'
p_6	'Dogs bark'
p_7	'Ralph is barking'
p_8	'Cats are nasty'
p_9	'Ralph barks'
p_{47}	'Howie is a cat'
p_{301}	'Ralph will catch a cat'
p_{312}	'Bill is afraid of dogs'
p_{317}	'Bill is walking quickly'
p_{4318}	'If Ralph is barking, then Ralph will catch a cat'
p_{4719}	'Ralph is barking'

Some of the semi-formal wffs of this realization are:

Ralph is a dog $\wedge$ $\neg$(four cats are sitting in a tree)

Bill is afraid of dogs $\wedge$ Ralph barks $\rightarrow$ Bill is walking quickly

Juney is barking loudly $\rightarrow$ Juney is barking

Ralph is a dog $\wedge$ dogs bark $\rightarrow$ Ralph barks

Actually, these are only abbreviations of semi-formal wffs, using the conventions for informally deleting parentheses. The second, for example, is an abbreviation of the realization of $(((p_{312}) \wedge (p_9)) \rightarrow (p_{317}))$. Note that not all variables need be realized. Nor need we assign distinct propositions to distinct variables. Here the realization of p_7 is the same as the realization of p_{4719}. And note well that a realization is linguistic, a formalized fragment of English.

The propositions that we assign to the variables are *atomic*, that is, they are the simplest propositions of the realization, ones whose internal structure will not be under consideration in this particular analysis. So the assignment of 'If Ralph is barking, then he will catch a cat' to p_{4318} is a bad choice, for it will not allow us to make a semantic analysis based on the form of that proposition in relation to 'Ralph is barking'. Generally speaking, the atomic propositions of a realization shouldn't include any English expressions that we've agreed to formalize with our formal connectives.

Before we can speak of the realizations of compound wffs as propositions we must decide upon interpretations of the formal connectives.

Exercises for Section D

1. Why do we introduce a formal language?
2. Identify which of the following are formal (unabbreviated) wffs:
   a. $(p_1) \vee \neg(p_2)$
   b. $((p_1) \rightarrow (p_2))$
   c. $((p_1 \vee p_2) \rightarrow p_2)$
   d. $(\neg(p_1)(p_2) \wedge (p_1))$
   e. $(\neg(\neg(p_1)) \vee \neg(p_1))$
   f. $((\neg(\neg(p_1))) \vee \neg(p_1))$
   g. $((\neg(\neg(p_1))) \vee (\neg(p_1)))$
3. Abbreviate the following wffs according to our conventions on abbreviations:
   a. $((((p_1) \rightarrow (p_2)) \wedge (\neg(p_2))) \rightarrow (\neg(p_1)))$
   b. $((((p_4) \wedge (p_2)) \vee (\neg(p_6))) \rightarrow ((p_7) \rightarrow (p_8)))$
4. Write the following informally presented wffs without abbreviations:
   a. $(p_1 \vee p_2)$
   b. $(p_0 \wedge \neg p_1)$
   c. $(p_0 \wedge \neg\neg p_1)$
   d. $(p_1 \rightarrow p_2)$
   e. $(p_1 \rightarrow (p_2 \rightarrow p_3))$
   f. $(\neg p_1)$
   g. $(\neg\neg p_1)$
   h. $(\neg\neg\neg\neg p_1)$
5. Give an example (using abbreviations) of a formula that is:
   a. A conjunction, the conjuncts of which are disjunctions of either atomic propositions or negated atomic propositions.

b. A conditional whose antecedent is a disjunction of negations and whose consequent is a conditional whose consequent is a conditional.

6. a. Calculate the length of the following wffs.

   i. $((((p_1) \rightarrow (p_2)) \wedge (\neg(p_2))) \rightarrow (\neg(p_1)))$

   ii. $((((p_4) \wedge (p_2)) \vee (\neg(p_6))) \rightarrow (\neg((p_7) \rightarrow (p_8))))$

   b. Give an unabbreviated example of a wff of length 5. For a wff of length 5 what is the minimum number of parentheses that can be used? The maximum? For a wff of length 10? (Hint: See Exercise 5 on p. 26 below.)

7. List all subformulas of each of the wffs in Exercise 6.a. What propositional variables appear in them?

8. Distinguish the following: ordinary language, the formal language, a semi-formal language.

9. a. Give the realization of the following, using (1) on p. 11:

   i. $((p_8 \wedge p_{4318}) \wedge p_7) \rightarrow p_1$
   ii. $(p_0 \wedge p_1) \rightarrow p_2$
   iii. $\neg(p_4 \wedge \neg p_5)$
   iv. $p_3 \rightarrow \neg\neg p_6$
   v. $\neg(p_{312} \wedge p_7) \wedge \neg p_{317}$
   vi. $p_{312} \wedge p_7 \rightarrow \neg p_{317}$

   b. The following wff is the realization of what formal wff in realization (1)?

   (Four cats are sitting in a tree ∧ four is a lucky number) →
   ¬(If Ralph is barking then he will catch a cat → Howie is a cat)

   c. Exhibit formal wffs of which the following could be taken to be realizations:

   i. Cats are nasty → Howie is nasty

   ii. ¬((Ralph is a dog ∨ ¬¬Ralph barks) ∨ Ralph is a puppet) →
   no number greater than 4 billion is a perfect square

E. Classical Propositional Logic

1. The classical abstraction and truth-functions

What are the simplest interpretations we can provide for the formal connectives that will be consonant with the assumptions we've made? We ignore every aspect of propositions other than their form and the single semantic property that makes them propositions.

The classical abstraction The only properties of a proposition that matter to logic are its form and its truth-value.

With the classical abstraction, the truth-value of a compound proposition can depend only on the connectives appearing in it and the truth-values of its parts. If that weren't the case, then the truth-values of 'Ralph is a dog' and of 'cats are nasty' wouldn't determine the truth-value of 'Ralph is a dog ∧ cats are nasty'. But if those truth-values don't, and we've agreed that no other property of these propositions matters, what could determine the truth-value of the compound proposition? If there is something nonfunctional, transcendent, that occurs when, say, $\wedge$ connects two sentences, how are we to reason? From the truth of one proposition how could we deduce the truth of another? Without some regularity reasoning cannot take place.

The Fregean assumption The truth-value of a proposition is determined by its form and the semantic properties of its constituents.

So if we join two atomic propositions with $\wedge$, $\vee$, $\rightarrow$, or place $\neg$ in front of one, the truth-value of the resulting proposition must depend on only the truth-values of those atomic propositions. The connectives must operate semantically as functions of the truth-values of the constituent propositions. This is what we mean when we say they are *truth-functional.*

Which truth-functions correspond to our connectives? I'll let you convince yourself that the only reasonable choices for $\neg$ and $\wedge$ are given by the following tables, where p and q stand for atomic propositions, and 'T' and 'F' stand for 'true' and 'false'.

p	$\neg$p
T	F
F	T

p	q	p$\wedge$q
T	T	T
T	F	F
F	T	F
F	F	F

That is, if p is T, then $\neg$p is F; if p is F, then $\neg$p is T. And p$\wedge$q is T if both p and q are T; otherwise it is F.

For $\vee$ there are two choices, corresponding to an inclusive or exclusive reading of 'or' in English. The choice is arbitrary, but it's customary to use the inclusive version, as in 'p or q or both'.

p	q	p$\vee$q
T	T	T
T	F	T
F	T	T
F	F	F

The hardest choice, however, is the table for $\rightarrow$. For that we use:

p	q	p$\rightarrow$q
T	T	T
T	F	F
F	T	T
F	F	T

The second row is the essence of 'if . . . then . . .', namely, a true conditional does not have a true antecedent and a false consequent. The first row is also fundamental, so long as we ignore all aspects of atomic propositions except their truth-values.

For the last two rows, consider that if both were to result in F, then $p \rightarrow q$ would be evaluated the same as $p \wedge q$. If we were to take the third row as T and the fourth as F, then $p \rightarrow q$ would be the same as q; were the third row F and the last T, then $p \rightarrow q$ would be the same as $q \rightarrow p$. Rejecting these counterintuitive choices, we are left with the table that is most generous in assigning T to conditional propositions: $p \rightarrow q$ is true so long as it's not the case that p is T and q is F.

An example from mathematics will illustrate why we want to classify conditionals with false antecedent as true. I think you'll agree the following is true of the counting numbers:

If x and y are odd, then the sum of x and y is even.

If we were to take either of the last two rows of the table for '$\rightarrow$' to be F, the formalization of this proposition would be false, for we could provide false instances of it: $4 + 8 = 12$, which is even, and $4 + 7 = 11$, which is not even. The formalization of 'if . . . then . . .' we have chosen allows us to deal with cases where the antecedent "does not apply" by treating them as vacuously true.

Note that I've used 'and', 'or', 'not', and 'if . . . then . . .' to explain the meaning of the formal connectives. This isn't circular. We are not defining or giving meaning to these English words, but to $\neg$, $\rightarrow$, $\wedge$, $\vee$. I must assume you understand the ordinary English connectives.

Given any atomic propositions p and q, we now have a semantic analysis of $\neg p$, $p \rightarrow q$, $p \wedge q$, and $p \vee q$. But what about:

(2) (Ralph is a dog $\wedge$ dogs bark) $\rightarrow$ Ralph barks

Here the antecedent is not a formless entity that is simply true or false: It contains a formal connective that must be accounted for. How shall we analyze this conditional compared to the following?

(3) (Ralph is a dog $\rightarrow$ dogs bark) $\rightarrow$ Ralph barks

Certainly the antecedents of (2) and (3), with the forms $p \wedge q$ and $p \rightarrow q$, are to be evaluated by different methods. But if both are evaluated as true or both as false, is there anything that can distinguish them semantically? We have agreed that the only semantic value of a proposition that we shall consider is its truth-value. So if two propositions have the same truth-value, they are semantically indistinguishable and should play the same role in any further semantic analysis.

The division of form and content If two propositions have the same semantic properties, then they are indistinguishable in any semantic analysis, regardless of their form.

Thus to evaluate (2), we first determine the truth-value of 'Ralph is a dog $\wedge$ dogs bark', and then only the truth-value of that proposition and the truth-value of 'Ralph barks' matter in determining the truth-value of the whole using the table for $\rightarrow$. We proceed in the same manner in determining the truth-value of (3), so that if 'Ralph is a dog $\wedge$ dogs bark' and 'Ralph is a dog $\rightarrow$ dogs bark' have the same truth-value, then the two conditionals (2) and (3) will have the same truth-value.

Adopting the division of form and content, the tables for $\neg$, $\rightarrow$, $\wedge$, and $\vee$ above apply to all propositions.

The classical truth-tables **(*classical evaluations of the connectives*)**

A	$\neg$A
T	F
F	T

A	B	A $\wedge$ B
T	T	T
T	F	F
F	T	F
F	F	F

A	B	A $\vee$ B
T	T	T
T	F	T
F	T	T
F	F	F

A	B	A $\rightarrow$ B
T	T	T
T	F	F
F	T	T
F	F	T

Suppose now that we are given a realization, say (1) above. We agree that the sentences assigned to the propositional variables are propositions. In addition we have a method for assigning truth-values to compound formulas in terms of the truth-values of their parts. Are we then to assume that every well-formed formula of the semi-formal language is a proposition? Consider:

$\neg\neg\neg\neg\neg\neg\neg\neg\neg\neg\neg\neg\neg\neg\neg\neg\neg$ (Ralph is a dog)

We can take this to be a formalization of a common language proposition, unlikely as it is that we would ever use it in our reasoning. But what if there are 613 $\neg$'s in front of '(Ralph is a dog)'?

It seems a harmless idealization to accept such a formula as a proposition, even though we might not ever use it. And to try to establish further criteria for what formulas of the semi-formal language are to be taken as propositions more than to say that they are well-formed (or perhaps shorter in length than some specific limit) involves us in a further semantic analysis not already taken into account by the assumptions we have adopted. So we simplify our work by adopting the following idealization.

Form and meaningfulness What is grammatical and meaningful is determined solely by form and what primitive parts of speech are taken as meaningful. In particular, given a semi-formal propositional language, every well-formed formula will be taken to be a proposition.

Thus, the semi-formal language of a realization is the collection of propositions formed from the realizations of the propositional variables on the pattern of the formulas of the formal language.

2. Models

Suppose we have a realization, say (1) above. We agree that the sentences assigned to the propositional variables, for example, 'Ralph is a dog' for p_0, are propositions, that is, each has a truth-value. But it is a further agreement to say which truth-value it has. When we do that we call all the assignments of specific truth-values together a *propositional valuation*, υ, and write $\upsilon(p) = T$ or $\upsilon(p) = F$ according to whether the atomic proposition p is taken to be true or false.

Since there is only one way to parse each semi-formal wff, the valuation υ plus the truth-tables determine the truth-value of every compound proposition of the semi-formal language, as you can show. So for any semi-formal wff A we can write $\upsilon(A) = T$ or $\upsilon(A) = F$ if A is assigned T or is assigned F by this method.

Models A *model* is a realization of the propositional variables as propositions, an assignment of truth-values to those atomic propositions, and the extension of that assignment to all formulas of the semi-formal language via the classical truth-tables. If $\upsilon(A) = T$, A is *true in the model*, and if $\upsilon(A) = F$, A is *false* in the model.

A formal wff is true or false in the model according to whether its realization is true or false.

Model Type I

$L(\neg, \rightarrow, \wedge, \vee, p_0, p_1, \ldots)$

↓ } realization

{real(p_0), real(p_1), . . . , propositions formed from these using $\neg, \rightarrow, \wedge, \vee$}

↓ } υ plus truth-tables

{T, F}

Exercises for Sections E.1 and E.2

1. Why do we take $A \rightarrow B$ to be true if A is false?
2. What is a truth-functional connective?
3. For each semi-formal wff below, find an assignment of truth-values to the atomic propositions in realization (1) that will make it have truth-value T, if that is possible.
   a. Ralph is barking $\rightarrow$ cats are nasty
   b. (Ralph is a dog $\vee$ dogs bark) $\wedge$ $\neg$(Juney is barking)
   c. Ralph is barking $\wedge$ cats are nasty $\rightarrow$ Ralph is barking
   d. Dogs bark $\vee$ $\neg$ (dogs bark) [as a realization of $p_6 \vee \neg(p_6)$]
   e. Dogs bark $\vee$ $\neg$ (dogs bark) [as a realization of $p_6 \vee \neg(p_3)$]
   f. $\neg$(cats are nasty $\rightarrow$ $\neg\neg$ (cats are nasty))
   g. ((Ralph is a dog $\vee$ dogs bark) $\wedge$ $\neg$(Ralph is dog)) $\rightarrow$ dogs bark

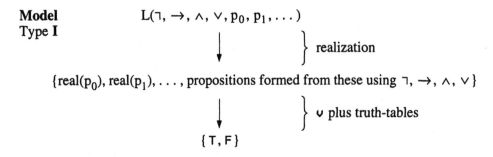

4. Distinguish between a realization and a model.
5. Show by induction on the length of wffs that in a model the valuation ν plus the truth-tables determine uniquely the truth-value of every compound proposition.

3. Validity and semantic consequence

Some propositions we know are true from our experience, such as 'The Eiffel Tower is in Paris.' The truth of others, though, follows solely from the assumptions we've made about the meaning of the logical connectives. For example, no matter what truth-value we assign to 'Ralph is a dog', '(Ralph is a dog) $\vee$ $\neg$(Ralph is a dog)' will be evaluated as true. Similarly, in any model 'Ralph is a dog $\wedge$ Bill is walking quickly $\rightarrow$ Ralph is a dog' is true. Turning to formal wffs for generality and precision, we can make the following definition.

Tautologies A formal wff is a *tautology* or *valid* means that in every model its realization is evaluated as true; in that case we write $\models A$.

A semi-formal proposition is a tautology means that it is the realization of some wff that is a tautology.

We can speak of a proposition in ordinary English being a tautology if there is a straightforward formalization of it into semi-formal English that is a tautology. We might say, for example, that 'Either Ralph is a dog or Ralph is not a dog' is a tautology, since it seems obvious that the right way to formalize it is as 'Ralph is a dog $\vee$ $\neg$(Ralph is a dog)'.

Similarly, 'Ralph is a dog $\wedge$ $\neg$(Ralph is a dog)' will be evaluated as false no matter what value is assigned to 'Ralph is a dog'. A proposition or formal wff that is evaluated as false regardless of the truth-values of its atomic constituents is called a *contradiction*. A contradiction is false, relative to the classical interpretation of the connectives, due solely to its form. If A is a tautology then $\neg A$ is a contradiction, and if A is a contradiction then $\neg A$ is a tautology.

To say that a *scheme*, such as $A \vee \neg A$, is a *tautology* or *valid* means that every result of replacing the metavariables by wffs is a valid wff. The replacement must be uniform; for example, that $A \rightarrow (B \rightarrow A)$ is a valid scheme requires that when we replace one A by a wff we use that same wff to replace the second A.

Form can determine relations among propositions, too. If the proposition 'Ralph is a dog $\wedge$ Howie is a cat' is true in a model, then each of its constituents must be true. If either 'Ralph is a dog' or 'Howie is a cat' is assigned T in a model, then 'Ralph is a dog $\vee$ Howie is a cat' must be assigned T. From the truth of one proposition or a collection of propositions, we can conclude the truth of another, based solely on their form. Using capital Greek letters Γ, Σ, and Δ for collections of propositions or wffs, we make the following definition for formal wffs.

Semantic consequence A wff B is a *semantic consequence* of a collection of wffs Γ, written $\Gamma \vDash B$, if and only if for every model in which every wff in Γ is true, B is also true. That is, it is not possible for all the wffs in Γ to be true and B to be false at the same time. We sometimes say the pair Γ, B is a semantic consequence.

Two wffs are *semantically equivalent* if each is a semantic consequence of the other: In every model they have the same truth-value.

The notion of semantic consequence formalizes the idea that one proposition *follows from* another or collection of others, relative to our understanding of the connectives. To assert that A follows from B is an *inference* (or *deduction* or *implication*), with Γ the *premises* and A the *conclusion*, and it is this that is our formalization of *therefore*, relative to our semantic assumptions. A *valid inference* is a semantic consequence. We often write an inference with a line separating the premises from the conclusion, as in:

$$\frac{A \rightarrow B\ ,\ A}{B}$$

This notation does not assume that the inference is actually valid.

Note that writing $\vDash A$ when A is a tautology is still appropriate, for if A is a tautology then it is a semantic consequence of the empty collection of wffs.

The same terminology can be applied to semi-formal propositions viewed as realizations of formal wffs, and then in turn to ordinary English propositions, as described for validity. For example, consider:

Either Ralph is a dog or Howie is a duck.
No way is Howie a duck.
So Ralph is a dog.

Arguing that 'either . . . or . . .' should be formalized as classical disjunction and 'no way' as classical negation, this is formalized as:

{Ralph is a dog $\vee$ Howie is a duck, $\neg$(Howie is a duck)} $\vDash$ Ralph is a dog

And that is a valid inference, as it has the form: $\{p_7 \vee p_{10}\, ,\, \neg\, p_{10}\} \vDash p_7$.

We can also talk about two propositions being semantically equivalent. For example, '$\neg$(Ralph is a dog $\wedge$ $\neg$(George is a duck))' is semantically equivalent to 'Ralph is a dog $\rightarrow$ George is a duck'. Semantic equivalence is a formalization of the idea that two propositions *mean the same* for all our logical purposes.

Let's review what we've done. We began by establishing a formal language that would make precise the form of the propositions we would study. We based that formalization on the four connectives : 'and', 'or', 'it's not the case that', and 'if . . . then . . .'. It was the semi-formal language in which we were primarily interested: English with the formal propositional connectives $\neg$, $\rightarrow$, $\wedge$, $\vee$.

Then we made the assumption that the only aspects of a proposition that are of concern to logic are its truth-value and its form as built up from sentence connectives. Making assumptions about the relationship of the semantic properties of a proposition to the properties of its constituents, we analyzed the truth of a compound proposition in terms of its parts by using the truth-tables. Then we explained by reference to the formal language what it means for a proposition to be true due solely to its propositional form, and what it means for one proposition to follow from another or a collection of others. Now we have a logic.

Classical propositional logic The definition of classical models, of classical validity, and of classical semantic consequence together constitute *classical propositional logic*. The name *classical propositional calculus* is also common, which we abbreviate as **PC**.

Exercises for Sections E.3

1. a. What is a tautology in the formal language?
   b. What is a tautology in the semi-formal language?
   c. What is a tautology in ordinary English?
2. Exhibit three formal classical tautologies and three formal classical contradictions.
3. a. What is a classical semantic consequence?
   b. Exhibit three formal classical semantic consequences.

F. Formalizing Reasoning

To see if classical propositional logic is a good formalization of reasoning, we need to see examples of propositions and inferences from ordinary language. I'll start with examples in English and then, if necessary, rewrite them in a way that's equivalent to the original for all our logical purposes and which can be easily formalized. Here 'for all our logical purposes' means relative to the assumptions of classical propositional logic. We can further abstract the semi-formal equivalent of a proposition into a wff of the formal language, and that is what we'll call its *(propositional) logical form.*

For example, consider the proposition:

(4) Ralph is a dog, but George is a duck.

Since none of the words or phrases we intend to formalize with the formal connectives are present, we could take (4) to be atomic. Alternatively, perhaps as a general agreement concerning the formalization of 'but', or perhaps because we think it correct to deduce from (4) both 'Ralph is a dog' and 'George is a duck', we could say that for all our logical purposes (4) is equivalent to:

Ralph is a dog, and George is a duck

This we would formalize as:

Ralph is a dog $\wedge$ George is a duck.

Thus the logical form of (4) depends on whether to view it as atomic or compound. But given that agreement, we can speak of its logical form as either p_6 or as:

(5) $p_0 \wedge p_1$

This last step, however, is misleading. We could also use $p_{36} \wedge p_{412}$. Perhaps we should use metavariables to speak of the logical form of (4) as, say, $p \wedge q$. But that further level of abstraction is no real simplification. Better is to keep in mind that talk of "the" logical form of a proposition presupposes an informal identification of formal wffs.

If we say that (5) is the logical form of (4), we needn't construe that to mean that (5) was waiting to be discovered as the skeleton of (4), though you may wish to see it that way. Rather, (5) reflects an agreement or recognition that the main properties of interest to logic of (4) are preserved by the rewriting into (5).

Relative to the assumptions of classical logic we can use the classical connectives to formalize other English connectives. Here are a few such formalizations:

(6)

English connective	*Formalization*	
A but B	$A \wedge B$	
A only if B	$A \rightarrow B$	
A if and only if B	$(A \rightarrow B) \wedge (B \rightarrow A)$	
A unless B	$\neg B \rightarrow A$	
neither A nor B	$\neg A \wedge \neg B$	
B provided that A	$A \rightarrow B$	
A or else B	$(A \vee B) \wedge \neg(A \wedge B)$	*exclusive disjunction*
B if A	$A \rightarrow B$	
if A, B	$A \rightarrow B$	
B in case A	$A \rightarrow B$	
both A, B	$A \wedge B$	

We usually abbreviate 'A if and only if B' as 'A *iff* B' and abbreviate its formalization as $A \leftrightarrow B$. We call '$\leftrightarrow$' the *biconditional.* We call 'if A, then B' the *left to right direction* of 'A iff B', and 'if B, then A' the *right to left direction.* The *contrapositive* of $A \rightarrow B$ is $\neg B \rightarrow \neg A$, which has the same truth-value. The *converse* of $A \rightarrow B$ is $B \rightarrow A$, which need not have the same truth-value.

Still, we need to take care with these and even with the four connectives we've agreed are the basis of our logic. Consider, for example, 'Bring me an ice cream cone and I'll be happy.' It would be wrong to formalize 'and' here as classical $\wedge$, for what is intended is 'If you bring me an ice cream cone, then I'll be happy', which requires $\rightarrow$.

What guide do we have in formalizing? In *Predicate Logic* I've motivated and discussed criteria we can use that are necessary for a semi-formal proposition to be a good formalization of an English proposition. There I've also analyzed many examples from ordinary speech. Here I will present just a few examples with some ideas about formalizing, leading to examples used in mathematics texts.

1. Ralph is a dog or he's a puppet.

Ralph is a dog $\vee$ Ralph is a puppet

$p_1 \vee p_2$

Analysis First we have to get rid of the indexical 'he'. We rewrite the proposition as 'Ralph is a dog or Ralph is a puppet.'

We've agreed to formalize 'or' as inclusive disjunction. Here that seems odd, for we suppose the alternatives are mutually exclusive. But that is an observation about the truth-values of the constituent propositions, not about the use of 'or'. We'll only take 'or' in the exclusive sense if the proposition is formulated using 'or else' or if the context demands it.

2. Ralph is a dog if he's not a puppet.

$\neg$(Ralph is a puppet) $\rightarrow$ Ralph is a dog

$\neg p_2 \rightarrow p_1$

Analysis We rewrite the proposition as 'Ralph is a dog if Ralph is not a puppet.' Should 'if' be formalized as a conditional? Is our rewrite equivalent to 'If Ralph is not a puppet, then Ralph is a dog' ? It seems so, and the formalization then follows.

3. Ralph is not a dog because he's a puppet.

Not formalizable.

Analysis Rewriting 'he' as 'Ralph', we have two propositions: 'Ralph is not a dog' and 'Ralph is a puppet'. Can we formalize 'because' as a propositional connective? Compare:

Ralph is a dog because Socorro is a seaport.

Ralph is a dog because Socorro is not a seaport.

Each of these is false, regardless of whether 'Ralph is a dog' and 'Socorro is a seaport' are true. So 'because' is not truth-functional, and hence we cannot formalize these or Example 3. Nor can we take the example as atomic, for from it we can conclude 'Ralph is not a dog', and an informal deduction that we recognize as valid would not be respected.

4. Three faces of a die are even numbered.
Three faces of a die are not even numbered.
***Therefore*, Ralph is a dog.**

Analysis The two premises appear to be contradictory, and from a contradiction any proposition follows. But both premises are true. The word 'not' applies to 'even numbered' only, and cannot be assimilated to a use of 'It is not the case that . . .'.

5. Ralph is a dog; he is not a puppet.

Ralph is a dog $\wedge$ Ralph is not a puppet.

$p_1 \wedge \neg p_2$

Analysis There is no word that could be construed as equivalent to 'and' in this example. But semi-colons, commas, and other punctuation are often used in English in place of connectives and should be formalized accordingly.

6. Every natural number is even or odd.

Every natural number is even or odd

p_1

Analysis The example is atomic. Propositional logic does not recognize the internal complexity of it. We cannot formalize 'or' by $\vee$, for the example is not equivalent to 'Every natural number is even or every natural number is odd.'

7. Suppose $\{s_n\}$ is monotonic. Then $\{s_n\}$ converges if and only if it is bounded.

$\{s_n\}$ is monotonic $\rightarrow$ ($\{s_n\}$ converges $\leftrightarrow$ $\{s_n\}$ is bounded)

$p_1 \rightarrow (p_2 \leftrightarrow p_3)$

Analysis In formalizing this theorem from a mathematical textbook on real analysis, (*Rudin*, *1964*, p. 47) I've changed the grammar significantly. It might seem we should formalize the example as:

$\{s_n\}$ is monotonic.
Therefore, $\{s_n\}$ converges $\leftrightarrow$ $\{s_n\}$ is bounded.

That would not be wrong, but when someone states a theorem in mathematics we normally understand it as asserting a claim. The words 'suppose', 'let', and others that we might take as indicating an inference are best formalized as indicating an antecedent of a conditional when used in mathematical texts.

The context of the discussion in that book indicates what $\{s_n\}$ is. But the form of the wff does not recognize that all three atomic propositions are about $\{s_n\}$.

8. Let A and B be sets of real numbers such that
(a) every real number is either in A or in B;
(b) no real number is in A and in B;
(c) neither A nor B is empty;
(d) if $\alpha \in A$, and $\beta \in B$, then $\alpha < \beta$.
Then there is one (and only one) real number γ such that $\alpha \leq \gamma$ for all $\alpha \in A$, and $\gamma \leq \beta$ for all $\beta \in B$.

{(A is a set of real numbers $\wedge$ B is a set of real numbers) $\wedge$ (every real number is either in A or in B $\wedge$ (no real number is in A and in B $\wedge$ (($\neg$(A is empty) $\wedge$ $\neg$(B is empty)) $\wedge$ ($\alpha \in A \wedge \beta \in B \rightarrow \alpha < \beta$)))) } $\rightarrow$ there is one and only one real number γ such that $\alpha \leq \gamma$ for all $\alpha \in A$ and $\gamma \leq \beta$ for all $\beta \in B$

$$((p_1 \wedge p_2) \wedge (p_3 \wedge (p_4 \wedge ((\neg p_5 \wedge \neg p_6) \wedge (p_7 \wedge p_8 \rightarrow p_9))))) \rightarrow p_{10}$$

Analysis I take 'Let' to indicate that the antecedent of a conditional follows, 'such that' to indicate that further conditions are conjoined, the semi-colons to be conjunctions, and 'neither . . . nor . . .' to be formalized according to our earlier convention at (6). The grouping of the conjuncts is not entirely arbitrary, though the association to the right in the formalizations of (a)–(d) is. I'll let you convince yourself that 'either . . . or' in (a), and 'and' and 'no' in (b) should not be formalized as propositional connectives here.

Despite the considerable complexity of the propositional form of this example (from *Rudin, 1964*, p. 9) note how much is ignored in the formalization. The use of variables, quantifiers, and mathematical terms is not recognized in propositional logic. To take account of those we need to look at the internal structure of propositions. But first we'll look more closely at classical propositional logic.

Exercises for Section F

1. What do we mean by 'the logical form of a proposition'?

2. Present and argue for formalizations of the following English connectives:
   a. not both A and B   b. when A, B   c. B just in case A   d. A, when B

3. a. Show that a conditional and its contrapositive have the same truth-value.
   b. Give an example of a conditional and its converse that have different truth-values.

4. Give two examples not from the text of English connectives that are not truth-functional and show why they are not truth-functional. (Hint: Compare Example 3.)

5. Formalize and discuss the following in the format above, or explain why they cannot be formalized in classical logic.

a. 7 is not even.
b. If 7 is even, then 7 is not odd.
c. Unless Milt has a dog, Anubis is the best-fed dog in Cedar City
d. Horses eat grass because grass is green.
e. It's impossible that $2 + 2 \neq 4$.
f. Anubis, you know, eats, sleeps, and barks at night.
g. You can't make an omelette without breaking eggs.

6. Formalize and discuss an example from a mathematics text.
7. Formalize each of the following inferences. Then evaluate it for validity in classical logic. To do that, attempt to determine if it is possible for the premises to be true and the conclusion false. If it is not possible, then the inference is valid.
   a. If Ralph is a cat, then Ralph meows. Ralph is not a cat. So Ralph does not meow.
   b. If Bob takes Mary Ann to Jamaica for the holidays, she will marry him. Mary Ann married Bob last Saturday. So Bob must have taken Mary Ann to Jamaica for the holidays.
   c. The students are happy if and only if no test is given. If the students are happy, the professor feels good. But if the professor feels good, he won't feel like lecturing, and if he doesn't feel like lecturing a test is given. Therefore, the students are not happy. (*Mates, 1965*)
   d. The government is going to spend less on health and welfare. If the government is going to spend less on health and welfare, then either the government is going to cut the Medicare budget or the government is going to slash spending on housing. If the government is going to cut the medicare budget, the elderly will protest. If the government is going to slash spending on housing, then the poor will protest. Therefore, the elderly will protest or the poor will protest.
   e. All men are mortal. Socrates is a man. Therefore, Socrates is mortal.
8. Make a list of each term or formal symbol that is introduced or defined in this chapter.

Proof by induction

Consider an example. We wish to prove that for every natural number n,

$$1 + 2 + \cdots + n = {}^1/_2\, n \cdot (n + 1)$$

We can test this for some small numbers, say 1, 2, 3, 4. But how are we to proceed for all numbers?

We've checked the result when $n = 1$. This the *basis of the induction.* Let's assume now that we've proved it for some number m:

$$1 + 2 + \cdots + m = {}^1/_2\, m \cdot (m + 1)$$

This is called the *induction hypothesis.* We try to use this hypothesis to prove that the theorem must be true for the next larger natural number, $m + 1$. We have:

$$1 + 2 + \cdots + m + (m + 1) = ({}^1/_2\, m \cdot (m + 1)) + (m + 1)$$

$$= \tfrac{1}{2}(m^2 + m) + \tfrac{1}{2}(2m + 2)$$
$$= \tfrac{1}{2}(m^2 + 3m + 2)$$
$$= \tfrac{1}{2}(m + 1) \cdot (m + 2)$$
$$= \tfrac{1}{2}(m + 1) \cdot ((m + 1) + 1)$$

We've shown the theorem is true for $m + 1$ if it is true for m. So we claim the theorem is true for every natural number.

In summary, the method of induction is:

> We show the statement is true for the number 1. Then we *assume* it's true for m, an arbitrary but fixed number, and show that it's true for $m + 1$. So it's true for all numbers.

We claim the statement is true for all numbers: It's true for 1; and if it's true for 1, then it's true for 2, so it's true for 2; since it's true for 2, it's therefore true for 3; and so on.

Those little words 'and so on' carry a lot of weight. We believe that the natural numbers are completely specified by their method of generation: add 1, starting at 0.

0 1 2 3 4 5 6 7 . . .

To prove a statement A by induction, we first prove it for some starting point in this list of numbers, usually 1, but just as well 0 or 47. We then establish that we have a method of generating proofs that is exactly analogous to the method of generating natural numbers: If A(n) is true, then A(n + 1) is true. We have the list:

A(0)	If A(0) , then A(1), so A(1).	If A(1) , then A(2), so A(2).	If A(2) , then A(3), . . . so A(3).

Then the statement is true for all natural numbers equal to or larger than our basis, whether the statement is for all numbers larger than 0 or all numbers larger than 47.

In essence we have only one idea: a process for generating objects one after another without end, in one case numerals or numbers, and in the other, proofs. We believe induction is a correct form of proof because the two applications of the single idea are matched up.

Another principle of induction we often use is the following:

> We show the statement is true for the number 1. Then we *assume* it is true for *all numbers less than or equal to* m, an arbitrary but fixed number, and show it's true for $m + 1$. So it's true for all numbers.

Exercises

Prove by induction the following.

1. The sum of the first n even numbers is $n \cdot (n + 1)$.
2. The sum of the first n odd numbers is n^2.
3. $1^2 + 2^2 + \cdots + n^2 = \tfrac{1}{6} n \cdot (n + 1) \cdot (2n + 1)$.
4. The minimum number of parentheses that can appear in a wff of length n is $2n$.
5. The maximum number of parentheses that can appear in a wff of length n is $2^{n+1} - 2$.

II Abstracting and Axiomatizing Classical Propositional Logic

---

In Chapter I we set out how we'll reason with propositions and combinations of those formed using propositional connectives. We developed a formal logic and saw examples of how it could be used to formalize ordinary and mathematical reasoning. In this chapter we'll see how to abstract that logic in order to be able to apply mathematics to the study of it. I'll then present a formalization of the notion of proof to axiomatize the logic.

A. The Fully General Abstraction

In classical propositional logic every atomic proposition is abstracted to just its truth-value. Nothing else matters. So if we have two models in which for each propositional variable p_i the one model assigns to p_i a proposition that is true iff the other model assigns one that is true, then the two models are indistinguishable. The propositions in one may be about mathematics and in the other about animal husbandry, but that cannot enter into our deliberations if we are using classical logic. It is appropriate to ignore differences between models of type **I** (p. 17) that do not matter to classical logic. We can simplify our models to:

Model Type **II**

$$\begin{array}{c} L(p_0, p_1, \ldots \neg, \rightarrow, \wedge, \vee) \\ \Big\downarrow \; \nu\text{, truth-tables} \\ \{T, F\} \end{array}$$

Here ν assigns truth-values to the variables p_0, $p_1, \ldots$ and is extended to all formal wffs by the classical truth-tables for $\neg$, $\rightarrow$, $\wedge, \vee$. We often name this abstracted version of a model by its valuation, ν.

Propositions, as I've presented them, are written or uttered. There can only be a finite number of them. But though finite, we can create more propositions at any time: the supply is unlimited. So rather than restricting our formal or semi-formal language to, say, all wffs using fewer than 47 symbols made up from p_0, $p_1, \ldots,$ p_{13} or the realizations of those in a particular model, it is more general and reflects better our assumption on form and meaningfulness (p. 16) to include all well-formed formulas in our logical analyses. This will also simplify our work considerably.

Let's denote by PV the collection of propositional variables. We can take this collection, as well as the formal and semi-formal languages, as potentially infinite collections, capable of being extended as needed. However, for some applications of mathematics to logic we will need to view these collections as completed infinite wholes, and I'll try to point out when we need that assumption. In any case, the definitions for the formal language given in the previous chapter are sufficiently mathematical to serve us.

In Chapter IX.B we'll look at the assumptions we make in reasoning about collections. For now I'll use the notation of set-theory informally, enclosing by parentheses of the sort '{ }' either the names of several things or a description of things to indicate that those things are to be taken as a collection. The symbol '$\in$' stands for 'is an element of'; '$\notin$' for 'is not an element of'; the symbol '$\varnothing$' stands for the empty collection consisting of no objects whatsoever.

For the semantics, a model of type **II** is now simply a function $\nu: \mathrm{PV} \rightarrow \{T, F\}$. It is extended inductively to all wffs by the truth-tables.

Now we make a powerful idealization.

The fully general abstraction Any function υ: PV$\rightarrow$ {T, F} that is extended to all wffs by the truth-tables is a model.

Now not only does it not matter how a model of type **II** arises, we assume any possible function gives a model. In particular, we assume that we can independently assign truth-values to all the variables in PV.

The fully general abstraction applies to models of type **II**, not type **I**. The simplicity and generality we get by making it is justified so long as we don't arrive at any contradiction with our previous, more fundamental assumptions and intuitions when we apply a metalogical result to actual propositions.

Platonists on the abstraction of models
Someone who holds a platonist conception of logic and views propositions as abstract things would take as facts about the world what I have called assumptions, idealizations, and generalizations. Words and sentences are understood as abstract objects that sometimes can be represented as inscriptions or utterances; they are not equivalences we impose on the phenomena of our experience. All of them exist whether they are ever uttered or not. The formal and informal language and any model are abstract objects complete in themselves, composed of an infinity of things. Though the distinction between a model of type **I** and **II** can be observed by a platonist, it does not matter whether a model is ever expressed or brought to our attention. It simply exists and hence, the platonist believes, the fully general abstraction is no abstraction but an observation about the world.

B. A Mathematical Presentation of PC

In this section I'll set out the definitions from Chapter I for classical propositional logic in terms of models of type **II**, stripped of all connection to ordinary language.

1. Models and the semantic consequence relation

A *model* for classical logic is a function υ: PV$\rightarrow$ {T, F} that is extended to all wffs by the classical tables for $\neg$, $\rightarrow$, $\wedge$, $\vee$. The extension is unique since every wff has a unique reading. The function υ is called a *valuation.*

A *wff* A *is true* (*or valid*) *in model* υ if $\upsilon(\text{A}) = \text{T}$, which we notate $\upsilon \models \text{A}$, and read as '$\upsilon$ validates A'; it *is false in* υ if $\upsilon(\text{A}) = \text{F}$, which we write $\upsilon \not\models \text{A}$.

Letting Γ, Σ, Δ stand for collections of wffs, we write $\Gamma \models \text{A}$, read as Γ *validates* A or A *is a semantic consequence of* Γ, if every model υ that validates all the wffs in Γ also satisfies $\upsilon(\text{A}) = \text{T}$. We write '$\text{A} \models \text{B}$' for $\{\text{A}\} \models \text{B}$, and '$\models \text{A}$' for $\varnothing \models \text{A}$, that is, A is true in every model. If $\models \text{A}$, we say that A is a (**PC**-) *tautology* or is (**PC**-) *valid.* We say that υ is a *model of* Γ, written $\upsilon \models \Gamma$, if $\upsilon \models \text{A}$ for every A in Γ. A scheme of wffs is tautological if every instance of the scheme is a tautology, and similarly for a scheme of inferences. *Two propositions* (or *wffs*) *are semantically equivalent* if each is a semantic consequence of the other: In every model they have the same truth-value.

We denote by **PC**, the (*classical*) *propositional calculus*, the collection of consequences $\{(\Gamma, A) : \Gamma \vDash A\}$. The relation $\vDash$ is called the *semantic consequence relation of* **PC** and is also written $\vDash_{\mathbf{PC}}$.

In what follows I will write, for example, Γ, A for $\Gamma \cup \{A\}$.

Theorem 1 Properties of the semantic consequence relation

a. *Reflexivity* $A \vDash A$.

b. If $\vDash A$, then $\Gamma \vDash A$.

c. If $A \in \Gamma$, then $\Gamma \vDash A$.

d. *Monotonicity* If $\Gamma \vDash A$ and $\Gamma \subseteq \Delta$, then $\Delta \vDash A$.

e. *Transitivity* If $\Gamma \vDash A$ and $A \vDash B$, then $\Gamma \vDash B$.

f. *The cut rule* If $\Gamma \vDash A$ and $\Delta, A \vDash B$, then $\Gamma \cup \Delta \vDash B$.

g. If $\Gamma \cup \{A_1, \ldots, A_n\} \vDash B$ and for each i, $\Gamma \vDash A_i$, then $\Gamma \vDash B$.

h. $\vDash A$ iff for every nonempty Γ, $\Gamma \vDash A$.

i. *The semantic deduction theorem* $\Gamma, A \vDash B$ iff $\Gamma \vDash A \rightarrow B$.

j. *The semantic deduction theorem for finite consequences*
$\Gamma \cup \{A_1, \ldots, A_n\} \vDash B$ iff $\Gamma \vDash A_1 \rightarrow (A_2 \rightarrow (\cdots \rightarrow (A_n \rightarrow B) \cdots)$.

k. *Alternate version of* (j)
$\Gamma \cup \{A_1, \ldots, A_n\} \vDash B$ iff $\Gamma \vDash (A_1 \wedge \cdots \wedge A_n) \rightarrow B$.

l. *Substitution* If $\vDash A(p)$, then $\vDash A(B)$, where $A(B)$ is the result of substituting B uniformly for p in A (i.e., B replaces every occurrence of p in A).

Proof I'll leave the proofs of all of these as exercises except for (h) and (l).

(h) The left to right direction follows from (b). For the other direction, if every nonempty Γ validates A, then in particular $\neg A \vDash A$. So by (i), $\vDash \neg A \rightarrow A$ and hence $\neg A \rightarrow A$ is true in every model. Therefore, A is true in every model.

(l) The only place that p can occur in A(B) is within B itself. Hence, given any model υ, once we have evaluated $\upsilon(B)$ the evaluation of $\upsilon(A(B))$ proceeds as if we were evaluating A(p) in a model w, where w and υ agree on all propositional variables except p and $w(p) = \upsilon(B)$. Since A(p) is true in every model, $w(A(p)) = T$, and so $\upsilon(A(B)) = T$. Hence, $\vDash A(p)$. ∎

Part (l) says that propositional variables are really variables that can stand for any proposition, a formal version of the division of form and content.

Exercises for Sections A and B.1

1. a. Explain the difference between a model of type **I** and a model of type **II**.
   b. Give an example of two very different models of type **I** that result in the same model of type **II**.
   c. Why is it appropriate to identify a model with its valuation?
2. What does it mean to say that v validates a wff A?
3. What does '$v \vDash \Gamma$' mean?
4. What does '$\Gamma \vDash A$' mean?
5. a. Prove parts (a)–(g) and (i) of Theorem 1.
   b. Prove parts (j) and (k) of Theorem 1. (Hint: Use induction.)

2. The choice of language for PC

Suppose we add to the language a new connective $\gamma(A_1, \ldots, A_n)$ that we intend to interpret as a truth-function. In that case we must specify its interpretation by a truth-table that for every sequence of n T's and F's assigns either T or F. Doing this we've added a *truth-functional connective* to the language.

But the connectives we have are already enough. Given any truth-functional connective γ, we can find a scheme $S(A_1, \ldots, A_n)$ that uses only $\neg, \rightarrow, \wedge, \vee$ such that for all wffs $A_1, \ldots, A_n$, $S(A_1, \ldots, A_n)$ is semantically equivalent to $\gamma(A_1, \ldots, A_n)$, which is what we mean by saying that $\{\neg, \rightarrow, \wedge, \vee\}$ is *truth-functionally complete*. Thus, except for the convenience of abbreviation, further truth-functional connectives add nothing of significance to our language and semantics.

Theorem 2 $\{\neg, \rightarrow, \wedge, \vee\}$ is truth-functionally complete.

Proof Let γ be a truth-functional connective. If γ takes only the value F, then $\gamma(A_1, \ldots, A_n)$ is semantically equivalent to $(A_1 \wedge \neg A_1) \vee \cdots \vee (A_n \wedge \neg A_n)$. Otherwise let the following be those sequences of T's and F's that are assigned T by the table for γ:

$$\alpha_1 = (\alpha_{11}, \ldots, \alpha_{1n})$$
$$\vdots$$
$$\alpha_k = (\alpha_{k1}, \ldots, \alpha_{kn})$$

Define for $i = 1, \ldots, k$ and $j = 1, \ldots, n$, $B_{ij} = \begin{cases} A_{ij} & \text{if } \alpha_{ij} = \mathsf{T} \\ \neg A_{ij} & \text{if } \alpha_{ij} = \mathsf{F} \end{cases}$

Set $D_i = B_{i1} \wedge \cdots \wedge B_{in}$, associating to the left (this corresponds to the ith row of the table). Then $D_1 \vee D_2 \vee \cdots \vee D_k$, associating to the left, is semantically equivalent to $\gamma(A_1, \ldots, A_n)$, as you can show. ■

As an example of this procedure, suppose we wish to define in **PC** a ternary connective γ with the following table:

A_1	A_2	A_3	$\gamma(A_1, A_2, A_3)$
T	T	T	T
T	T	F	F
T	F	T	F
T	F	F	T
F	T	T	T
F	T	F	T
F	F	T	F
F	F	F	F

Then:

$\alpha_1 = (T,T,T)$ $\alpha_3 = (F,T,T)$

$\alpha_2 = (T,F,F)$ $\alpha_4 = (F,T,F)$

$D_1 = (A_1 \wedge A_2) \wedge A_3$

$D_2 = (A_1 \wedge \neg A_2) \wedge \neg A_3$

$D_3 = (\neg A_1 \wedge A_2) \wedge A_3$

$D_4 = (\neg A_1 \wedge A_2) \wedge \neg A_3$

And $((D_1 \vee D_2) \vee D_3) \vee D_4$ is semantically equivalent to $\gamma(A_1, A_2, A_3)$.

By noting the following semantic equivalences in **PC** we can show that *each of* $\{\neg, \rightarrow\}$, $\{\neg, \wedge\}$, *and* $\{\neg, \vee\}$ *is truth-functionally complete.*

$A \rightarrow B \equiv \neg(A \wedge \neg B)$

$A \rightarrow B \equiv \neg A \vee B$

$A \wedge B \equiv \neg(A \rightarrow \neg B)$

$A \wedge B \equiv \neg(\neg A \vee \neg B)$

$A \vee B \equiv \neg A \rightarrow B$

$A \vee B \equiv \neg(\neg A \wedge \neg B)$

Since every truth-functional connective is definable in **PC** from $\neg$ and $\rightarrow$, the connectives $\wedge$ and $\vee$ are in some sense superfluous. We can define a language $L(\neg, \rightarrow, p_0, p_1, \ldots)$ and formulate classical propositional logic in it, defining the connectives $\wedge$ and $\vee$ as above. We feel that we have the same logic because we have a translation λ from the language $L(\neg, \rightarrow, \wedge, \vee, p_0, p_1, \ldots)$ to $L(\neg, \rightarrow, p_0, p_1, \ldots)$ via:

$\lambda(p_i) = p_i$

$\lambda(\neg A) = \neg(\lambda(A))$ $\lambda(A \wedge B) = \neg(\lambda(A) \rightarrow \lambda(\neg B))$

$\lambda(A \rightarrow B) = \lambda(A) \rightarrow \lambda(B)$ $\lambda(A \vee B) = \neg\lambda(A) \rightarrow \lambda(B)$

This translation is faithful to our models. That is, given any $\upsilon: PV \rightarrow \{T, F\}$, if we first extend it to all wffs of $L(\neg, \rightarrow, \wedge, \vee, p_0, p_1, \ldots)$ and call that υ_1, and then extend it to $L(\neg, \rightarrow, p_0, p_1, \ldots)$ and call that υ_2, then $\upsilon_1 \models A$ iff $\upsilon_2 \models \lambda(A)$.

Similarly, we have the same logic if we formalize it using $\{\neg, \wedge\}$ or $\{\neg, \vee\}$ or any other truth-functionally complete set of connectives. We can say, loosely, that **PC** may be *formalized in any of several languages*, and speak of a *formalization of* **PC** *in the language of* $L(\neg, \rightarrow, p_0, p_1, \ldots)$. But these phrases are meant only to refer to the translations and equivalences we've established and are not meant to suggest that there is some formless body of truths comprising **PC**.

Normal forms

The proof of Theorem 2 suggests how we can take any wff and find a wff in a particularly simple form that is equivalent to it.

Normal forms A *disjunctive normal form for a wff* A is any formula B that (i) is semantically equivalent to A, (ii) uses exactly the same variables as A, and (iii) is $B_1 \vee B_2 \vee \cdots \vee B_n$ and each B_i is of the form $C_1 \wedge \cdots \wedge C_m$ for some m, where each C_j is either a propositional variable or the negation of one.

It is a *conjunctive normal form* for A if it satisfies (i) and (ii) and is a conjunction of disjunctions of variables or negations of variables.

For example:

$((p_1 \wedge p_2) \wedge (\neg p_1)) \vee (((\neg p_6 \wedge \neg p_8) \wedge p_9) \vee p_2)$ is in disjunctive normal form

$((p_1 \vee p_2) \vee (\neg p_1)) \wedge (((\neg p_6 \vee \neg p_8) \vee p_9) \wedge p_2)$ is in conjunctive normal form

There is no unique normal form for a wff since $(A \wedge B) \wedge C$ is semantically equivalent to $A \wedge (B \wedge C)$, and $(A \vee B) \vee C$ is equivalent to $A \vee (B \vee C)$, and A is equivalent to both $A \vee A$ and to $A \wedge A$. So for formulas such as $p_1 \vee (\neg p_2 \vee p_3)$ and $(\neg p_{17} \wedge p_{31}) \wedge (p_6 \wedge p_8)$ we can say they are in either disjunctive or conjunctive normal form.

The normal form theorem Given any wff A we can find a disjunctive normal form for A and a conjunctive formal form for A.

Proof Any wff A determines a truth-function: Suppose the propositional variables appearing in A are $\{q_1, \ldots, q_n\}$. We can make a table with 2^n rows corresponding to the ways we can assign T or F to the variables. We can then calculate the truth value of A under each assignment. The proof of Theorem 2 establishes a disjunctive normal form for A. By repeated use of the following equivalences we can obtain a conjunctive normal form for A:

$A \vee (B \wedge C)$ is semantically equivalent to $[(A \vee B) \wedge (A \vee C)]$

$A \wedge (B \vee C)$ is semantically equivalent to $[(A \wedge B) \vee (A \wedge C)]$ ■

3. The decidability of tautologies

There is a simple procedure to determine whether any given wff is a tautology. I'll set it out here, noting particularly that our idealizations have not created problems.

a. In any model, the truth-value of any wff depends on only the truth-values of the propositional variables appearing in it.

b. Given any assignment of truth-values to a finite number of propositional variables, the assignment can be extended to all variables (for example, take F to be the value assigned to all other propositions).

c. Therefore, a wff is not a tautology iff there is an assignment of truth-values to the variables appearing in it for which it takes the value F. This is because if there is such an assignment, it can be extended to be a model in which the wff is false; if there is no such assignment, then the wff must be a tautology.

d. For any assignment of truth-values to the variables appearing in a wff we can effectively calculate the truth-value of the wff by repeatedly using the classical tables for $\neg$, $\to$, $\wedge$, $\vee$.

Thus, we can decide whether a wff is a **PC**-tautology by listing all possible truth-value assignments to the variables appearing in it—if there are *n* variables, there are 2^n assignments—and for each assignment mechanically checking the resulting truth-value for the wff. The wff is a tautology iff it always receives the value T. In Chapter 19 of *Epstein and Carnielli, 1989* this method is set out precisely. Note that the decision procedure for the validity of wffs also gives a decision procedure for whether wffs are semantically equivalent: A is semantically equivalent to B iff $A \leftrightarrow B$ is a tautology.

This decision procedure is practically unusable for wffs containing 83 or more variables: We would have to check some 2^{83} assignments. There are many shortcuts we can take, but to date no one has come up with a method of checking validity that can be run on a computer in less than (roughly) exponential time relative to the number of variables appearing in the wff; it has been conjectured that there is no such faster method.

One particular way of using (a)–(d) above is worth noting: We attempt to falsify the wff. If we can falsify the wff, it's not a tautology; otherwise it is. This is particularly useful for $\to$-wffs because only one line of the table yields F, so $A \to B$ can be falsified iff there is an assignment that makes both A true and B false. Generally we look at schema rather than particular wffs. For example:

	$\neg(A \wedge B)$	$\to$	$(B$	$\to$	$A)$
is false iff	T			F	
which is iff	$A \wedge B$		B		A
are	F		T		F

And this is the falsifying assignment. Hence, the scheme is not tautological.

Similarly,	$((A \wedge B) \to C)$	$\to$	$(A$	$\to$	$(B$	$\to$	$C))$	is false
iff	T			F				
iff			A			$B \to C$		
are			T			F		
iff					B		C	
are					T		F	

But if A is T, B is T, and C is F, then $(A \wedge B) \to C$ is F, so there is no way to falsify the scheme. Hence, the scheme is a tautology.

We can also determine if an inference is valid by using the deduction theorem (Theorem 1.k): $A_1, \ldots, A_n \models B$ is valid iff $(A_1 \wedge \cdots \wedge A_n) \rightarrow B$ is a tautology.

Exercises for Sections B.2 and B.3

1. Determine whether the following are classical tautologies. Here p, q, and r are used in place of p_1, p_2, and p_3.
   a. $[\neg(p \wedge q) \wedge p] \rightarrow \neg q$
   b. $[(p \rightarrow q) \wedge (\neg p \rightarrow r)] \rightarrow (q \vee r)$
   c. $(q \vee \neg q) \rightarrow q$
   d. $q \rightarrow (q \vee \neg q)$
   e. $\neg(\neg p \vee \neg\neg p) \vee p$
   f. $((p \leftrightarrow \neg q) \leftrightarrow \neg p) \leftrightarrow q$
   g. $[q \leftrightarrow (r \rightarrow \neg p)] \vee [(\neg q \rightarrow p) \leftrightarrow r]$
2. Evaluate whether the following schema are classically valid inferences.
   a. $\dfrac{(A \wedge \neg\neg A) \vee B}{\neg B}$
   b. $\dfrac{A \rightarrow \neg B, \ B \wedge \neg C}{A \rightarrow C}$
   c. $\dfrac{A \rightarrow \neg\neg B, \ \neg C \vee A, \ C}{B}$
3. Prove that $A \leftrightarrow B$ is a tautology iff A is semantically equivalent to B.
4. Prove that the following pairs are semantically equivalent.
   a. $A \vee (B \wedge C)$ and $(A \vee B) \wedge (A \vee C)$
   b. $A \wedge (B \vee C)$ and $(A \wedge B) \vee (A \wedge C)$ } *distribution laws*
   c. $\neg(A \wedge B)$ and $(\neg A \vee \neg B)$
   d. $\neg(A \vee B)$ and $(\neg A \wedge \neg B)$ } *De Morgan's laws*
5. Show that the following are schema of **PC**-tautologies:
   a. *Exportation* $((A \wedge B) \rightarrow C) \rightarrow (A \rightarrow (B \rightarrow C))$
   b. *Importation* $(A \rightarrow (B \rightarrow C)) \rightarrow ((A \wedge B) \rightarrow C)$
   c. *Double negation* $\neg\neg A \leftrightarrow A$
   d. *Clavius' law (consequentia mirabilis)* $(\neg A \rightarrow A) \rightarrow A$
   e. *Transitivity of the conditional* $((A \rightarrow B) \wedge (B \rightarrow C)) \rightarrow (A \rightarrow C)$
   f. *Contraposition* $(A \rightarrow B) \leftrightarrow (\neg B \rightarrow \neg A)$
   g. *Principle of contradiction, reductio ad absurdum* $(A \rightarrow B) \rightarrow ((A \rightarrow \neg B) \rightarrow \neg A)$
6. The *Sheffer stroke* ('nand') is the connective: $A \mid B$ is T iff not both A and B are T. Show that it by itself is truth-functionally complete.
7. Show that $\{\rightarrow, \wedge, \vee\}$ is not truth-functionally complete. (Hint: Consider wffs made from these connectives using only p_0.)

C. Formalizing the Notion of Proof

1. Reasons for formalizing

Historically, logic proceeded from observations that there were propositions true solely because of their form, to an investigation of those forms, to symbolization and generality, and then finally to laying out a few forms from which all other logically acceptable ones could be derived. That approach dates back to the Stoics, and to Aristotle too, though he was concerned more with the form of acceptable inferences than valid wffs. The modern versions came to fruition in the work of *Gottlob Frege, 1879* and *Whitehead and Russell, 1910–13*.

Logic in this tradition was seen as one language for all reasoning. It was only in the nineteenth century, with the use of models for non-Euclidean geometry, that mathematicians considered the idea that a set of formal propositions could have more than one interpretation. This led to the explicit statement of the truth-tables, though they had been implicit in logical analyses since the ancient Greeks.

In 1921 *Emil L. Post* showed that the propositional wffs derivable as theorems in Whitehead and Russell's system of logic coincide with the classical tautologies of the truth-tables. Since we now know that we can use the truth-tables to decide whether a wff is a **PC**-tautology, why should we bother to show that we can derive the **PC**-tautologies from a few acceptable valid wffs?

For one thing, the decision procedure takes an unrealistic amount of time, so for all practical purposes we have no decision procedure for wffs of even moderate length. Worse, when we consider the internal form of propositions we will find that there is no mechanical way to determine whether a wff is true in all models. We will need a way to derive tautologies syntactically.

Moreover, formalizing the notion of derivability, of when one proposition can be proved from others, is itself important not only for logic but for mathematics and science. The semantics of **PC** gave us a notion of semantic consequence. We want a syntactic notion of consequence.

Finally, by approaching logic in terms of the forms and not the meanings of wffs we'll have another way to isolate our assumptions. We start with an informal notion of validity and take as fundamental the validity of a few simple wffs and some few basic rules that lead always from valid wffs to valid wffs. Then we can say: If you accept these you will accept our logic. That is, unless you accept other wffs that aren't derivable in our system. To convince you that there are no other acceptable wffs we need to return to our formalization of the meaning of wffs and prove that all the valid wffs, that is, the tautologies, can be derived within our system.

Though there are many different ways to formalize the notion of proof, I will use just one in this text, what is usually called Hilbert-style proof theory.

2. Proof, syntactic consequence, and theories

We have the formal language $L(\neg, \to, \wedge, \vee, p_0, p_1, \ldots)$. Whenever I say 'proposition' below I mean one in the semi-formal language of a model of L.

To formalize the notion of proof, we begin with a collection of propositions that we take to be self-evidently true, the *axioms*. We then set out a collection of rules that allow us to derive consequences; the rules together with the axioms comprise an *axiom system*. In axiomatizing logic itself, we'll want to take as axioms propositions that we believe are self-evidently true due to their form only. The best way to describe those is by using formal wffs or schema. The axiom system is then in the formal language. To axiomatize a particular discipline such as geometry or physics we would then add propositions specific to that area as axioms.

Rules A *rule* is a direct consequence relation: Given a collection of propositions of specified form, the *premises*, another proposition, the *conclusion*, is taken to be a consequence. Rules are given in the formal language and are usually presented as schema, where any collection of wffs of the form $\{A_1, \ldots, A_n\}$ can be taken as premises, and a wff of the form B as consequence.

Typically we write the premises of a rule above the consequence, separated by a line. For instance, we have *modus ponens* $\dfrac{A, A \to B}{B}$ and *adjunction* $\dfrac{A, B}{A \wedge B}$.
When we choose a rule, we believe that it is self-evident that when the premises are instantiated with true propositions, the conclusion will be true, too.

Proofs There is a *proof* or *derivation of* A *from an axiom system* means that there is a sequence $A_1, \ldots, A_n$ such that A_n is A and each A_i is either an axiom or is a result of applying a rule of the system to some of the preceding A_j's. In that case A is a *theorem* and we write $\vdash A$.

A *proof of a proposition* A *from a collection of propositions* Σ is a proof sequence as above except that we also allow that any A_i can be a proposition in Σ. We say that A is then a *syntactic consequence of* Σ and write $\Sigma \vdash A$.

We sometimes read $\Sigma \vdash A$ as A is *deducible* from Σ. We write $A \vdash B$ for $\{A\} \vdash B$, and $\Sigma, A \vdash B$ for $\Sigma \cup \{A\} \vdash B$. Note that $\varnothing \vdash A$ is the same as $\vdash A$. We write $\Sigma \nvdash A$ for 'there is not a proof of A from Σ'.

The relation '$\vdash$' is our formalization of the syntactic notion of a proposition following from one or more propositions. In this we are again primarily interested in the syntactic forms, so we will generally take Σ to be a collection of formal wffs and then pass back and forth between wffs and propositions as we did for the notions of truth and validity in Chapter I. Thus Σ, Γ, Δ and subscripted versions of these can *stand variously for collections of wffs, semi-formal propositions, or ordinary language propositions* depending on the context.

Note well that we have not defined the notions of proof and consequence here. What we've done, relative to any particular axiom system, is give a definition of '$\vdash$', which we read colloquially as 'is a theorem' or 'has as a consequence' or 'proves'. I have to assume that you already have an idea of what it means to prove something, and it is that which we are formalizing, as we need to use the informal notion in proving theorems about the formal or semi-formal languages. We prove those (informal) theorems in the language of this book, the *metalanguage*, which is English supplemented with various technical notions.

I'll leave to you to prove the following basic properties of syntactic consequence relations.

Theorem 3 Properties of syntactic consequence relations

a. *Reflexivity* $A \vdash A$.

b. If $\vdash A$, then $\Gamma \vdash A$.

c. If $A \in \Gamma$, then $\Gamma \vdash A$.

d. *Monotonicity* If $\Gamma \vdash A$ and $\Gamma \subseteq \Delta$, then $\Delta \vdash A$.

e. *Transitivity* If $\Gamma \vdash A$ and $A \vdash B$, then $\Gamma \vdash B$.

f. *The cut rule* If $\Gamma \vdash A$ and $\Delta, A \vdash B$, then $\Gamma \cup \Delta \vdash B$.

g. If $\Gamma \cup \{A_1, \ldots, A_n\} \vdash B$ and for each $i = 1, \ldots, n$, $\Gamma \vdash A_i$, then $\Gamma \vdash B$.

h. *Compactness* $\Gamma \vdash A$ iff there is some finite collection $\Delta \subseteq \Gamma$ such that $\Delta \vdash A$.

Theorem 3.g shows that with these definitions of 'consequence' and 'theorem', once we prove a theorem we can use it to prove further theorems: If Γ is a collection of theorems and $\Gamma \vdash A$, then $\vdash A$.

When we think of scientific theories, we conceive of them as not just their premises, but their consequences too, and that is how we will define theories here.

Theories Σ is a *theory* means that it is closed under syntactic deduction: if $\Sigma \vdash A$, then A is in Σ. The collection of all syntactic consequences of Σ is the *theory of* Σ, which we write as $\mathrm{Th}(\Sigma) = \{A : \Sigma \vdash A\}$. Thus, Σ is a theory iff $\mathrm{Th}(\Sigma) = \Sigma$.

Some theories are better than others, as any physicist will tell you. As logicians we're especially interested in theories that are consistent, that is, ones from which no contradiction can be deduced.

Consistency and completeness

Γ is *consistent* iff for every A, either $\Gamma \nvdash A$ or $\Gamma \nvdash \neg A$.

Γ is *complete* iff for every A, either $\Gamma \vdash A$ or $\Gamma \vdash \neg A$.

A collection Σ is *inconsistent* means it is not consistent. The collection of all wffs is complete and inconsistent. But *a consistent and complete theory gives as full a description of the world, relative to the atomic propositions of the language, as is possible in our logic.*

3. Soundness and completeness

Given a semantics and a syntax, if every theorem is a tautology then we say that the *axiomatization is sound.* That is, for all A, if $\vdash$A then $\models$A. This is the minimal condition we impose on any pair of semantics and syntax that we want to claim characterize the same pre-formal logical intuitions.

Given a sound axiomatization of a semantics, if we also have that every tautology is a theorem, that is for all A, if $\models$A then $\vdash$A, then we say that the *axiomatization is complete.* A proof of that is called a *completeness theorem.*

The strongest correlation we can get is when the semantic and syntactic consequence relations are the same. That is, for all Γ and A, $\Gamma \models$ A iff $\Gamma \vdash$ A. In that case we say that we have *strongly complete semantics for the axiomatization*, or a *strongly complete axiomatization of the semantics.* A proof of the equivalence is called a *strong completeness theorem* for the logic.

Exercises for Section C

1. Prove that if Γ is a collection of theorems and $\Gamma \vdash$ A, then $\vdash$A. (Hint: Induct on the length of a proof of A from Γ.) Does it matter whether Γ is finite or not?
2. Why is it right to identify $\vdash$A and $\varnothing \vdash$A?
3. Show that if Γ is a theory, then every axiom is in Γ.
4. Prove Theorem 3.
5. What does it mean to say an axiomatization is sound? Complete? Strongly complete?
6. We could rephrase several parts of Theorem 3 and other statements about '$\vdash$' in terms of theories, which is what many texts do. For example, Theorem 3.c can be stated s: $\Gamma \subseteq \mathrm{Th}(\Gamma)$. Restate parts (d) and (e) of Theorem 3 similarly.

D. An Axiomatization of PC

1. The axiom system

How are we to devise an axiomatization for **PC**? Generally we have some idea, some strategy for how to prove an axiom system complete or strongly complete. So we begin a proof with no axioms in hand and, say, one rule, *modus ponens.* Then each time we need an axiom (scheme) or an additional rule to make the completeness proof work, we add it to the list. The list might get a bit long that way, but we can be sure it's complete if we've added enough axioms to make the proof go through. We can always go back and try to simplify the list by showing that one of the axioms is superfluous by deducing it from the others. This is how I got the axiomatization in this section. I'll highlight uses of the axiom schema by putting them in boldface so you can see where they're needed.

To simplify our work, we'll start by axiomatizing **PC** using only $\neg$ and $\rightarrow$ as connectives, taking all the definitions for the formal language as before, deleting all clauses involving $\wedge$ and $\vee$.

Classical Propositional Logic, PC *in* $L(\neg, \rightarrow)$

Each formal wff that is an instance of one of the following schema is an axiom.

1. $\neg A \rightarrow (A \rightarrow B)$
2. $B \rightarrow (A \rightarrow B)$
3. $(A \rightarrow B) \rightarrow ((\neg A \rightarrow B) \rightarrow B)$
4. $(A \rightarrow (B \rightarrow C)) \rightarrow ((A \rightarrow B) \rightarrow (A \rightarrow C))$

rule $\dfrac{A, A \rightarrow B}{B}$

Here are two examples of derivations, both of which we will need in the completeness proof below. I'll write 'axiom' as shorthand for 'axiom scheme'.

Lemma 4

a. $\vdash A \rightarrow A$

b. $\{A, \neg A\} \vdash B$

Proof On the right-hand side I give a justification for each step of the derivation.

(a)	1. $\vdash A \rightarrow ((A \rightarrow A) \rightarrow A)$	(an instance of) **axiom 2**
	2. $\vdash A \rightarrow (A \rightarrow A)$	**axiom 2**
	3. $\vdash (A \rightarrow ((A \rightarrow A) \rightarrow A)) \rightarrow ((A \rightarrow (A \rightarrow A)) \rightarrow (A \rightarrow A))$	**axiom 4**
	4. $\vdash (A \rightarrow (A \rightarrow A)) \rightarrow (A \rightarrow A)$	*modus ponens* using (1) and (3)
	5. $\vdash A \rightarrow A$	*modus ponens* using (2) and (4)
(b)	1. $\vdash \neg A \rightarrow (A \rightarrow B)$	**axiom 1**
	2. $\neg A$	premise
	3. $(A \rightarrow B)$	*modus ponens* using (1) and (2)
	4. A	premise
	5. B	*modus ponens* on (3) and (4) ■

Theorem 5

a. ***The syntactic deduction theorem*** $\Gamma, A \vdash B$ iff $\Gamma \vdash A \rightarrow B$

b. ***The syntactic deduction theorem for finite consequences***
$\Gamma \cup \{A_1, \ldots, A_n\} \vdash B$ iff $\Gamma \vdash A_1 \rightarrow (A_2 \rightarrow (\cdots \rightarrow (A_n \rightarrow B) \cdots))$

Proof (a) From right to left is immediate, since *modus ponens* is our rule.

To show that if $\Gamma, A \vdash B$, then $\Gamma \vdash A \rightarrow B$, suppose that $B_1, \ldots, B_n$ is a proof of B from $\Gamma \cup \{A\}$. I'll show by induction that for each i, $1 \leq i \leq n$, $\Gamma \vdash A \rightarrow B_i$.

Either $B_1 \in \Gamma$ or B_1 is an axiom, or B_1 is A. In the first two cases the result follows by using **axiom 2**. If B is A, we could just add $A \rightarrow A$ to our list of axioms. But I showed in Lemma 4.a, using **axioms 2** and **4**, that $\vdash A \rightarrow A$.

Now suppose that for all $k < i$, $\vdash A \rightarrow B_k$. If B_i is an axiom, or $B_i \in \Gamma$, or B_i is A, we have $\vdash A \rightarrow B_i$ as before. The only other case is when B_i is a consequence by *modus ponens* of B_m and $B_j = B_m \rightarrow B_i$ for some $m, j < i$. Then by induction $\Gamma \vdash A \rightarrow (B_m \rightarrow B_i)$ and $\Gamma \vdash A \rightarrow B_m$, so by **axiom 4**, $\Gamma \vdash A \rightarrow B_i$. I'll leave part (b) as an exercise. ■

Theorem 6 Theories, consistency, and completeness

a. If Γ is a theory, then every axiom is in Γ.

b. If Γ is consistent and $\Delta \subseteq \Gamma$, then Δ is consistent.

c. Γ is consistent iff every finite subset of Γ is consistent.

d. Γ is consistent iff there is some B such that $\Gamma \nvdash B$.

e. Γ is a complete theory iff for every A, either $A \in \Gamma$ or $\neg A \in \Gamma$.

f. If Γ is complete, then for every A such that $\Gamma \nvdash A$, for every B, we have $\Gamma, A \vdash B$.

g. If Γ is consistent, then $\Gamma \cup \{A\}$ or $\Gamma \cup \{\neg A\}$ is consistent.

h. $\Gamma \nvdash A$ iff $\Gamma \cup \{\neg A\}$ is consistent.

i. Γ is a complete and consistent theory iff for every A, exactly one of $A, \neg A \in \Gamma$.

Proof I'll leave (a), (b) and (e) to you.

(c) I'll prove the negated version. Suppose Γ is inconsistent. Then there is some A such that $\Gamma \vdash A$ and $\Gamma \vdash \neg A$. Hence, by Theorem 3.h, there is some finite $\Sigma \subseteq \Gamma$ and finite $\Delta \subseteq \Gamma$ such that $\Sigma \vdash A$ and $\Delta \vdash \neg A$. Then $\Sigma \cup \Delta \subseteq \Gamma$ and is finite and inconsistent. The other direction is immediate.

(d) I'll show the negated equivalence. If Γ is inconsistent, then for some A, $\Gamma \vdash A$ and $\Gamma \vdash \neg A$. So by **axiom 1**, for any B, $\Gamma \vdash B$. The other direction is easy.

(f) If $A \notin \Gamma$, then $\neg A \in \Gamma$, so $\Gamma \cup \{A\}$ is inconsistent, and we're done by (d).

(g) Suppose that both $\Gamma \cup \{A\}$ or $\Gamma \cup \{\neg A\}$ are inconsistent. Then for any B, by part (d), $\Gamma \cup \{A\} \vdash B$ and $\Gamma \cup \{\neg A\} \vdash B$. Hence by the syntactic deduction theorem and **axiom 3**, for every B, $\Gamma \vdash B$. So Γ is inconsistent.

(h) If $\Gamma \vdash A$, then $\Gamma \cup \{\neg A\}$ is inconsistent. On the other hand, if $\Gamma \cup \{\neg A\}$ is inconsistent, then by part (d), $\Gamma \cup \{\neg A\} \vdash A$. Hence by the syntactic deduction theorem, $\Gamma \vdash \neg A \to A$. By Lemma 4.a, $\Gamma \vdash A \to A$. Hence by **axiom 3**, $\Gamma \vdash A$.

(i) Suppose for every A, exactly one of A, $\neg A$ is in Γ. By (e) Γ is a complete theory, and so it must be consistent. The other direction is immediate. ■

Exercises for Section D.1

1. Exhibit derivations that establish the following:
   a. $\vdash (\neg A \to A) \to A$
   b. $\vdash A \to (B \to B)$
   c. $\vdash A \to \neg\neg A$
   d. $\vdash \neg\neg A \to A$ (Hint: Use the deduction theorem.)
2. Prove Theorem 5.b. (Hint: Use induction on *n*.)
3. a. Establish $A \vdash \neg A \to B$.
   b. Establish $\{A \to B, B \to C\} \vdash A \to C$.
   c. Use the algorithm implicit in the proof of Theorem 5.a to convert the derivation in (a) to a proof of $\vdash A \to (\neg A \to B)$.
4. What changes would need to be made to the material in this section if we replace axiom scheme 1 by $A \to (\neg A \to B)$ and kept the other axioms and rule ?

2. A completeness proof

Rather than showing directly that every valid wff is a theorem, I'll show the contrapositive: If a wff is not a theorem, then it is not valid. First I'll establish the informal remark I made above that a complete and consistent collection of wffs is as full a description of the world, relative to the atomic propositions, as we can give in our logic.

Lemma 7 Γ is a complete and consistent theory iff there is a model ν such that $\Gamma = \{A: \nu(A) = \mathsf{T}\}$.

Proof I'll let you show that the set of wffs true in a model is a complete and consistent theory.

Suppose now that Γ is a complete and consistent theory. Define a function ν: Wffs $\to \{\mathsf{T}, \mathsf{F}\}$ by setting $\nu(A) = \mathsf{T}$ iff $A \in \Gamma$. To show that ν is a model we need to show that it evaluates the connectives correctly.

If $\nu(\neg A) = \mathsf{T}$, then $\neg A \in \Gamma$,
so $A \notin \Gamma$ by consistency,
so $\nu(A) = \mathsf{F}$.

If $\nu(A) = \mathsf{F}$, then $A \notin \Gamma$,
so $\neg A \in \Gamma$ by completeness,
so $\nu(\neg A) = \mathsf{T}$.

Suppose $\upsilon(A \rightarrow B) = T$. Then $A \rightarrow B \in \Gamma$. If $\upsilon(A) = T$, we have $A \in \Gamma$, so by *modus ponens*, $B \in \Gamma$, as Γ is a theory. Hence $\upsilon(B) = T$. Conversely, suppose $\upsilon(A) = F$ or $\upsilon(B) = T$. If the former then $A \notin \Gamma$, so $\neg A \in \Gamma$, and by **axiom 1**, $A \rightarrow B \in \Gamma$; so $\upsilon(A \rightarrow B) = T$. If the latter then $B \in \Gamma$, and as Γ is a theory, by **axiom 2**, $A \rightarrow B \in \Gamma$; so $\upsilon(A \rightarrow B) = T$. ■

Given a wff that is not a theorem, I'll show now that there is a complete and consistent collection of wffs that does not contain it. Using that and the previous Lemma, it will be easy to prove the completeness of our axiomatization.

Lemma 8 If $\nvdash D$, there is some complete and consistent theory Γ such that $D \notin \Gamma$.

Proof Let $A_0, A_1, \ldots$ be our numbering of the wffs of the formal language. Define:

$$\Gamma_0 = \{\neg D\}$$

$$\Gamma_{n+1} = \begin{cases} \Gamma_n \cup \{A_n\} & \text{if this is consistent} \\ \Gamma_n & \text{otherwise} \end{cases}$$

$$\Gamma = \bigcup_n \Gamma_n$$

We have that Γ_0 is consistent by Lemma 7.h. So by construction, each Γ_n is consistent. Hence Γ is consistent, for if not some finite $\Delta \subseteq \Gamma$ is inconsistent by Theorem 6.c, and Δ being finite, $\Delta \subseteq \Gamma_n$ for some n. In that case Γ_n would be inconsistent.

Γ is complete, because if $\Gamma \nvdash A$, then by Theorem 6.h, $\Gamma \cup \{\neg A\}$ is consistent, and hence by construction, $\neg A \in \Gamma$, and so $\Gamma \vdash \neg A$.

Finally, Γ is a theory, since if $\Gamma \vdash A$, then $\Gamma \cup \{A\}$ is consistent, and hence by construction, $A \in \Gamma$. ■

Theorem 9

***a. Completeness of the axiomatization of* PC** $\vdash A$ iff $\models A$.

b. Finite strong completeness For finite Γ, $\Gamma \vdash A$ iff $\Gamma \models A$.

Proof (a) First, the axiomatization is sound: Every axiom is a tautology, as you can check, and if A and $A \rightarrow B$ are tautologies, then so is B. Hence if $\vdash A$, then A is a tautology.

Now suppose that $\nvdash A$. We will show that A is not a tautology. Since $\nvdash A$, by Lemma 8 there is a complete and consistent Γ such that $A \notin \Gamma$. By Lemma 7 there is some model υ such that Γ is the set of wffs true in υ. So $\upsilon(A) = F$, and A is not a tautology. Thus, if A is a tautology, it is a theorem.

(b) Suppose $\Gamma = \{A_1, \ldots, A_n\}$. If $\{A_1, \ldots, A_n\} \vdash A$ then by the syntactic deduction theorem for finite consequences, $\vdash A_1 \rightarrow (A_2 \rightarrow (\cdots \rightarrow (A_n \rightarrow A) \cdots)$. Hence, by (a), $\models A_1 \rightarrow (A_2 \rightarrow (\cdots \rightarrow (A_n \rightarrow A) \cdots)$, and so $\{A_1, \ldots, A_n\} \models A$ (Theorem 1.j). ■

We have shown that for classical propositional logic the syntactic and semantic formalizations of 'follows from' result in the same relation on wffs and hence on propositions. Moreover, these metalogical notions can be identified with the theoremhood or validity of the corresponding conditional.

Corollary 10 The following are equivalent:

$A \vDash B \quad A \vdash B \quad \vDash A \to B \quad \vdash A \to B$

Proof The equivalence of the first two is Theorem 9.b. The equivalence of the first and third is the semantic deduction theorem. The equivalence of the second and fourth is the syntactic deduction theorem. ■

In Lemma 8 we showed that given a non-theorem, there exists a complete and consistent collection that does not contain it. But that proof is not constructive, for we did not show how to determine whether $\Gamma_n \cup \{ A_n \}$ is consistent. Though Theorem 9 tells us a derivation for a tautology such as $(p_1 \to (p_2 \to p_3)) \to (p_2 \to (p_1 \to p_3))$ exists in our system, we haven't shown how to produce it. In an appendix below I give a constructive proof of completeness that sets out a method for producing derivations of tautologies.

In any case, to prove the following strong completeness theorem in which Γ can be infinite, which I'll show is an easy extension of the method of proof above, we must use nonconstructive methods (see *Henkin, 1954*).

Lemma 11

a. If Σ is consistent and $\Sigma \nvdash D$, then there is a complete and consistent set Γ such that D is not in Γ and $\Sigma \subseteq \Gamma$.

b. Every consistent collection of wffs has a model.

Proof Part (a) follows just as for Lemma 8, except that we take Γ_0 to be $\Sigma \cup \{\neg D\}$. Part (b) then follows by Lemma 7. ■

***Theorem 12 Strong completeness of the axiomatization of* PC**

For all Γ and A, $\Gamma \vdash A$ iff $\Gamma \vDash A$

Proof If $\Gamma \vdash A$, then given any model of Γ all the wffs in Γ are true, as are the axioms. Since the rule is valid, A must be true in the model, too. Hence $\Gamma \vDash A$. Suppose $\Gamma \nvdash A$. Then proceed as in the proof of Theorem 9 using Lemma 11. ■

The most striking consequence of this nonconstructive method is the following, which cannot be proved by constructive means.

Theorem 13 Compactness of the semantic consequence relation

a. $\Gamma \vDash A$ iff there is some finite collection $\Delta \subseteq \Gamma$ such that $\Delta \vDash A$.

b. Γ has a model iff every finite subset of Γ has a model.

Proof Part (a) follows from Theorem 12 and Theorem 3.h.

(b) If Γ has a model, then so does every finite subset of it. Now suppose every finite subset of Γ has a model. If Γ is not consistent, then for some finite $\Delta \subseteq \Gamma$, we have $\Delta \vdash A$ and $\Delta \vdash \neg A$. Hence by strong completeness, $\Delta \models A$ and $\Delta \models \neg A$. But then Δ cannot have a model, which is a contradiction. So Γ has a model. ■

Exercises for Section D.2

1. Prove that if υ is a model and Γ is the collection of wffs true in υ, then Γ is complete and consistent.

2. Use the completeness theorem to establish:
   a. $\vdash \neg(A \to B) \to (A \to \neg B)$
   b. $\vdash ((A \to B) \to A) \to A$

3. Prove Lemma 8 using at stage n of the construction:
$$\Gamma_{n+1} = \begin{cases} \Gamma_n \cup \{A_n\} & \text{if this is consistent} \\ \Gamma_n \cup \{\neg A_n\} & \text{otherwise} \end{cases}$$

4. Show that axiom scheme 3 can be replaced by $(\neg A \to A) \to A$.
(Hint: Search through the entire completeness proof beginning with Section D.1 to see where axiom scheme 3 is used.)

5. Prove: $\{A_1, \ldots, A_n\} \vdash B$ iff $\vdash \neg((A_1 \wedge \cdots \wedge A_n) \wedge \neg B)$.

6. Prove that if $\vdash A$, then there is a proof $A_1, \ldots, A_n$ of A in **PC** such that all the propositional variables that appear in $A_1, \ldots, A_n$ appear in A. (Hint: If A is a theorem, it's valid. So restrict the formal language to the variables in A.)

3. Independent axiom systems

Simpler axiom systems are better because the syntactic agreements necessary to establish the logic are then fewer and more persuasive. In particular, we prefer systems in which no axiom is superflous.

Given an axiom system we say that an *axiom is independent* of the others if we can't prove it from the system that results by deleting that axiom. An *axiom scheme is independent* if there is at least one instance of that scheme that cannot be proved if the scheme is deleted. A *rule* is independent if there is at least one wff that can be proved using the rule that cannot be proved if the rule is deleted. *An axiom system is independent* if each axiom (scheme) and rule is independent. We often say an axiom is independent when we mean that the scheme is. An axiom (scheme) that is a consequence of the others in a system is superfluous, though it may serve to shorten proofs.

In a series of exercises below I ask you to show that the axiomatization of **PC** given above is independent.

4. Derived rules and substitution

A *derived rule* is a scheme of consequences $\{A_1, \ldots, A_n\} \vdash B$. By using the deduction theorem and the completeness theorem we can obtain a derived rule from each of our axioms: Take the antecedent as premise and consequent as conclusion. For example, we have the derived rule: $\dfrac{\neg A}{A \rightarrow B}$. Some tautologies are more familiar in the form of rules:

modus tollens $\dfrac{\neg B,\ A \rightarrow B}{\neg A}$ *reductio ad absurdum* $\dfrac{A \rightarrow B,\ A \rightarrow \neg B}{\neg A}$

I'll let you show (Exercise 1 below) that if we can prove A from Σ using a derived rule as well as the original rules, then $\Sigma \vdash A$.

With the style of proof we use, it is customary to minimize the number of rules taken as primitive at the expense of additional axioms. There are other formalizations of the notion of proof, however, that use only rules, and in some cases those give clearer derivations.

Our axiom system uses schema. However, some authors prefer to present axiom systems using what they call the *rule of substitution*:

$$\frac{\vdash A(p)}{\vdash A(B)}$$

By Theorem 1 and the finite strong completeness theorem, this holds in **PC**. It shows that propositional variables are really variables: If $\vdash A(p)$, then no matter what proposition 'p' is to stand for, simple or compound, we have a theorem. But there is a drawback to using this as a proof method. We must formulate it to apply only to the derivation of theorems and not to syntactic consequences as in:

$$\frac{A(p)}{A(B)}$$

If we were to use this latter, every wff would be a syntactic consequence of $\{p_1\}$, which we certainly don't want. We cannot use the rule of substitution to derive consequences from an arbitrary collection of wffs. Moreover, the rule of substitution only makes sense for the formal language, not for the semi-formal language of propositions. Indeed, by our original definition it isn't really even a rule.

One derived rule we often do employ, though, is the following, a syntactic version of the division of form and content.

Corollary 14 Substitution of logical equivalents If C(B) is the result of substituting B for some but not necessarily all occurrences of the subformula A in C, then we have the derived rule in **PC**, $\dfrac{A \leftrightarrow B}{C(A) \leftrightarrow C(B)}$.

Proof Given evaluations of $\upsilon(A)$ and $\upsilon(B)$, the evaluations of $\upsilon(C(A))$ and $\upsilon(C(B))$ can differ only if $\upsilon(A) \neq \upsilon(B)$. So if $\upsilon \vDash A \leftrightarrow B$, then $\upsilon \vDash C(A) \leftrightarrow C(B)$. Hence, $A \leftrightarrow B \vDash C(A) \leftrightarrow C(B)$, and the corollary follows. ■

5. An axiomatization of PC in $L(\neg, \rightarrow, \wedge, \vee)$

Originally we defined **PC** in $L(\neg, \rightarrow, \wedge, \vee)$, so we should give an axiomatization in that language, too. We can start with the axioms for **PC** in $L(\neg, \rightarrow)$. Then in the completeness proof for $L(\neg, \rightarrow)$ in Section D the only place the other connectives have to be considered is in the proof of Lemma 7. We need additional axioms to ensure that $\wedge$ and $\vee$ are evaluated correctly.

PC *in* $L(\neg, \rightarrow, \wedge, \vee)$

axiom schema

1. $\neg A \rightarrow (A \rightarrow B)$
2. $B \rightarrow (A \rightarrow B)$
3. $(A \rightarrow B) \rightarrow ((\neg A \rightarrow B) \rightarrow B)$
4. $(A \rightarrow (B \rightarrow C)) \rightarrow ((A \rightarrow B) \rightarrow (A \rightarrow C))$
5. $A \rightarrow (B \rightarrow (A \wedge B))$
6. $(A \wedge B) \rightarrow A$
7. $(A \wedge B) \rightarrow B$
8. $A \rightarrow (A \vee B)$
9. $B \rightarrow (A \vee B)$
10. $(A \rightarrow C) \rightarrow ((B \rightarrow C) \rightarrow ((A \vee B) \rightarrow C))$

rule $\dfrac{A,\ A \rightarrow B}{B}$

Theorem 15 This axiom system is strongly complete for **PC**.

Proof The proof is as for the system in $L(\neg, \rightarrow)$ of Section D.1, except that we supplement the proof of Lemma 7:

$$\begin{aligned} \upsilon(A \wedge B) = \mathsf{T} \quad &\text{iff} \quad \upsilon(A) = \mathsf{T} \text{ and } \upsilon(B) = \mathsf{T} \\ &\text{iff} \quad A \in \Gamma \text{ and } B \in \Gamma \\ &\text{iff} \quad (A \wedge B) \in \Gamma \text{ using \textbf{axioms 5, 6} and \textbf{7}} \end{aligned}$$

For disjunction, we have that if $A \in \Gamma$ or $B \in \Gamma$, then $(A \vee B) \in \Gamma$ by **axioms 8** and **9**. So if $\upsilon(A) = \mathsf{T}$ or $\upsilon(B) = \mathsf{T}$, then $\upsilon(A \vee B) = \mathsf{T}$. In the other direction, if $\upsilon(A \vee B) = \mathsf{T}$, then $(A \vee B) \in \Gamma$. If both $A \notin \Gamma$ and $B \notin \Gamma$, then by Theorem 6.f, for any C, $\Gamma \cup \{A\} \vdash C$ and $\Gamma \cup \{B\} \vdash C$. So by the syntactic deduction theorem, for any C, $\Gamma \vdash A \rightarrow C$ and $\Gamma \vdash B \rightarrow C$. Hence by **axiom 10,** $\Gamma \vdash (A \vee B) \rightarrow C$. As $(A \vee B) \in \Gamma$, this means that for every C, $\Gamma \vdash C$, which is a contradiction on the consistency of Γ. Hence, $A \in \Gamma$ or $B \in \Gamma$, so that $\upsilon(A) = \mathsf{T}$ or $\upsilon(B) = \mathsf{T}$. ■

Corollary 16 $\Gamma \cup \{A_1, \ldots, A_n\} \vdash B$ iff $\Gamma \vdash (A_1 \wedge \cdots \wedge A_n) \rightarrow B$.

Exercises for Sections D.3–D.5

1. Prove: If there is a sequence $A_1, \ldots, A_n$ where each A_i is either an axiom, is in Σ, or is a consequence of some of the previous A_i using a rule of the system or a derived rule, then $\Sigma \vdash A_n$. (Hint: Each time a derived rule is employed, insert into the proof the derivation that justifies the rule.)

2. *Independence of the axiomatization of* **PC** *in* L(¬, →) (*R. Flagg, 1978*)
Axiom scheme 1 Consider the two tables below for evaluating ¬ and → .

A	¬A
0	0
1	0

A → B	B = 0	B = 1
A 0	0	1
1	0	0

a. Show that no matter what values are given to the propositional variables, each instance of axiom scheme 2, 3, or 4 will be assigned value 0 by these tables. (For example, consider scheme 2. Suppose A is given value 1 and B is given value 0. Then (A→B) is given value 0, and B→(A→B) is given 0.)

b. Show that if these tables assign value 0 to C and to C→D, then D must have value 0, and hence every consequence of axiom schema 2, 3, and 4 can receive only value 0.

c. Show that axiom scheme 1 can be assigned value 1 by these tables. (Hint: In shorthand, ¬0→(0→1) = 1.) Conclude that axiom scheme 1 is independent.

d. *Axiom scheme* 2 Use the following tables and proceed as for axiom scheme 1 to show axiom scheme 2 is independent, except that here every consequence of axiom schema 1, 3, and 4 takes value 1. (Hint: Assign 0 to A and 1 to B for scheme 2.)

A	¬A
0	2
1	2
2	1

A → B	B = 0	B = 1	B = 2
A 0	1	2	2
1	1	1	2
2	1	1	1

e. *Axiom scheme* 3 Use the following tables and proceed as for axiom scheme 1 to show that axiom scheme 3 is independent. (Hint: Assign 1 to A and 1 to B for scheme 3.)

A	¬A
0	1
1	1

A → B	B = 0	B = 1
A 0	0	1
1	0	0

f. *Axiom scheme* 4 Use the following tables and proceed as for axiom scheme 1 to show that axiom scheme 4 is independent. (Hint: For axiom scheme 4 assign 1 to A, 3 to B, 2 to C.)

A	¬A
0	3
1	2
2	1
3	0

A → B	B = 0	B = 1	B = 2	B = 3
A 0	0	1	2	3
1	0	0	2	0
2	0	1	0	3
3	0	0	0	0

g. Show that each scheme is independent of all others plus $A \to A$. (Hint: Show what value each table gives to $A \to A$.)

h. Show that if we define an axiom scheme to be independent of others if all instances of it are independent of all instances of all others, then axiom scheme 1 is not independent of the other schema. (Hint: Consider the wff $\neg A \to (A \to (C \to C))$.)

3. Give an axiomatization of **PC** in the language $L(\neg, \wedge)$. (Hint: Use $\dfrac{\neg(A \wedge \neg B),\ A}{B}$.)

4. Prove or disprove: The axiomatization of **PC** in $L(\neg, \to, \wedge, \vee)$ is independent.

5. We say Σ is *closed under* a rule if whenever wffs that can serve as the premises of the rule are in Σ, then so is their consequence.

   The closure of a collection of wffs under a rule
   Given a collection of wffs Γ and a rule $\{A_1, \ldots, A_m\} \vdash B$,
   i. Assign the number 1 to each wff in Γ.
   ii. If the maximum of the numbers assigned to $B_1, \ldots, B_m$ in this procedure is n, and $B_1, \ldots, B_m$ are wffs of the correct form to be premises of the rule, and the application of the rule to these wffs in this order results in C, then C is assigned the number $n + 1$.
   iii. A wff is in the closure of Γ iff it is assigned some number $n \geq 1$ by (i) or (ii).

a. List all wffs that are assigned 2 in the closure of $\{p, q, p \vee r\}$ under the rule of adjunction.

b. Is $\{p_1, p_2 \to p_3, p_4 \to p_5\}$ closed under *modus ponens*?

c. Is $\{p_1, p_2, p_1 \wedge p_2, p_2 \wedge p_1\}$ closed under the rule of adjunction?

d. Is $\{p_1, p_1 \to p_2, p_2, p_2 \to p_3\}$ closed under *modus ponens*?

e. A platonist would say that the *closure* of Γ under a rule is the smallest collection of wffs Δ such that $\Delta \supseteq \Gamma$ and Δ is closed under the rule. Explain the virtues and defects of that definition compared to the inductive one given above.

A constructive proof of completeness of PC

The first axiomatization of classical logic in terms that are recognizably the same as ours was by *Russell and Whitehead, 1910–13*, in their *Principia Mathematica*. Their work was indebted to an earlier axiomatization by *Frege, 1879*. For Frege, Whitehead, and Russell, and many others at the time, the question of completeness simply did not arise. Their goal was to show that all logic, and indeed all mathematics, could be developed within their system; hence they did not look outside their formal systems for justification in terms of meanings of their formulas. *Dreben and van Heijenoort, 1986*, pp. 44–47, and *Goldfarb, 1979*, discuss their views.

The first use of semantic notions to justify Whitehead and Russell's system, giving a completeness proof, was by Bernays, though that was not published until *Bernays, 1926* (see *Dreben and van Heijenoort, 1986*). The first published completeness proof for Whitehead and Russell's system, and long the most influential, was by *Post, 1921*. He showed that every wff is semantically equivalent to one in disjunctive normal form.

Supplementing the system of *Principia Mathematica* with further axioms, he then showed how given any tautology A it was possible to produce a derivation of the equivalent disjunctive normal form B of A and a derivation of $B \to A$. Joining those derivations he then had a proof of A. Only later was it shown that his further axioms could be proved in the system of *Principia Mathematica.*

Post's proof has a very clear advantage over the proof in Section D: Given a tautology it shows how to produce a derivation for it in the formal system. However, Post's proof has a disadvantage: The strong completeness of **PC** cannot be deduced from it, since the proof of strong completeness requires substantial infinitistic nonconstructive assumptions (see *Henkin, 1954*) as in the proof I gave, which is due to *Łos, 1951*, based on the work of Adolf Lindenbaum (see *Tarski, 1930*, Theorem 12). There are several other constructive proofs of completeness, surveyed in *Surma, 1973 B,* all of which share the same advantage and disadvantages.

The axiomatization of **PC** given in Section D was first presented no later than 1969 by H. A. Pogorzelski (see *Flagg, 1978*).

The constructive proof of completeness of **PC** that I'll present here is due to *László Kalmár, 1935*. I will use Lemma 4 and the syntactic deduction theorem of Section D above, since the proofs of those show how to produce the derivations.

Lemma A

a. $\vdash A \to \neg\neg A$

b. $\{A, \neg B\} \vdash \neg(A \to B)$

Proof (a) As before, I'll put the justification to the right of each step.

1. $(\neg A \to (A \to \neg\neg A)) \to [(\neg\neg A \to (A \to \neg\neg A)) \to (A \to \neg\neg A)]$ — axiom 3
2. $\neg A \to (A \to \neg\neg A)$ — axiom 1
3. $(\neg\neg A \to (A \to \neg\neg A)) \to (A \to \neg\neg A)$ — *modus ponens* on (1) and (2)
4. $\neg\neg A \to (A \to \neg\neg A)$ — axiom 2
5. $A \to \neg\neg A$ — *modus ponens* on (3) and (4)

(b) It's immediate that $\{A, \neg B, A \to B\} \vdash \neg B$ and $\{A, \neg B, A \to B\} \vdash B$. Hence by Lemma 4.b, $\{A, \neg B, A \to B\} \vdash \neg(A \to B)$. So by the syntactic deduction theorem, $\{A, \neg B\} \vdash (A \to B) \to \neg(A \to B)$. So $\{A, \neg B\} \vdash \neg(A \to B) \to \neg(A \to B)$ by Lemma 4.a and Theorem 3.b. An instance of axiom 3 is $[(A \to B) \to \neg(A \to B)] \to [(\neg(A \to B) \to \neg(A \to B)) \to \neg(A \to B)]$. Hence, by applying *modus ponens* twice, we get $\{A, \neg B\} \vdash \neg(A \to B)$. ∎

In the proof of part (b) I've shown that from our previous work we can conclude there is such a derivation. The proofs of the earlier lemmas on which this one depends show how to produce a derivation.

Lemma B Let C be any wff and $q_1, \ldots, q_n$ the propositional variables appearing in C. Let υ be any valuation. Define, for all $i \leq n$:

$$Q_i = \begin{cases} q_i & \text{if } \upsilon(q_i) = \mathsf{T} \\ \neg q_i & \text{if } \upsilon(q_i) = \mathsf{F} \end{cases}$$

If $\upsilon(C) = \mathsf{T}$, then $\{Q_1, \ldots, Q_n\} \vdash C$. If $\upsilon(C) = \mathsf{F}$, then $\{Q_1, \ldots, Q_n\} \vdash \neg C$.

The proof is by induction on the length of C. If C is (p_i), then, as you can show, we need only that $\vdash p_i \rightarrow p_i$ and $\vdash \neg p_i \rightarrow \neg p_i$, which we have from Lemma 4.a .

Now suppose it is true for all wffs shorter than C. Let $\Gamma = \{Q_1, \ldots, Q_n\}$ as above. If C is $\neg$A, suppose $\upsilon(C) = T$. Then $\upsilon(A) = F$. So by induction $\Gamma \vdash \neg A$. If $\upsilon(C) = F$, then $\upsilon(A) = T$, so by induction $\Gamma \vdash A$. By the previous lemma and Lemma 4.b, $\Gamma \vdash A \rightarrow \neg\neg A$, so using *modus ponens*, $\Gamma \vdash \neg\neg A$, as we wish.

If C is $A \rightarrow B$, first suppose $\upsilon(C) = T$. Then $\upsilon(A) = F$, or $\upsilon(B) = T$. If $\upsilon(A) = F$, then $\Gamma \vdash \neg A$, so by axiom 1, $\Gamma \vdash A \rightarrow B$. If $\upsilon(B) = T$, then axiom 2 allows us to conclude that $\Gamma \vdash A \rightarrow B$. Finally, if $\upsilon(C) = F$, then $\upsilon(A) = T$ and $\upsilon(B) = F$. So $\Gamma \vdash A$ and $\Gamma \vdash \neg B$. Hence by the previous lemma, $\Gamma \vdash \neg(A \rightarrow B)$. ∎

Theorem C

a. ***Completeness of the axiomatization of*** **PC** $\vdash A$ iff $\models A$.

b. ***Finite strong completeness*** For finite Γ, $\Gamma \vdash A$ iff $\Gamma \models A$.

Proof (a) As before, if $\vdash A$, then $\models A$.

Now suppose A is valid, that is, for every valuation υ, $\upsilon(A) = T$. Let $q_1, \ldots, q_n$ be the propositional variables appearing in A. Consider the valuation υ such that for all *i*, $\upsilon(q_i) = T$. By Lemma B, we have $\{q_1, \ldots, q_n\} \vdash A$. Hence, by the syntactic deduction theorem, $\{q_1, \ldots, q_{n-1}\} \vdash q_n \rightarrow A$. Similarly, as the valuation $\upsilon(q_i) = T$ for $i < n$ and $\upsilon(q_i) = F$ yields $\upsilon(A) = T$, we get $\{q_1, \ldots, q_{n-1}\} \vdash \neg q_n \rightarrow A$. Thus, using axiom scheme 3, we can obtain $\{q_1, \ldots, q_{n-1}\} \vdash A$. Repeating this argument *n* times, we have $\vdash A$.

Part (b) follows by the syntactic deduction theorem, as in the proof of Theorem 9.b. ∎

In the proof of Theorem C I didn't exhibit the derivation of the given tautology, but you can trace through the earlier proofs on which it depends to produce one.

III The Language of Predicate Logic

A. Things, the World, and Propositions

Propositional logic allows us to justify our intuition that certain inferences are valid due solely to the form of the propositions. For example:

Ralph is a dog and George is a duck.
Therefore, Ralph is a dog.

This can be justified as valid by arguing that the first proposition has the form of two propositions joined by 'and', which we can analyze via our formal models of $\wedge$.

But now consider:

(1) All dogs bark.
Ralph is a dog.
Therefore, Ralph barks.

This is clearly valid: It's not possible for the premises to be true and the conclusion false. And certainly the form of the propositions guarantees that.

But what is the form that guarantees the validity of this inference? Propositional logic is no help to us, for the propositional form is simply: p, q therefore r. Nor can propositional logic be used to justify our intuition that the following is valid:

8 is divisible by 2.
Therefore, some number is divisible by 2.

We want to analyze why we consider these to be valid, based on general principles that apply to many other examples. To model more kinds of inferences and to justify further formulas or inferences as valid (or invalid), we have to consider the internal structure of propositions.

To look at all possible forms of propositions would be too complicated. How, then, shall we parse the internal structure of propositions? Roughly speaking, the syntactic structures we'll look at are based on considering nouns to be the most important part of propositions. The reason we concentrate on nouns is because of a particular view of the world, a view that is part but not all of the way we "see" in our language: The world is made up of things.

We have some intuition about what a thing is, some general notion—not entirely precise, but in many cases of *individual* things we can agree: tables, chairs, people, dogs, are individual things. In many cases we don't agree: real numbers, infinite sets, wisdom, atoms—are these things?

We use our basic shared notion of what an individual thing is to motivate our work. In the end, what a thing is will be circumscribed by the agreements we use to establish a formal syntactic and semantic analysis. Differences about our idea of things within that framework may determine, in part, which logical forms we consider valid. We'll make the following assumptions.

- The world is made up (to some extent) of individual things.
- There are many propositions and deductions that can be understood as being about individual things.
- We shall be concerned with propositions and deductions only insofar as they can be construed as being about individual things.

In brief, we summarize our assumptions with the following.

Things, the world, and propositions The world is made up of individual things; propositions are about individual things.

Repetitive use of the word *thing* makes for tedious reading. For variety I'll also use *object*, *individual*, *entity*, and *element*.

Our assumption that the world is made up of things stands in contrast to other ways of seeing. From the view that the world is all flux and becoming, that we cannot step into the same river twice, it follows that processes, not things, are fundamental. But the assumption that the world is made up of things is sufficiently pervasive in our ordinary reasoning to serve as the basis for our formal work for now. In my *Predicate Logic* and in an appendix to my *Five Ways of Saying "Therefore"* I discuss the nature of this assumption and the limitations it imposes.

B. Names and Predicates

If propositions are about individual things, then words that pick out or purport to pick out specific individual things are of special importance. Such words are called *names*, for example, 'Socrates', 'Richard L. Epstein', 'Santa Claus'.

Which words we accept as names depends partly on what (we believe) exists, and partly on the roles names play in our reasoning. Is 7 a name? π? The United Nations? In the examples here I'll use names of objects that I'm familiar with rather than names of only well-known people and objects that carry many irrelevant associations, hoping you'll trust that they pick out specific objects.

An example of the simplest kind of proposition that involves a name is:

Ralph is sitting.

We have a name, 'Ralph', and what grammarians call a 'predicate', namely 'is sitting'. There's no further for our analysis to go.

Consider now a proposition with two names:

(2) Dick loves Juney.

We can view this as a name, 'Dick', and a predicate, 'loves Juney'. Or we can view the proposition as being about Juney, and hence it is composed of a name, 'Juney', and what's asserted about her, 'Dick loves'. Or we can view it as about both Dick and Juney, composed of two names and a predicate, 'loves'. This strains the usual reading of the word 'predicate' you learned in school, but we can say that a predicate arises by deleting either or both of the names in (2).

Consider another example using three names:

Dick and Juney are the parents of Ralph.

There are seven ways to parse this proposition depending on whether we consider it to be about one, two, or all three of the objects named in it. For instance, we can view it as a proposition about Dick and Ralph, and hence the predicate that arises is '— and Juney are the parents of —'. Here I've used dashes where I deleted names in order to make it clear where the names go. In any proposition that contains names we can obtain a predicate by deleting one or more of the names. Reversing this, we have a definition of the fundamental part of our grammar.

Predicate A *predicate* is any incomplete (English) phrase with specified gaps such that when the gaps are filled with names of things the phrase becomes a proposition.

This definition seems to say that no matter what names fill in the gaps in a predicate, a proposition results. But if '— loves —' is a predicate, what are we to make of '7 loves Juney'?

We have two choices. We may say that when we took '— loves —' to be a predicate we weren't thinking of names of all things as being suitable to fill in the blanks, but only, say, names of people and animals. That is, we can restrict what kinds of things we are talking about.

Alternatively, we can assume there is a homogeneity to all things: Whatever can be asserted about one thing can be asserted about any other, though of course the assertion may be false. In that case, '7 loves Juney' wouldn't be nonsensical, but false. In either case, what predicates we reason with depends in part on what things we are talking about.

In order to reduce ambiguity, let's require that each name-place of a predicate has a separate dash associated with it, and that for any particular predicate the number of dashes won't vary. We don't want '— is the brother of —' to be a predicate if by that we mean the latter blank can be filled with as many names as we wish. The number of blanks in a predicate is its *arity*: if a predicate has one blank it is *unary*, if two it is *binary*, if three it is *ternary*, ... Predicates other than unary ones are often called *relations*.

When it is clear where and how many dashes occur in a predicate we may informally leave the dashes off, saying 'barks' or 'is green' or 'is the father of' are predicates. Some writers leave the word 'is' to be understood too, saying, for example, that 'green' is a predicate.

This definition of 'predicate' recognizes two primitive grammatical categories: propositions and names. What other grammatical categories shall we recognize?

C. Propositional Connectives

We take propositions as fundamental, and so we continue to recognize the ways of forming new propositions from others as in propositional logic. As before, we'll concentrate on formalizations of 'and', 'or', 'if . . . then . . .', and 'not', symbolized by $\wedge$, $\vee$, $\rightarrow$, $\neg$.

An atomic proposition in propositional logic was one whose internal structure was not taken into account. Thus 'Ralph sits' was atomic. But now we recognize the structure of that proposition as made up of a name and a predicate. Nonetheless, let us continue to classify it as *atomic*, in the sense that it contains no formal symbol of our logic.

Consider now an example:

Ralph is a dog and Ralph barks.

It would be wrong to classify this as atomic, for then we could not justify the deduction from it of 'Ralph barks'. We should formalize it as:

Ralph is a dog $\wedge$ Ralph barks

If we delete the names in this semi-formal proposition, we obtain:

(3) — is a dog $\wedge$ — barks

This, by our definition, is a predicate. Predicates, just as propositions, can be classified as *atomic* or *compound*, according to whether they result in atomic or compound propositions when the blanks are filled with names, where a *compound proposition* is one whose structure is taken into account in terms of the logical symbols we'll use.

There is an ambiguity in the use of dashes in (3). I intended both gaps to be filled with the same name, but I had no way to indicate that. We need place-holders for cross-referencing: the letters 'x', 'y', 'z' will do for now. So 'x is a dog $\wedge$ x barks' becomes a proposition, though not an atomic one, when both x's are replaced with the same name.

D. Variables and Quantifiers

So far we have considered only propositions about specific named things. For such propositions, propositional logic is often adequate to formalize our reasoning. We are still not in a position to analyze argument (1), for we have not considered how to talk about things in general. The use of place-holders suggests a way.

Sometimes we use place-holders to indicate generality. For instance, reasoning about the arithmetic of the real numbers, we might assert '$(x = y) \rightarrow (y = x)$', meaning no matter what numbers x and y refer to, $(x = y) \rightarrow (y = x)$. Here '$x$' and '$y$' are acting more like 'this' and 'that', not simply as place-holders for names, for not every real number has a name that we could substitute for 'x' or 'y'. The reference of 'x' and 'y', as much as the reference of 'this' and 'that', can vary in context. Hence, we call a letter used for generality a *variable.* What we mean by 'the reference for 'x' can vary', and why this doesn't reintroduce the problems associated with indexicals discussed in Chapter I.B, we'll discuss in the next section. For now we're primarily interested in the form of how we can express generality.

Confusion will follow if we trust to context to make it clear whether we are taking 'x is a dog' as a predicate, or as a formalization of 'it is a dog', awaiting a reference for 'it' in order to become a proposition, or as a proposition about all things. We can avoid this ambiguity by requiring that we be explicit when we intend to talk about all things. For example:

For any thing, x, x is a dog.

Here we clearly have a proposition, a sentence that is true or false.

Similarly, for our example from arithmetic we should say:

For any thing, x, and any thing, y, $x = y \rightarrow y = x$.

The use of 'x' and 'y' to express generality has been made explicit, though it is implicit that the things we intend 'x' and 'y' to vary over are real numbers.

The use of quotation marks to indicate that we are talking about the letters 'x' and 'y' is becoming tedious. As with other formal symbols, let's assume that letters can name themselves and that we can figure out by context when we're using them and when we're mentioning them.

There are several words and phrases in English that we use to indicate that we are making an assertion about all things: 'for each thing', 'anything', 'everything', 'all things', 'given anything', 'no matter what thing'. Though their syntax varies—'any', 'each', 'every' take the singular, while 'all' takes the plural—let's use a single symbol, $\forall$, called the *universal quantifier*, to formalize uses of them that can be assimilated to that of 'for any thing'.

Thus we can formalize 'For any thing, x, x is a dog' as '$\forall x$, x is a dog'. But what of 'For any thing, x, and any thing, y, $x = y \rightarrow y = x$'? Is 'and' to be formalized as $\wedge$? No, for it does not connect two propositions. Rather than use an unformalized 'and', let's agree that successive uses of $\forall$ that apply to the same phrase can be treated as a list by being marked off by commas. So we'll have '$\forall x$, $\forall y$, $x = y \rightarrow y = x$'. Actually, we'd do well to dispense with commas in favor of the less ambiguous parentheses that clearly mark the beginning and end of phrases. So we will have semi-formal propositions such as '$\forall x$ (x is a dog)' and '$\forall x\, (\forall y\, (x = y \rightarrow y = x))$'.

We use other words and phrases in English to refer to unnamed objects in some kind of generality: 'some', 'many', 'lots of', 'all but seven', 'a few', 'exactly two', 'there is a', ... At this stage let's choose just one group of these as basic: those that we can reasonably interpret as meaning 'there is at least one thing'. This includes 'there is a', 'there exists a', and, depending on context, 'some' and 'for some'. We will formalize these with the symbol $\exists$, called the *existential quantifier*.

Thus if we have a predicate 'x is a dog' and wish to treat x as a variable to express generality and not simply as a place-holder, we can make a proposition by saying 'There is a thing, x, such that x is a dog'. The words 'such that' mark off the clause; since we've agreed to use parentheses for that purpose, it's appropriate to formalize this proposition as '$\exists x$ (x is a dog)'.

Similarly, we can transform the predicate '$x = y \rightarrow y = x$' into a proposition by indicating that x and y are to be taken as a specific pair of numbers, though which pair we don't say: 'For some thing, x, and some thing, y, $x = y \rightarrow y = x$'. We can symbolize this as '$\exists x\, (\exists y\, (x = y \rightarrow y = x))$'. Or we might mix quantifiers to formalize the proposition of arithmetic that there is a smallest number as '$\exists y\, (\forall x\, (y \leq x))$'.

We have taken as basic two ways to transform a predicate into a proposition by indicating that the place-holder(s) are to be taken to express some kind of generality: the use of the phrase 'for all' and the phrase 'there is a'. Either of these phrases we call an (*informal*) *quantification*; the formalized versions of these we also call *quantifications*.

I must assume that you understand the phrases 'there is a', 'for all', 'for some', 'there exists', etc., just as I had to assume you understood 'and' when I introduced the symbol $\wedge$. We shall replace these phrases with formal symbols to which we shall give fairly explicit and precise meaning, based on our understanding of English. Then we'll be able to claim that their use generates propositions from predicates.

E. Compound Predicates and Quantifiers

How do the propositional connectives interact with the quantifiers? Consider:

$\forall x$ (x is a dog) $\wedge$ $\exists x$ (x barks)

The connective joins two propositions; no new analysis is needed.

But what about:

(4) x is a dog $\wedge$ x barks

This is a predicate, a compound predicate: when a name replaces the place-holders, a compound proposition results. For example:

(5) Ralph is a dog $\wedge$ Ralph barks

Suppose we wish to transform (5) into a proposition by a quantification:

(6) $\exists x$ (x is a dog $\wedge$ x barks)

We can take this as a formalization of 'There is something that is a dog and that barks', which is a proposition. But what is the role of $\wedge$ in (6)? Its use in (4) is justified by saying that (4) is merely the skeleton of a proposition: When names replace the variable we will have a proposition, such as (5), in which $\wedge$ joins propositions. But (6) is already meant to formalize a proposition: the variable is not supposed to be replaced by a name. In the next chapter we will see a semantic analysis of quantification that depends, in the end, on replacing variables by either proper or temporary names, so that the use of $\wedge$ in (6) will be justified as the propositional connective.

Once we specify how we will interpret the formal quantifiers, they can be used generally to transform any predicate, atomic or compound, into a proposition. For example, from 'Ralph barks $\vee$ y meows', we can form both:

$\exists y$ (Ralph barks $\vee$ y meows) $\forall y$ (Ralph barks $\vee$ y meows)

From '$x > y \vee x = 0$' we can form any of:

$\forall x\, (\forall y\, (x > y \vee x = 0))$

$\forall x\, (\exists y\, (x > y \vee x = 0))$

$\exists x\, (\forall y\, (x > y \vee x = 0))$

$\exists x\, (\exists y\, (x > y \vee x = 0))$

This is getting too complicated to follow without a formal language to guide us in the construction of semi-formal predicates and propositions. Before we turn to that, however, let's summarize what we've done.

F. The Grammar of Predicate Logic

In our analysis of the structure of propositions, the parts of speech we have considered are:

propositions	names
propositional connectives	predicates
phrase markers	quantifications

Predicates are dependent on our choice of what atomic propositions and names we take. Atomic propositions, however, though primitive in the sense of containing no formal symbols of our logic, are no longer structureless. They are parsed as composed of names and what's left over when the names are deleted.

Compound propositions are formed from predicates and names by using propositional connectives, which we formalize as before, and quantifications. The formalization of quantifications will use variables and quantifiers, to which we will give interpretations once we've agreed on the precise forms we're considering. Parentheses will be used to mark off phrases. Compound propositions, now, can have no part which is atomic, for example, '$\exists x$ (x is a dog $\wedge$ x barks)'.

This list was motivated by the assumption 'Things, the world, and propositions'. But it doesn't exhaust the ways we can look at the structure of propositions that reflect that assumption. Nonetheless, we'll stop with this list for now. It will provide us with enough structure to formalize many kinds of inferences that we could not previously analyze. This is the grammar of *predicate logic.*

Exercises for Sections A–F

1. Show all possible ways to parse the following into predicates and names.
   a. Dick has two dogs, Ralph and Juney.
   b. Juney barks at raccoons.
   c. Juney and Ralph are Dick's dogs.
2. Parse each of the following propositions into a predicate and name(s). Explain your choice in terms of which words stand for things.
   a. Juney and Ralph went to a movie.
   b. Juney and Ralph are dogs.
   c. $8 < 9$.
   d. Ralph hit Juney with a newspaper.
   e. Ralph hit Juney with Dick's copy of *Predicate Logic.*
   f. Paris is in France.
   g. Ralph is a dog or he's a puppet.

h. Juney isn't a puppet.
i. The set of natural numbers contains π.
j. The set of natural numbers is contained in the set of real numbers.
k. Green is pleasanter than grey.

3. Which of the predicates in Exercise 2 should be formalized as compound?

4. a. Discuss whether '7 loves Juney' is a proposition.
   b. Discuss whether 'Ralph is divisible by Juney' is a proposition.

5. Transform the following into universal propositions:
   a. $x < y$ iff $x + 1 \leq y$.
   b. If x is a dog, then x barks.

6. Find as many ways as possible to use quantifiers to make the following into true propositions about the natural numbers.
   a. $x < y$
   b. $x < y \wedge \neg(x < z \wedge z < y)$

G. A Formal Language for Predicate Logic

Predicates and names are the building blocks of the propositions we'll study, but we have no list of all predicates and names nor any method for generating them, for English is not a fixed, formal, static language. Therefore, let us introduce *predicate symbols* $P_0, P_1, \ldots$ to stand for predicates we wish to take as atomic, and *name symbols* $c_0, c_1, \ldots$ to stand for names. We call these 'symbols' rather than 'variables' in order not to confuse them with place-holders/variables. We need an ample supply of those, too, which we do call *variables* or *individual variables*: $x_0, x_1, \ldots$. Note again that I'm using formal symbols to name themselves in order to reduce the use of quotation marks.

The ellipses '...' are to indicate that the list goes on indefinitely. We need not take more than, say, 47 predicate and name symbols and 6 variables. But it's simpler to assume that there are always enough when we need them. In later work we might assume that the collection of predicate symbols, the collection of name symbols, and/or the collection of variables are completed infinite totalities, but for now that's not needed.

Each predicate in English has an arity, the number of gaps that must be filled with names to make it a proposition (unary, binary, ternary, ...). So really we need predicate symbols $P_0^1, P_0^2, P_0^3, \ldots, P_1^1, P_1^2, P_1^3, \ldots$. That is, we need a predicate symbol labeled with 0 for each arity, a predicate symbol labeled with 1 for each arity, ... Note that already this is an idealization, for it's unlikely that we'd ever consider reasoning with a predicate having 4,317,232 blanks. Because the additional indices make formulas hard to read, I will generally indicate the arity of a predicate symbol only when it might not otherwise be clear, or when I am making a formal definition, or when I want to illustrate a formal definition.

To talk about the formal language we will need *metavariables*. We can use $t, u, v, t_0, t_1, \ldots, u_0, u_1, \ldots, v_0, v_1, \ldots$ to stand for what we call *terms*, that is, any variable or name symbol. We'll use $\mathrm{A}, \mathrm{B}, \mathrm{C}, \mathrm{A}_0, \mathrm{A}_1, \ldots, \mathrm{B}_0, \mathrm{B}_1, \ldots, \mathrm{C}_0, \mathrm{C}_1, \ldots$ to stand for any formulas. By letting $i, j, k, m, n, i_0, i_1, \ldots, j_0, j_1, \ldots, k_0, k_1, \ldots, m_0, m_1, \ldots, n_0, n_1, \ldots$ stand for natural numbers, we can talk about any of the predicate symbols, name symbols, or variables. For example, P_i^n can stand for any predicate symbol. I'll use x, y, z, w, and $y_0, y_1, \ldots$ as metavariables for variables as well as informally as variables.

In English the blanks can appear anywhere in a predicate, for example, '— is a dog' or '— and — are the parents of —'. In the formal language, it's better to use a uniform symbolization, writing the blanks, as marked by variables, following the predicate symbol, for example, $\mathrm{P}_2^3(x_6\ x_1\ x_{14})$. To make it easier to read I'll put *commas* between the variables, as in $\mathrm{P}_2^3(x_6, x_1, x_{14})$. The commas are a technical aid in the formal language and are not intended to formalize anything.

The formal language of predicate logic

Vocabulary *predicate symbols* $\mathrm{P}_0^1, \mathrm{P}_0^2, \mathrm{P}_0^3, \ldots, \mathrm{P}_1^1, \mathrm{P}_1^2, \mathrm{P}_1^3, \ldots, \mathrm{P}_2^1, \mathrm{P}_2^2, \mathrm{P}_2^3, \ldots$

name symbols $c_0, c_1, \ldots$ } *terms*

variables $x_0, x_1, \ldots$ } *terms*

propositional connectives $\neg, \rightarrow, \wedge, \vee$

quantifiers $\forall, \exists$

Punctuation *parentheses* (,) *comma* ,

***An inductive definition of* wff** (*well-formed formula*)

i. For every $i \geq 0$, $n \geq 1$, and terms t_j, $1 \leq j \leq n$, $(\mathrm{P}_i^n(t_1, \ldots, t_n))$ is an *atomic wff* to which we assign the number 1.

ii. If A is a wff to which the number n is assigned, each of $(\neg\mathrm{A})$ and $(\forall x_i\,\mathrm{A})$ and $(\exists x_i\,\mathrm{A})$, for any i, is a *compound* wff to which we assign $n+1$.

iii. If A and B are wffs and the maximum of the numbers assigned to A and to B is n, then each of $(\mathrm{A}\rightarrow\mathrm{B})$, $(\mathrm{A}\wedge\mathrm{B})$, $(\mathrm{A}\vee\mathrm{B})$ is a *compound* wff to which we assign $n+1$.

iv. A concatenation of symbols is a *wff* iff it is assigned some number $n \geq 0$.

A wff of the form $(\forall x\,\mathrm{B})$ is a *universal wff*, the *universal generalization of* B *with respect to* x. A wff of the form $(\exists x\,\mathrm{B})$ is an *existential wff*, the *existential generalization of* B *with respect to* x.

We generally name a formal language by listing its vocabulary. So a name for the formal language we've just defined would be:

$L(\neg, \rightarrow, \wedge, \vee, \forall, \exists, x_0, x_1, \ldots, P_0^1, P_0^2, P_0^3, \ldots, P_1^1, P_1^2, P_1^3, \ldots, c_0, c_1, \ldots)$

That is a long name. Since we'll always use the same variables, we can leave those out. We can also leave the superscripts on the predicate symbols to be understood. So an abbreviated name for this formal language is:

$$L(\neg, \rightarrow, \wedge, \vee, \forall, \exists, P_0, P_1, \ldots, c_0, c_1, \ldots)$$

We may define a formal language with other propositional connectives, fewer (but at least one) or more predicate symbols, fewer or more name symbols, or no name symbols at all. The definitions of this chapter generalize easily and I'll assume those whenever we need them.

Note that we don't require x to appear in A in order to form $(\forall x\ A)$ or $(\exists x\ A)$. Superfluous quantification, as in '$(\forall x_1$ (Ralph is a dog))', is a (harmless) technical device that allows us to deal uniformly with all wffs; it is not meant to formalize anything from our ordinary reasoning.

The use of parentheses is important to mark clearly how we parse a wff, as we'll see in the next section. But in informal discussions we can make formulas look less complicated by adopting the following.

Conventions on deleting parentheses The parentheses around atomic wffs and the outer parentheses around the entire wff can be deleted. Parentheses between successive quantifiers at the beginning of a wff may be deleted. As for propositional logic, $\neg$ binds more strongly than $\wedge$ and $\vee$, which bind more strongly than $\rightarrow$. And $\forall x$, $\exists x$ bind more strongly than any of those. We can use [ ] in place of the usual parentheses. A conjunction or disjunction without parentheses is understood as associating the conjuncts or disjuncts to the left.

Thus we can take $\forall x_1 \exists x_2 \forall x_3 P_1(x_1, x_2, x_3)$ to abbreviate:

$$(\forall x_1 (\exists x_2 (\forall x_3 (P_1^3(x_1, x_2, x_3)))))$$

And we can take $\neg \forall x_7 P_0(c_{47}) \rightarrow \exists x_3 P_1(x_3, c_{16}) \wedge P_{13}(x_3)$ as an informal abbreviation of:

$$((\neg(\forall x_7 (P_0^1(c_{47})))) \rightarrow ((\exists x_3 (P_1^2(x_3, c_{16}))) \wedge (P_{13}^1(x_3))))$$

H. The Structure of the Formal Language

We first need to establish that there is only one way to read each wff. The proof is very similar to the one for the language of propositional logic.

Theorem 1 Unique readability of wffs
There is one and only one way to parse each wff.

Proof To each symbol α of the formal language assign an integer $\lambda(\alpha)$ according to the following chart:

$\neg$	$\rightarrow$	$\wedge$	$\vee$	(	)	x_i	c_i	P_i^n	$\forall$	$\exists$
0	0	0	0	-1	1	1	1	$-n$	-1	-1

To the concatenation of symbols $\alpha_1\, \alpha_2 \cdots \alpha_n$ assign the number:

$$\lambda(\alpha_1\, \alpha_2 \cdots \alpha_n) = \lambda(\alpha_1) + \lambda(\alpha_2) + \cdots + \lambda(\alpha_n)$$

We proceed by induction on the number of occurrences of $\forall, \exists, \neg, \rightarrow, \wedge, \vee$ in A to show that for any wff A, $\lambda(A) = 0$.

If there are no occurrences, then A is atomic. That is, for some $i \geq 0$, $n \geq 1$, A is $(P_i^n(t_1, \ldots, t_n))$. Then $\lambda(\text{'('}) = -1$, $\lambda(P_i^n) = -n$, $\lambda(t_1) = \cdots = \lambda(t_n) = 1$, and $\lambda(\text{')'}) = 1$. Adding, we have $\lambda(A) = 0$.

The inductive stage of the proof, and then the proof that initial segments and final segments of wffs have value $\neq 0$, and then that there is only one way to parse each wff, is done almost exactly as for the propositional language, pp. 8–9. ■

By Theorem 1 we know that every wff has exactly one number assigned to it in the inductive definition of wffs. We call that number the *length* of the wff. Now when we want to show that all wffs have some property we can use induction on the length of wffs as we did for propositional logic.

The definition of one formula being a *subformula* is as for the propositional language, with the addition of:

If C has length $n + 1$ then: If C has the form $(\forall x\, A)$ or $(\exists x\, A)$, its subformulas are C, A, and the subformulas of A.

We also define inductively what it means for a *symbol to appear* or *occur in a wff*:

The symbols that appear in the atomic wff $(P_i^n(t_1, \ldots, t_n))$ are: P_i^n and $t_1, \ldots, t_n$.

The symbols that appear in $(\neg A)$ are '$\neg$' and those that appear in A.

The symbols that appear in $(A \rightarrow B)$, $(A \wedge B)$, $(A \vee B)$ are, respectively, $\rightarrow$, $\wedge$, $\vee$, and all those that appear in either A or in B.

The symbols that appear in $(\forall x\, A)$ are x, $\forall$, and those that appear in A; and similarly for $(\exists x\, A)$, reading '$\exists$' for '$\forall$'.

Wffs are always read from left to right, so when we say the *symbol* β *follows* the symbol α in a wff we mean that β is to the right of α.

It is important that we can number all wffs. But which particular numbering we use doesn't matter. In an exercise below I ask you to give a numbering, using the numbering of propositional wffs on p. 10 as a guide.

I. Free and Bound Variables

When we use a quantifier to change the role of a variable from place-holder to one that is used to express generality, we need to be clear about which variable or variables are affected. We often refer to the concatenations '$\forall x$' and '$\exists x$' as *quantifiers*. We say:

The *scope* of the initial quantifier $\forall x$ in $(\forall x\, A)$ is A.

The *scope* of the initial quantifier $\exists x$ in $(\exists x\, A)$ is A.

A particular *occurrence of x in* A is *bound in* A if it immediately follows an occurrence of the symbol '$\forall$' or '$\exists$' or lies within the scope of an occurrence of '$\forall x$' or '$\exists x$'. If an occurrence of x in A is not bound, it is *free in* A. Note that a variable can have both free and bound occurrences in a single wff. For example:

(7) $$(\forall x_1\, P_2(x_1)) \rightarrow (\exists x_2\, P_7(x_1, x_2) \vee \exists x_{12}\, \forall x_1\, P_3(x_1))$$

In this wff the scope of the first occurrence of $\forall x_1$ is $P_2(x_1)$, the scope of the second $\forall x_1$ is $P_3(x_1)$, the scope of $\exists x_2$ is $P_7(x_1, x_2)$, and the scope of $\exists x_{12}$ is $\forall x_1\, P_3(x_1)$. Every variable is bound except for the third occurrence of x_1 (reading from the left)

In a wff $\forall x\, A$ every free occurrence of x in A is *bound by* the initial $\forall x$; in $\exists x\, A$, the initial $\exists x$ *binds* every free occurrence of x in A. For example, in (7) $\exists x_2$ binds the occurrence of x_2 in $P_7(x_1, x_2)$, the first $\forall x_1$ binds the second occurrence of x_1, no quantifier binds the third occurrence of x_1, and $\exists x_{12}$ binds no variable. In $\forall x_1 \exists x_1 P_2(x_1)$, $\exists x_1$ binds the last appearance of x_1 in the wff and $\forall x_1$ binds nothing, for x_1 is not free in the scope of $\forall x_1$.

The wffs that are symbolic versions of propositions are those in which no occurrence of a variable is free. They are called *closed wffs* or (*formal*) *sentences.* A wff that is not closed is *open* or an *open expression* or *open sentence*; these correspond to predicates. Example (7) is open, while its antecedent is closed. An *atomic sentence* is a closed atomic wff, that is, an atomic wff of the form $P_i^n(c_{j_1}, \ldots, c_{j_n})$ for some i, n, and $j_1, \ldots, j_n$. Note that A, B, C, ... are used to range over all wffs, not just those corresponding to propositions.

Mathematicians often treat an open expression such as '$x \geq y \vee y \geq x$' as if it were about all objects. We won't do that, but we can convert any open expression into one that is a proposition about all objects by prefixing it with appropriate universal quantifiers. I'll write '$\equiv_{\text{Def}}$' to mean 'equivalent by definition'.

The (universal) closure of A Let $x_{i_1}, \ldots, x_{i_n}$ be a list of all the variables that occur free in A, where $i_1 < \cdots < i_n$.

The (*universal*) *closure* of A is $\forall \ldots A \equiv_{\text{Def}} \forall x_{i_1} \cdots \forall x_{i_n} A$

The *existential closure* of A is $\exists \ldots A \equiv_{\text{Def}} \exists x_{i_1} \cdots \exists x_{i_n} A$.

J. The Formal Language and Propositions

The formal language codifies and clarifies the grammar of predicate logic. Using it as a guide we can formalize English propositions into a semi-formal language where $\neg$, $\rightarrow$, $\wedge$, $\vee$, $\forall$, $\exists$ and variables replace their English counterparts. Once we fix on how to interpret these formal symbols, it will be legitimate to view a formal expression as a proposition, for example:

$$\forall x_1\ (x_1 \text{ is a dog} \wedge x_1 \text{ is bigger than Ralph} \rightarrow \exists x_2\ (x_2 \text{ is smaller than } x_1))$$

The terminology we've adopted to discuss the formal language can then be carried over to apply to such expressions. For example, we can say that the second occurrence of x_2 in the example is bound by the quantifier $\exists x_2$ and that the predicate 'is bigger than' appears in the proposition. In this section I'll explain how the formal language can be used to specify which semi-formal expressions can be considered propositions.

First, we *realize* the name symbols as names and the predicate symbols as atomic predicates, that is, predicates whose internal structure will not be considered. For example, we could assign 'Ralph' to 'c_0', 'Juney' to 'c_1', 'is a dog' to 'P_0^1', 'is a cat' to 'P_1^1', and so on.

But what do we mean by 'and so on'? If we want every predicate and name symbol to be realized, then we must have an explicit method for generating the predicates and names we wish to assign to the symbols. Generally, though, we're interested in only a few atomic predicates and names in any discussion, so we realize only some of the predicate and name symbols. Let's make the convention that *the formal language is restricted to just those predicate and name symbols that we realize*, without explicitly restricting the formal language each time. At least one predicate symbol must be realized in order to have any propositions at all.

The *realization of a formal wff* is what we get by replacing the name symbols and predicate symbols in the formula with their realizations. A *realization* is then the collection of expressions that are realizations of the formal language. For example:

(8) $L(\neg, \rightarrow, \wedge, \vee, \forall, \exists, P_0, P_1, \ldots, c_0, c_1, \ldots)$

$\downarrow$

$L(\neg, \rightarrow, \wedge, \vee, \forall, \exists$; 'is a dog', 'is a cat', 'eats grass', 'is a wombat', 'is the father of'; 'Ralph', 'Bon Bon', 'Howie', 'Juney')

A realization is a *semi-formal language*, though to call this 'a language' we need to fix on some interpretation of the formal symbols. The predicates interpret the predicate symbols in order from left to right, unary predicate symbols first, then binary, then (if any) ternary, and so on. So 'is a dog' is the realization of P_0^1, which we can notate as 'is a dog' = real(P_0^1), and 'is a wombat' = real(P_3^1), and 'is the father of' = real(P_0^2). Similarly, the names, set off by a semicolon, realize the name symbols in order from left to right, so 'Howie' realizes c_2, which we notate as 'Howie' = real(c_2). The expressions or formulas of the semi-formal language are the realizations of the formal wffs. For example,

Ralph is a dog.

This is an expression of the semi-formal language. It is the realization of:

$P_0^1(c_0)$

We say that real($P_0^1(c_0)$) = 'Ralph is a dog'.

Similarly, we have the following expressions of the semi-formal language:

$\neg$(Ralph is a dog) $\rightarrow \forall x_0$ (x_0 is a cat)

x_{32} is a dog

These are realizations of, respectively,

$\neg P_0^1(c_0) \rightarrow \forall x_0\, (P_1^1(x_0))$

$P_0^1(x_{32})$

A semi-formal language is linguistic, a formalized fragment of English. Because it is linguistic I have used quotation marks to indicate that I am mentioning pieces of language and not using them. That is why real($P_0^1(c_0)$) = 'Ralph is a dog'. Let us adopt the convention that predicates and names in a presentation of a realization are to be understood to be pieces of language. So we can write (8) as:

$$L(\neg, \rightarrow, \wedge, \vee, \forall, \exists, P_0, P_1, \ldots, c_0, c_1, \ldots)$$
$$\downarrow$$
L($\neg, \rightarrow, \wedge, \vee, \forall, \exists$; is a dog, is a cat, eats grass, is a wombat, is the father of; Ralph, Bon Bon, Howie, Juney)

Which English propositions can be formalized into semi-formal English depends not only on the syntactic assumptions we have made, but also on how we shall interpret the formal symbols, which we'll consider in the next chapter.

Exercises for Sections G–J

1. a. Distinguish between the use of a letter as a place-holder and as a variable.
   b. Distinguish between the roles of variables, predicate and name symbols, and metavariables.

2. Give an example (abbreviated if you wish) of a formal wff that is:
   a. Atomic, open, and contains name symbols.
   b. A universal sentence.
   c. A universal open expression.
   d. A sentence that is the universal generalization of an existential generalization of a conditional whose antecedent is a universal formula.

e. A formula that is open but which has a closed subformula.
f. A compound sentence no part of which is a sentence.

3. For each of the following list which of (i)–(iv) apply:

   i. It is an unabbreviated wff. iii. It is existential.
   ii. It is atomic. iv. It is universal.

   a. $P_1^1(x_1)$
   b. $(P_2^2(x_2))$
   c. $(P_2^1(x_1))$
   d. $\exists x_2\,(P_2^1(x_2))$
   e. $(\forall x_1\,(P_{47}^1(x_2)))$
   f. $((P_3^1(x_4)) \rightarrow (\neg(P_4^2(x_2, x_1))))$
   g. $\neg P_5^2(x_1, x_2) \rightarrow (P_1^2(x_2, x_1) \vee \neg P_1^2(x_2, x_1))$

4. Convert each of the following abbreviations into an unabbreviated semi-formal wff. Give its length and list its subformulas and the terms that appear in it.

   a. $\forall x_1$ (x_1 is a dog)
   b. $\forall x_1$ (x_1 is a dog) $\wedge$ Ralph barks $\rightarrow$ Juney howls $\vee$ Bon Bon is a horse
   c. $\neg\exists x_{13}$ (x_{13} is a dog $\wedge$ Ralph is a dog $\rightarrow$ Juney barks)
   d. $\forall x_1$ ($\neg$ Juney barks $\rightarrow$ x_1 is a dog)

5. For each of the following (abbreviations of) wffs:

   i. Identify the scope of each quantifier.
   ii. State which occurrences of variables are free and which are bound.
   iii. State whether the wff is a sentence or an open expression, as well as whether it is atomic.
   iv. Give the universal closure of the wff.

   a. $P_1(c_8)$
   b. $P_8(x_1)$
   c. $(\exists x_2\, P_1(x_1)) \vee \neg P_2(c_3)$
   d. $\forall x_1\,(\neg P_2(x_1) \vee P_1(x_3)) \rightarrow \exists x_3\, P_4(x_1, x_3)$
   e. $\forall x_2\, P_1(x_1, x_2, x_3) \rightarrow \exists x_2\, \exists x_3\, P_1(x_1, x_2, x_3)$
   f. $P_1(x_1) \vee \neg P_1(x_1)$
   g. $\forall x_1\, P_1(x_1) \wedge \neg\exists x_2\, P_2(x_2)$

6. As for Exercise 5, do (i)–(iv) for the following semi-formal wffs:

   a. x_1 is an uncle of x_2 $\vee$ $\exists x_1$ (x_1 is a woman)
   b. Ralph is an uncle $\vee$ $\exists x_1$ (x_1 is a dog)
   c. $\forall x_2\, \neg$(x_2 is a father $\vee$ $\exists x_2$ (x_2 is related to Ralph))
   d. $\exists x_1$ (x_1 is in the nucleus of x_2 $\wedge$ x_2 is an atom) $\rightarrow$ $\forall x_2$ (x_1 is smaller than x_2)

7. Using the atomic predicates '— is a dog', '— is the brother of —', and '— eats meat', exhibit a compound 7-ary predicate.

8. Give a numbering of all wffs of the formal language of predicate logic. (Hint: Use the numbering of wffs of the language of propositional logic, p. 10, as a guide.)

IV The Semantics of Classical Predicate Logic

Our goal now is to give meanings to the *logical* or *syncategorematic* vocabulary, which consists of the connectives $\neg$, $\rightarrow$, $\wedge$, $\vee$, the quantifiers, $\forall$, $\exists$, and the variables, $x_0, x_1, \ldots$. We need to settle on interpretations that will not vary from proposition to proposition, nor from realization to realization. Predicates and names—the *nonlogical* or *categorematic* parts of speech—are what give content to propositions, and we need to explain how our interpretations of these connect to the meanings of the logical vocabulary in determining the truth or falsity of propositions.

A. Names

We saw that names play a central role in predicate logic. In this section we'll adopt two agreements to govern our use of names in reasoning. Any word that purports to pick out an individual thing and that satisfies those agreements can then be used as a name, whether or not we classify it as a proper name in our ordinary speech.

If I write 'Richard L. Epstein has two dogs' you'd naturally think the proposition is about me. But there is also a person who teaches linguistics who's called 'Richard L. Epstein'. This kind of ambiguity is common in English because we use a limited number of words as proper names, making it easier to recognize when a word is being used as a name. But in formal reasoning we can and should avoid the confusion that results from using a word in more than one way, as we did in deciding to treat words as types. We make the following proviso:

A name can pick out at most one thing.

This agreement does not rule out using different names for the same individual, which may be essential to discovering there is only one object being discussed.

So a name purports to pick out one thing. But need it actually pick out something? Consider:

Santa Claus is a dog.	false
Santa Claus is not a dog.	true
Santa Claus has a dog.	true? false?

Our intuition begins to falter when names do not "really" pick out things.

Or consider:

Theaetetus was an intelligent disciple of Socrates.
Therefore, Theaetetus was a disciple of Socrates.

This inference looks valid. But what does it mean to say it is valid if there was no Theaetetus? If 'Theaetetus' does not name anything, what do we mean by saying that these propositions are true or are false? We're not concerned here with a character in a story, but whether an actual person existed as described in Plato's dialogue.

Consider further 'Theaetetus did not exist.' If we want to reason with this proposition, an awkward question arises: If the proposition is true, what is it that didn't exist? That would only be a confusion. We would have agreed that names needn't actually pick out things, yet would have fallen into the rather natural assumption that they do.

For now, let's require that a name picks out at least one object. The analyses will be simpler. But we will be unable to reason within our semi-formal languages about whether Theaetetus existed. So in sum we have the following.

Names refer If we use a phrase or a word as a name, we assume that there is exactly one thing it picks out.

We say a name *refers to* or *denotes* what it picks out, and call that thing its *reference* or *denotation.*

In classical propositional logic we ignored all properties of propositions except those that could not be dispensed with in analyzing reasoning: form and truth-value. For classical predicate logic we will ignore all properties of names except those that are essential to any reasoning. Since names have no form in predicate logic, we make the following abstraction.

The classical abstraction of names The only property of a name that matters to logic is whether it refers, and if it does refer then what object it refers to.

B. Predicates

1. A predicate applies to an object

Atomic propositions connect atomic predicates to named things. Given a predicate such as '— is a dog', it is true or false of the thing named 'Ralph' according to whether the proposition formed with that predicate and name, 'Ralph is a dog', is true or false.

If we were to restrict logic to reasoning about only named objects, objects for which we have names in the semi-formal language, then we would not need to say anything more about the relation of language (predicates) to things than we already have. But we will want to reason about things that aren't named, such as all the pigs in Denmark, or about things it would be inconvenient to list names of, such as all pickup trucks in the U.S. on November 4, 2004, or about things that we cannot name except in some idealized sense, such as atoms. To make sense of what it means to say that '$\forall x_1$ (x_1 is a dog)' is true or false, we need some notion of what it means to say that '— is a dog' is true or false of a particular object, named or unnamed.

The words 'true' and 'false' apply to propositions. To connect a predicate to an object, named or unnamed, we shall have to form a proposition. Let's consider an example, restricting our discussion in this section to unary predicates.

Suppose I wish to communicate to you that the lamp on my desk weighs less than 2 kg. Of course, the lamp has no given name with which we are both familiar. If you were in the room with me I might point to the lamp and say, 'That lamp weighs less than 2 kg.' I use the words 'That lamp' (plus my gesture) to pick out one thing, and hence by the standards we have established, I use those words as a name. Or if I were writing to you, I might write 'The brass lamp on my desk weighs less than 2 kg,' using the phrase 'The brass lamp on my desk' to pick out one object, and hence as a name. I name the object, however temporarily, and state a proposition using that name and the predicate '— weighs less than 2 kg'. The connection is reduced to supplying a name, that is, naming plus a proposition.

If every time we wish to assert a connection between a predicate and an unnamed object we must coin a new name and form a new semi-formal proposition not previously available to us, then our semi-formal language will be always unfinished. We will not be able to get agreement in advance on the vocabulary we shall reason with.

So, rather than accepting a proliferation of new names, let's use variables as temporary names: They await supplementation to pick out one object, just like the words 'this' or 'that'. For example, I could say '*x* weighs less than 2 kg' while I emphatically point to the lamp and utter '*x*'. Or I could write to you '*x* weighs less than 2 kg' and add 'By '*x*' I mean the lamp on my table.' I still have to indicate what '*x*' is to refer to, by pointing or describing in words, but that way of indicating is separated from the semi-formal language.

When we stipulate by some means what '*x*' is to refer to and agree that '— weighs less than 2 kg' is a predicate, then we can agree to view '*x* weighs less than 2 kg' as true or false. That is, when we supplement the sentence '*x* weighs less than 2 kg' with an indication of what '*x*' is to refer to, we have a proposition, indeed an atomic proposition. *The connection between atomic predicates and unnamed objects is reduced to naming and truth-values for atomic propositions.*

But why the variable '*x*' ? We could just as easily have used the variable '*y*', so that '*y* weighs less than 2 kg' would provide "the" connection between the predicate and that object. Yet '*y* weighs less than 2 kg' is a different sentence, and hence a different proposition from '*x* weighs less than 2 kg', even though we've agreed to use both '*x*' and '*y*' to refer to the same object. Indeed, we might even consider two utterances of '*x* weighs less than 2 kg' as different propositions if for the first I say ''*x*' is to refer to the brass lamp on my table,' and after the second I point emphatically at the lamp and say, 'That's what '*x*' refers to.'

Still, if '*x* weighs less than 2 kg' is true (respectively, false) when '*x*' is meant to indicate my lamp, then '*y* weighs less than 2 kg' must be true (respectively, false) too when '*y*' is taken to refer to my lamp. Variables have form, but no content except what we might assign to them via an indication of a particular reference. The method of assignment of reference is necessarily outside our semi-formal language, connecting as it does a piece of language to the world. Hence these propositions should be considered equivalent: They have the same semantic properties, since we cannot take account of their differences within the language.

But what if an object already has a name? Or several names? For example, the person named 'Marilyn Monroe' was also known by the names 'Norma Jean Baker' and 'Norma Jean Mortenson'. Under what circumstances should we say that the predicate 'was blonde' is true of that person? When '*x* was blonde' is true, where '*x*' is meant to indicate that person? Or when 'Marilyn Monroe was blonde' is true? Or when 'Norma Jean Baker was blonde' is true?

It doesn't seem to matter, for all these should have the same truth-value. But we can't argue for that as before, for in 'Marilyn Monroe was blonde' the method of indicating reference is noted within the semi-formal language. Shall we insist none-theless that what is true of an object is independent of how it is named? Consider:

Stanisław Krajewski thinks that Marilyn Monroe was blonde.

Stanisław Krajewski thinks that Norma Jean Baker was blonde.

The first was true, the second false when I first asked Stanisław Krajewski.

Let us make the following definition: Given a unary predicate P and term t, if t is a name or is a variable supplemented with an indication of what object it is to refer to, then $P(t)$ is a *predication*. We say that P is *predicated of the object according to the reference provided by* t. The predication is *atomic* if P is atomic. Viewing a variable as supplemented with an indication of what it refers to as a name, I have argued that predications are propositions. Now let us make a further restriction that is characteristic of classical logic and which follows because of the classical abstraction of names.

> Every atomic predicate is *extensional*. That is, given any terms t and u that refer to the same object (possibly through some temporary indication of reference), then $P(t)$ and $P(u)$ have the same truth-value.

If we can't agree that a predicate such as '— thinks that — was blonde' is extensional, we shall have to exclude it from our realizations.

Now we can define what it means for a predicate to be true of an object independently of the method of indicating reference, for the truth-value of an atomic predication no longer depends on the name or variable used to refer to the object.

> A unary atomic predicate P *applies to* or *is true of* a particular object if given a variable x with an indication that x is to refer to the object, or given a name a of the object, then $P(x)$, respectively $P(a)$, is true.

The use of variables as temporary names has reintroduced the use of indexicals into our reasoning. Just as the truth-value of 'it weighs less than 2 kg' depends on what 'it' refers to, so two utterances of 'x weighs less than 2 kg' can be taken to be two different propositions depending on what we take 'x' to stand for. Some such controlled use of indexicals seems necessary if we are to reason about unnamed objects. We can avoid confusion if we remember that an open wff such as 'x weighs less than 2 kg' is not a proposition of the semi-formal language, but only a tool to connect that language to the things of which we speak.

2. Predications involving relations

Consider the atomic predicate '— is bigger than —'. What shall we mean when we say this predicate is true of:

If we let 'x' stand for the left-hand object and 'y' stand for the right-hand object, then 'x is bigger than y' is true. If we let 'x' stand for the right-hand object and 'y' the left-hand one, then 'x is bigger than y' is false. It's not enough to speak of a *relation* (an n-ary predicate for $n > 1$) applying to objects; we have to speak of the relation applying to the objects in a certain ordering, saying which blank is to be filled with the name of which object.

Predications Given any n-ary predicate P and terms $t_1, \ldots, t_n$ along with an indication of what each variable among those terms is to refer to, then $P(t_1, \ldots, t_n)$ is a *predication*; we say that P is *predicated of* the objects referred to by the terms according to those references. The predication is *atomic* if P is atomic. Every predication is a proposition.

Note that if we let both 'x' and 'y' refer to the same object, say me, then 'x has the same blood type as y' will be a proposition by this definition and assumption. We could require that distinct variables always stand for distinct things in a predication, but that would require an analysis of what we mean by two things being distinct, which in turn is greatly facilitated by allowing distinct variables to refer to the same object. Moreover, we allow different proper names to refer to the same object, as with the example of 'Marilyn Monroe' and 'Norma Jean Baker'. So it seems advisable and reasonable to allow distinct variables in a predication to refer to the same object.

Within one predication, however, we don't want a single variable to refer in different places to distinct objects. We don't want 'x is the father of x' to be a proposition if someone suggests that the first 'x' is to refer to Juney and the second to Ralph. *Our agreement that a name can refer to only one thing will be extended to apply to variables used to pick out objects within a single predication.*

Now we can state the assumption about the extensionality for all predicates.

The extensionality of atomic predicates Every atomic predicate in the semi-formal language is *extensional.* That is, given any terms $t_1, \ldots, t_n$ and $u_1, \ldots, u_n$ such that each term that is a variable is supplemented with an indication of what it is to refer to, and for each i, $1 \leq i \leq n$, t_i and u_i refer to the same object, then $P(t_1, \ldots, t_n)$ has the same truth-value as $P(u_1, \ldots, u_n)$.

After we define truth-conditions for compound formulas, I'll show in the next chapter that this assumption extends to compound predicates, too.

So now if we use 'is a dog' along with 'Ralph' and 'Marilyn Monroe' in a realization, we shall assume that the following all have the same truth-value:

'x is the dog of y'	where 'x' is to refer to Ralph and 'y' to Marilyn Monroe

'x is the dog of Marilyn Monroe'	where 'x' is to refer to Ralph
'x is the dog of Norma Jean Baker'	where 'x' is to refer to Ralph
'Ralph is the dog of y'	where 'y' is to refer to Marilyn Monroe
'Ralph is the dog of Marilyn Monroe'	
'Ralph is the dog of Norma Jean Baker'	
'z is the dog of w'	where 'z' is to refer to Ralph and 'w' to Marilyn Monroe

To introduce some terminology about predications we need to be able to refer to the objects picked out by the variables and names in a predication. In writing I have no choice but to use names or descriptions, so I'll take $a, b, c, a_0, a_1, \ldots,$ $b_0, b_1, \ldots$ as *metavariables* ranging over names in the semi-formal language and in ordinary speech.

We cannot speak of an n-ary relation applying to objects without specifying which blank of the predicate is to be filled with a name of which object. Labeling the blanks, however, we can make the following definition.

A predicate is true of $P(y_1, \ldots, y_n)$ *applies to* or *is true of* $a_1, \ldots, a_n$ *in that order* means that if for each i we let y_i refer to a_i, then $P(y_1, \ldots, y_n)$ is true.

Normally we label the blanks in a predicate in order from left to right, so we can say that '— was the teacher of —' applies to Socrates and Plato, understanding the order of the objects to be that in which we mention them, first Socrates, then Plato. Some common synonyms are the following (where 'in that order' is assumed):

$P(y_1, \ldots, y_n)$ *applies to* $a_1, \ldots, a_n$.

$P(y_1, \ldots, y_n)$ *is true of* $a_1, \ldots, a_n$.

$P(y_1, \ldots, y_n)$ *holds of* $a_1, \ldots, a_n$.

$a_1, \ldots, a_n$ *satisfy* $P(y_1, \ldots, y_n)$.

$a_1, \ldots, a_j$ *stand in relation* $P(y_1, \ldots, y_n)$ to $a_{j+1}, \ldots, a_n$.

So Socrates stands in the relation '— was the teacher of —' to Plato, though Plato does not stand in that relation to Socrates. When a predication is not true we say it is *false of* or *is not satisfied* (etc.) of the objects, in the order given in the predication.

The story of how reference is stipulated, that is, on what basis a variable is actually assigned an object as reference, is crucial to any particular application of predicate logic. But that is a big subject, so I'll just refer you to the discussion of it in my *Predicate Logic*.

In the end, then, the question of an atomic predicate applying to an object can be reduced to two semantic notions we already decided to rely on: reference and the truth or falsity of atomic propositions.

Note that we now have three roles for variables:

'x is a dog'	place-holder
'x is a dog' where 'x' is to refer to Ralph	temporary name
'$\forall x$ (x is a dog)'	generalizing

We now make an abstraction of predicates to just those properties we've considered so far

The classical abstraction of predicates The only properties of an atomic predicate that matter to logic are its arity and for each (sequence of) object(s) under consideration whether it applies to the object(s) or does not apply.

The platonist conception of predicates and predications

A platonist conceives of propositions as abstract objects, some of which can be represented or expressed in language. In accord with this view, the platonist says that predicates or properties or qualities are abstract things, some of which can be represented or expressed as, or simply correspond to, linguistic predicates as I have defined them. But a property or relation exists independently of our conceiving of it or representing it. Thus we have an inscription 'is blue' that stands for a linguistic type (taken as an abstract thing) that corresponds (etc.) to the property of being blue, or "blueness". There may be properties and relations that do not correspond to any linguistic predicate in our language.

Different linguistic expressions may represent the same platonic predicate, for example, 'x is a bachelor' and 'x is a man and x is not married'. Where I would undertake a semi-formal analysis of the linguistic expressions to decide whether we should treat them as semantically equivalent, the platonist goes through much the same procedure to determine whether the expressions denote the same predicate.

'A unary predicate applies to an object' is understood by a platonist to mean (or is simply true just in case) the object has the property. For example, the predicate denoted by '— is a dog' applies to Ralph just in case Ralph has the property of being a dog; the predicate expressed by 'is a horse' is true of Bon Bon just in case the quality of horsieness belongs to Bon Bon. How we indicate what a variable is to refer to is of no concern to a platonist, nor what name we use, for an object has the property or it does not, and it is an accident how or whether we refer to the object with a variable or name.

Nor is the platonist concerned with what variable is to refer to what object in the predication of a relation. The "concreta" Socrates and Plato "participate in" the relation expressed by 'was the teacher of' in the order represented by the sequence (Socrates, Plato). And sequences of objects are abstract objects, not our ways of assigning reference. Sometimes it's said that the sequence (Socrates, Plato) satisfies the predicate represented by 'was the teacher of', and that is if and only if Socrates stands in the relation of being the teacher of to Plato.

I take as primitive the truth or falsity of atomic propositions, and that is used to explain what is meant by a predicate applying to an object or sequence of objects. On the platonist conception, the possession of a property by an object or sequence of objects is taken as primitive, and that is used to explain the truth or falsity of atomic propositions.

Properties and relations are abstract things; sequences, even sequences of just one object, are abstract things; and a proposition is sometimes said to be composed of these. What exactly that composition is seems unclear, since it can't be juxtaposition, for abstract things aren't like concrete things that can be placed side by side, nor can it be explained by saying that the composition is the objects having the property, because that doesn't account for false propositions.

The platonist view of predicates is recommended by its adherents for giving an explanation of truth independent of us and our capabilities. But if so, it does it at the cost of separating logic from methods of reasoning. To use logic we will still have to indicate what variable is to refer to what thing and how we will specify sequences.

Exercises for Sections A and B

1. List the categorematic and syncategorematic parts of the following:

    a. $\forall x_1 ((\text{Ralph is a dog} \land \lnot (x_1 \text{ is a cat})) \to \text{Ralph likes } x_1)$

    b. Richard L. Epstein taught in João Pessoa, Brazil

    c. Plato knew Socrates $\to$ Plato is dead

    d. $\forall x_1 (x_1 \text{ belongs to Richard L. Epstein} \to x_1 \text{ was made in America})$

2. Describe the three roles that variables play. How can we distinguish which role is intended?

3. a. State precisely what it means to predicate the phrase 'is round' of: △

    b. State precisely what it means for the predicate 'stands to the left of' to apply to:

    △ ○

    c. State precisely what it means to say that the predicate '— and — are smaller than —' is true of: △ ○ ▢

    d. State precisely what it means to say that Marilyn Monroe stands in the relation '— is more honest than —' to Richard Nixon.

4. a. Are the following semantically equivalent?

    i. x is a dog   where 'x' is meant to stand for Ralph

    ii. y is a dog   where 'y' is meant to stand for Ralph

    iii. Ralph is a dog

    b. What role do (i)–(iii) play in predicating '— is a dog' of Ralph?

    c. Instead of (i) why don't I write the following?

    x is a dog   where 'x' is meant to stand for 'Ralph'

5. Classify the following predicates as extensional or not, supporting your choices with examples and arguments.

a. '— is a dog'
b. '— is a domestic canine'
c. '— believes that — is a dog'
d. '— sees —'
e. '— smells sweet'
f. '— was thinking about —'
g. '— knows that — is the father of —'
h. '— has diameter < 3 cm'

6. Here is a well-known argument for rejecting the view that predicates are linguistic.
   i. There are only finitely many English words.
   ii. So there are at most countably many English sentences.
   iii. Let $Q_1, Q_2, \ldots$ be a list of all predicates that can be derived from English sentences and that have only a single gap to be filled by a name.
   iv. Each of these predicates either applies or does not apply to any other (e.g., '— is a predicate containing fewer than 437 letters' applies to '—is a dog').
   v. So the following phrase is a predicate: '— is a predicate on our list that does not apply to itself'. Call it Q.
   vi. So Q must be on our list.
   vii. But then we have that Q applies to itself iff Q is a predicate on our list that does not apply to itself.
   viii. This is a contradiction, so Q cannot be on our list.
   ix. But our list was supposed to contain all predicates. So predicates aren't linguistic.

   Show this argument fails. (Hint: The adjective 'countable' applies only to fixed collections of objects or fixed ways of generating objects. Yet the very example of a new meaning to 'our list' shows that English isn't fixed in that way.)

C. The Universe of a Realization

A realization (Chapter III.J) is a particular formalized fragment of English. Specific predicates and names replace the predicate and name symbols of the formal language.

By choosing the names that realize the name symbols we have implicitly chosen some objects to be under discussion: those things that are named. In choosing the predicates that realize the predicate symbols we also have some implicit idea of what objects we are discussing, since we have the general agreement that whenever the blanks in a predicate are filled with names of things we are talking about, the resulting sentence becomes a proposition.

We should be explicit and state as precisely as we can what things are under discussion. Those things comprise the *universe of the realization* (some authors use the term *domain*). For example, consider the realization:

(1) $L(\neg, \rightarrow, \wedge, \vee, \forall, \exists, P_0, P_1, \ldots, c_0, c_1, \ldots)$

$\downarrow$

L(¬, →, ∧, ∨, ∀, ∃; is a dog, is a cat, eats grass, is a wombat, is the father of; Ralph, Bon Bon, Howie, Juney)

We might take the universe to be all animals, living or toy, or all things in the

U.S. over 3 cm tall, or everything made of cloth or flesh, or the collection composed of just Ralph, Bon Bon, Howie, and Juney. There are many choices.

We could even choose the universe to be all things. That choice is said by some to be the only appropriate one, for logic is supposed to be completely neutral with respect to what things there are. In that case, we would have to argue that the choice of predicates imposes no restrictions on what things we are discussing: '*x* loves Juney' is simply false when '*x*' is taken to refer to the number 7.

But there are serious difficulties in taking the universe to be all things. We must be able to distinguish one thing from all others and give some explanation of what it means to assign that thing as reference to a variable. To give such an explanation amounts to coming up with a single idea of what it means to be a thing, which is what we hoped to clarify by our development of predicate logic. Worse, though, the notion of all things may be incoherent, leading to paradoxes when we try to make it precise, as we'll see in Chapter IX.B.

Therefore, though we need not reject unequivocally that the collection of all things could be the universe of a realization, the problems inherent in working with such a collection make the choice of other "more restricted" universes attractive. Each choice will impose some sense, to be explained by us, of what it means to assign reference to a variable.

So, for example, we can choose as universe for realization (1) the collection of all animals, living or toy.

(2) $L(\neg, \rightarrow, \wedge, \vee, \forall, \exists; P_0, P_1, \ldots, c_0, c_1, \ldots)$

$\downarrow$

$L(\neg, \rightarrow, \wedge, \vee, \forall, \exists;$ is a dog, is a cat, eats grass, is a wombat, is the father of; Ralph, Bon Bon, Howie, Juney)

universe: all animals, living or toy

The universe is not part of the semi-formal language, though it is an essential part of the realization. An expression such as '$\exists x$ (x is a wombat)' cannot be considered a proposition unless we know what things we are talking about, even if it be all things. Compare (2) to:

(3) $L(\neg, \rightarrow, \wedge, \vee, \forall, \exists; P_0, P_1, \ldots, c_0, c_1, \ldots)$

$\downarrow$

$L(\neg, \rightarrow, \wedge, \vee, \forall, \exists;$ is a dog, is a cat, eats grass, is a wombat, is the father of; Ralph, Bon Bon, Howie, Juney)

universe: all animals, living or toy, that are in the U.S. and aren't in a zoo

Though the predicate and name symbols in (3) have the same realizations as in (2), this is a distinct realization: the universe differs. In (2) there is some object that satisfies 'x is a wombat'; in (3) there is none.

If we have names in the semi-formal language, then since names refer we have to have at least one object in the universe. If only predicate symbols are realized, though, it would seem there need not be anything in the universe. But if we're talking about nothing, then the whole story of how we evaluate predications makes no sense. When we reason, we assume we're talking about something.

Nonempty universe The universe of a realization contains at least one thing.

We won't investigate whether there exist any unicorns by taking as universe all unicorns, but by taking as universe all animals and considering '$\exists x$ (x is a unicorn)'.

D. The Self-Reference Exclusion Principle

Our goal is to give fairly precise meanings to the logical grammar of our semi-formal languages, and in doing so state conditions under which a semi-formal proposition is true. But certain problematic sentences stand in our way. Consider the following sentence, which I'll call 'β':

β is not true.

On the face of it this is going to get us into trouble:

(4) β is true iff 'β is not true' is true
iff β is not true

Such contradictions threaten to undermine all our efforts at defining truth-values of compound propositions in terms of the semantic properties of the primitives of the language.

Some logicians claim that not all propositions are true or false but may be indeterminate, while others deny that the sentence β, the *liar paradox*, is a proposition. Others question the assumption that equiform propositions necessarily have the same semantic properties, claiming that the problem in (4) arises from the equivocal use of a quotation name and sentence types.

By far the simplest solution, though, and the one we'll use here, is due to Alfred Tarski. It amounts to denying such sentences the status of propositions, casting them outside the scope of our formal methods.

What exactly do we mean by 'such sentences'? Tarski said the problem is the possibility of self-reference. In *Tarski, 1933*, he proposed that we should do logic only in semi-formal languages in which there can be no self-reference. Thus, within the semi-formal language we cannot talk about the semi-formal language, but only of objects outside it.

How can we formulate Tarski's restriction? Perhaps it's enough to exclude names of parts of our language from the language itself. But we still could get paradoxes by using descriptions:

> $\forall x$ (x is a sentence $\wedge$ x appears in a logic text $\wedge$ x is displayed $\wedge$
> x is followed by 14 asterisks $\rightarrow \neg$(x is true)). * * * * * * * * * * * * * *

There is, I'm fairly sure, only one sentence that satisfies the antecedent of the last displayed sentence, namely the sentence itself. So we have a version of the liar paradox.

It seems that problems appear with sentences that involve predicates of the semantics of the language, such as 'is true' or 'refers to'. The simplest, surest way to enforce Tarski's restriction is to cast out all predicates about the syntax or semantics of the language. So we shall adopt the following.

The self-reference exclusion principle We exclude from consideration in our logic sentences that contain words or predicates that refer to the syntax or semantics of the language for which we wish to give a formal analysis of truth, namely, formalized English.

In terms of restrictions on realizations we can make this principle more precise:

> No name symbol can be realized as a name of any wff or part of a wff of the semi-formal language. No predicate symbol can be realized as a predicate that can apply to wffs or parts of wffs of the semi-formal language.

Still, this principle requires judgment in its use. Shall we exclude 'is green' or 'is loud' from semi-formal languages on the grounds that a predicate can be written in green ink or a proposition when spoken can be loud? To the extent that these predicates enter into our analysis of the truth-value and meaning of propositions of the semi-formal language, the answer is yes; otherwise, no.

Exercises for Sections C and D

1. a. Specify two other universes that we could use with realization (1).
   b. How exactly did you specify those universes?
2. Why do we assume that the universe for a realization is not empty?
3. Why do we adopt the self-reference exclusion principle?
4. Give two predicates that you could apply to part of the semi-formal language, one which should be excluded by the self-reference exclusion principle, and one which need not.

E. Models

A realization establishes the propositions with which we shall reason. When we assign semantic properties to those propositions we will have a model.

For propositional logic we begin with a realization and assign truth-values to the atomic propositions. Then we extend the assignment of truth-values to all propositions of the realization by means of the truth-tables.

We will follow roughly the same procedure here, assigning semantic properties to the logical parts of the semi-formal language and devising an explanation of the nonlogical parts in a way that allows us to assign truth-values to all propositions of the realization. I'll take several pages to motivate this; the key definitions of 'model' and 'semantic consequence' are summarized at the end of the chapter.

1. The assumptions of the realization

We begin with a realization with a nonempty universe. For this realization we have already said that we will agree to the following.

a. Every name has a reference. That is, we agree to view each as denoting (picking out, referring to, having as reference) one object of the universe. We do not necessarily agree on which object the name picks out.

b. The formal symbols '$\forall$' and '$\exists$' are meant to formalize 'for all things in the universe' and 'there is something in the universe'.

c. Any one object of the universe can be distinguished (perhaps in some ideal sense) from any other in order to recognize it as a referent of a name or assign it as temporary reference to a variable in a predication.

d. Every atomic predicate can be predicated of any object(s) in the universe. That is, when a name or names or variables supplemented with indications of what each is to refer to in the universe fill in the blanks, we have a proposition.

For example, in realization (2), (a) means that we agree that each of 'Ralph', 'Bon Bon', 'Howie' and 'Juney' has a reference, though what reference that is we needn't yet specify. Assumption (d) means that 'is a dog', 'is the father of ', . . . can be predicated of any objects in the universe to obtain a proposition, for example, 'Ralph is a dog', 'x is the father of Ralph' where 'x' is to refer to Rin Tin Tin, etc.

When we establish a realization we assume the meaningfulness of its parts. For propositional logic we assumed that every syntactically acceptable composition of meaningful parts, every well-formed formula of the semi-formal language, is meaningful. But here, consider:

$\forall x_1$ (Ralph is a dog)

As a formalization of 'For every object, Ralph is a dog' this seems so odd that we may have doubts about accepting it as a proposition. Still, 'No matter what x_1 is, Ralph is a dog' (another sentence the displayed wff could formalize) makes some sense. It's like saying something irrelevant before making the point.

If we were to put further restrictions on what formulas of the semi-formal language are to be taken as meaningful more than to say that they are well-formed (or perhaps shorter in length than some specified limit), then we are embarked on a semantic analysis not taken into account by the assumptions we've adopted. So we adopt just as for propositional logic the assumption of *form and meaningfulness*:

> What is grammatical and meaningful is determined solely by form and what primitive parts of speech are taken as meaningful. In particular, given a semi-formal language, every well-formed formula will be taken as meaningful, and every closed formula will be taken as a proposition.

Exactly what we mean by an assignment of semantic properties to the primitive categorematic parts of the language is what we need to determine now, after which our goal will be to give explanations of the logical parts of the language in order to extend any such assignment of semantic properties to compound wffs.

2. Interpretations

We have a realization. Now we wish to *interpret* the names and predicates as having specific semantic properties.

e. For each name a specific object of the universe will be taken as reference (denotation). It is not for us to say what means are used to establish the reference of a name, only that we have some method.

f. The method whereby we indicate which object of the universe is to be taken as (temporary) reference of which variable is also semantically primitive. It is not for us as logicians to specify it, but only to assume that such a method is given, as indeed we already have in selecting a universe for the realization. The method itself does not provide any particular references for variables. It only specifies how such references can be provided.

For example, were the universe all objects on my desk, we might take the method to be picking up an object and saying 'This is what 'x' is to refer to'. For a universe of all atoms comprising the objects on my desk a more complicated story will be needed.

We denote the particular ways of assigning references according to the method we have chosen with the meta-variables $\sigma, \sigma_0, \sigma_1, \ldots, \tau, \tau_0, \tau_1, \ldots$. For any one predication or proposition only the variables appearing in it will be of concern to us. It is convenient, however, to imagine that each variable $x_0, x_1, \ldots$ is (or can be) assigned a reference by each σ.

Further, for '∀' to model 'for all' we must assume that the method of assigning references allows us to pick any object of the universe as the reference of any variable. That is, we must have available to us all possible assignments of references. A problem of having to speak of references for all variables at once can be avoided by reducing assignments to assigning reference to one variable at a time.

Completeness of the collection of assignments of references There is at least one assignment of references. For every assignment of references σ, and every variable x, and every object of the universe, either σ assigns that object to x or there is an assignment τ that differs from σ only in that it assigns that object to x.

Though this version seems less weighted with assumptions about the nature of naming, we need to remember that when describing a general method of assigning references for the interpretation of a realization, we have to show why, without being circular, the method ensures that we have a complete collection of assignments.

Given these ways to assign semantic values to terms we can assign semantic values to the atomic formulas.

g. A truth-value is assigned to each atomic proposition. It is not for us as logicians to specify why or how.

h. Atomic predications are also semantically primitive. That is, given any n-ary predicate P and any terms of the realization $t_1, \ldots, t_n$, where the variables among these are assigned reference, a truth-value is given for $P(t_1, \ldots, t_n)$. If a predicate realizes more than one predicate symbol, we require that the truth-value assigned to it doesn't depend on which symbol it is interpreting.

Perhaps there is only one model for a realization. That is, 'Ralph is a dog' is true or is false, though we might not know which. But we can allow ourselves various interpretations of the predicates and names as, so to speak, hypothetical models, since we may not yet agree on what the "correct" model is.

We can express these assumptions in more concise mathematical notation, so long as we remember that the notation carries no more assumptions than we've made so far.

Interpretations We have a complete collection of assignments, where an *assignment of references to variables* is σ: Variables $\rightarrow$ Universe. For a uniform way to write references for both variables and names, we take these assignments to give references to names, too. Thus, an *assignment of references* is:

σ: Terms $\rightarrow$ Universe

And all assignments σ and τ satisfy for every name a, $\sigma(a) = \tau(a)$.

A *valuation on atomic propositions* is:

υ: Atomic Propositions $\rightarrow$ {T, F}

A *valuation on atomic wffs based on* σ is an extension of the valuation υ to all atomic wffs:

υ_σ: Atomic Wffs $\rightarrow$ {T, F}

That is, for each atomic proposition A, $\upsilon_\sigma(\mathrm{A}) = \upsilon(\mathrm{A})$. The collection of such valuations must satisfy the following condition.

Consistency and extensionality of predications
For all atomic wffs $Q(t_1, \ldots, t_n)$ and $Q(u_1, \ldots, u_n)$ and any assignments σ and τ, if for all $i \leq n$, $\sigma(t_i) = \tau(u_i)$, then $\upsilon_\sigma(Q(t_1, \ldots, t_n)) = \upsilon_\tau(Q(u_1, \ldots, u_n))$.

We can write σ('Ralph') for the object assigned in the interpretation to the name 'Ralph', and $\sigma(x_{26})$ for the object assigned by σ to 'x_{26}'. Remembering the convention that symbols and semi-formal wffs name themselves, we can write σ(Ralph) for σ('Ralph') and υ(Ralph is a dog) rather than υ('Ralph is a dog').

Now we have to describe how to extend valuations to all propositions (closed formulas) of the semi-formal language.

3. The Fregean assumption and the division of form and content

The Fregean assumption in propositional logic says that the truth-value of a proposition is determined by its form, as given by the connectives appearing in it, and the semantic properties of its constituents. For predicate logic shall we say that the truth-value of the whole is determined by the form of the proposition, as given by the logical vocabulary appearing in it, and the properties of its categorematic parts?

On one point this fails to grasp the structure of propositions, for predicates have form, unlike atomic propositions in propositional logic. A binary predicate is different from a ternary one. So the form of a proposition depends on the predicates appearing in it, too.

But even adjusting for that difference, the proposed assumption is not right. The truth-value of '$\forall x$ (x is a dog)' depends not only on the predicate '— is a dog', the quantifier, and the variable, but on what things there are.

Thus, the truth-value of a compound proposition in predicate logic depends on its form, the properties of its constituents, and what things there are. Is there anything further on which the truth-value of a compound proposition could depend? Any other factor would be extraneous to the nature of propositions as we have chosen to consider them and must be ignored if our reasoning is to be regular and take into account only what we have explicitly chosen to take into account.

The Fregean assumption The truth-value of a compound proposition is determined by its form, the semantic properties of its parts, and what things are being discussed.

One of the properties of a part of a proposition is its form, for example '¬(Ralph is a dog)' in '¬(Ralph is a dog) → cats are nice'. However, we want to make a sharp distinction between the role that form plays and the semantic properties of propositions and their constituents. If we don't, then we will have confusions between what is semantic and what is syntactic. The assumption that served us for propositional logic will serve for predicate logic, too, though here we must be careful to speak not only of propositions but of any part of a proposition.

The division of form and content If two propositions or parts of propositions have the same semantic properties, then they are indistinguishable in any semantic analysis, regardless of their form.

4. The truth-value of a compound proposition: discussion

In this section I'll show how we can extend valuations based on assignments of references to all closed formulas. In doing so we'll see how we are justified in using propositional connectives between open formulas (Chapter III, p. 59).

Let's consider realization (2) again:

$$L(\neg, \rightarrow, \wedge, \vee, \forall, \exists; P_0, P_1, \ldots, c_0, c_1, \ldots)$$
$$\downarrow$$

L(¬, →, ∧, ∨, ∀, ∃; is a dog, is a cat, eats grass, is a wombat, is the father of; Ralph, Bon Bon, Howie, Juney)

universe: all animals, living or toy

Let's assume further that we have agreed on what we mean by assigning reference to a variable, say pointing with a finger or describing in English, and that we have a valuation υ and valuations υ_σ for the atomic predications here.

How shall we extend the valuation υ to '$\forall x_1$ (x_1 is a dog)'?

$\upsilon(\forall x_1$ (x_1 is a dog)$) = \mathsf{T}$ iff (a) (informally) everything is a dog

iff (b) no matter what thing (in the universe) 'x_1' refers to, 'x_1 is a dog' is true

iff (c) for any assignment of references σ, $\upsilon_\sigma(x_1$ is a dog$) = \mathsf{T}$

For (c) to reflect our intuition of (a), we must have enough assignments of references. For every element of the universe there must be some assignment that assigns that element to 'x_1'; that is, the collection of assignments must be complete. Further, the valuations must be consistent with respect to variables.

Similarly:

$\upsilon(\exists x_1\,(x_1 \text{ is a dog})) = \mathsf{T}$ iff (a) (informally) something is a dog

iff (b) there is something (in the universe) such that if 'x_1' is supplemented with an indication that it is to refer to that thing, then 'x_1 is a dog' is true

iff (c) there is some assignment of references σ, such that $\upsilon_\sigma(x_1 \text{ is a dog}) = \mathsf{T}$

Consider now:

(5) $\upsilon(\forall x_1\, \exists x_2\,(x_2 \text{ is the father of } x_1)) = \mathsf{T}$

iff no matter what thing 'x_1' refers to, '$\exists x_2\,(x_2$ is the father of $x_1)$' is true

iff no matter what thing 'x_1' refers to, there is something such that if the use of 'x_2' is supplemented with an indication that 'x_2' is to refer to that thing, 'x_2 is the father of x_1' is true

This example illustrates the inductive procedure whereby we can assign truth-values to compound propositions. For any assignment of references to 'x_1' and 'x_2', the predication 'x_2 is the father of x_1' has been assigned a truth-value by our valuations. Therefore, for each assignment of reference to 'x_1' we can assign a truth-value to '$\exists x_2\,(x_2$ is the father of $x_1)$'. Therefore, we can assign a truth-value to '$\forall x_1\, \exists x_2\,(x_2$ is the father of $x_1)$'.

Rephrasing (5) in mathematical notation requires care. If in the first equivalence we say 'for every assignment of references, σ', then saying 'there is some assignment of references, τ' at the second step will not work, since σ has already stipulated what 'x_1' is to refer to. The following recasting of (5) avoids this problem.

(6) $\upsilon(\forall x_1\, \exists x_2\,(x_2 \text{ is the father of } x_1)) = \mathsf{T}$

iff for every assignment of references σ, $\upsilon_\sigma(\exists x_2\,(x_2 \text{ is the father of } x_1)) = \mathsf{T}$

iff for every assignment of references σ, there is an assignment of references τ that differs from σ at most in what it assigns as reference to 'x_2', such that $\upsilon_\tau(x_2 \text{ is the father of } x_1) = \mathsf{T}$

That is, σ assigns reference to 'x_1', and we ignore what it assigns as reference to 'x_2', leaving that assignment to τ. The formulation of what it means for a collection of assignments to be complete ensures that (6) captures the intuition of (5).

This rephrasing may seem an awkward way to express (5), but the alternative is less appealing. Instead of assuming that σ assigns references to all variables, we could assume it assigns reference only to 'x_1'; and τ could assign reference only to 'x_2'. Then we could talk about $\upsilon_{\sigma+\tau}$ assigning a truth-value to 'x_2 is the father of x_1'. When we consider that many variables may be involved in the analysis of a proposition, the prospect of writing $\upsilon_{\sigma_1 + \cdots + \sigma_{417}}$ and/or taking $\sigma_1 + \cdots + \sigma_{417}$ looks even unhappier.

The phrasing in (6) allows for a greater uniformity of expression. It is the best of a bad lot, and the most common. We can adopt it so long as we keep in mind that *at each stage of an analysis of the truth-value of a compound proposition we will have to account for assigning reference to at most a finite number of occurrences of a single variable not assigned reference before.*

In the examples above we unpack the use of quantifiers one at a time. But consider the example with which we began our analysis of quantification in Chapter III.D:

$$\forall x\, \forall y\, (x = y \rightarrow y = x)$$

According to that earlier discussion:

'$\forall x\, \forall y\, (x = y \rightarrow y = x)$' is true
  iff for anything, x, and anything, y, $x = y \rightarrow y = x$

There are no stages: something is asserted about any pair of numbers. We could say:

(7) $\upsilon(\forall x_1\, \forall x_2\, (x_1 = x_2 \rightarrow x_2 = x_1)) = \mathsf{T}$
  iff for every assignment of references, σ, $\upsilon_\sigma(x_1 = x_2 \rightarrow x_2 = x_1) = \mathsf{T}$

But to do so will make it more difficult to give an inductive definition of the procedure of assigning truth-values. If we strictly follow the structure of the semi-formal proposition, dealing with one quantifier or one propositional connective at each stage, the inductive procedure will be clear and simple to express. We shall have to verify, however, that it gives the same result as (7):

$\upsilon(\forall x_1\, \forall x_2\, (x_1 = x_2 \rightarrow x_2 = x_1)) = \mathsf{T}$
  iff for every assignment of references σ_1, $\upsilon_{\sigma_1}(\forall x_2\, (x_1 = x_2 \rightarrow x_2 = x_1)) = \mathsf{T}$
  iff for every assignment of references σ_1, and every assignment of references σ_2 that differs at most in what it assigns as reference to 'x_2', $\upsilon_{\sigma_2}(x_1 = x_2 \rightarrow x_2 = x_1) = \mathsf{T}$
  iff for every assignment of references σ, $\upsilon_\sigma(x_1 = x_2 \rightarrow x_2 = x_1) = \mathsf{T}$

The last equivalence follows from the completeness of our collection of assignments of references.

Consider now how propositional connectives enter into our evaluations:

$\upsilon(\forall x_1\ (x_1$ is a dog $\rightarrow \neg(x_1$ is a cat$)))=\mathsf{T}$

iff no matter what thing 'x_1' refers to,
'x_1 is a dog $\rightarrow \neg(x_1$ is a cat)' is true

iff for every assignment of references σ,
$\upsilon_\sigma(x_1$ is a dog $\rightarrow \neg(x_1$ is a cat$))=\mathsf{T}$

When 'x_1' has been supplemented with an indication of what it is to refer to, 'x_1 is a dog $\rightarrow \neg(x_1$ is a cat)' is a proposition to which our propositional semantics apply. *Assignments of references, then, provide the link that justifies the use of propositional connectives between open wffs.*

The same remarks apply when propositional connectives appear at different stages of the analysis:

$\upsilon(\forall x_1 [x_1$ is a dog $\vee \exists x_2\ (x_2$ is the father of $x_1 \wedge x_2$ is a cat$)])=\mathsf{T}$

iff for every assignment of references σ,
$\upsilon_\sigma(x_1$ is a dog $\vee \exists x_2\ (x_2$ is the father of $x_1 \wedge x_2$ is a cat$))=\mathsf{T}$

To reduce the right-hand side we use the propositional semantics, since with the reference for 'x_1' provided by σ, both 'x_1 is a dog' and '$\exists x_2\ (x_2$ is the father of $x_1 \wedge x_2$ is a cat)' are propositions.

What happens if an atomic proposition is a constituent of a quantified proposition?

$\upsilon(\exists x_1\ (x_1$ is a cat $\vee$ Ralph is a dog$))=\mathsf{T}$

iff there is some assignment of references σ, such that
$\upsilon_\sigma(x_1$ is a cat $\vee$ Ralph is a dog$)=\mathsf{T}$

Now we can apply the propositional semantics. Here 'Ralph is a dog' is atomic and σ plays no role in establishing its truth-value, so υ_σ (Ralph is a dog) = υ (Ralph is a dog).

In this way, too, we can deal with superfluous quantifiers. For example:

$\upsilon(\forall x_1$ (Ralph is a dog$))=\mathsf{T}$

iff for every assignment of references σ, υ_σ(Ralph is a dog$)=\mathsf{T}$

iff υ(Ralph is a dog$)=\mathsf{T}$

Superfluous quantification also occurs when a variable is already bound.

$\upsilon_\sigma(\exists x_1\ \forall x_1\ (x_1$ is a man$))=\mathsf{T}$

iff there is some assignment of references σ_1 that differs from σ at most in what it assigns x_1, such that $\upsilon_{\sigma_1}(\forall x_1\ (x_1$ is a man$))=\mathsf{T}$

iff there is some assignment of references σ_1 that differs from σ at most in what it assigns x_1, such that for every assignment of references σ_2 that differs from σ_1 at most in what it assigns x_1, $\upsilon_{\sigma_2}(x_1 \text{ is a man}) = \mathsf{T}$

The choice of what σ_1 assigns to x_1 does not affect what σ_2 assigns to x_1. So continuing we have:

iff for every assignment of references σ_2 that differs from σ at most in what it assigns x_1, $\upsilon_{\sigma_2}(x_1 \text{ is a man}) = \mathsf{T}$

iff $\upsilon_{\sigma}(\forall x_1\, (x_1 \text{ is a man})) = \mathsf{T}$

A superfluous quantifier does not affect the truth-value of the formula to which it is applied.

One last example illustrates how to deal with a variable that appears both free and bound in a formula.

$\upsilon(\forall x_1\, (x_1 \text{ is a dog} \vee \exists x_1\, (x_1 \text{ eats grass}))) = \mathsf{T}$

iff for every assignment of references σ, $\upsilon_{\sigma}(x_1 \text{ is a dog} \vee \exists x_1\, (x_1 \text{ eats grass})) = \mathsf{T}$

iff for every assignment of references σ, $\upsilon_{\sigma}(x_1 \text{ is a dog}) = \mathsf{T}$ or $\upsilon_{\sigma}(\exists x_1\, (x_1 \text{ eats grass})) = \mathsf{T}$

iff for every assignment of references σ, $\upsilon_{\sigma}(x_1 \text{ is a dog}) = \mathsf{T}$ or there is some τ that differs from σ at most in what it assigns as reference to 'x_1' such that $\upsilon_{\tau}(x_1 \text{ eats grass}) = \mathsf{T}$

Phrasing this inductive procedure to apply generally is not hard if we keep in mind the examples of this section and the reasons for the technical choices.

5. Truth in a model

We began with a formal language. We then realized (some of) the predicate and name symbols and agreed on a universe, obtaining a semi-formal language. We then provided an interpretation of the semi-formal language, where for every atomic wff of the semi-formal language $P(t_1, \ldots, t_n)$ and every assignment of references σ, $\upsilon_{\sigma}(P(t_1, \ldots, t_n))$ is defined via the interpretation. When we add a specific method of extending the valuation to all propositions (closed formulas) of the semi-formal language will be a *model*.[1]

[1] Surveying different logics in *Predicate Logic*, I distinguished sharply between interpretations and models, since it is only in the definition of model that a particular logic is selected. Since we are considering only classical predicate logic in this volume, I will use the words *model* and *interpretation* interchangeably.

The extension of valuations based on assignments of references to all formulas
We define v_σ: Wffs $\to$ {T, F} by extending the valuations v_σ to all formulas of the semi-formal language simultaneously for all assignments of references σ:

$\mathsf{v}_\sigma(\neg A) = \mathsf{T}$ iff $\mathsf{v}_\sigma(A) = \mathsf{F}$

$\mathsf{v}_\sigma(A \wedge B) = \mathsf{T}$ iff $\mathsf{v}_\sigma(A) = \mathsf{T}$ and $\mathsf{v}_\sigma(B) = \mathsf{T}$

$\mathsf{v}_\sigma(A \vee B) = \mathsf{T}$ iff $\mathsf{v}_\sigma(A) = \mathsf{T}$ or $\mathsf{v}_\sigma(B) = \mathsf{T}$

$\mathsf{v}_\sigma(A \to B) = \mathsf{T}$ iff $\mathsf{v}_\sigma(A) = \mathsf{F}$ or $\mathsf{v}_\sigma(B) = \mathsf{T}$

$\mathsf{v}_\sigma(\forall x\, A) = \mathsf{T}$ iff for every assignment of references τ that differs from σ at most in what it assigns as reference to x, $\mathsf{v}_\tau(A) = \mathsf{T}$

$\mathsf{v}_\sigma(\exists x\, A) = \mathsf{T}$ iff for some assignment of references τ that differs from σ at most in what it assigns as reference to x, $\mathsf{v}_\tau(A) = \mathsf{T}$

We say that σ *satisfies* or *validates* A if $\mathsf{v}_\sigma(A) = \mathsf{T}$, and write $\mathsf{v}_\sigma \models A$ or more commonly, $\sigma \models A$.

For this definition to work we need that for each σ, v_σ assigns a unique truth-value to each wff. That follows by induction on the length of a wff using the unique readability of wffs, as I'll leave for you to show.

The definition of 'σ satisfies A' is meant to formalize (or can be understood as meaning) that A is true of the objects assigned by σ as references to the free variables and names in A. I'll continue to speak of an object or objects satisfying a wff if it is clear which objects are assigned to which variables as reference.

Truth in a model
The valuation v is extended to all closed wffs A by setting:

(8) $\mathsf{v}(A) = \mathsf{T}$ iff for every assignment of references σ, $\mathsf{v}_\sigma(A) = \mathsf{T}$.

If $\mathsf{v}(A) = \mathsf{T}$ we say that A is *true in the model,* and we write $\mathsf{v} \models A$. If A is not true, then $\mathsf{v}(A) = \mathsf{F}$, and we write $\mathsf{v} \not\models A$. We often read $\mathsf{v} \models A$ as 'v validates A'.

I'll use the letters M and N and subscripted versions of these to denote models. That is, M is composed of:

—A realization.

—A universe, denoted U or U_M.

—The collection of assignments of references for the realization and universe.

—The collection of valuations built on those.

—The valuation v, also written v_M.

Given a model M, the *formal wff* A *is true in* M if the realization of A is true in M. We also say that A is *valid in* M, or M *validates* A, and write $\upsilon(\text{A}) = \mathsf{T}$, or $\mathsf{M} \models \text{A}$. We say that A is *false* in M if it is not true in M, and write $\upsilon(\text{A}) = \mathsf{F}$, or $\mathsf{M} \not\models \text{A}$. We say a model is *a model for a collection of closed wffs* of the formal or semi-formal language if every wff in the collection is true in the model; we write $\mathsf{M} \models \Gamma$ to mean that for every A in Γ, $\mathsf{M} \models \text{A}$.

For example, $\forall x_7\, \text{P}_0(x_7)$ is false in the model based on realization (2), where the "obvious" truth-values are assigned to predications, because its realization '$\forall x_7$ (x_7 is a dog)' is false in the model. And that's because Fred Kroon is an animal and hence is in the universe, so letting 'x_7' stand for him, 'x_7 is a dog' is false: Fred Kroon is not now and never has been a dog.

Schematically we have:

***A predicate logic model* Type I**

$$\text{L}(\neg, \rightarrow, \wedge, \vee, \forall, \exists;\ \text{P}_0, \text{P}_1, \ldots, \text{c}_0, \text{c}_1, \ldots)$$

$\downarrow$ realization

L($\neg, \rightarrow, \wedge, \vee, \forall, \exists$; realizations of predicate symbols; of name symbols)
universe: specified in some manner

$\downarrow$ assignments of references, σ, τ, . . . ; assignments of truth-values to atomic predications, υ_σ, υ_τ, . . . ; truth-tables for $\neg, \rightarrow, \wedge, \vee$; evaluation of the quantifiers

$\{\mathsf{T}, \mathsf{F}\}$

For any closed wff every υ_σ will give the same truth-value, since there are no free variables. So we have the following.

Theorem 1

a. Given a wff A and assignments of references σ and τ, if for every variable x free in A, $\sigma(x) = \tau(x)$, then $\upsilon_\sigma(\text{A}) = \upsilon_\tau(\text{A})$.

b. If A is closed, then $\upsilon_\sigma(\text{A}) = \mathsf{T}$ for every assignment of references σ iff $\upsilon_\sigma(\text{A}) = \mathsf{T}$ for some assignment of references σ.

Proof I'll show (a) and leave (b) to you. We proceed by induction on the length of A. If A is atomic, the theorem follows by the requirement in the definition of valuations that predications be consistent.

Suppose now that the theorem is true for all wffs shorter than A. If A has the form $\neg\text{B}$, $\text{B} \rightarrow \text{C}$, $\text{B} \wedge \text{C}$, or $\text{B} \vee \text{C}$, then the theorem follows by induction, since the propositional semantics take no note of assignments of references.

Suppose now that A has the form ($\exists x$ B). Let σ and τ be as in the theorem. If $\upsilon_\sigma(\exists x\, \text{B}) = \mathsf{T}$, then there is some γ that differs from σ at most in what it assigns x, such that $\upsilon_\gamma(\text{B}) = \mathsf{T}$. Fix such a γ, and define ρ to be the same as τ except that

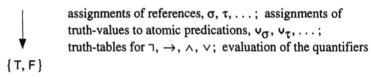

$\rho(x) = \gamma(x)$. Then ρ and γ agree on all variables free in B, so by induction $\upsilon_\rho(\mathrm{B}) = \mathsf{T}$. And hence $\upsilon_\tau(\exists x\, \mathrm{B}) = \mathsf{T}$. The same argument works with the roles of σ and τ reversed, so $\upsilon_\sigma(\exists x\, \mathrm{B}) = \upsilon_\tau(\exists x\, \mathrm{B})$.

The proof for A of the form $\forall x\, \mathrm{B}$ follows similarly. ■

When I talk informally of the *truth-conditions of* a proposition A or the *circumstances under which* A *is true*, I mean the expression that results when the semantic analysis $\upsilon(\mathrm{A}) = \mathsf{T}$ *iff*... is carried out.

For those who wish to adopt the convention that an open formula such as '$x = y \rightarrow y = x$' is a proposition, viewing each free variable as expressing generality, we could extend definition (8) to all wffs, using the following.

Theorem 2

a. For every wff A with exactly *n* free variables $y_1, \ldots, y_n$, $\upsilon(\forall y_1 \ldots \forall y_n\, \mathrm{A}) = \mathsf{T}$ iff for every σ, $\sigma \vDash \mathrm{A}$.

b. For every wff A with exactly *n* free variables $y_1, \ldots, y_n$, $\upsilon(\exists y_1 \ldots \exists y_n\, \mathrm{A}) = \mathsf{T}$ iff for some σ, $\sigma \vDash \mathrm{A}$.

Proof I'll do part (a) and leave (b) for you. The proof is by induction on the number of free variables. If $n = 1$, the theorem is true by definition. Suppose now that the theorem is true for *n*, and let A have exactly $x, y_1, \ldots, y_n$ free. Then:

$\upsilon(\forall y_1 \ldots \forall y_n\, \forall x\, \mathrm{A}) = \mathsf{T}$

iff for every σ, $\sigma \vDash \forall x\, \mathrm{A}$ (by induction)

iff for every σ, for every τ that differs from σ at most in what it assigns x, $\tau \vDash \mathrm{A}$

iff for every σ, $\sigma \vDash \mathrm{A}$ ■

Still, it is unwise to read open wffs as true or false, for confusions then follow. For example, we expect a proposition and its negation to have opposite truth-values, yet 'x is even' and '$\neg(x$ is even)' would both be false if the universe is taken to be all natural numbers and 'is even' is given its usual interpretation. The wff that has the opposite truth-value to that of 'x is even' is '$\neg\forall x\, (x$ is even)'.

6. The relation between ∀ and ∃

In one class I taught there were only a few students: Eleonora, Angela, Geovanni, and Chico. If I wish to say that all of them are Brazilian I can use the proposition:

$\forall x\, (x$ is Brazilian)

But equally well I could use the name of each student, asserting:

Eleonora is Brazilian $\wedge$ Angela is Brazilian $\wedge$ Geovanni is Brazilian $\wedge$ Chico is Brazilian

Similarly, instead of asserting $\exists x$ (x is male), I could assert:

Eleonora is male $\vee$ Angela is male $\vee$ Geovanni is male $\vee$ Chico is male

For a model with a small universe all of whose objects are named, we can dispense with quantification. We have:

$\forall \equiv$ conjunction $\quad \exists \equiv$ disjunction

From propositional logic we have $A \wedge B \leftrightarrow \neg(\neg A \vee \neg B)$. So, letting B stand for '— is Brazilian', we have:

$$B(\text{Eleonora}) \wedge B(\text{Angela}) \wedge B(\text{Geovanni}) \wedge B(\text{Chico}) \leftrightarrow$$
$$\neg(\neg B(\text{Eleonora}) \vee \neg B(\text{Angela}) \vee \neg B(\text{Geovanni}) \vee \neg B(\text{Chico}))$$

The line before the biconditional is equivalent to $\forall x\, B(x)$, and the second line is equivalent to $\neg \exists x \neg B(x)$. For a small universe that we can inspect, this equivalence seems right: A predicate is true of everything iff it is not the case that we can find something of which it is not true.

A tenet of classical mathematics is that the same relationship between $\forall$ and $\exists$ holds for any universe. Classical logic, with the goal of formalizing classical mathematics, adopts the same.

The classical relation between $\forall$ ***and*** $\exists$ In any model, for any wff A and any assignment of references σ, $\sigma \vDash \forall x A$ iff $\sigma \vDash \neg \exists x \neg A$.

This is a separate assumption that is not a consequence of our earlier agreements establishing classical predicate logic. It reflects the view that reasoning about infinite collections or unsurveyable collections of unnamed things is the same as reasoning about small collections each of whose objects can be named. Not all reasoning need generalize from finite collections, but the classical logician feels confident that the assumptions of classical predicate logic, and in particular this relationship between $\forall$ and $\exists$, do. For those who disagree, as some mathematicians do (see Chapter 26 of *Epstein and Carnielli*), the option is either to reject this last assumption of the equivalence of $\forall$ and $\neg\exists\neg$, or else hold that classical predicate logic is applicable only to finite (surveyable) collections.

Given the classical relation between $\forall$ and $\exists$, we also have the following.

Theorem 3 In every model M, for any wff A, and for every assignment of references σ, $\sigma \vDash \exists x A$ iff $\sigma \vDash \neg \forall x \neg A$.

Proof We have that $\sigma \vDash \forall x \neg A$ iff $\sigma \vDash \neg \exists x \neg\neg A$ by the classical relation between $\forall$ and $\exists$. Since for any wffs C, D, $\upsilon_\sigma(C) = \upsilon_\sigma(D)$ iff $\upsilon_\sigma(\neg C) = \upsilon_\sigma(\neg D)$. We have $\upsilon_\sigma(\neg \forall x \neg A) = \upsilon_\sigma(\neg\neg \exists x \neg\neg A)$. So $\upsilon_\sigma(\neg \forall x \neg A) = v_\sigma(\exists x A)$. ■

F. Validity and Semantic Consequence

We can now define validity and semantic consequence for predicate logic in the same manner as for propositional logic. First we look at the formal language.

A purely formal closed wff A is *valid* iff it is true in every model. To classify a wff as valid does not necessarily mean that we can survey all models or that the models comprise some fixed collection. General agreements about the nature of models normally suffice. We say that a *proposition in a semi-formal language is valid* if it is the realization of a valid wff. Valid formal wffs or semi-formal propositions are called *tautologies*.

For example, the following is valid:

(9) $\forall x_7\, (P_{43}(x_7) \vee \neg P_{43}(x_7))$

And then so is: $\forall x_7$ ((x_7 is a dog) $\vee$ $\neg$(x_7 is a dog))

We can say a proposition in ordinary English is a tautology if there is a formalization of it on which we feel certain we'll all agree and which is valid. For example,

Everything is either a dog or it isn't.

This is a tautology because (9) is an "obvious" formalization of it.

Now we can set out a formalization of the notion that a proposition B follows from a collection of propositions Γ iff it is not possible for all the wffs in Γ to be true and B to be false at the same time.

Semantic consequence A purely formal wff B is a *semantic consequence* of a collection of formal wffs Γ, written $\Gamma \vDash B$, iff B is true in every model in which all the wffs in Γ are true.

A semi-formal proposition B is a semantic consequence of a collection of semi-formal propositions Γ iff B can be taken as the realization of some purely formal wff B*, and the propositions in Γ can be taken as realizations of some collection of formal wffs Γ^* such that $\Gamma^* \vDash B^*$.

We sometimes read '$\Gamma \vDash B$' as 'Γ validates B', and write '$\Gamma \nvDash B$' for 'Γ does not validate B'. Extending the notion of semantic consequence to propositions in ordinary English requires us to agree on formalizations of those.

If B is a semantic consequence of Γ, then we say as in propositional logic that the *inference* Γ *therefore* B is *valid* or a *valid deduction.* However, our formalization of semantic equivalence is more complex than for propositional logic, for we have not only models but assignments of references within those.

Semantically equivalent wffs A is *semantically equivalent* to B iff for every model and every assignment of references σ, $\sigma \vDash A$ iff $\sigma \vDash B$.

For propositions, that is closed wffs with an interpretation, by Theorem 1.b, as for propositional logic, A and B are semantically equivalent iff $A \models B$ and $B \models A$.

Theorem II.1.(a)–(k) on the semantic consequence relation of propositional logic (p. 30) holds for predicate logic, too, as you can show. I'll also let you show the following.

Theorem 4 If A is semantically equivalent to B, then both: $\models \forall \ldots (A \leftrightarrow B)$ and $\models \exists \ldots (A \leftrightarrow B)$.

Exercises for Sections E and F

1. Explain why we are justified in reading $\neg$ as a propositional connective in '$\exists x_1 \neg(x_1$ is a dog)'.
2. Explain what it means for a collection of assignments of references to be complete.
3. Distinguish between a realization and a model.
4. Prove that for each σ, υ_σ assigns a unique truth-value to each wff.
5. Give an analysis of the truth-conditions for the following propositions.
   a. $\forall x_1$ (x_1 loves Juney $\vee \neg$(x_1 has a heart))
   b. $\forall x_1 \neg [\exists x_2$ (x_1 loves x_2) $\wedge \neg\exists x_2$ (x_1 loves x_2)]
   c. $\forall x_1$ (x_1 is a dog $\rightarrow \neg$(x_1 eats grass))
   d. $\forall x_1 \forall x_2$ (x_1 is the father of x_2 $\vee$ $\exists x_2$ (x_2 is a clone))
   e. $\forall x_1 \exists x_2$ (x_2 is the father of x_1)
   f. $\exists x_2 \forall x_1$ (x_2 is the father of x_1)
   g. $\exists x_1 \forall x_2$ (x_2 is the father of x_1)
6. Explain in your own words why for any closed formula A and any assignment of references σ, $\upsilon_\sigma(A) = T$ iff $\upsilon_\tau(A) = T$ for every assignment of references τ.
7. Prove Theorem 1.b and Theorem 2.b.
8. Explain how we can understand an open formula to be a proposition. Why is it a bad idea to do so?
9. Distinguish semantic consequence for formal wffs, semi-formal wffs, and ordinary language sentences.
10. Prove Theorem II.1.(a)–(k) for semantic consequence for predicate logic models.
11. Classify the following as valid or invalid.
    a. Ralph is a dog. *Therefore,* $\exists x_1$ (x_1 is a dog).
    b. $\forall x_1$ (x_1 is a bachelor) *Therefore,* Ralph is a bachelor.
    c. $\forall x_1$ (x_1 is a bachelor) *Therefore,* $\neg$ (Ralph is married).
    d. Ralph is a dog $\vee \neg\exists x_1$ (x_1 barks). Juney barks. *Therefore,* Ralph is a dog.
    e. Juney is barking loudly. *Therefore,* Juney is barking.

f. $\exists x_1$ ($x_1 > 5 \wedge x_1$ is the sum of 3 perfect squares) $\rightarrow$
$\neg \forall x_1$ ($\neg (x_1 > 5) \vee \neg(x_1$ is the sum of 3 perfect squares)).

g. $\exists x_1$ ($x_1 > 5 \wedge x_1$ is the sum of 3 perfect squares) $\rightarrow$
$\neg \forall x_1$ ($x_1 \leq 5 \vee \neg(x_1$ is the sum of 3 perfect squares)).

12. Consider a realization whose universe has exactly four objects, as in Section E.6, each of which is named in the semi-formal language: 'Eleonora', 'Angela', 'Geovanni', 'Chico'.
    a. Give two formalizations of each of the following, one using quantifiers and one not:
       Everyone is a student.
       No one failed the exam.
    b. Do the same for a realization with 4318 objects in the universe.

Summary: The definition of a model

The realization

a. Every name has a reference. We need not agree on which object the name picks out.

b. The formal symbols '$\forall$' and '$\exists$' formalize 'for all things in the universe' and 'there is something in the universe'.

c. Any one object of the universe can be distinguished from any other in order to recognize it as a referent of a name or assign it as temporary reference to a variable in a predication.

d. Every atomic predicate can be predicated of any object(s) in the universe. That is, when a name or names or variables supplemented with indications of what each is to refer to in the universe fill in the blanks, we obtain a proposition.

The interpretation

e. For each name a specific object of the universe is taken as reference.

f. The method whereby we indicate which object of the universe is to be taken as (temporary) reference of which variable is semantically primitive.

Completeness of the collection of assignments of references
There is at least one assignment of references. For every assignment of references σ, and every variable x, and every object of the universe, either σ assigns that object to x or there is an assignment τ that differs from σ only in that it assigns that object to x.

g. A truth-value is assigned to each atomic proposition.

h. Predications are semantically primitive. That is, given any n-ary predicate P and any terms of the realization $t_1, \ldots, t_n$, where the variables among these are assigned reference, a truth-value is given for $P(t_1, \ldots, t_n)$. If a predicate realizes more than one predicate symbol, the truth-value assigned to it doesn't depend on which symbol it is interpreting.

In mathematical notation
We have a complete collection of assignments, where an *assignment of references* is σ: Terms $\rightarrow$ Universe. For every name a and all assignments σ and τ, $\sigma(a) = \tau(a)$.

A *valuation on atomic propositions* is ν : Atomic Propositions $\rightarrow$ {T, F}.

A *valuation on atomic wffs based on* σ is an extension of the valuation ν to all atomic wffs, ν_σ : Atomic Wffs $\rightarrow$ {T, F}

Consistency and extensionality of predications
For all atomic wffs $Q(t_1, \ldots, t_n)$ and $Q(u_1, \ldots, u_n)$ and for any assignments σ and τ, if for all $i \leq n$, $\sigma(t_i) = \tau(u_i)$, then $\nu_\sigma(Q(t_1, \ldots, t_n)) = \nu_\tau(Q(u_1, \ldots, u_n))$.

The model

The extension of valuations based on assignments of references to all formulas
Define ν_σ: Wffs $\rightarrow$ {T, F} by extending the valuations ν_σ to all formulas of the semi-formal language simultaneously for all assignments of references σ:

$\nu_\sigma(\neg A) = T$ iff $\nu_\sigma(A) = F$

$\nu_\sigma(A \wedge B) = T$ iff $\nu_\sigma(A) = T$ and $\nu_\sigma(B) = T$

$\nu_\sigma(A \vee B) = T$ iff $\nu_\sigma(A) = T$ or $\nu_\sigma(B) = T$

$\nu_\sigma(A \rightarrow B) = T$ iff $\nu_\sigma(A) = F$ or $\nu_\sigma(B) = T$

$\nu_\sigma(\forall x\, A) = T$ iff for every assignment of references τ that differs from σ at most in what it assigns as reference to x, $\nu_\tau(A) = T$

$\nu_\sigma(\exists x\, A) = T$ iff for some assignment of references τ that differs from σ at most in what it assigns as reference to x, $\nu_\tau(A) = T$

σ *satisfies* or *validates* A if $\nu_\sigma(A) = T$, notated as $\nu_\sigma \models A$ or $\sigma \models A$.

The classical relation between $\forall$ *and* $\exists$ In any model, for any wff A and any assignment of references σ, $\sigma \models \forall x A$ iff $\sigma \models \neg \exists x \neg A$.

Truth in a model The valuation ν is extended to all closed wffs A by setting $\nu(A) = T$ iff for every assignment of references σ, $\nu_\sigma(A) = T$. If $\nu(A) = T$ we say that A is *true in the model*, and we write $\nu \models A$. If A is not true, then $\nu(A) = F$, and we write $\nu \not\models A$.

Semantic consequence A formal wff B is a *semantic consequence* of a collection of formal wffs Γ, written $\Gamma \models B$, iff B is true in every model in which all the wffs in Γ are true.

Semantic equivalence A is *semantically equivalent* to B iff for every model and every assignment of references σ, $\sigma \models A$ iff $\sigma \models B$.

V Substitutions and Equivalences

In the last chapter we gave a formal definition of truth in a model. Here we'll see that the assumptions that motivated that definition are respected in our models. At the same time, I'll draw out some consequences for our further work.

A. Evaluating Quantifications

1. Superfluous quantifiers

I'll leave the proof of the following two theorems to you, using Theorem IV.2.

Theorem 1 Superfluous quantifiers
If x is not free in A, then $\forall x$ A and $\exists x$ A are both semantically equivalent to A.

Theorem 2 Commutativity of quantifiers
$\forall x \forall y$ A is semantically equivalent to $\forall y \forall x$ A.
$\exists x \exists y$ A is semantically equivalent to $\exists y \exists x$ A.

Alternating quantifiers, that is, a pair $\forall x \exists y$ or $\exists y \forall x$, cannot in general be reversed to obtain an equivalent proposition. For example, $\forall x \exists y\, (x < y)$ is true about the natural numbers, where '<' is given its usual interpretation, but reversing the quantifiers yields a false proposition $\exists y \forall x\, (x < y)$.

2. Substitution of terms

Suppose we have a model in which the universe is all people older than twenty, and 'is a man', 'is a woman' have their usual interpretations. Then the following is false:

(1) $\forall x$ (x is a man) $\vee$ $\forall x$ (x is a woman)

It is not equivalent to the following, which is true:

(2) $\forall x$ (x is a man $\vee$ x is a woman)

In (1) the two occurrences of '$\forall x$' serve different purposes. The first (reading from the left) governs the 'x' in 'x is a man', but has no effect on the 'x' in 'x is a woman'. We could have used a different variable in the second disjunct:

(3) $\forall x$ (x is a man) $\vee$ $\forall z$ (z is a woman)

This is equivalent to (1). In (2), however, we cannot replace one or two of the occurences of 'x' with 'z', for then we would have a free variable, and hence no proposition. For example:

(4) $\forall x$ (x is a man $\vee$ z is a woman)

In our ordinary language it is often difficult to tell what is governed by a phrase such as 'for some' or 'for all'. But our formal and semi-formal languages allow us to be precise. Recall from Chapter III that in $\forall x$ A the *scope* of the initial $\forall x$ is A, and in $\exists x$ A the *scope* of the initial $\exists x$ is A. Thus in (1) the 'x' in 'x is a woman' is in the scope of the second '$\forall x$', but not in the scope of the initial '$\forall x$'. In (2) the 'x' in 'x is a woman' is in the scope of the initial '$\forall x$'.

Recall also from Chapter III that a particular *occurrence of* x *in* A *is bound in* A if it immediately follows an occurrence of the symbol '$\forall$' or '$\exists$' or else lies within the scope of an occurrence of '$\forall x$' or '$\exists x$'. If an occurrence of x in A is not bound, it is *free in* A. Thus, for example, each occurrence of 'z' in (3) is bound, and in (4) the single occurrence of z is free. In $\forall x$ A (respectively $\exists x$ A) every free occurrence of x in A is *bound by* the initial quantifier.

Consider the proposition:

(5) Given anything, there's something distinct from it.

In the next chapter we'll discuss how to formalize 'is the same as' with a predicate '=', and '— is distinct from —' as its negation, written '$\neq$'. Using those, we can formalize (5) as:

(6) $\forall x \exists y (y \neq x)$

This is true in every model that has at least two things in the universe.

The particular variables we use to formalize (5) do not matter. For instance, (6) is equivalent to $\forall w \exists z (z \neq w)$. But we cannot replace 'y' with 'x' in $(y \neq x)$ in

(6) and get an equivalent proposition, for by Theorem 1, $\forall x\, \exists y\ (x \neq x)$ is equivalent to $\forall x\ (x \neq x)$, and $\forall x\, \exists x\ (x \neq x)$ is equivalent to $\exists x\ (x \neq x)$. Both are false in all models, and hence are not equivalent to (6). It is crucial that semantic analyses of (6) allow for distinct references to be assigned to 'x' and 'y' in '$x \neq y$' when the universe contains distinct objects.

The moral of these examples is: *Which variable we use doesn't matter so long as we introduce no new cross-referencing*. The problem is how to recognize when we have introduced new cross-referencing. The examples suggest that we must avoid substitutions that convert free variables into bound ones. To make that precise, we need some definitions.

> A(x) means x occurs free in A (other variables may also be free in A).
>
> The term t is *free for an occurrence of* x in A (i.e., free to replace it) iff
> i. The occurrence of x is free.
> ii. If t is y, then the occurrence does not lie within the scope of some occurrence of $\forall y$ or $\exists y$.
>
> The term t is *free for* x in A iff it is free for every free occurrence of x in A.
>
> A(t/x) means the wff that results from *replacing every free occurrence of* x (if any) in A with t, unless we specifically say that t replaces some but not necessarily all occurrences of x for which it is free.

Note that every name is free for every free occurrence of a variable, and every variable is free for every occurrence of itself.

Theorem 3 If y is free for x in A, then in any model:

a. If for every σ, $\sigma \vDash$A(x), then for every σ, $\sigma \vDash$A(y/x).

b. If for some σ, $\sigma \vDash$A(x), then for some σ, $\sigma \vDash$A(y/x).

Proof The proof is by induction on the length of A. Suppose A is atomic, say $P(t_1, \ldots t_k, x, t_{k+1}, \ldots, t_n)$. Suppose that for some σ, $\sigma \nvDash P(y/x)$. Define τ by $\tau(t_i) = \sigma(t_i)$ for all i, $1 \leq i \leq n$, and $\tau(x) = \sigma(y)$. This is possible because x does not appear in P(y/x) and the collection of assignments of references for the model is complete. By the consistency of predications, $\tau \nvDash$P(x). So if all σ, $\sigma \vDash$P(x), then for all σ, $\sigma \vDash$P(y/x).

The cases when A is $B \wedge C$, $B \vee C$, $B \rightarrow C$, or $\neg B$ are straightforward.

Suppose, then, that A is $\forall z$ B. If z is not free in B, then we are done by induction. If z is y, then y is not free for x in A. If z is x, then A(y/x) is A, since no occurrence of x in A is free. Finally, if z is neither x nor y, and z is free in B, the proof is similar to that for atomic wffs, reading "does not appear free" for "does not appear". The case when A is $\exists z$ B is similar. ■

We can't improve Theorem 3 to get 'iff'. Consider:

If for all σ, $\sigma \models x = y$, then for all σ, $\sigma \models y = y$.

This is true in every model, since the consequent is always true. The converse, however, is true only in a model if it has just one object in the universe:

If for all σ, $\sigma \models y = y$, then for all σ, $\sigma \models x = y$.

The variable y may appear free in A, so the substitution A(y/x) may introduce new cross-referencing. If y is not free in A, however, the same argument as in the proof of Theorem 3 works in the other direction, and we do have equivalences.

Theorem 4 If y is free for x in A, but y is not itself free in A, then

a. x is free for y in A(y/x).

b. $\forall x\,\mathrm{A}(x)$ is semantically equivalent to $\forall y \mathrm{A}(y/x)$.

c. $\exists x\,\mathrm{A}(x)$ is semantically equivalent to $\exists y\,\mathrm{A}(y/x)$.

3. The extensionality of predications

The following theorem shows that compound predications are also extensional.

Theorem 5 Suppose x is free in A and t is free for x. Suppose also that t is substituted for some but not necessarily all occurrences of x for which it is free. Then for any assignment of references σ,

if $\sigma(x) = \sigma(t)$, then $\sigma \models \mathrm{A}(x)$ iff $\sigma \models \mathrm{A}(t/x)$.

Proof I'll prove this by induction on the length of A. If A is atomic, it is true by the assumption of the consistency and extensionality of atomic predications.

Suppose A is of the form $\mathrm{B} \wedge \mathrm{C}$. If x is free in B and C, then we have by induction that for every σ, if $\sigma(x) = \sigma(t)$, then $\sigma \models \mathrm{B}(x)$ iff $\sigma \models \mathrm{B}(t/x)$ and $\sigma \models \mathrm{C}(x)$ iff $\sigma \models \mathrm{C}(t/x)$ where t is substituted for some but not necessarily all occurrences of x for which it is free. Hence, we have $\sigma \models (\mathrm{B} \wedge \mathrm{C})(x)$ iff $\sigma \models (\mathrm{B} \wedge \mathrm{C})(t/x)$. I'll leave to you the cases where x is not free in one of B or C.

The cases when A is $\mathrm{B} \vee \mathrm{C}$, $\mathrm{B} \rightarrow \mathrm{C}$, or $\neg\mathrm{B}$ are done similarly.

Suppose now that A is of the form $\forall z \mathrm{B}$. If z is not free in B, then we are done by Theorem 1 and induction. So suppose z is free in B. Then we have by induction that for every σ, if $\sigma(x) = \sigma(t)$, then $\sigma \models \mathrm{B}(z,x)$ iff $\sigma \models \mathrm{B}(z, t/x)$. Hence, since z cannot be t (as t is free for x in A), for every σ, if $\sigma(x) = \sigma(t)$, $\sigma \models \forall z\,\mathrm{B}(z, x)$ iff $\sigma \models \forall z\,\mathrm{B}(z, t/x)$.

The case when A is of the form $\exists z \mathrm{B}$ is done similarly. ∎

Exercises for Section A

1. Show directly the following without using Theorem 1 or Theorem 2.
   a. $\nu(\forall x\, \forall y\, (x \text{ is a cousin of } y)) = \mathsf{T}$ iff for every σ, $\nu_\sigma(x \text{ is a cousin of } y) = \mathsf{T}$
   b. $\nu(\forall x\, \forall y\, (x \text{ is a cousin of } y)) = \nu(\forall y\, \forall x\, (x \text{ is a cousin of } y))$

c. $\nu(\exists x\, \exists y\, (x \text{ is a cousin of } y)) = \nu(\exists y\, \exists x\, (x \text{ is a cousin of } y))$

2. Prove Theorem 2. (For ease of notation you can write $\sigma \underset{x}{\neq} \tau$ for 'σ disagrees with τ only on the variable x', and similarly $\sigma \underset{x,y}{\neq} \tau$.)

3. Either show that the following two formulas are equivalent or provide a model in which one is true and the other false: $\forall x\, \exists y\, (x = y)$ and $\exists y\, \forall x\, (x = y)$

4. For each formula A below give $A(x_2/x_1)$ and $A(c_0/x_1)$.
   a. $P_1(c_8)$
   b. $P_8(x_1)$
   c. $\exists x_2\, P_1(x_1) \vee \neg P_2(c_3)$
   d. $\forall x_1\, (\neg P_2(x_1) \vee P_1(x_3)) \rightarrow \exists x_3\, P_4(x_1, x_3)$
   e. $\forall x_2\, P_1(x_1, x_2, x_3) \rightarrow \exists x_2\, \exists x_3\, P_1(x_1, x_2, x_3)$
   f. $P_1(x_1) \vee \neg P_1(x_1)$
   g. $\forall x_1\, P_1(x_1) \wedge \neg \exists x_2\, P_2(x_2)$

5. For each wff A below write out $A(y/x)$ and determine whether y is free for x in A.
   a. $\forall x\, (x \text{ loves } y)$
   b. $\forall x\, (x \text{ loves } y) \vee (x = y)$
   c. $\forall y\, (x \text{ loves } y)$
   d. $(x \text{ loves } y) \vee \exists z\, \exists y\, (\neg(y \text{ loves } z))$
   e. $(x \text{ loves } y) \vee \exists z\, \exists x\, (\neg(y \text{ loves } x))$
   f. $\exists y\, (y \neq y \vee x = x)$
   g. $\forall z\, (z = y \vee \exists x\, (\neg(x \text{ loves } y) \wedge x \text{ is a parent of } y)$

6. For each formula A below give $A(x_2/x_1)$ and $A(\text{Ralph}/x_1)$.
   a. x_1 is an uncle of x_2 $\vee$ $\exists x_1$ (x_1 is a woman)
   b. Ralph is an uncle $\vee$ $\exists x_1$ (x_1 is a dog)
   c. $\forall x_2\, \neg$(x_2 is a father $\vee$ $\exists x_2$ (x_2 is related to Ralph))
   d. $\forall x_1\, \forall x_2$ (x_1 is a dog $\vee$ x_2 is a cat $\rightarrow$ Juney barks)
   e. $\exists x_1$ (x_1 is in the nucleus of x_2 $\wedge$ x_2 is an atom) $\rightarrow$ $\forall x_2$ (x_1 is smaller than x_2)

7. a. Give an example of a formula $A(x)$ such that y is free for x in A and the following is not valid: $\forall \ldots \forall x\, A(x) \leftrightarrow \forall \ldots \forall y\, A(y)$.
   b. As for (a) for $\forall \ldots \exists x\, A(x) \leftrightarrow \forall \ldots \exists y\, A(y/x)$.

8. For each of the following, show that it is a tautology or exhibit a model in which it fails.
   a. $\forall x\, \exists y\, (x \text{ is barking at } y) \rightarrow \forall x\, \exists z\, (x \text{ is barking at } z)$
   b. $\exists x\, \forall y\, (x \text{ is barking at } y) \leftrightarrow \exists z\, \forall y\, (z \text{ is barking at } y)$
   c. $\forall y\, \forall x\, (x \text{ is barking at } y) \rightarrow \forall y\, (y \text{ is barking at } y)$
   d. $\forall y\, \forall x\, (x \text{ is barking at } y) \leftrightarrow \forall y\, (y \text{ is barking at } y)$
   e. $\forall x\, \forall y\, (x \text{ is barking at } y) \rightarrow \forall y\, (y \text{ is barking at } y)$
   f. $\forall y\, (\forall x\, (x \text{ is barking at } y) \rightarrow y \text{ is barking at } y)$
   g. $\exists x\, \forall y\, (x \geq y) \rightarrow \exists x\, \forall x\, (x \geq x)$
   h. $\forall z\, \exists x\, (\forall y\, (y \text{ is barking at } y) \rightarrow \neg(z \text{ is barking at } x)) \leftrightarrow$
   $\forall z\, \exists y\, (\forall y\, (y \text{ is barking at } y) \rightarrow \neg(z \text{ is barking at } x))$

9. If y is free for x in A, show:
   a. $\models \forall \ldots \forall x\, A(x) \rightarrow \forall \ldots \forall y\, A(y/x)$
   b. $\models \forall \ldots (\forall x\, A(x) \rightarrow \forall y\, A(y/x))$
   c. $\models \forall \ldots (\forall y\, \forall x\, A(x) \rightarrow \forall y\, A(y/x))$

10. If y is free for x in A, show:
    a. $\models \forall \ldots \exists x\, A(x) \rightarrow \forall \ldots \exists y\, A(y/x)$
    b. $\models \forall \ldots (\exists x\, A(x) \rightarrow \exists y\, A(y/x))$
    c. $\models \forall \ldots (\exists y\, \exists x\, A(x) \rightarrow \exists y\, A(y/x))$
11. Prove Theorem 4.

B. Propositional Logic within Predicate Logic

Since the propositional connectives in predicate logic are evaluated the same as in propositional logic, **PC** applies within predicate logic. That is, if A has the form of a propositional tautology, it's valid. But what do we mean by 'propositional form'?

Consider $\exists x$ (x is a dog) $\vee \neg \forall y$ (y is a cat). The propositional form of this is $A \vee \neg B$. But what about:

(7) (x is a dog) $\rightarrow \neg$ (y is a cat)

We'd like to speak of the propositional form of this wff, even though 'x is a dog' and 'y is a cat' are not propositions.

A propositional form of a wff *A propositional form of a wff* A is a wff B of the language of propositional logic such that there is some way to assign wffs (possibly open) to the propositional variables in B that results in A, where the same variable is assigned the same wff throughout B.

There is never a unique propositional form for a wff. For example, $p_4 \rightarrow \neg p_1$ and $p_8 \rightarrow \neg p_{17}$ are both propositional forms of (7). Even taking propositional forms schematically, there are two propositional forms for:

($\forall x$ (x is a dog) $\rightarrow \neg \exists y$ (y is a cat)) $\vee \neg$($\forall x$ (x is a dog) $\rightarrow \neg \exists y$ (y is a cat))

They are: $(A \rightarrow B) \vee \neg(A \rightarrow B)$ and also $(A \vee \neg A)$. Sometimes I'll speak loosely of *the* propositional form of a wff when there is one form that is of obvious interest.

I'll let you prove the following lemma.

Lemma 6 Suppose A is a wff which has propositional form $\alpha(q_1, \ldots, q_n)$, and A arises by substituting the predicate logic wffs $D_1, \ldots, D_n$ for $q_1, \ldots, q_n$. For any model and assignment of references σ, define the propositional valuation υ by $\upsilon(q_i) = T$ iff $\sigma \models D_i$. Then $\sigma \models A$ iff $\upsilon(\alpha) = T$.

Theorem 7

a. If A has a propositional form of a **PC**-tautology, then in every model for every assignment of references σ, $\sigma \models A$.

b. If A has a propositional form of a **PC**-tautology and $y_1, \ldots, y_n$ are the variables free in A, then for any quantifiers $Q_1 \ldots Q_n$, $\models Q_1 y_1 \ldots Q_n y_n A$.

c. Suppose A has a propositional form α and B has a propositional form β using the same propositional variables, and $\models_{\mathbf{PC}} \alpha \leftrightarrow \beta$. If B and A both arise by the same substitution of predicate wffs for those variables, then A is semantically equivalent to B, and $\models \forall \ldots (A \leftrightarrow B)$.

Proof (a) Let A have propositional form α, where α is a **PC**-tautology. Suppose $\sigma \not\models A$. Since α is a tautology, α must have length ≥ 1. Suppose α is $\beta \rightarrow \gamma$. Then A must have the form $B \rightarrow C$. Then $\sigma \models B$ and $\sigma \not\models C$. In that case, for the propositional valuation υ as in Lemma 6, $\upsilon(\beta) = \mathsf{T}$ and $\upsilon(\gamma) = \mathsf{F}$, which contradicts that $\beta \rightarrow \gamma$ is a tautology. So A does not have form $\beta \rightarrow \gamma$. I'll let you show that A also cannot have form $\neg\beta$, $\beta \wedge \gamma$, or $\beta \vee \gamma$. So there is no σ such that $\sigma \not\models A$.

I'll leave the rest of the theorem to you. ■

Exercises for Section B

1. Prove Lemma 6.
2. a. Prove Theorem 7.c.
   b. Prove: Suppose A has a propositional form α and B has a propositional form β using the same propositional variables, and $\models_{\mathbf{PC}} \alpha \rightarrow \beta$. If B and A both arise by the same substitution of predicate wffs for those variables, then $\models \forall \ldots (A \rightarrow B)$.
3. Show that the following is false: If A and B have propositional forms that are equivalent in **PC**, then $\models \forall \ldots A \leftrightarrow \forall \ldots B$. (Hint: Consider a very simple propositional form.)
4. For each of the following either justify that it is valid based on its propositional form, or explain why Theorem 7 does not apply.
   a. $\forall x$ (x is a dog) $\vee \neg \forall x$ (x is a dog)
   b. $\forall x$ (x is a dog $\vee \neg$ (x is a dog))
   c. [$\forall x$ (x is a dog) $\vee$ (y is a cat)] $\leftrightarrow$ [$\neg$($\neg \forall x$ (x is a dog) $\wedge \neg$ (y is a cat))]
   d. $\neg \forall x$ (x is a dog $\vee$ Ralph is a cat) $\leftrightarrow \exists x \neg$ (x is a dog $\vee$ Ralph is a cat)
   e. [$\neg\neg \forall x$ (x is a dog) $\rightarrow \exists y$ (y is a cat)] $\rightarrow$
      [$\neg$($\neg \forall x$ (x is a dog) $\wedge \exists y$ (y is a cat))]
   f. $\forall y$ ($\exists x$ (x is a dog) $\wedge$ (y is a cat)) $\leftrightarrow$ $\forall y \neg$($\exists x$ (x is a dog) $\rightarrow$ (y is a cat))
   g. $\forall x$ (Ralph is a dog $\rightarrow \neg \exists x$ (x is a cat)) $\vee \neg \forall x$ (Ralph is a dog $\rightarrow \neg \exists x$ (x is a cat))
   h. $\forall x \forall y \forall z$ (x is the daughter of y and z $\rightarrow$ (x is female $\wedge$ (y is female $\vee$ z is female)) $\rightarrow \neg$ (x is the daughter of y and z $\rightarrow$ (x is female $\wedge$ (y is female $\vee$ z is female))))
   i. $\forall x \forall y$ (((x is the uncle of y)$\rightarrow$ (x is male)) $\rightarrow$ (x is the uncle of y)) $\rightarrow$ (x is the uncle of y))
   j. $\forall x \exists y$ (($y < x$) $\vee$ ($x = 0$))

C. Distribution of Quantifiers

Theorem 8

a. For any wffs A and B, and for any model and any assignment of references σ, if $\sigma \models \forall x\,(A \to B)$, then $\sigma \models \forall x\, A \to \forall x\, B$.

b. $\forall x\,(A \wedge B)$ is semantically equivalent to $\forall x\, A \wedge \forall x B$.

c. $\exists x\, A \vee \exists x B$ is semantically equivalent to $\exists x\,(A \vee B)$.

Proof Suppose (a) is false. Then there is some σ such that $\sigma \models \forall x\,(A \to B)$, but $\sigma \not\models \forall x\, A \to \forall x B$. So $\sigma \models \forall x\, A$, but $\sigma \not\models \forall x B$. So there is some τ that differs from σ only in what it assigns to x, such that $\tau \not\models B$. But for any such τ, $\tau \models A$. So $\tau \not\models A \to B$. Hence, $\sigma \not\models \forall x\,(A \to B)$, a contradiction. So (a) is true.

Parts (b) and (c) I'll leave to you. ■

Corollary 9

a. $\models \forall \ldots (\forall x (A \to B) \to (\forall x A \to \forall x B))$

b. If A is semantically equivalent to B, then $\models \forall \ldots A \leftrightarrow \forall \ldots B$.

Proof Part (a) is from Theorem 8.a, and (b) follows by Theorem IV.4. ■

Not all distributions are valid. For example, $\exists x (x < 1 \to x < 0)$ is true for the usual interpretation of the predicates on the universe of natural numbers because 2 satisfies the predicate, since the antecedent is false. But $\exists x (x < 1) \to \exists x\, (x < 0)$ is false. Nor can we get an equivalence for Theorem 8.a, since $\forall x (x < 2) \to \forall x\, (x < 1)$ is true of the natural numbers, while $\forall x (x < 2 \to x < 1)$ is false.

Here is a list of equivalences that we will need. I'll leave the proofs to you.

Theorem 10 The Tarski-Kuratowski algorithm If A and B are any wffs and x is not free in B, each pair of formulas below are semantically equivalent.

a.	$\forall x\, A$	$\neg \exists x \neg A$	*i.*	$(\exists x\, A) \vee B$	$\exists x\, (A \vee B)$
b.	$\exists x\, A$	$\neg \forall x \neg A$	*j.*	$B \vee (\exists x\, A)$	$\exists x\, (B \vee A)$
c.	$\neg \forall x\, A$	$\exists x \neg A$	*k.*	$(\forall x\, A) \vee B$	$\forall x\, (A \vee B)$
d.	$\neg \exists x\, A$	$\forall x \neg A$	*l.*	$B \vee (\forall x\, A)$	$\forall x\, (B \vee A)$
e.	$(\exists x\, A) \wedge B$	$\exists x\, (A \wedge B)$	*m.*	$(B \to \exists x\, A)$	$\exists x\, (B \to A)$
f.	$B \wedge (\exists x\, A)$	$\exists x\, (B \wedge A)$	*n.*	$(B \to \forall x\, A)$	$\forall x\, (B \to A)$
g.	$(\forall x\, A) \wedge B$	$\forall x\, (A \wedge B)$	*o.*	$(\exists x\, A \to B)$	$\forall x\, (A \to B)$
h.	$B \wedge (\forall x\, A)$	$\forall x\, (B \wedge A)$	*p.*	$(\forall x\, A \to B)$	$\exists x\, (A \to B)$

Corollary 11 Suppose $Q_1, \ldots Q_n$ are quantifiers. Let Q_i' be $\exists$ if Q_i is $\forall$, and let Q_i' be $\forall$ if Q_i is $\exists$. Then for any wff D:

$$\neg\, Q_1 y_1 \ldots Q_n y_n \,\mathrm{D} \text{ is semantically equivalent to } Q_1' y_1 \ldots Q_n' y_n \neg \mathrm{D}.$$

Exercises for Section C

1. Show that $\forall \ldots \forall x\, (\mathrm{A} \vee \mathrm{B}) \to \forall x \mathrm{A} \vee \forall x \mathrm{B}$ is not valid.
2. For each of the following, show that it is valid or give a model in which it fails.
   a. $\forall x\, \exists y\, (x \text{ loves } y) \to \forall x\, \exists z\, (x \text{ loves } z)$
   b. $\exists x\, \forall y\, (x \text{ loves } y) \leftrightarrow \exists z\, \forall y\, (z \text{ loves } y)$
   c. $\forall y\, \forall x\, (x \text{ loves } y) \to \forall y\, (y \text{ loves } y)$
   d. $\forall y\, \forall x\, (x \text{ loves } y) \leftrightarrow \forall y\, (y \text{ loves } y)$
   e. $\forall x\, \forall y\, (x \text{ loves } y) \to \forall y\, (y \text{ loves } y)$
   f. $\forall y\, (\forall x\, (x \text{ loves } y) \to y \text{ loves } y)$
   g. $\exists x\, \forall y\, (x \geq y) \to \exists x\, \forall x\, (x \geq x)$
   h. $\forall z\, \exists x\, (\forall y\, (y \text{ loves } y) \to \neg (z \text{ loves } x)) \leftrightarrow$
   $\forall z\, \exists y\, (\forall y\, (y \text{ loves } y) \to \neg (z \text{ loves } x))$
3. Prove Theorem 10
4. Prove Corollary 11.

Prenex normal forms

In **PC** each wff has both a disjunctive and a conjunctive normal form. I'll show here how we can obtain normal forms for predicate logic wffs, too.

For any formula A we can find another formula B equivalent to it in which all the quantifiers occur at the beginning of the formula. For example, letting '≈' stand for 'is semantically equivalent', via the Tarski-Kuratowski algorithm we have:

$$\begin{aligned} \forall x\, (\exists y\, \mathrm{P}_1(x, y) \to \exists z\, \mathrm{P}_2(x, z)) &\approx \forall x \exists z\, (\exists y\, \mathrm{P}_1(x, y) \to \mathrm{P}_2(x, z)) \\ &\approx \forall x \exists z \forall y\, (\mathrm{P}_1(x, y) \to \mathrm{P}_2(x, z)) \end{aligned}$$

Sometimes we must make a substitution of one variable for another in order to effect such an equivalence.

(*) $\exists x\, \mathrm{P}(x) \to \exists x\, \mathrm{Q}(x) \approx \forall x\, (\mathrm{P}(x) \to \exists x\, \mathrm{Q}(x))$ by the T-K algorithm

$\approx \forall z\, (\mathrm{P}(z) \to \exists x\, \mathrm{Q}(x))$ by Theorem 3

$\approx \forall z\, \exists x\, (\mathrm{P}(z) \to \mathrm{Q}(x))$ by the T-K algorithm

---

Prenex forms A wff in which no quantifier appears is *quantifier-free*. A formula is in *prenex form* if it has the form:

$$\underbrace{Q_1 y_1 \ldots Q_n y_n}_{\textit{prefix}}\ \mathrm{C}$$

where for each $i \leq n$, Q_i is either $\forall$ or $\exists$ and C, the *matrix*, is quantifier-free.

We allow $n = 0$. If Q_i is different from Q_{i+1}, we say there is an *alternation of quantifiers* at that point.

Even identifying alphabetic variants, a wff may have more than one prenex form. For example, we could reduce (*) to a diferent prenex form:

$\exists x\, P(x) \rightarrow \exists x\, Q(x)$	$\approx \exists x\, (\exists x\, P(x) \rightarrow Q(x))$	by the T-K algorithm
	$\approx \exists z\, (\exists x\, P(x) \rightarrow Q(z))$	by Theorem 3
	$\approx \exists z\, \forall x\, (P(x) \rightarrow Q(z))$	by the T-K algorithm

---

Prenex normal form A *prenex normal form for a wff* A is any formula B such that:

i. B is in prenex form;

ii. A and B have the exactly the same free variables;

iii. B is semantically equivalent to A; and

iv. Every variable that appears in the prefix of B appears also in the matrix of B, and no variable appears twice in the prefix of B.

It is a *disjunctive* (*conjunctive*) *prenex normal form* if its matrix is in disjunctive (conjunctive) propositional normal form.

The same terminology applies to semi-formal wffs as realizations of formal wffs. For example, a prenex normal form of $\exists x$ (x is a cat)$\rightarrow\neg\forall y$ (y is lovable) is $\forall x\, \exists y$ ((x is a cat $\rightarrow \neg$(y is lovable)). And a disjunctive normal form for it is $\forall x\, \exists y$ ($\neg$(x is a cat) $\vee$ $\neg$(y is lovable)).

Normal form theorem Given any wff A we can find a a conjunctive prenex normal form for it and a disjunctive prenex normal form for it.

Proof We proceed by induction on the length of A to show that first A has a prenex normal form. If A is atomic or is quantifier free, we are done. So suppose theorem is true for all wffs shorter than A.

Suppose A has the form $\neg$C. Then by induction there is a wff C* in prenex normal form for C. By Corollary 11, we are then done.

Suppose A has the form C$\wedge$D. Then by induction C and D have prenex normal forms $Q_1 y_1 \cdots Q_n y_n$ C* and $Q_{n+1} z_1 \cdots Q_{n+m} z_m$ D*. By Theorem 3 we can substitute so that we may assume that no y_i is a z_i. Then by parts (e)–(h) of the Tarski-Kuratowski algorithm, $Q_1 y_1 \cdots Q_n y_n Q_{n+1} z_1 \cdots Q_{n+m} z_m$ (C*$\wedge$D*) is semantically equivalent to C$\wedge$D, and is in prenex form, and satisfies (ii) and (iv). The case when A has the form C$\vee$D is done similarly.

Suppose that A is of the form C$\rightarrow$D. Then C$\rightarrow$D is semantically equivalent in **PC** to $\neg$(C$\wedge\neg$D). So we can use Theorem 7 and what we've already established.

If A has form $\forall x$ C, then by induction there is a wff C* in prenex normal form for C. Hence, by Theorem 1, Theorem 4, and Theorem IV.4, we are done. If A is of the form $\exists x$ C the proof is the same.

Finally, by applying Theorem 7 and the normal form theorem for propositional logic to a prenex normal form for A, we can obtain both a conjunctive normal form for A and a disjunctive normal form for A. ∎

Exercises

1. Find a prenex normal form for each of the following formulas.
   a. $\exists x\, P_1(x) \wedge P_2(x)$
   b. $\forall x\, (P_1(x) \rightarrow \forall y\, (P_2(x, y) \rightarrow \neg \forall z\, P_3(y, z)))$
   c. $\exists x\, P_1(x, y) \rightarrow (P_2(x) \rightarrow \neg \exists z\, P_3(x, z))$
   d. $\forall x\, P_1(x) \leftrightarrow \exists x\, P_2(x)$
   e. $\forall x\, P_1(x) \leftrightarrow \exists x\, P_1(x)$
   f. $(\exists x\, P_1(x) \wedge \exists x\, P_2(x)) \rightarrow P_3(x)$
   g. $\exists x\, P_1(x) \rightarrow [\exists y\, (\forall x\, P_2(x, y) \rightarrow \forall y\, P_2(x, y)) \rightarrow \forall x\, (P_1(x) \leftrightarrow \exists y\, P_2(y))]$
   h. $\exists x$ (x is a cat) $\rightarrow \neg \forall y$ (y is lovable)
   i. $\forall x$ (x is a dog $\wedge\ \exists y$(y is a master of x) $\rightarrow \neg \forall z$ (x is hated by z))
2. Find a disjunctive prenex normal form for (b) and (h) of Exercise 1.
3. Prove that if B is a prenex normal form for A, then
   a. $\vDash \forall \ldots (A \leftrightarrow B)$
   b. $\vDash \forall \ldots A \leftrightarrow \forall \ldots B$

D. Names and Quantifiers

We've assumed that in every model a name is a name of something. So from the truth of a wff that uses a name we can conclude that something exists.

Theorem 12 Existential generalization For every wff $A(x)$ and name (symbol) c,

a. In any model, for every σ, if $\sigma \vDash A(c/x)$ then $\sigma \vDash \exists x\, A(x)$.

b. $\vDash \forall \ldots (A(c/x) \rightarrow \exists x\, A(x))$.

But consider:

Socrates is dead

This is true. So it seems we are committed to accepting as true:

$\exists x$ (x is dead)

What kind of existence can we mean by '$\exists$' if the dead are said to exist?

It is not necessary to become entangled in metaphysical or religious doctrines: '$\exists x$ (x is dead)' is true because 'Socrates is dead' is true. In predicate logic we understand propositions as timelessly true or false. Reference, too, must be construed as timeless. What a name refers to simply *is* in a way that cannot take into account its coming into being and passing away, as we could, perhaps, were we to model reasoning that focuses on processes.

Since names refer, we also know that if we are talking about everything, then in particular we are talking about whatever is named. So we have the following.

Theorem 13 Universal instantiation For every wff $A(x)$ and name (symbol) c,

a. In any model, if for every σ, $\sigma \vDash \forall x\, A(x)$ then $\sigma \vDash A(c/x)$.

b. $\vDash \forall \ldots (\forall x\, A(x) \rightarrow A(c/x))$.

The wff $A(c/x)$ is called an *instantiation* or *particularization* of $\forall x\, A(x)$. Theorem 13 is also called *universal specification* or *particularization*. For example:

$\vDash$ $\forall x$ (x is mortal) $\rightarrow$ Socrates is mortal

$\vDash$ $\forall x$ (x is kind $\vee$ $\neg\forall y$ (y is a dog $\rightarrow$ x loves y)) $\rightarrow$
Marilyn Monroe is kind $\vee$ $\neg\forall y$ (y is a dog $\rightarrow$ Marilyn Monroe loves y)

If we combine universal instantiation with existential generalization, we obtain the following for any wff A in which only x is free.

Theorem 14 All implies exists $\forall x\, A(x) \vDash \exists x\, A(x)$

The validity of this scheme of inferences appears to depend on our having names available in the semi-formal language. But really it follows because we have assumed that the universe of a realization is nonempty. Because of that, we can validly assert that something exists, using any wff that has a propositional logic tautology form, for example, $\vDash \exists x_1\, (P_1(x_1) \vee \neg P(x_1))$.

How did we get into the position of logic guaranteeing the existence of something? The assumption that logic is about things led to a referential interpretation of the quantifiers, which depends on there being at least one object in the universe that variables can take as reference. We have no truth-conditions for wffs of the form $\forall x\, A(x)$ when the universe is empty. *All implies exists* simply reflects that we are always reasoning about some objects.

E. The Partial Interpretation Theorem

For **PC** the truth-value of a wff depends only on the assignment of truth-values to the propositional variables appearing in it. Similarly, for classical predicate logic the truth-value of a formula depends only on the semantic properties of the predicates and names appearing in it, as well as the range of the variables. So if two models agree on all the categorematic parts of a sentence and have the same universe, they should yield the same truth-value for the sentence. To prove this, we introduce some notation.

Models agree on a language For any collection of wffs Γ of a formal or semi-formal language L, let $L(\Gamma)$ stand for the language L restricted to just those names (symbols) and predicates (symbols) appearing in wffs in Γ. We write $L(A)$ for $L(\{A\})$. Two models M and N *agree on* $L(\Gamma)$ when the models have the same universe and the same interpretations of the names and the predicates that are in L(Γ). That is, the following all hold:

i. $U_M = U_N$.
ii. For every name c appearing in Γ, for every σ on M and τ on N, $\sigma(c) = \tau(c)$.
iii. For every n-ary predicate P appearing in Γ, whenever $\sigma(y_i) = \tau(y_i)$ for each $i \leq n$, $\sigma \vDash P(y_1, \ldots, y_n)$ iff $\tau \vDash P(y_1, \ldots, y_n)$.

I'll let you prove the next theorem, using induction on the length of wffs.

Theorem 15 The partial interpretation theorem
If M and N agree on L(A), then $M \vDash A$ iff $N \vDash A$.

Theorem 16 Substitution of wffs (the division of form and content verified)
Let C(A) be a wff that contains A as a subformula. Let B be another wff with the same variables free as A has. Let C(B) be the result of replacing some but not necessarily all occurrences of A in C(A) by B.

a. In any model, if for all σ, $\sigma \vDash A$ iff $\sigma \vDash B$, then for all σ, $\sigma \vDash C(A)$ iff $\sigma \vDash C(B)$.

b. $\vDash \forall \ldots ((A \leftrightarrow B) \rightarrow (C(A) \leftrightarrow C(B)))$.

c. $\vDash \forall \ldots (A \leftrightarrow B) \rightarrow \forall \ldots (C(A) \leftrightarrow C(B))$.

Proof (a) We proceed by induction on the number of connectives and quantifiers occuring in C that do not occur in A. If there are none, then C is A and the result is immediate.

Suppose the theorem is true for fewer connectives and quantifiers than occur in C. Suppose also that we are given a model in which for all σ, $\sigma \vDash A$ iff $\sigma \vDash B$. If C(A) has the form $\neg E(A)$, then for all σ, $\sigma \vDash \neg E(A)$ iff $\sigma \nvDash E(A)$ iff $\sigma \nvDash E(B)$ iff $\sigma \vDash \neg E(B)$. I'll leave the cases for the other propositional connectives to you.

Suppose now that C(A) has the form $\forall x E(A)$. Then given any σ,

$\sigma \vDash \forall x E(A)$ iff for any τ that differs from σ at most in what it assigns x, $\tau \vDash E(A)$

iff for any τ that differs from σ at most in what it assigns x, $\tau \vDash E(B)$

iff $\sigma \vDash \forall x E(B)$

The case when C(A) has the form $\exists x E(A)$ is done similarly.

I'll let you prove part (b). Part (c) then follows by Theorem 8 on the distribution of quantifiers. ■

Exercises for Sections D and E

1. Give an example where $\models \exists x\, A(x)$, but $\not\models A(c/x)$.
2. Prove :
   a. $\forall \ldots A(c/x) \models \forall \ldots \exists x\, A(x)$.
   b. $\models \forall \ldots A(c/x) \rightarrow \forall \ldots \exists x\, A(x)$.
3. Prove :
   a. $\forall \ldots \forall x A(x) \models \forall \ldots A(c/x)$.
   b. $\models \forall \ldots \forall x A(x) \rightarrow \forall \ldots A(c/x)$.
4. For the realization described at the beginning of Chapter IV.F.4, write out explicitly the language of { $\forall x_1$ (x_1 is a dog), $\forall x_1 \exists x_2$ (x_2 is the father of x_1), $\exists x_1$ (x_1 is a cat $\vee$ Ralph is a dog) }).
5. Prove Theorem 15 and give an example of it.
6. Give an example of Theorem 16.

VI Equality

Essential to our analyses of the truth of a wff is our ability to distinguish one object from another as the reference of a name or a variable. Though this is a semantic notion, we can formalize it within the language of predicate logic without creating self-referential paradoxes.

A. The Equality Predicate

The phrases '— is identical to —', '— is the same as —', and '— equals —' are roughly synonymous in our ordinary speech. Sometimes 'is' can also be understood as 'is the same as', as in '7 · 2 is 14' or 'Marilyn Monroe was Norma Jean Baker'. We can abstract from what is common to these locutions or grammatical variations of them such as '— and — are the same'. We will replace these predicates with a new formal symbol, '=', which we call the *equality predicate*. So we'll formalize:

> 7 + 3 equals 12.
> The morning star is identical to the evening star.
> Howie and Ralph are the same.

as:

> 7 + 3 = 12.
> The morning star = the evening star.
> Howie = Ralph.

We can use '$\neg$(— = —)' to formalize '— is different from —', since that is the antonym of '— is the same as —'. I'll abbreviate '$\neg$(— = —)' as '— ≠ —'.

The predicate '=' is meant to be interpreted the same across all models, formalizing our notion of '— is the same object of the universe as —'. Hence, it is syncategorematic, part of the logical vocabulary.

The equality predicate The equality predicate '=' is part of the formal language, the definition of which now includes an additional clause for all terms t, u:

$(t = u)$ is an *atomic wff* for any terms t and u.

I'll continue to use '=' informally in our metalogical discussions when there's not much chance of confusion.

B. The Interpretation of '=' in a Model

When we choose a realization, we specify a universe. With the specification of a universe we have an implicit idea of identity, implicit criteria for distinguishing any one object of the universe from any other. That's what we will use to interpret '='.

Interpretation of the equality predicate For any assignment of references σ in a model, $\sigma \vDash t = u$ iff σ assigns the same object of the universe to both t and u.

For example, if the universe is all objects on my desk, then we interpret '=' as 'is the same object on my desk as'; if the universe is all real numbers, then we interpret '=' as 'is the same real number as'; if the universe consists of Ralph, all canaries, and the Eiffel Tower, we interpret '=' as 'is the same Ralph, canary, or Eiffel Tower as'.

In classical predicate logic we reason with only extensional predicates (pp. 74, 85 and Theorem V.5). So for any semi-formal language, we have the following.

(1) For any assignment of references σ, if $\sigma \vDash x = y$, then for every open wff A, $\sigma \vDash \mathrm{A}(x)$ iff $\sigma \vDash \mathrm{A}(y/x)$, where y replaces some but not necessarily all occurrences of x in A for which it is free.

We have to restrict ourselves to substituting y for occurrences of x for which it is free to avoid cross-referencing problems. Now we have the following.

Theorem 1 Principle of substitution of equals
$\vDash \forall \ldots (x = y \rightarrow (\mathrm{A}(x) \leftrightarrow \mathrm{A}(y/x)))$ where y replaces some but not necessarily all occurrences of x in A for which it is free.

I'll let your show that this principle can be extended to apply to names, too. That is, for any term t, if $\sigma \vDash x = t$, then $\sigma \vDash \mathrm{A}(x)$ iff $\sigma \vDash \mathrm{A}(t/x)$ where t replaces some but not necessarily all occurrences of x for which it is free.

Now we can form a tautology that contains no categorematic part.

Theorem 2 Principle of identity $\vDash \forall x\,(x = x)$

Can we make the implicit notion of identity for a universe any more explicit?

C. The Identity of Indiscernibles

Suppose I'm talking about my dog and you're talking about a dog you saw last week and we wonder whether "they" are the same. What do we do? I describe mine: It's 45 cm tall, it's black and white, it's a border collie . . . . If the dog you saw fits the same description for a sufficiently large number of predicates, we conclude it's the same dog. We might be mistaken because we haven't considered enough predicates: Perhaps you didn't notice enough about the dog you saw, or there's no way to test whether it responds to the name 'Juney' since you didn't call it that when you saw it. But we nonetheless believe that if they were different we could distinguish them.

If there is a difference, something must make a difference. Depending on what things we are discussing, that difference may be as immediate as my touching the one and not the other, or as remote from experience as one satisfying and one not satisfying the equation $\pi x^4 - x^7 = 93$. In a platonist vein it's often said:

> If two things are distinct, then there must be a property that one has which the other does not.

In terms of identity rather than difference this would be the following.

Identity of indiscernibles If every property applies to both x and y or to neither x nor y, then x is identical to y.

The principle of identity of indiscernibles purports to capture true, universal identity, a single analysis for all objects: What cannot be distinguished is one, not two. But 'cannot be distinguished' must mean any property, whether expressible in language or not. What we may not be able to distinguish verbally we might be able to distinguish physically, or may reside in Plato's heaven as distinct, or God or the gods may be able to distinguish.

Such an absolute notion of identity, independent of us, our language, and our capabilities cannot be of use to us as logicians, for what cannot be expressed in language cannot be part of reasoning. Even were you to have access to Plato's heaven and return with a certainty that there are two not one dogs, so long as that knowledge remains inchoate, incapable of expression in language, it remains outside the domain of logic.

This absolute identity becomes relevant to logic when we understand 'property' as a predicate or informal open sentence. In that case there are significant restrictions on how we may interpret 'every property'. When we realize the formal language we realize the predicate and name symbols as particular predicates and names, and then only those predicates and semi-formal open sentences of the realization may enter into our deliberations. The principle of identity of indiscernibles reconstructed in any particular discussion in which we use predicate logic, that is, in any particular realization and model, gives us the converse of (1).

The predicate logic criterion of identity For any assignment of references σ, $\sigma \vDash x = y$ iff for every open wff A, $\sigma \vDash A(x)$ iff $\sigma \vDash A(y/x)$, where y replaces some but not necessarily all occurrences of x in A for which it is free.

Is the predicate logic criterion of identity meant to take into account wffs that contain '=' itself? Is the property 'is identical to' meant to be one of those covered by the principle of identity of indiscernibles? For platonic identity, yes. 'Every property' means every property, including the property of being identical to itself. But as a tool in trying to make precise and explicit our idea of identity, no. Given a realization, we wish to add '=' and ask if its evaluation is determined by the predicates we already have in hand.

For example, suppose we wish to reason about the objects on my desk and we use the following language and model:

(2) $L(\neg, \rightarrow, \wedge, \vee, \forall, \exists$; '— is brown', '— is green', '— is red', '— is a book', '— is a pencil')
universe: all objects on my desk at the time I am writing this
predicates are given their ordinary interpretations

Among the objects on my desk are an eraser, a pencil, some books, my black eyeglass case, and a blue pen. So the following is true using the implicit identity.

(3) $\exists x \exists y\, (x \neq y \wedge (x \text{ is brown} \leftrightarrow y \text{ is brown}) \wedge (x \text{ is green} \leftrightarrow y \text{ is green})$
$\wedge\ (x \text{ is red} \leftrightarrow y \text{ is red}) \wedge (x \text{ is a book} \leftrightarrow y \text{ is a book})$
$\wedge\ (x \text{ is a pencil} \leftrightarrow y \text{ is a pencil}))$

But it is false using the predicate logic criterion of identity because we don't have enough predicates to discern between the black eyeglass case and the blue pen. We could add the predicate '— is black' to the language, with its ordinary interpretation, and then (3) is false according to the predicate logic criterion of identity, too. But then whether a wff is true or false in a model is not determined by just the predicates appearing in it: *The predicate logic criterion of identity does not satisfy the partial interpretation theorem.*

Even if we do add enough predicates to be able to distinguish each object as different from all others, those predicates do not summarize or express the implicit idea of what it means to be an individual thing on my desk, the idea we had when we chose the realization. They serve only to flesh out the predicate logic criterion of identity in this realization to stand as an explicit substitute for that implicit notion, a substitute that correctly classifies distinct things as distinct.

Whatever idea we have of what it means to say something is a *thing* of the universe is difficult if not impossible to express. In some cases, such as the universe of objects on my desk, we may be able to distinguish each object by enumerating enough predicates in our realizations. Then we can replace the implicit notion of identity with the explicit predicate logic criterion of identity. In other cases, such as the universe of all pigs in Denmark, we feel confident we could find enough predicates to distinguish each object, though we might not incorporate enough in a particular model. In some cases, such as the universe of real numbers, it is not possible to add enough predicates to any one realization to distinguish them all, as there are too many objects (Chapter IX.B). A constructivist points to this as evidence that the universe of all real numbers is ill-defined and unacceptable; a platonist points to this as evidence of the limitations of our formal methods: there are enough properties, we just cannot represent them all in language.

In the case of the universe of all things, the problem of making explicit our implicit notion of identity is acute. To give a criterion of identity is neither more nor less than to say what is a thing. Because of the difficulty of doing that, many feel it is illegitimate to use such a universe.

In those cases where the implicit identity of the universe satisfies the predicate logic criterion of identity, the implicit notion has been made explicit. The explicit criterion might not be what we meant by identity, but satisfying that criterion might be the best we can do in making our background explicit. Our general goal should be to make the implicit identity of the universe satisfy the predicate logic criterion of identity by employing sufficient predicates to distinguish distinct things. But that goal, depending on what we believe there is, might not always be attainable.

D. Equivalence Relations

By identifying indiscernibles we can convert any model into one in which the predicate logic criterion of identity holds. I'll show that here in a more general way that we'll need in Chapter XB.1; if you wish you can skip this section until then.

Equivalence relations and congruence relations A binary relation $R(x,y)$ is an *equivalence relation* iff it satisfies the following three conditions:

For every x, $R(x,x)$. *reflexivity*

For every x and y, if $R(x,y)$, then $R(y,x)$. *symmetry*

For every x,y,z, if $R(x,y)$ and $R(y,z)$, then $R(x,z)$ *transitivity*

Taken as an interpretation of a binary connective in a model, R is a *congruence relation* iff it is an equivalence relation and satisfies the condition on consistency and extensionality of predications with respect to R, namely:

For all atomic wffs $Q(t_1, \ldots, t_n)$ and $Q(u_1, \ldots, u_n)$ and assignments of reference σ and τ, if for all $i \leq n$, $R(\sigma(t_i), \tau(u_i))$, then

$$\sigma \vDash Q(t_1, \ldots, t_n) \text{ iff } \tau \vDash Q(u_1, \ldots, u_n).$$

I'll leave the proof of the following lemma to you.

Lemma 3

a. Any equivalence relation, $\simeq$, on the universe of a model partitions the universe into distinct nonempty classes. That is, (i) every object in the universe is in some class, (ii) no object is in two classes, and (iii) every class has something in it. Each class consists of objects identified by the relation, that is, x and y are in the same class iff $x \simeq y$.

b. Any partition of the universe into distinct nonempty classes yields an equivalence relation: $x \simeq y$ iff x and y are in the same class.

c. The interpretation of '=' as the predicate logic criterion of identity in a model is a congruence relation.

d. For any congruence relation, $\simeq$, on the universe, for all wffs $A(t_1, \ldots, t_n)$ and $A(u_1, \ldots, u_n)$ and any σ and τ, if for all $i \leq n$, $\sigma(t_i) \simeq \tau(u_i)$, then $\sigma \vDash A(t_1, \ldots, t_n)$ iff $\tau \vDash A(u_1, \ldots, u_n))$.

There can be more than one congruence relation on a model. For example, for the language with just one unary predicate, let $\mathsf{M} = \langle \{1, 2, 3, 4, 5, 6\}; P \rangle$, where P is true of 1, 3, and 5, and is false of 2, 4, and 6. Then both of the following are congruence relations on M:

$\simeq_1$ defined by the classes $\{1, 3\}$, $\{2, 4\}$, $\{5\}$, $\{6\}$

$\simeq_2$ defined by the classes $\{1, 3, 5\}$, $\{2, 4, 6\}$

A congruence relation $\simeq$ on a model is *maximal* iff there is no congruence relation on the model that strictly contains it, where $\simeq$ *strictly contains* $\simeq^*$ iff $\simeq$ is not identical to $\simeq^*$ and whenever $\mathsf{a} \simeq^* \mathsf{b}$ then $\mathsf{a} \simeq \mathsf{b}$. For example, $\simeq_2$ strictly contains $\simeq_1$ above. I'll let you show that *every model has just one maximal congruence relation: the predicate logic criterion of identity.*

Theorem 4 Let M be a model in which '=' is interpreted as a congruence relation $\simeq$, with a complete collection of assignments of references and valuations for that. Then there is a model $\mathsf{M}/_{\simeq}$ for the same language satisfying:

1. $\mathsf{M}/_{\simeq}$ uses the implicit identity of the universe to interpret '='.
2. The universe of $\mathsf{M}/_{\simeq}$ is no larger than the universe of M.
3. For every wff A, $\mathsf{M} \vDash \mathrm{A}$ iff $\mathsf{M}/_{\simeq} \vDash \mathrm{A}$.

Proof Let M be as in the hypothesis of the theorem.

$$\mathrm{L}(\neg, \rightarrow, \wedge, \vee, \forall, \exists, =, \mathrm{P}_0, \mathrm{P}_1, \ldots, c_0, c_1, \ldots)$$
$$\downarrow$$
$$\mathrm{L}(\neg, \rightarrow, \wedge, \vee, \forall, \exists, \simeq, Q_0, Q_1, \ldots, a_0, a_1, \ldots)$$

universe: U

For each object d in U, define [d] to be {e : e ∈ U and e ≃ d}. Denote by 'U$/_{\simeq}$' the collection of those classes. We read '$/_{\simeq}$' as 'modulo ≃'.

To get a realization for the language, for each name a_i of M take the name '$[a_i]$', and for each predicate Q_i add 'modulo ≃' to get the phrase $Q_i/_{\simeq}$:

$$L(\neg, \rightarrow, \wedge, \vee, \forall, \exists; \text{ '— is the same equivalence class as —'}, Q_0/_{\simeq}, Q_1/_{\simeq}, \ldots, [a_0], [a_1], \ldots)$$

universe: U$/_{\simeq}$

For each assignment of references σ of M define an assignment of references:

For every term t, $\sigma/_{\simeq}(t) \equiv_{\text{Def}} [\sigma(t)]$.

I'll let you show that if $\sigma(t) \simeq \sigma(u)$, then $\sigma/_{\simeq}(t)$ is the same as $\sigma/_{\simeq}(u)$, so this definition does not depend on the representative of the equivalence class on the right-hand side.

Since the collection for M is complete, the collection of these assignments is complete for this realization.

Finally, to create a model, M$/_{\simeq}$, define for each $\sigma/_{\simeq}$:

$$\sigma/_{\simeq} \vDash Q_i/_{\simeq}(u_1, \ldots, u_n) \text{ iff } \sigma \vDash Q_i(t_1, \ldots, t_n)$$

where if u_i is x, then $t_i = x$, and if u_i is $[a_i]$, then $t_i = a_i$.

Because ≃ is a congruence relation, if $[a_i]$ is the same as $[a_j]$ it doesn't matter whether we take a_i or a_j in this definition.

We now want to show that for every wff B of the semi-formal language, M⊨B iff M$/_{\simeq}$⊨B. That will follow from the stronger claim:

(4) For every formal wff B and every σ on M, σ⊨B iff $\sigma/_{\simeq}$⊨B.

The proof of (4) is by induction on the length of B. If B is atomic, then (4) is true by definition. Suppose (4) is true for all wffs shorter than B. If B is of the form C∧D, C∨D, C→D, or ¬C, then (4) follows easily by induction. So suppose B is of the form $\forall x$ C. Then by induction:

$\sigma \vDash \forall x\, C$ iff for every τ that differs from σ at most in what it assigns x, $\tau \vDash C(x)$

iff for every $\tau/_{\simeq}$ that differs from $\sigma/_{\simeq}$ at most in what it assigns x, $\tau/_{\simeq} \vDash C(x)$

iff $\sigma/_{\simeq} \vDash \forall x\, C(x)$

The case where B is of the form $\exists x$ C is done similarly. Hence, for every B, M⊨B iff M$/_{\simeq}$⊨B. ■

Exercises for Chapter VI

1. Show that the principle of substitution of equals for all terms follows from the principle of substitution of equals for variables.
2. If we were to use the predicate logic criterion of identity to interpret '=' in a model with a universe of 5 objects, what would be the minimal number of unary predicates needed to be able to distinguish all those objects? For a universe with n objects?
3. Prove Lemma 3.
4. Consider a model $<\{1, 2, 3, \ldots\}; \equiv (\text{mod } 2), \equiv (\text{mod } 3)>$ where

   $n \equiv (\text{mod } j)\ m$ iff $[m > n$ and j divides $(m - n)$, or $n > m$ and j divides $(n - m)$, or $m = n]$

   Prove that $\equiv (\text{mod } 12)$ is a congruence relation on this model, but is not maximal and does not satisfy the predicate logic criterion of identity.
5. Prove that every model has exactly one maximal congruence relation: the predicate logic criterion of identity.
6. For $\mathsf{M} = <\mathsf{U}\ ; Q_1, \ldots, Q_n>$ where each Q_i is unary, what is the greatest number of classes there can be for a maximal congruence relation on M? (Hint: Induct on n.) If each Q_i is binary?

VII Examples of Formalization

In this text we're interested in formalizations of reasoning from mathematics. But the general methods of formalizing can be introduced more clearly for ordinary reasoning, which will also alert us to some limitations of predicate logic. A full discussion of formalizing ordinary reasoning with criteria for what constitutes a good formalization can be found in my *Predicate Logic.*

A. Relative Quantification

1. All dogs bark.

$\forall x$ (x is a dog $\rightarrow$ x barks)

predicates: '— is a dog', '— barks'
names: none

$\forall x_1 \, (P_0(x_1) \rightarrow P_1(x_1))$

Analysis The grammatical subject of this example is 'dogs'. That is not a name of an individual nor a pronoun that we can construe as a place-holder. There is no one thing in the world called 'dogs' that barks. Rather, the example is equivalent to:

(a) Every dog barks.

This is a relative quantification; it's quantifying over all dogs, rather than all things. We can rewrite (a) as:

(b) Everything which is a dog is a thing which barks.

Now we have two predicates, 'is a dog' and 'barks', with a quantification that governs both. A formalization of (b) needs to connect these. Consider:

(c) $\forall x$ (x is a dog $\rightarrow$ x barks)

This models:

For each thing, if it is a dog then it barks.

All objects of the universe that are not dogs are irrelevant to the truth-value of (c), for if 'x' does not refer to a dog, then 'x is a dog $\rightarrow$ x barks' is vacuously true. This is just what we want, and indeed it was in part to be able to model propositions such as this that we chose this classical reading of '$\rightarrow$' (pp. 14–15).

Suppose, though, that the universe of a model contains no dogs. Then (c) is true. Is Example 1 true, too? Some say 'no'. They would formalize Example 1 as:

(d) $\exists x$ (x is a dog) $\wedge$ $\forall x$ (x is a dog $\rightarrow$ x is barks)

But most logicians do not want to ascribe existential assumptions to propositions unless compelled to do so by the role of the proposition in inferences, and they do not think that the inference 'All dogs bark, *therefore*, there is a dog' is valid. The standard now is to adopt (c) as a formalization of Example 1, though in some contexts we may want to use (d).

The same formalization is also used for the following:

Dogs bark. Every dog barks. Each dog barks. Any dog barks.

2. Some cats are nasty.

$\exists x$ (x is a cat $\wedge$ x is nasty)

predicates: '— is a cat', '— is nasty'
names: none

$\exists x_1 (P_0(x_1) \wedge P_1(x_1))$

Analysis This is another relative quantification. Rewriting this example in the manner of Example 1 we get:

Some thing which is a cat is a thing which is nasty.

But we should not formalize this as:

$\exists x$ (x is a cat $\rightarrow$ x is nasty)

That would be true in a universe that contained only nice cats and a dog, for any dog makes the conditional 'x is a cat $\rightarrow$ x is nasty' true when taken as reference for 'x'. But Example 2 is false for that universe. We want instead:

$\exists x$ (x is a cat $\wedge$ x is nasty)

This is also a good formalization of the following:

There is a cat that is nasty. Some cat is nasty.
There exists a nasty cat. There is at least one nasty cat.

Some say Example 2 is true only if there are at least two cats in the universe. We'll see how to formalize that in Section D below. But the convention nowadays is to understand 'some' as 'there is at least one', unless context demands otherwise.

3. Dogs are nicer than cats.

$\forall x$ (x is a dog $\rightarrow \forall y$ (y is a cat $\rightarrow x$ is nicer than y))

predicates: '— is a dog', '— is a cat', '— is nicer than —'
names: none

$\forall x_1\,(P_0(x_1) \rightarrow \forall x_2\,(P_1(x_2) \rightarrow P_1^2(x_1,x_2))$

Analysis In this example we have quantifications relative to both dogs and cats. Universal quantification seems appropriate, so we can use the method of Example 1 sequentially. The same formalization can also be used for:

All dogs are nicer than all cats. Any dog is nicer than any cat.

Note that the formalization is semantically equivalent to:

$\forall x\, \forall y$ ((x is a dog $\wedge$ y is a cat) $\rightarrow x$ is nicer than y)

4. Some cat is smarter than some dog.

$\exists x$ (x is a cat $\wedge \exists y$ (y is a dog $\wedge$ x is smarter than y))

predicates: '— is a dog', '— is a cat', '— is smarter than —'
names: none

$\exists x_1\,(P_0(x_1) \wedge \exists x_2\,(P_1(x_2) \wedge P_1^2(x_1,x_2))$

Analysis This is just the sequential application of the method of Example 2. It can also be used for the formalization of:

Some cats are smarter than some dogs.
There are a cat and a dog such that the cat is smarter than the dog.
There exist a cat and a dog where the cat is smarter than the dog.
There's a cat that's smarter than a dog.

Note that it is semantically equivalent to:

$\exists x\, \exists y$ ((x is a cat $\wedge$ y is a dog) $\wedge$ x is smarter than y)

5. Every dog hates some cat.

$\forall x$ (x is a dog $\rightarrow \exists y$ (y is a cat $\wedge$ x hates y))

predicates: '— is a dog', '— is a cat', '— hates —'
names: none

$\forall x_1\,(P_0(x_1) \rightarrow \exists x_2\,(P_1(x_2) \wedge P_1^2(x_1,x_2))$

Analysis We have first a universal relative quantification and then an existential one. So we use the methods of Examples 1 and 2 sequentially. If the universe contains no dogs, then Example 5, just as 'Dogs bark', is true. If the universe does contain dogs, then for each one of those there must be at least one cat that it hates for the formalization to be true.

6. **Some boy loves all dogs.**

$\exists x\ (x \text{ is a boy} \wedge \forall y\ (y \text{ is a dog} \rightarrow x \text{ loves } y))$

predicates: '— is a boy', '— is a dog', '— loves —'
names: none

$\exists x_1\ (P_0(x_1) \wedge \forall x_2\ (P_1(x_2) \rightarrow P_1^2(x_1, x_2))$

Analysis Sequentially applying the methods of Examples 1 and 2 yields the formalization.

7. **All black dogs bark.**

$\forall x\ (x \text{ is black} \wedge x \text{ is a dog} \rightarrow x \text{ barks})$

predicates: '— is black', '— is a dog', '— barks'
names: none

$\forall x_1\ ((P_0(x_1) \wedge P_1(x_1)) \rightarrow P_2(x_1))$

Analysis Here the relative quantification is over black dogs. But we cannot treat '— is a black dog' as atomic, for the following is a valid inference that would not be respected by that choice:

Juney is a black dog.
Therefore, Juney is a dog.

Still, we cannot always treat an adjective or adjectival phrase as a separate predicate as in the formalization of this example. Consider:

Cheema is a small St. Bernard.

It would be wrong to formalize this as:

Cheema is small $\wedge$ Cheema is a St. Bernard.

That would justify as valid:

Cheema is a small St. Bernard.
Therefore, Cheema is small.

And that deduction is not valid, for the premise would be true if Cheema were 1 m tall rather than the normal height of 1.6 m for a St. Bernard, and a dog 1 m tall is a big dog. The adjective 'small' is being invoked relative to other St. Bernards only.

8. If Ralph is a dog, then anything is a dog.

Ralph is a dog → ∀*x* (*x* is a dog)

predicates: '— is a dog'
name: 'Ralph'

$P_0(c_0) \rightarrow \forall x_1 P_0(x_1)$

Analysis The formalization is straightforward, reading 'any' as ∀.

9. If anything is a dog, then Ralph is a dog.

∃*x* (*x* is a dog) → Ralph is a dog

predicates: '— is a dog'
name: 'Ralph'

$\exists x_1 P_0(x_1) \rightarrow P_0(c_0)$

Analysis Reading 'any' as 'all' here is wrong. The natural reading of 'anything' is 'something' in this example. Generally, we read:

'If S, then anything P' as S → ∀*x* P

'If anything S, then P' as ∃*x* S → P

10. If something weighs over 30 lbs., it cannot be sent by parcel post.

∀*x* (*x* weighs over 30 lbs. → ¬(*x* can be sent by parcel post))

predicates: '— weighs over — lbs.', '— can be sent by parcel post'
name: '30'

$\forall x_1 (P_0^2(x_1, c_0) \rightarrow \neg P_1(x_1))$

Analysis We cannot invariably formalize uses of 'some' as '∃'. Here the proposition is a regulation applying to all things, not to at least one.

B. Adverbs, Tenses, and Locations

11. (a) Juney is barking loudly.
***Therefore*, Juney is barking.**

Not formalizable.

Explanation: The example is clearly valid. But we can't view the premise as compound as in 'Juney is a black dog'. The only categorematic parts we recognize are names and what we get when we take away names from propositions, that is, predicates. To justify the validity of Example 11 we would need to take actions or processes, what we use verbs for in English, as equally fundamental with things.

Example 11 is outside the scope of predicate logic, and 'Juney is barking loudly' is not formalizable.

Some disagree. They say that our language is our guide, and we normally talk about actions as things. But if it's natural to speak of actions as things, we ought to be able to give a reading of (a) that will justify the validity of Example 11 in predicate logic. Consider:

(b) The barking of Juney is loud.

But that's not right: proposition (a) is talking of the moment when it's uttered, whereas (b) means that Juney's bark is normally or always loud. So consider:

There is a barking of Juney and it is loud.

We can formalize this as:

(c) $\exists x$ (x is a barking of Juney $\wedge$ x is loud)

From (c) we can conclude:

$\exists x$ (x is a barking of Juney)

And that's how we would have to construe 'Juney is barking' in order to justify the validity of Example 11. It only seems that 'Juney is barking' is atomic. When we "realize" that actions are things we'll understand that 'Juney is barking' is really a complex existential proposition. All verbs except 'to be' will have to disappear from our formalizations in favor of noun clauses and existential assumptions.

To force arguments involving adverbs into the forms of predicate logic is to deny other views of the world, other ways of parsing experience, in favor of thing-oriented expressions only. That's not good logic, as I discuss in my *Predicate Logic.*

12. A cat is never your friend.

$\forall x \forall y$ (x is a cat $\wedge$ y is a person $\rightarrow \neg$(x is a friend of y))

predicates: '— is a cat', '— is a person', '— is a friend of —'
names: none

$\forall x_1 \forall x_2 (P_1(x_1) \wedge P_2(x_2) \rightarrow \neg P^2_{4318}(x_1, x_2))$

Analysis Is this a proposition? Aren't 'never' and 'your' too ambiguous? Perhaps it's just an exclamation. I think it is a proposition, because when I say it some people disagree, telling me, 'That's not true.'

I understand Example 12 as a statement of a natural law, akin to 'Electrons have spin' or 'Atoms never disintegrate'. In that sense, it is atemporal: 'is never' can be rewritten as 'is not ever', which, reading 'is' atemporally, we can rewrite as 'is not'. The pronoun 'your' I understand to introduce generality, as in 'anyone's', rather than referring to the person to whom the example might be addressed. Hence, rewriting we have:

A cat is not a friend of any person.

The initial word 'A' now should be taken in the sense of 'all' if the example is to be understood as a natural law. The formalization then follows as in Example 3.

13. A pencil is lying on the desk of Richard L. Epstein.

$\exists x$ (x is a pencil and x is lying on the desk of Richard L. Epstein)

predicates: '—is a pencil', '— is lying on —'
names: 'the desk of Richard L. Epstein'

$\exists x_1 ((P_0(x_1) \wedge P_1^2(x_1, c_0))$

Analysis As I am writing now, this example is false. I stop for a moment and lay my pencil down; it is then true. Is Example 13 a proposition? Is it two propositions?

Compare Example 13 to 'I am an American.' We could say that is true when I say it, false when Fred Kroon says it. But we don't. We say that we won't recognize the sentence as a proposition until the reference of the indexical is made explicit.

Similarly, in Example 13 the words 'is lying on' are indexical when we consider the time of utterance. Can we make that phrase more precise? Perhaps we could formalize Example 13 as:

$\exists x$ (x is a pencil at 4:31 a.m., April 24, 2005 $\wedge$ x is lying on the desk of Richard L. Epstein at 4:31 a.m., April 24, 2005)

This might work for Example 13. But to implement it in general, we'd need a lot of unrelated atomic predicates in place of '— is a pencil —':

— is a pencil at 4:31 a.m., April 24, 2005; — is a pencil at 4:32 a.m., April 24, 2005; . . .

Perhaps we should take as atomic: '— is a pencil at —'. We could expand every predicate to include a blank for a time label. If we did that we'd have to treat '4:31 a.m., April 24, 2005' as the name of a thing, so our semi-formal languages would have to include time names, and every universe would need to contain moments of time as things.

What are these moments of time? Exactly how precise do we need to be with our time indexing? The year? The day? The hour? The minute? The second? The nanosecond? And what precise time label is appropriate to fix the time of:

Juney was barking.

Continuous tenses demand time intervals. Yet suppose we formalize this as:

Juney was barking between 4:16 a.m. and 6:00 a.m., April 24, 2005.

Then we shall have to assume a great deal about the nature of time to be able to conclude:

Juney was barking between 4:31 a.m. and 5:31 a.m., April 24, 2005.

The problem with taking moments of time to formalize tensed propositions is that we must then take instants of time as things, on a par for the basis of reasoning with numbers and baseball bats. One wonders whether an instant of time is concrete like a sandwich, or abstract like a real number. I have no intuition about instants of time, how to tell one from another, what would be true of of an instant of time and what false, beyond what's needed to make reasoning involving tenses come out "right". Yet these things would be so fundamental, they would have to be in the universe of every model.

The assumptions of predicate logic, however, seem to force us in the direction of postulating some things to account for tenses. The reasoning we're formalizing is based on the assumption that the world is made up of things. Unless we expand the grammatical categories of predicate logic to take account of verbs and tenses in some way such as assuming that the world is made up, in part, of processes, we cannot adequately model reasoning involving tenses. The focus and methods of predicate logic are askew to concerns of reasoning with tenses, for tenses are aspects of verbs.

One of the main motives for the development of modern predicate logic was to formalize reasoning in mathematics and, secondarily, science. Claims of mathematics are meant as timeless: we don't ask when '$2 + 2 = 4$' is true. The resources of predicate logic are best applied to timeless propositions. 'A pencil is lying on the desk of Richard L. Epstein' is timelessly false, if we consider the time of which it is to be understood.

But we don't want to restrict predicate logic to reasoning about only timeless abstract objects. Electrons, if we are to believe the scientists, are physical things, yet we don't ask when 'Electrons have negative charge' is true. A law of science is not meant to be true today only: 'have' is to be understood not as in the present, but as timeless. The reasoning of scientific experiments can be formalized in predicate logic: mass, energy, instants of time, locations can all be represented numerically, and reasoning about them becomes reasoning about timeless propositions of numbers. Within the domain of physics, incorporating assumptions about the nature of time seems quite appropriate.

C. Qualities, Collections, and Mass Terms

14. Intelligence is good.

Not formalizable?

Analysis This not equivalent to: $\forall x$ (x is intelligent $\rightarrow$ x is good). It may be that intelligence is good while some intelligent people are bad.

Rather, we should rewrite Example 14 as:

Intelligence is a good thing.

Then it becomes clear that 'intelligence' is being used to designate a thing; the word is being used as a name. It may be possible to "reduce" Example 14 or the more problematical 'Blue is a color' to talk about objects satisfying '— is intelligent' or '— is blue', but the conversion is hardly obvious and the rewriting is awkward. Such rewritings would ignore the limitations imposed by our assumptions.

Whether we take 'intelligence', 'blue', etc., as names depends on what we believe exists. For platonists these denote things, and the naming is by directing our intellects to those abstract objects: 'blue' does not denote something blue, it is the color, beyond our world of sensation and becoming. But if we take 'intelligence' as a name of something, it must be primitive. Logic alone will yield no connection between 'intelligence' and '— is intelligent', which also is atomic.

15. The flock of sheep in Alan's corral is more homogeneous than the collection consisting of Ralph, all canaries, and the Eiffel Tower.

Not formalizable.

Analysis There is no obvious way to "reduce" this to talk about objects satisfying predicates such as '— is a sheep' and '— is a canary'. The predicate '— is homogeneous' applies to collections, not to objects in the collections. In this example 'the flock of sheep in Alan's corral' is meant to denote a single thing; it is used as a name. In that case it is primitive, so our logic recognizes no connection between it and the predicate '— is a sheep'.

Worse, consider:

(a) The collection of all things that do not belong to themselves belongs to itself.

(b) The property of not being possessed by itself is possessed by itself.

Possesses, belongs to are recastings of the notion of *satisfaction, applies to* in terms of properties and collections; (a) and (b) are recastings of the liar paradox. Seen in this light, we must exclude these words from realizations according to the self-reference exclusion principle.

With some artifice we might be able to transform a particular proposition involving collections or qualities and what belongs to or is possessed by them into a semi-formal proposition. But in general *we cannot formalize talk of properties or collections along with talk of the objects that are possessed by or belong to them.* So the following are not formalizable:

Every pack of dogs has a leader.
Marilyn Monroe had all the qualities of a great actress.
Ralph has some quality that distinguishes him from every other thing.

In Chapter XIV we'll consider extensions of our work to formalize these.

16. **(a) The puddle in front of R. L. Epstein's home is water.**
(b) Water is wet.
Therefore,
(c) The puddle in front of R. L. Epstein's home is wet.

Not formalizable.

Analysis Only a thing can make '*x* is wet' true. So in (b), 'water' is used as a name. But 'water' used as a name of a thing seems odd if not actually wrong, for water is wet in a different sense than Juney is wet after a rainstorm. Water is an undifferentiated mass that we can treat as a thing only by dividing it up: a glass of water, a puddle of water, the water in my bathtub. Water is not abstract, like beauty or wisdom, but what kind of physical thing can be in no one place? Water in one place, in the puddle in front of my home, is *that* water, not water in general.

In (a) we're asserting something about the puddle. There 'is water' is a predicate. The problem is not that *mass terms* do not fit into our categories of names and predicates; it's that they fit into both categories.

Generally, mass terms are not accepted when we reason in predicate logic, though we do accept 'is made of water', 'is made of green cheese', 'is an instance of fire' as atomic predicates, and 'the snow on R. L. Epstein's front yard', 'the water in Lake Erie' as names, so long as doing so doesn't conflict with the roles these play in inferences. No method of extending predicate logic to incorporate mass terms has gained general acceptance. The perspective of the world being made up (in part) of undifferentiated substances seems too much at odds with the view that the world is made up of individual things.

Reasoning with mass terms can be formalized within Aristotelian logic, as discussed in Chapter V of *Predicate Logic*.

D. Finite Quantifiers

17. **There are at least two dogs.**

$\exists x\, \exists y\, (x \neq y \wedge (x \text{ is a dog} \wedge y \text{ is a dog}\,)$

predicates: '— is a dog'
name: none

$\exists x_1\, \exists x_2\, (x_1 \neq x_2 \wedge (\mathrm{P}_0(x_1) \wedge \mathrm{P}_0(x_2)))$

Analysis The formalization is true in a model iff there are at least two dogs in the universe. For this to be a good formalization, we must view 'two' as part of a quantifier, syncategorematic, rather than as a name and also view '=' as syncategorematic.

Since we have plenty of variables available, we can also model 'There are at least 3', 'There are at least 4', ... We first give an inductive definition of formulas $\mathrm{E}_n(y_1, \ldots, y_{n+1})$, where $y_1, \ldots, y_n$ are distinct variables:

$$\mathrm{E}_1(y_1) \equiv_{\mathrm{Def}} y_1 = y_1 .$$

$$\mathrm{E}_2(y_1, y_2) \equiv_{\mathrm{Def}} y_1 \neq y_2$$

$$\mathrm{E}^1_{n+1}(y_1, \ldots, y_{n+1}) \equiv_{\mathrm{Def}} \mathrm{E}_{n+1}(y_1, \ldots, y_n) \wedge (y_1 \neq y_{n+1})$$

$$\mathrm{E}^{m+1}_{n+1}(y_1, \ldots, y_{n+1}) \equiv_{\mathrm{Def}} \mathrm{E}^m_{n+1}(y_1, \ldots, y_{n+1}) \wedge (y_{m+1} \neq y_{n+1})$$

$$\mathrm{E}_{n+1}(y_1, \ldots, y_{n+1}) \equiv_{\mathrm{Def}} \mathrm{E}^n_{n+1}(y_1, \ldots, y_{n+1})$$

Then for $n \geq 2$, for any model and any assignment of references σ,

$\sigma \models \mathrm{E}_n(y_1, \ldots, y_n)$ iff σ assigns distinct objects to $y_1, \ldots, y_n$

$\exists y_1 \ldots \exists y_n\, \mathrm{E}_n(y_1, \ldots, y_n)$ is true in a model iff
there are at least n objects in the universe.

For any wff $\mathrm{A}(x)$, if $y_1, \ldots, y_n$ are the first n variables in increasing order that do not appear in A for $n \geq 2$, we define:

$$\exists_{\geq 1} x\, \mathrm{A}(x) \equiv_{\mathrm{Def}} \exists x\, \mathrm{A}(x)$$

$$\exists_{\geq n} x\, \mathrm{A}(x) \equiv_{\mathrm{Def}} \exists y_1 \cdots \exists y_n (\mathrm{E}_n(y_1, \ldots, y_n) \wedge (\mathrm{A}(y_1/x) \wedge \cdots \wedge \mathrm{A}(y_n/x)))$$

Then if x is the only variable free in A:

$\exists_{\geq n} x\, \mathrm{A}(x)$ is true in a model iff
there are at least n distinct objects in the universe that satisfy $\mathrm{A}(x)$.

If we construe 'there are at least n' as syncategorematic, we can use these compound quantifiers in any formalization. We can produce semi-formal equivalents to 'There is at least one dog that barks', 'There are at least two dogs that bark', 'There are at least three dogs that bark', ...

18. There are at most two nice cats.

$\neg \exists_{\geq 3} x$ (x is a cat $\wedge$ x is nice)

predicates: '— is a dog', '— is nice'
name: none

$\neg \exists_{\geq 3} x_1\, (\mathrm{P}_0(x_1) \wedge \mathrm{P}_1(x_1))$

Analysis The formalization is true in a model iff there are at most two nice cats in the universe.

For every wff A with variable x free, and for every $n \geq 1$, we can define:

$$\exists_{< n+1} x\, \mathrm{A}(x) \equiv_{\mathrm{Def}} \neg \exists_{\geq n} x\, \mathrm{A}(x)$$

If x is the only variable free in A, then $\exists_{< n+1} x\, \mathrm{A}(x)$ is true in a model iff there are at most n distinct objects in the universe that satisfy $\mathrm{A}(x)$.

19. There are exactly forty-seven dogs that don't bark.

$\exists_{\geq 47} x$ (x is a dog $\wedge$ $\neg$(x barks)) $\wedge$ $\exists_{\leq 47} x$ (x is a dog $\wedge$ $\neg$(x barks))

predicates: '— is a dog', '— barks'
name: none

$$\exists_{\geq 47} x_1\ (P_0(x_1) \wedge \neg P_1(x_1)) \wedge \exists_{\leq 47} x_1\ (P_0(x_1) \wedge \neg P_1(x_1))$$

Analysis For any $n \geq 1$ and any wff A, we define:

$$\exists !_n x\ A(x) \equiv_{\text{Def}} \exists_{\geq n} x\ A(x) \wedge \exists_{< n+1} x\ A(x)$$

If x is the only variable free in A, then:

$\exists !_n x\ A(x)$ is true in a model iff there are exactly n things in the universe that satisfy $A(x)$

For $n = 1$ we usually take a simpler formula to formalize 'There is a unique x such that' or 'There exists uniquely an x such that':

$$\exists ! x\ A(x) \equiv_{\text{Def}} \exists x\ A(x) \wedge \forall x\, \forall y\ (A(x) \wedge A(y) \rightarrow x = y)$$
where y is the least variable that does not appear in A

20. There are exactly forty-seven things.

$\exists !_{47} x\ E_{47}(x_1, \ldots, x_{47})$

predicates: '='
name: none

Analysis With $\exists !_n x\ E_n(x_1, \ldots, x_n)$ we have a formula that is true in a model iff there are exactly n things in the universe.

21. All but two bears in the London Zoo are brown.
There are forty-seven bears in the London Zoo.
***Therefore,* there are forty-five brown bears in the London Zoo.**

$\exists !_2 x$ ((x is a bear $\wedge$ x is in the London Zoo) $\wedge$ $\neg$(x is brown))
$\exists !_{47} x$ (x is a bear $\wedge$ x is in the London Zoo)

$\exists !_{45} x$ ((x is a bear $\wedge$ x is brown) $\wedge$ (x is in the London Zoo))

predicates: '— is a bear', '— is in the London Zoo', '— is brown', '='
names: none

$\exists !_2\, x_1\ ((P_1(x_1) \wedge P_2(x_1)) \wedge \neg P_3(x_1))$
$\exists !_{47}\, x_1\ (P_1(x_1) \wedge P_2(x_1))$

$\exists !_{45}\, x_1\ ((P_1(x_1) \wedge P_3(x_1)) \wedge P_2(x_1))$

Analysis 'All but two bears are brown' means 'Two bears are not brown and all the rest are brown'. That is, there are exactly two bears that are not brown.

By viewing 'two', 'forty-seven', 'forty-five' as parts of syncategorematic expressions, a problem in arithmetic can be converted to an investigation of logical validity. We can convert counting things into counting variables.

22. There are no nice cats.

$\neg\exists x$ (x is a cat $\wedge$ x is nice)

predicates: '— is a cat', '— is nice'
name: none

$\neg\exists x_1\ (P_0(x_1) \wedge P_1(x_1))$

Analysis This is true in a model iff there is no cat in the universe that is nice.

Generally, we formalize uses of 'no' as in 'No cats are nice' or 'No horse eats meat' with '$\neg\exists x$'. Some prefer to formalize 'No cats are nice' as:

$\forall x$ (x is a cat $\rightarrow$ $\neg$(x is nice))

This is equivalent to the formalization above.

Similarly, we can model 'nothing' as 'no thing' or 'there is not a thing'. For example, we can formalize: 'Nothing both barks and meows.' as: $\neg\exists x$ (x barks $\wedge$ x meows)

23. Nothing is friendlier than a dog.

$\forall y$ (y is a dog $\rightarrow$ $\neg\exists x$ ($\neg$(x is a dog) $\wedge$ x is friendlier than y))

predicates: '— is friendlier than —', '— is a dog'
names: none

$\forall x_1\ (P_1(x_1) \rightarrow \neg\exists x_2\ (\neg P_1(x_2) \wedge P_0^2(x_2,x_1)))$

Analysis We've agreed to formalize 'nothing' using $\neg\exists x$. The problem is what to do with 'a dog'. I understand that to mean 'any dog', and use the method of Example 1. But what should be the scope of the universal quantifier?

Suppose we take 'nothing' to have precedence over 'any'. Then we'd have:

$\neg\exists x\ \forall y$ (y is a dog $\rightarrow$ x is friendlier than y)

But that's wrong. We don't mean by this example that nothing is friendlier than all dogs at once. Rather, given any dog there is nothing friendlier than it:

$\forall y$ (y is a dog $\rightarrow$ $\neg\exists x$ (x is friendlier than y))

But that can't be right either, for it would be true only if all dogs are equally friendly. Example 23 is equivalent to:

Nothing that is not a dog is friendlier than a dog.

24. Only Ralph barks.

Ralph barks $\wedge \neg\exists x\ (x \neq \text{Ralph} \wedge x \text{ barks})$

predicates: '— barks', '='
names: 'Ralph'

$$P_1(c_0) \wedge \neg\exists x_1\ (\neg(x_1 = c_0) \wedge P_1(x_1))$$

Analysis If this example is true, then Ralph barks and nothing else barks. That is, 'Ralph barks and nothing that is not Ralph barks.' We've agreed to formalize 'nothing' as $\neg\exists x$, and here, since 'Ralph' is a name, we take the 'is' as '='.

25. Only dogs bark.

$\exists x\ (x \text{ is a dog} \wedge x \text{ barks}) \wedge \neg\exists x\ (\neg(x \text{ is a dog}) \wedge x \text{ barks})$

predicates: '—is a dog', '— barks'
names: none

$$\exists x_1\ (P_1(x_1) \wedge P_2(x_1)) \wedge \neg\exists x_1\ (\neg P_1(x_1) \wedge P_2(x_1))$$

Analysis We proceed as in Example 24 and get:

Dogs bark and nothing that is not a dog barks.

The problem is whether to take 'Dogs' as 'All dogs' or 'Some dogs'. Compare, 'Only rich people have Rolls Royces.' It's not that all rich people own Rolls Royces, but rather that some do and no one else does. We want 'only' in the sense of 'some'.

26. Not only dogs bark.

$\exists x\ (x \text{ is a dog} \wedge x \text{ barks}) \wedge \neg(\neg\exists x\ (\neg(x \text{ is a dog}) \wedge x \text{ barks}))$

predicates: '—is a dog', '— barks'
names: none

$$\exists x_1\ (P_1(x_1) \wedge P_2(x_1)) \wedge \neg(\neg\exists x_1 (\neg P_1(x_1) \wedge P_2(x_1)))$$

Analysis Example 26 is not the negation of Example 25, for the following inference is valid:

Not only dogs bark.
Therefore, dogs bark.

Here 'not' is part of a compound quantifier 'not only', rather than a negation of the whole. We can rewrite this example as 'Dogs bark and not: nothing that is not a dog barks.' As for Example 25, we use $\exists$ to govern 'Dogs'.

E. Examples from Mathematics

27. Through any two points there is exactly one line.

$\forall x \, \forall y \, ((x \text{ is a point} \wedge y \text{ is a point}) \wedge x \neq y \rightarrow$
$\exists ! z \, (z \text{ is a line} \wedge (z \text{ is through } x \wedge z \text{ is through } y)))$

predicates: '— is a point', '— is a line', '— is through —', '='
names: none

$\forall x_1 \, \forall x_2 \, ((P_1(x_1) \wedge P_1(x_2)) \wedge \neg(x_1 = x_2) \rightarrow$
$\exists ! x_3 \, (P_2(x_3) \wedge (P_0^2(x_3, x_1) \wedge P_0^2(x_3, x_2)))$

Analysis I understand 'any two' as 'any two distinct things', for the example would be false in its normal interpretation in Euclidean geometry if we were to allow the same object to be referred to by the two variables we use in 'x is a point $\wedge$ y is a point'. Mathematicians sometimes say 'any two' when they mean any things referred to by two variables, as in 'Any two points are identical or determine a line', which we would formalize as '$\forall x \, \forall y \, (x = y \vee x$ and y determine a line)'. Sometimes you can find them really struggling with language, saying 'For any two not necessarily distinct . . .'.

28. Every number is divisible by 2.

$\forall x \, (x \text{ is a number} \rightarrow x \text{ is divisible by } 2)$

predicates: '—is a number', '— is divisible by —'
names: '2'

$\forall x_1 \, (P_6(x_1) \rightarrow P_1^2(x_1, c_1))$

Analysis What objects is the proposition about? Numbers and 2. I take '2' as a name, though if no other numerals are used in the propositions we are formalizing, and 'is divisible by' only appears with '2', we could take 'is divisible by 2' as an atomic predicate.

A mathematician might ask why we don't formalize the proposition as:

(a) $\forall x \, (x \text{ is divisible by } 2)$

If we take the proposition within the context of formal assumptions about numbers, then (a) would be acceptable. Otherwise, we must consider numbers to be just one kind of object that a universe could contain.

29. Every number is even.

Formalization as for Example 28.

Analysis The word 'even' in this sense is shorthand for 'is divisible by 2'. It has no other meaning. Either predicate could be taken as an abbreviation of the other in formalizations of arithmetic. Hence they can have the same formalizations. The only choice is whether to take 'is divisible by 2' or 'is even' for both.

30. There are 2 prime numbers that differ by 2.

$\exists x\, \exists y\, ((x \neq y \wedge ((x \text{ is a number} \wedge y \text{ is a number}) \wedge$
$(x \text{ is a prime} \wedge y \text{ is a prime}))) \wedge x \text{ differs from } y \text{ by } 2)$

predicates: '— is a prime', '— differs from — by —', '='
'— is a number —'

names: '2'

$\exists x_1\, \exists x_2\, ((\neg(x_1 = x_2) \wedge ((P_1(x_1) \wedge P_1(x_2)) \wedge (P_2(x_1) \wedge P_2(x_2))))$
$\wedge\ P_0^3(x_1, x_2, c_0))$

Explanation: Here, one occurrence of '2' is formalized as syncategorematic and one occurrence as categorematic.

31. Given any line and any point not on that line, there is a line parallel to that given line through that point.

$\forall x\, \forall y\, (x \text{ is a line} \wedge y \text{ is a point} \wedge \neg(y \text{ lies on } x) \rightarrow$
$\exists z (z \text{ is a line} \wedge (z \text{ is parallel to } x \wedge y \text{ lies on } z)))$

predicates: '— is a line', '— is a point', '— lies on —',
'— is parallel to —

names: none

$\forall x_1\, \forall x_2\, ((P_1(x_1) \wedge P_2(x_2) \wedge \neg P_6^2(x_2, x_1)) \rightarrow$
$\exists x_3\, (P_1(x_3) \wedge (P_4^2(x_3, x_1) \wedge P_6^2(x_2, x_3)))$

Analysis I understand the mathematician's use of the phrases 'Given any . . .' and 'Let x be any . . .' to indicate universal quantification relative to the objects described (compare Example 8 of Chapter I). In this example, the relative quantification for the first point and line should extend over the whole proposition, since they recur in the second part. But as mention of another line enters only in the consequent, I've restricted the scope of the relative quantification for that predicate to the consequent.

I used the predicate '— lies on —' instead of '— is through —', because we use 'x passes through y', 'y lies on x', 'x is through y', 'y is on x' for variation of style, understanding them to be equivalent.

32. Every nonempty collection of natural numbers has a least element.

Not formalizable.

Analysis This proposition is fundamental to reasoning about the natural numbers, but we have no way to formalize it. We would have to take both natural numbers and collections of natural numbers as things and use 'belongs to' or 'has' as a predicate, which we are prevented from doing by the self-reference exclusion principle, as discussed in Example 15. In Chapter XIV we'll consider a way to extend the grammar of our formal and semi-formal languages to formalize this.

Exercises for Chapter VII

1. Show that neither $\forall x\, (x \text{ is a dog} \vee x \text{ barks})$ nor $\forall x\, (x \text{ is a dog} \wedge x \text{ barks})$ are suitable to formalize 'All dogs bark'.

2. Show that neither of the following are suitable to formalize 'Every dog hates some cat.'
   $\forall x\, \exists y\, ((x \text{ is a dog} \wedge y \text{ is a cat}) \rightarrow x \text{ hates } y)$
   $\forall x\, \exists y\, ((x \text{ is a dog} \wedge y \text{ is a cat}) \wedge x \text{ hates } y)$

3. Show that the formalization of 'Some boy loves all dogs' given in Example 6 is equivalent to: $\exists x\, \forall y\, (x \text{ is a boy} \wedge (y \text{ is a dog} \rightarrow x \text{ loves } y))$.

4. Show that (a) and (b) in Example 15 are paradoxical.

5. Show that the following are not good formalizations of 'Some boys love all dogs':
   $\exists x\, \forall y\, ((x \text{ is a boy} \wedge y \text{ is a dog}) \rightarrow x \text{ loves } y)$
   $\exists x\, \forall y\, ((x \text{ is a boy} \wedge y \text{ is a dog}) \wedge x \text{ loves } y)$

6. Formalize each of the following in the manner of the examples of this chapter or explain why it cannot be formalized.
   a. Every dog is friendly.
   b. Every dog is friendly or feral.
   c. All dogs that aren't friendly are feral.
   d. Some teachers are nasty.
   e. Some nasty teachers yell.
   f. A scout is reverent.
   g. A lady is present.
   h. If anyone can be president, then Groucho Marx can be president.
   i. Ralph was snoring heavily.
   j. Atoms have diameter less than $^1/_2$ cm.
   k. Every pack of dogs has a leader.
   l. New Mexico is in the United States.

7. Formalize each of the following.
   a. Everyone loves someone.
   b. Everyone loves someone other than himself.
   c. Any horse is smarter than a cat.
   d. Everyone knows someone who is famous.
   e. No three people in Socorro have the same birthday.
   f. There are at least seven but fewer than forty-two cats that have not clawed someone.
   g. Of every three people in Socorro, two were born in New Mexico.

8. Formalize each of the following. You may use formal symbols and assume the propositions are given in the context of a formalization of the appropriate part of mathematics.
   a. Every number is even or odd.
   b. Every odd is not even.
   c. Suppose x is an irregular heptagon. Then x cannot be circumscribed by a circle.
   d. Any two lines either meet in a point or are parallel, where every line is considered parallel to itself.
   e. Suppose that parallel lines meet. Then similar triangles are congruent.
   f. Every pair of prime numbers that differ by 2 are less than 19.
   g. If 2 is added to a number, then both the original number and the result are odd, or both are even.
   h. Every set of real numbers that is bounded above has a least upper bound.
   i. There are two pairs of primes each of which differs by two.
      (Hint: We cannot speak both of a pair as a thing and of the things in the pair as belonging to the pair (Example 15), and two pairs need not be four distinct things.)
9. Formalize each of the following. Evaluate whether the inference is valid.
   a. Horses are animals. *Therefore,* something is an animal.
   b. Gold is valuable. *Therefore,* something is valuable.
   c. All but 4,318 horses do not live in Utah. There are 18,317,271 horses.
      Therefore, there are 18,312,956 horses that do not live in Utah.
   d. Only Ralph and Juney are dogs. Both Ralph and Juney bark.
      Therefore, all dogs bark.
   e. Every pair of real numbers has a least upper bound. π and 47 are real numbers.
      Therefore, π and 47 have a least upper bound.
   f. Every set of real numbers that has an upper bound has a least upper bound.
      Therefore, the square roots of all numbers less than 47 have a least upper bound.

VIII Functions

A. Functions and Things

In attempting to place the infinitesimal calculus on an intellectually solid foundation in the 19th century, the notion of limit was refined along with a new theory of collections viewed as sets. In this view there was no longer room for the conception of functions as processes. The function *sin x*, for example, was no longer a process, represented as a continuous curve, but a relationship between real numbers, where 0 is related to 0; $\pi/2$ is related to 1; π is related to 0; and so on. A function was reduced to or reconceived of as a set of ordered pairs: (input, output). Anomalies arose, however, as in the theorem that there is an everywhere continuous, nowhere differentiable function, which is incompatible with viewing functions as processes.

Nonetheless, the mathematical utility of this view of functions in the context of set theory led to its adoption in modern mathematics. This view is compatible with and indeed part of the motive for the development of mathematical logic, in which the world is conceived as made up of things as opposed to processes.

But even though a function such as *sin x* was "really" no more than a relation, no one began to write *sin*(x, y) in place of *sin* $x = y$. The notation of functions was not only convenient but essential for the intelligibility of mathematical theorems. What, though, was the status of a notation such as *sin x*? Such notations serve as name-makers. Given the name 0, we have the name *sin* 0; given the name $\pi/2$, we have the name *sin* $\pi/2$.

We have name-making devices in ordinary language, too. Consider:

The wife of Richard Nixon.

The phrase 'the wife of—' creates a new name when the blank is filled with a name. But if we take 'the wife of —' as a name-making phrase, what are we to make of:

the wife of Pope John Paul II
the wife of x where 'x' is to refer to the eraser on my desk
the wife of King Henry VIII

We could say that we intended the blank in 'the wife of —' to be filled with only names of people who have exactly one wife. But how are we to enforce that?

Compare the phrase '— is the wife of —'. We can take this as a predicate relative to the universe consisting of all people and get:

Stanisław Krajewski is the wife of Richard L. Epstein.
x is the wife of Marilyn Monroe where 'x' stands for Fred Kroon
Marilyn Monroe is the wife of x where 'x' stands for Betty Grable

We don't rule these out as propositions, because to do so would complicate our reasoning. We classify them as false. Falsity is the default value for the application of predicates. But what default value can we assign to 'the wife of Marilyn Monroe'? Some method must be found if we are to use 'the wife of —' as a name-making device, because we require that names refer.

In Chapter XXI we'll look at such possibilities. But for now we'll avoid that issue by requiring that a phrase used as a name-maker must produce a name of an object when the name of any object is put into the blank.

Such phrases that always produce a name are rare in ordinary speech. But in mathematics they are common. For example, for the universe of natural numbers we have 'x^2', 'the smallest prime greater than x', 'the smallest number divisible by 3 and greater than x'; for the universe of non-negative real numbers, we have '$\sqrt{x}$'. But for the universe of natural numbers '$\sqrt{x}$' cannot serve as a name-maker because '$\sqrt{3}$' is not a name of a natural number. What we accept as a name-maker depends on what things we are talking about.

As there are predicates of several variables, so there are functions of several variables. For example, for the universe of natural numbers, we have addition, $x + y$. When 'x' and 'y' are replaced by names or are supplemented with an indication of what they are to refer to, we get a name, for example:

$2 + 3$
$4{,}318 + 78{,}162{,}298{,}489{,}728{,}213{,}906$
$x + 17$ where 'x' stands for eleven

But unlike predicates, we can iterate functions. For example, $(4^2 \cdot 15) + 6$ is a name, a name of an object we also call '246'. We now have two names for the one object two hundred and forty-six, and we can create indefinitely more. With the use of functions the equality predicate plays a substantial role, so I'll assume that it is in every language we consider in this chapter.

In the same way that we defined 'predicate' in Chapter II, we can define 'function' in accord with the view that the world is made up of things.

Functions A *function* is any incomplete phrase with specified gaps such that when the gaps are filled with names of things, the phrase becomes a name.

In this definition 'name' means also a variable supplemented with an indication of what object it is to refer to.

The lowercase roman letters f, g, h are commonly used in mathematics to stand for functions, and I'll adopt those as metavariables, along with $g_0, g_1, \ldots$ and $h_0, h_1, \ldots$.

As a predicate is unary, binary, ternary, or n-ary according to how many blanks it has, so is a function. If a phrase serves as a function for a particular universe, we say it is a function *on* that universe. Functions are sometimes said to be defined on a *domain*, but that terminology is usually adopted for partial functions. For example, for the universe of all real numbers, $\sqrt{x}$ is said to be defined on or have as domain the nonnegative reals.

We begin with some functions as primitive, for example $\cdot$ and $+$. We can then form compound functions from these, for example $(x + y) \cdot z$. As with predicates, this may give functions of higher arity than what we started with. To make this clearer, we need a formal language.

B. A Formal Language with Function Symbols and Equality

The definitions here extend those of Chapter III. To define the formal language $\text{L}(\neg, \rightarrow, \wedge, \vee, \forall, \exists, =; \text{P}_0, \text{P}_1, \ldots, \text{f}_0, \text{f}_1, \ldots, \text{c}_0, \text{c}_1, \ldots)$, we first add to the vocabulary of our previous formal language (p. 62):

equality symbol: $=$

function symbols: $\text{f}_0^1, \text{f}_0^2, \text{f}_0^3, \ldots, \text{f}_1^1, \text{f}_1^2, \text{f}_1^3, \ldots, \text{f}_2^1, \text{f}_2^2, \text{f}_2^3, \ldots$

The subscript on f_i^n is the number of the symbol, the superscript is the arity.

Terms and their depth are defined inductively.

Terms Every name symbol and every variable is a term. These are the *atomic terms*, and they have *depth* 0.

If $t_1, \ldots, t_m$ are terms of depth $\leq n$, and at least one of them has depth n, then $\text{f}_i^m(t_1, \ldots, t_m)$ is a term and has *depth* $n + 1$.

A concatenation of symbols is a term iff it is a term of depth n for some $n \geq 0$.

A term that is not atomic is *compound*.

In an exercise below I suggest how to prove the unique readability of terms in much the same ways as we showed the unique readability of wffs.

We define inductively what it means for a symbol to *appear* or *occur in a term*:

For every $i \geq 0$, the only symbol that appears in x_i is x_i, and the only symbol that appears in c_i is c_i.

For every $i \geq 0$ and $n \geq 1$ and terms $t_1, \ldots, t_n$, the terms that appear in $f_i^n(t_1, \ldots, t_n)$ are $f_i^n(t_1, \ldots, t_n)$, and those terms that occur in any one of $t_1, \ldots, t_n$.

A term is *closed* if no variable appears in it; otherwise it is *open*.

I will use f, g, $f_0, f_1, \ldots, g_0, g_1, \ldots$ as metavariables for function symbols.

The base stage of the inductive definition of a *well-formed-formula* (*wff*) (p. 62) is now changed to:

Atomic wffs:

$(P_i^n(t_1, \ldots, t_n))$ where $0 \leq i$, and $1 \leq n$, and t_j is a term for $1 \leq j \leq n$

$(t_1 = t_2)$ where t_1, t_2 are terms

The *unique readability of wffs* follows as before (pp. 63–64), modifying the definition of λ to read 'for every term t, $\lambda(t) = 1$'. In an exercise below I ask you to define a numbering of the wffs of this language.

We define inductively what it means for a *term to appear* or *occur in* a wff:

The terms that appear in $P_i^n(t_1, \ldots, t_n)$ are those terms that appear in $t_1, \ldots, t_n$. The terms that appear in $(t_1 = t_2)$ are those terms that appear in t_1 and t_2.

The terms that appear in $\neg$A are those that appear in A. The terms that appear in A$\rightarrow$B, A$\wedge$B, A$\vee$B are those that appear in A or in B.

The terms that appear in $\forall x_i$ A or $\exists x_i$ A are x_i and those that appear in A.

The definitions of the following are as in Chapter III:

length of a wff	*bound variable*	*open wff*
subformula	*quantifier binding a variable*	*closure of a wff*
free variable	*closed wff* or *sentence*	

A term t is *free for an occurrence of x in* A (meaning free to replace it) means that both:

i. That occurrence of x is free.
ii. For any y that appears in t, the occurrence of x does not lie within the scope of $\forall y$ or $\exists y$.

Note that a closed term is free for every free occurrence of a variable.

A term t is *free for x in* A iff it is free for every free occurrence of x in A. As before, A(t/x) means the wff that results from replacing every free occurrence of x in A with t if x appears free in A, otherwise A itself, unless we specifically say that t may replace some but not necessarily all occurrences of x for which it is free. Similarly, $\mathrm{f}(t_1/y_1, \ldots, t_n/y_n)$ is the term we get by replacing every occurrence of y_i by t_i in $\mathrm{f}(y_1, \ldots, y_n)$.

Sometimes we need to talk about replacing a term (rather than a variable) by another term. I'll write A(t/u) to mean that t replaces all occurrences of u, unless noted otherwise. Similarly, $\mathrm{f}(t_1/u_1, \ldots, t_n/u_n)$ stands for the term that results from replacing each u_i by t_i in $\mathrm{f}(u_1, \ldots, u_n)$

C. Realizations and Truth in a Model

The only change necessary to the notion of *realization* (Chapter IV.E) for this formal language is to say that we realize each f_i^n as an n-ary function. A realization need not realize any function symbols, nor if it does realize some need it realize all. In the presentation of a realization, functions are listed after the predicates and before the names. The notion of a *universe* for a realization is as before.

Expressions in the realization inherit the terminology of the wffs they realize. For example, a closed term of the semi-formal language is one realizing a closed term of the formal language.

A function, just as a predicate, requires names (or variables acting as temporary names) to yield a value. But unlike a predicate, when the gaps in a function are filled with names we get a name rather than a proposition.

The value of a function Given any n-ary function f and terms $t_1, \ldots, t_n$ where each variable (if any) appearing in each term is supplemented with an indication of what it is to refer to, the object referred to by $f(t_1, \ldots, t_n)$ is called the *value of the function applied to the objects* referred to by $t_1, \ldots, t_n$ in that order. If $t_1, \ldots, t_n$ are atomic, $f(t_1, \ldots, t_n)$ is an *atomic application* of the function f.

Just as for predications, our agreement that a name can refer to only one thing is extended to apply to variables used as demonstratives in an application of a function. That is, in evaluating $f(t_1, \ldots, t_n)$, each occurrence of a variable x in $t_1, \ldots, t_n$ must be assigned the same reference. For example, taking the universe to be the natural numbers with the usual interpretation of '+', the value of + applied to 7 and 3 in that order is 10. And the value of '$x^2 + (x + y)$', where both occurrences of 'x' are assigned 6 as reference and 'y' is assigned '3' is 45.

The definition of predication given in Chapter IV.E.2 remains the same, with this new definition of 'term'. Comparable to our abstraction of predicates, we have the following.

The classical abstraction of functions The only properties of a function that matter to logic are its arity and, for every (sequence of) term(s) consistently supplemented with an indication of what each variable (if any) is to refer to, the value of the function applied to the object(s) referred to by those terms.

What does the value of a function depend on? Since we've decided that we shall ignore all semantic aspects of names and functions other than the references they provide, we adopt a Fregean assumption for functions:

> The value of a function applied to particular terms (as consistently supplemented with an indication of what each variable is to refer to) is determined by the semantic properties of its parts: the references of those terms.

So our assignments of references must satisfy the following.

The extensionality of functions For any function f of the realization and terms $t_1, \ldots, t_n$ and $u_1, \ldots, u_n$, and any assignments of references σ, τ, if for all $i \leq n$, $\sigma(t_i)$ refers to the same object as $\tau(u_i)$, then $\sigma(f(t_1, \ldots, t_n))$ refers to the same object as $\tau(f(y_1, \ldots, y_n))$.

I will let you show that *if we take as primitive the assignments of references of applications of functions to variables, then the condition on extensionality of functions uniquely determines the assignments of references to all terms.*

Now that we have assignments of references for all terms, the definitions establishing our models from Chapters IV and VI can be used. The definitions of *truth in a model, validity*, etc. are as before, too, with the understanding that the word 'term' now has the broader meaning given above.

The theorems and lemmas proved in Chapters IV and V, and the principle of substitution and the principle of identity of Chapter VI.B continue to hold, as you can verify. In particular, since a closed term is a (possibly compound) name, existential generalization and universal instantiation hold for closed terms as well as atomic names.

D. Examples of Formalization

The examples here are from arithmetic, where the mathematician's ordinary language is so close to a semi-formal one that I have presented only semi-formal propositions. These examples contain commonly accepted abbreviations or semi-formal predicates such as $+$, $\cdot$, x^2, x^y, following the mathematician's implicit conventions on deleting parentheses.

In the list of categorematic terms used in the formalization I've listed only atomic names and not compound ones such as '2 + 3'.

1. $\forall x\, \forall y\, (x + y = y + x)$

predicates: '='
functions: '+'
names: none

$\forall x_1\, \forall x_2\, (\, f_1^2(x_1, x_2) = f_1^2(x_2, x_1))$

Analysis Addition is a binary function. The proposition is an identity of terms.

2. $\forall x\, (x + 0 = x)$

predicates: '='
functions: '+'
names: '0'

$\forall x_1\, (\, f_1^2(x_1, c_7) = x_1)$

Analysis Again we have an identity of terms, only here one of the terms contains a name. The application of the function, however, is atomic.

3. $2 + 3 > 7$

predicates: '>'
functions: '+'
names: '2', '3', '7'

$P_1^2(\, f_1^2(c_0, c_1), c_2)$

Analysis Here we have a compound closed term: '2 + 3' acts as a name.

4. $(2 + 3^3) \cdot 7 > 4000 + 18$

predicates: '>'
functions: '+', '·', '$\_\_^{-}$'
names: '2', '3', '7', '4000', '18'

$P_1^2(\, f_2^2(\, f_1^2(c_0, f_3^2(c_1, c_1)), c_2), f_1^2(c_3, c_4))$

Analysis All terms are closed. Here we begin to see the complexity we can get by applying functions to terms involving terms. I have viewed '3^3' as a substitution in the function x^y, rather than in x^3 or x^x.

5. $\forall x\, \forall y\, \exists z\, (((2x)^2 + (2y)^2) = 2 \cdot z)$
Therefore,
$\exists z\, ((2 \cdot 3)^2 + (2 \cdot 5)^2 = 2 \cdot z)$

predicates: '='
functions: '+', '·', '—²'
names: '2', '5', '3'

$$\frac{\forall x_1 \forall x_2 \exists x_3 \left(f_1^2(f_3^2(f_2^2(c_0,x_1), c_0),\ f_3^2(f_2^2(c_0,x_2), c_0)) = f_2^2(c_0,x_3)\right)}{\exists x_3 \left(f_1^2(f_3^2(f_2^2(c_0,c_1), c_0),\ f_3^2(f_2^2(c_0,c_2), c_0)) = f_2^2(c_0,x_3)\right)}$$

Analysis The example contains nonatomic applications of functions, for example, $(2x)^2$, $(2x)^2 + (2y)^2$, $(2 \cdot 3)^2$. The example is a valid inference, an instance of universal instantiation (twice).

6. $\forall x\, \forall y\, \exists z\, (((2x)^2 + (2y)^2) = 2 \cdot z)$
   Therefore,
   $\exists z\, (8 + 3 = 2 \cdot z)$

predicates: '='
functions: '+', '·', '—²'
names: '2', '3', '8'

$$\frac{\forall x_1 \forall x_2 \exists x_3 \left(f_1^2(f_3^2(f_2^2(c_0,x_1), c_0),\ f_3^2(f_2^2(c_0,x_2), c_0)) = f_2^2(c_0,x_3)\right)}{\exists x_3 \left(f_1^2(c_1,c_2) = f_2^2(c_0,x_3)\right)}$$

Analysis The deduction is not valid. This is not an application of universal instantiation: '8' has not been substituted for a variable, but for a compound term.

7. $\forall x\, \exists y\, (x + y = 0)$
   Therefore,
   $\exists y\, ((y + 1) + y = 0)$

predicates: '='
functions: '+'
names: '0', '1'

$$\frac{\forall x_1\, \exists x_2 \left(f_1^2(x_1, x_2) = c_7\right)}{\exists x_2 \left(f_1^2(f_1^2(x_2, c_1), x_2) = c_7\right)}$$

Analysis The deduction is invalid: on the usual interpretation of the functions and names on the universe of integers, the premise is true, but the conclusion is false. The term $(y + 1)$ is not free to be substituted for x in $\exists y\, (x + y = 0)$.

Exercises for Sections A–D

1. For each formula below decide whether it is a term, and if so whether it is atomic or compound, open or closed, and give an unabbreviated realization of it from arithmetic.

   a. (c_0) b. $f_1^2(c_0, c_1)$ c. $f_2^2(c_1, f_1^2(c_1, c_2), c_3)$

d. $f_2^2(f_1^2(x_1, c_0), f_1^1(c_1))$ e. $f_4^1(f_1^2(x_1, x_2))$ f. $f_3^2(f_1^2(x_1, x_2), c_0), x_1)$

g. $f_1^3(f_1^2(x_1, x_2, x_3))$ h. $f_1^3(f_1^1(c_1), f_1^1(c_1), f_1^1(c_1))$

2. For each of the following:
   i. If there is a commonly accepted reading of the sequence as a term, restore parentheses to make that explicit.
   ii. State whether the term is atomic or compound, open or closed.
   iii. Give a formal term of which it could be the realization.
   iv. If the term is open, state a universe for which the term serves as a function.

   a. $2+3$ d. x^{2yz} g. $\int_1^x e^t\, dt$

   b. $(2^2+3)+4$ e. $(x^2+1)\cdot 3-(4\cdot 5+6)$ h. $\vec{x}\times\vec{y}\times\vec{z}$

   c. $x^2+2\cdot y+x$ f. $\int_1^3 e^t\, dt$ i. $\sqrt{x}$

3. For each wff below, state whether the term listed is free to be substituted for x in the wff, and make the substitution if it is.
   a. 3 in $x^2+(2\cdot x)=1$
   b. 3 in $\forall x\,(x^2+(2\cdot x)=1)$
   c. $3+x$ in $\exists y\,(x+y=0)$
   d. $(x^2+(2\cdot y))$ in $\forall z\,\exists x\,(x+y=z)\rightarrow\exists y\,(x+y=z)$
   e. $x+y$ in $(x>y)\rightarrow\forall y\,(y^2>1)$
   f. $x+y$ in $\forall y\,(y>3)\vee\exists y\,(x>y)$
   g. $x+y$ in $\forall x\,(y\le x)$
   h. $x+y$ in $\forall y\,(y\le x)$

4. *The unique readability of terms*
   Let γ be the following assignment of numbers to the symbols:

x_i	c_i	f_i^n	(	)	for each $i\ge 0$ and $n\ge 0$
1	1	$2-n$	2	-2	

   To a concatenation of these symbols $\alpha_1\,\alpha_2\cdots\alpha_n$ assign the number $\gamma(\alpha_1\,\alpha_2\cdots\alpha_n)=\gamma(\alpha_1)+\gamma(\alpha_2)+\cdots+\gamma(\alpha_n)$. Show:
   a. For every term t, $\gamma(t)=1$.
   b. If β is an initial segment of a term other than the entire term, then $\gamma(\beta)<1$.
   c. If β is a final segment of a term other than the entire term, then $\gamma(\beta)>1$.
   d. No proper initial or final segment of a term is a term.
   e. There is one and only one way to parse each term.
   f. Each term has a unique number assigned as its depth.

5. Provide a numbering of wffs. (Hint: Use the numbering of wffs of the language of propositional logic on p. 10 as a guide.)

6. Prove: For a realization if for every n-ary function f and variables $y_1,\ldots,y_n$, every assignment of references σ provides a reference $\sigma(f(y_1,\ldots,y_n))$, then the condition on extensionality of functions uniquely determines the references for all terms.

7. Prove the partial interpretation theorem for languages with function symbols.

8. For languages with function symbols verify Theorem V.3 on the substitution of terms.

E. Translating Functions into Predicates

To accommodate talk of functions within predicate logic we've viewed functions as relations. So we should be able to replace any talk of functions with talk of the corresponding relations, so long as we make explicit our assumption that the application of the function yields a reference.

Consider, for example:

2 + 11 is prime.

We can use the predicate 'x is the sum of y and z' to convert this to:

$\exists ! x$ (x is the sum of 2 and 11) $\wedge$ $\forall x$ (x is the sum of 2 and 11 $\rightarrow$ x is prime)

Eliminating functions from compound wffs requires decisions and care about the scope of the existence and uniqueness conditions. For example,

(1) $\forall x \exists y\, ((y + 1 = x) \vee x = 0)$

Here it seems reasonable to take the least scope possible:

$\forall x \exists y$ ([$\exists ! z$ (z is the sum of y and 1) $\wedge$
$\forall z$ (z is the sum of y and 1 $\rightarrow z = x$)] $\vee\ x = 0$)

It would seem that we only need to enforce a rule of taking the minimal scope to give a general method for eliminating functions in favor of predicates. But an easier, more regular translation can be accomplished by instead taking an axiom for the uniqueness condition. For (1) we'd translate with respect to:

$\forall x \forall y \exists ! z$ (z is the sum of x and y)

to get: $\forall x \exists y$ ($\forall z$ (z is the sum of y and 1 $\rightarrow z = x$) $\vee\ x = 0$).

I'll set that translation out now. It is the most complicated piece of work in this book, and if you wish you can skip it until we need it in Chapter XII.

We start with a formal language L(f) with as many predicate and name symbols as you wish but just one n-ary function symbol, f. Let P be an $n + 1$-ary predicate symbol that does not appear in L(f). Let L(P) be the language L(f) with f deleted and P added.

I'll define a mapping * from L(f) to L(P). We begin by defining * on atomic wffs by induction on the maximum depth m of any term in the wff.

*Definition of * for atomic wffs*

If A is an atomic wff and the maximum depth of any term in A is 0, then A* = A.

Suppose * has been defined for all atomic wffs with terms of depth no greater than m. Suppose also that the maximum depth of any term in A is $m + 1$, and $u_1, \ldots, u_s$ are all the terms in A of depth $m + 1$, where

u_i is $f(t_1^i, \ldots, t_n^i)$. Let '$\bigwedge_i B_i$' stand for the conjunction of the indexed wffs, associating to the left. Define:

$$D \equiv_{Def} \bigwedge_i (y_1^i = t_1^i)^* \wedge \cdots \wedge (y_n^i = t_n^i)^* \wedge \bigwedge_i P(y_1^i, \ldots, y_n^i, z_i) \rightarrow A(z_1/u_1, \ldots, z_s/u_s)^*$$

where $y_1^i, \ldots, y_n^i, z_i$ are the least variables in order that are not in A and not in $\{ y_1^j, \ldots, y_n^j, z_j : j < i\}$.

Finally, let $w_1, \ldots, w_r$ be the list of all the variables in D in alphabetical order, omitting the variables that appear in A. Define:

(2) $$A^* = \forall w_1 \ldots \forall w_r\ D$$

(This definition is legitimate since by hypothesis the wffs $(y_1^i = t_1^i), \ldots, (y_n^i = t_n^i)$, and $A(z_1/u_1, \ldots, z_s/u_s)$ have no term of depth greater than m.)

Extend * *to all wffs of* L(f) *homophonically*:

$(\neg A)^*$	$= \neg A^*$	$(A \rightarrow B)^* =$	$A^* \rightarrow B^*$
$(A \wedge B)^*$	$= A^* \wedge B^*$	$(\forall x\, A)^*$	$= \forall x\, A^*$
$(A \vee B)^*$	$= A^* \vee B^*$	$(\exists x\, A)^*$	$= \exists x\, A^*$

Note that the variables free in A* are exactly those free in A. Define now:

The function assumption $\alpha_f \equiv_{Def} \forall x_1 \ldots \forall x_n \exists x_{n+1} P(x_1, \ldots x_n, x_{n+1})$

Correlated to this mapping of the language L(f) to L(P), we can define a mapping, also denoted *, from models of L(P) that satisfy α_f to models of L(f).

$$\begin{array}{ccc} L(f) & \xrightarrow{\ *\ } & L(P) \\ \text{models} & \xleftarrow{\ *\ } & \text{models of } \alpha_f \end{array}$$

(3) Delete the interpretation **P** of P.
Interpret the function symbol f by :
$\sigma(f(x_1, \ldots, x_n)) = \mathbf{b}$ iff $\sigma(x_1) = \mathbf{a}_1, \ldots, \sigma(x_n) = \mathbf{a}_n$ and $\mathbf{P}(\mathbf{a}_1, \ldots, \mathbf{a}_n, \mathbf{b})$.

Lemma 1 This mapping of models is onto. That is, given any model **N** of L(f), there is a model **M** of L(P) such that $\mathbf{M} \vDash \alpha_f$ and $\mathbf{M}^* = \mathbf{N}$.

Proof Let **N** be a model of L(f). Define the model **M** of L(P) by modifying **N**:

Delete the interpretation **f** of f.

Interpret the predicate symbol P by: $\sigma \vDash P(x_1, \ldots, x_n, x_{n+1})$ iff
$\sigma(x_1) = \mathbf{a}_1, \ldots, \sigma(x_n) = \mathbf{a}_n$, and $\sigma(x_{n+1}) = \mathbf{f}(\mathbf{a}_1, \ldots, \mathbf{a}_n)$.

I'll let you show that $\mathbf{M} \vDash \alpha_f$ and $\mathbf{M}^* = \mathbf{N}$. ∎

Lemma 2 For any A in L(f), for any model M of L(P), M*⊨A iff M⊨A*.

Proof To prove this, first define a mapping from assignments of references on M to assignments of references on M* by taking σ^* to be σ adding the evaluation of the function by (3) above. Note that the mapping is onto, that is, every assignment of references on M* is σ^* for some σ on M. I'll now show that for any A,

(4) $\sigma \vDash A^*$ iff $\sigma^* \vDash A$

If we can prove this for atomic wffs, we'll be done, since the translation is extended to all wffs homophonically. For atomic wffs we'll proceed by induction on the depth of terms in A. If A is an atomic wff of depth 0, then (4) is true since the evaluations are identical. Suppose now that (4) is true for atomic wffs every term of which has depth $\leq m$. Then:

$\sigma \vDash A^*$ iff for all τ that differ from σ only in what they assign the variables in the prefix at (2):
if $\tau \vDash \bigwedge_i (y_1^i = t_1^i)^* \wedge \cdots \wedge (y_n^i = t_n^i)^*$ and
$\tau \vDash \bigwedge_i \mathrm{P}(y_1^i, \ldots, y_n^i, z_i)$, then $\tau \vDash \mathrm{A}(z_1/u_1, \ldots, z_s/u_s)^*$

iff (by induction) for all τ that differ from σ only in what they assign the variables in the prefix at (2):
if $\tau^* \vDash \bigwedge_i (y_1^i = t_1^i) \wedge \cdots \wedge (y_n^i = t_n^i)$ and
$\tau^* \vDash \bigwedge_i \mathrm{f}(y_1^i, \ldots, y_n^i) = z_i$, then $\tau^* \vDash \mathrm{A}(z_1/u_1, \ldots, z_s/u_s)$

iff for all τ that differ from σ only in what they assign the variables in the prefix at (2):
if $\tau^* \vDash \bigwedge_i \mathrm{f}(t_1^i, \ldots, t_n^i) = z_i$, then $\tau^* \vDash \mathrm{A}(z_1/u_1, \ldots, z_s/u_s)$

iff for all τ that differ from σ only in what they assign the variables in the prefix at (2), $\tau^* \vDash \mathrm{A}(u_1, \ldots, u_s)$

iff $\sigma^* \vDash A$ since the variables in the prefix at (2) do not appear in A.

To complete the proof of the lemma, we have:

M⊨A* iff for all σ on M, $\sigma \vDash A^*$

iff for all σ^* on M*, $\sigma^* \vDash A$

iff M*⊨A since the mapping of assignments of references is onto. ■

Theorem 3 Let $\Gamma^* = \{B^* : B \in \Gamma\}$. Then $\Gamma \vDash A$ iff $\Gamma^* \cup \{\alpha_f\} \vDash A^*$.

Proof $\Gamma^* \cup \{\alpha_f\} \vDash A^*$

iff for every model M of L(P) such that M⊨α_f if M⊨Γ^*, then M⊨A*

iff (by Lemma 2) for every model M* of L(f), if M*⊨Γ, then M*⊨A

iff (by Lemma 1) for every model N of L(f), if N⊨Γ, then N⊨A ■

Corollary 4 There is a mapping * from the language $L(f_1, \ldots, f_n)$ to $L(P_1, \ldots, P_n)$ such that for any collection of wffs Γ and A of $L(f_1, \ldots, f_n)$, $\Gamma \vDash A$ iff $\Gamma^* \cup \{\alpha_f\} \vDash A^*$.

Proof Eliminate one function symbol at a time using Theorem 3. ■

The extension of Theorem 3 to languages with infinitely many function symbols requires no new ideas, but it is terribly complicated to say correctly. Fortunately we won't need it in this text.

Exercises for Section E

1. Eliminate the function symbols from the following according to the translation given in Section E. (Hint: When there are two function symbols, eliminate them sequentially.)
   a. $x + y = 2$
   b. $x + (y + z) = 2$
   c. $x + ((y + z) + x) = 2$
   d. $x + (y + z) = (x + y) + z$
   e. $\forall x \, \forall y \, (x + y = y + x)$
   f. $\forall x \, \forall y \, \forall z \, (x + (y + z) = (x + y) + z)$
   g. $x^z + y = 2$
   h. $\forall x \, \forall y \, \exists z \, (x^z + y = 2^y)$
   i. $\forall x \, \forall y \, \exists z \, (((2x)^2 + (2y)^2) = 2z)$
2. Complete the proof of Lemma 1.
3. Devise a mapping ° of L(f) to L(P) as described at the beginning of Section E, where the existential assumption is taken to have minimal scope for each wff instead of translating relative to the assumption α_f. Show that it satisfies $\Gamma \vDash A$ iff $\Gamma^\circ \vDash A^\circ$. (Hint: Modify the translation * for atomic wffs.)
4. *Translating names into predicates*
   Let L(c) be a language with as many predicate and function symbols as you wish but just one name symbol, c. Let P be a unary predicate symbol that does not appear in L(c). Let L(P) be the language L(c) with c deleted and P added.
   Define a mapping # from L(c) to L(P) inductively on the length of wffs:
   If A is atomic and c does not appear in A, then A# = A.
   If A is atomic and c appears in A, then
   $A(c)^\# = \forall x \, (P(x) \rightarrow A(x/c))$ x is the least variable not appearing in A
   Extend # to all wffs homophonically.
   Define the *name assumption* $\alpha_c \equiv_{Def} \exists! x \, P(x)$
   a. Prove: There is an onto mapping # from models M of L(P) that satisfy $M \vDash \alpha_c$ to models of L(c) such that $M^\# \vDash A$ iff $M \vDash A^\#$.

b. Prove: For any collection of wffs Γ and any A of L(c), $\Gamma \models A$ iff $\Gamma^{\#} \cup \{\alpha_c\} \models A^{\#}$.

c. Eliminate the name symbols from the following according to the translation.

i. $\forall x\ 0 < x$

ii. $\forall x\ \forall y\ (x + y > 0 \vee (x = 0 \wedge y = 0))$

iii. $0 < 1$

iv. $\forall x\ (x < 1 \rightarrow x = 0)$

IX The Abstraction of Models

In this chapter we'll further abstract the semantics of predicate logic in order to be able to use mathematics to study predicate logic. To do that, we'll have to consider more carefully what it means to classify a collection as a thing.

A. The Extension of a Predicate

Consider again the model discussed in Chapter IV.E:

$L(\neg, \rightarrow, \wedge, \vee, \forall, \exists; P_0, P_1, \ldots, c_0, c_1, \ldots)$

↓ realizations of name and predicate symbols; universe

$L(\neg, \rightarrow, \wedge, \vee, \forall, \exists;$ is a dog, is a cat, eats grass, is a wombat, is the father of; Ralph, Dusty, Howie, Juney)
universe: all animals, living or toy

↓ assignments of references; valuations of atomic wffs; evaluation of $\neg, \rightarrow, \wedge, \vee \forall, \exists$

$\{T, F\}$

Recall the classical abstraction of predicates: the only properties of a predicate that matter to logic are its arity and for each (sequence of) object(s) under consideration whether it applies to that (sequence of) object(s) or not. Thus, if we were to take a second model that differs from the one above only in that it realizes

P_1^1 as '— is a domestic feline —' and realizes P_0^2 as '— is the male parent of —', then the differences between the two models would be immaterial for classical logic, since 'is a domestic feline' and 'is a cat' apply to the same things in the universe, and 'is the father of' and 'is the male parent of' each are true of the same pairs from the universe. So for classical logic we could identify the two models, as they have the same universe and validate the same formal wffs.

To come up with a way to talk about such identifications of models and to simplify what we need to pay attention to in presentations of models, let's call the collection of sequences of the universe of which a predicate is true the *extension* of the predicate *relative to* the universe. Thus, relative to a universe of all objects with diameter greater than 3 cm on the moon 'is a dog' and 'is a cat' have the same extension. Relative to all objects with diameter greater than 3 cm in my house, 'likes brocolli' and 'is a human' have the same extension. For the universe of all living things with the usual interpretation of the predicates 'x is a step-brother to y' has the same extension as 'x is human $\wedge$ x is male $\wedge$ $\exists! z$ (z is a parent of x $\wedge$ z is a parent of y)'.

Collections of objects and sequences of objects are central to the discussion of the extensions of predicates. But if collections are to play a role in our abstraction of classical models, then, comparable to the classical abstraction of predicates, how we specify a collection cannot matter: $\mathsf{A} = \mathsf{B}$ iff (an object is in A iff it is in B). So relative to all living animals, the collection of objects x such that x is domestic and x is a canine is the same as the collection of objects x such that x is a dog.

In Chapter VIII.E we saw how to view functions as special kinds of predicates. Similarly, given a function f in a realization, we define the *extension* of f to be the collection of all $n+1$-tuples of the universe $\mathsf{b}_1, \ldots, \mathsf{b}_n, \mathsf{b}_{n+1}$ where b_{n+1} is the value of f applied to the objects $\mathsf{b}_1, \ldots, \mathsf{b}_n$ in that order.

Now, given a model, we write:

U is the universe of a model viewed as a collection.

P_i^n is the realization of P_i^n

P_i^n is the extension of P_i^n viewed as a collection of n-tuples.

f_i^n is the realization of f_i^n

f_i^n is the extension of f_i^n viewed as a collection of $n+1$-tuples.

a_i is the realization of c_i

a_i is the object assigned to a_i in the model.

In the model above, the realization of P_0^2 is '—is the father of —', and the extension of '—is the father of —' is the collection of pairs of animals the first of which is the father of the second. The realization of c_0 is 'Ralph', and the object assigned to the name 'Ralph' is what's pictured on p. 522.

Schematically, a model is then:

I $L(\neg, \rightarrow, \wedge, \vee, \forall, \exists, =; \mathrm{P}_i^n, \mathrm{f}_i^n, \ldots, \mathrm{c}_0, \mathrm{c}_1, \ldots)$ where $i, n \geq 0$

$\downarrow$ realizations of name and predicate symbols; universe

$L(\neg, \rightarrow, \wedge, \vee, \forall, \exists, =; P_i^n, f_i^n, \ldots, a_0, a_1, \ldots)$ where $i, n \geq 0$

$\downarrow$ assignments of references; valuations of atomic wffs

$$< \mathsf{U}; \mathsf{P}_0^1, \mathsf{P}_1^1, \ldots, \mathsf{P}_i^n, \ldots, \mathsf{f}_0^1, \mathsf{f}_1^1, \ldots, \mathsf{f}_i^n, \ldots, \mathsf{a}_0, \mathsf{a}_1, \ldots >$$

$\downarrow$ evaluation of $\neg, \rightarrow, \wedge, \vee, \forall, \exists, =$

$\{\mathsf{T}, \mathsf{F}\}$

This is a model of type **I**.

The classical abstraction of predicates requires that we ignore any differences between two models that appear only at the second level (realization). If two models have the same interpretation, they will validate exactly the same formal wffs. So we can make a *classical abstraction of the model*:

II $L(\neg, \rightarrow, \wedge, \vee, \forall, \exists, =; \mathrm{P}_i^n, \mathrm{f}_i^n, \ldots, \mathrm{c}_0, \mathrm{c}_1, \ldots)$ where $i, n \geq 0$

$\downarrow$ interpretation

$$< \mathsf{U}; \mathsf{P}_0^1, \mathsf{P}_1^1, \ldots, \mathsf{P}_i^n, \ldots, \mathsf{f}_0^1, \mathsf{f}_1^1, \ldots, \mathsf{f}_i^n, \ldots, \mathsf{a}_0, \mathsf{a}_1, \ldots >$$

$\downarrow$ assignments of references; valuations of atomic wffs; evaluation of $\neg, \rightarrow, \wedge, \vee\ \forall, \exists, =$

$\{\mathsf{T}, \mathsf{F}\}$

Here the mth collection of sequences, reading from the left, interprets P_m^n.

The extensional evaluation of predicates and functions

$\sigma \models \mathrm{P}_i^n(t_1, \ldots, t_n)$ iff the sequence $\sigma(t_1), \ldots, \sigma(t_n)$ is in P_i^n

$\sigma(\mathrm{f}_i^n(t_1, \ldots, t_n)) = \mathsf{b}$ iff the sequence $\sigma(t_1), \ldots, \sigma(t_n), \mathsf{b}$ is in f_i^n

---

Using these evaluations we have a model of type **II** for the formal language. I'll use $\mathsf{M}, \mathsf{N}, \mathsf{M}_0, \mathsf{M}_1, \ldots, \mathsf{N}_0, \mathsf{N}_1, \ldots$ as variables for type **II** models, too.

In a model of type **II** there are no longer any semi-formal propositions to assign truth-values. Rather, a *formal wff is to be considered a proposition under the interpretation*, for example, $\forall x_1\, \exists x_2\, (\mathrm{P}_1^2(x_1, x_2) \vee \mathrm{P}_{17}^1(\mathrm{c}_{36}))$. With the meaning given to the symbols of an open formal wff by a model of type **II**, a wff is *satisfied* or not by an assignment of references, and a closed formal wff is *true* or *false*.

For propositional logic, two type **I** models that result in different type **II** models must disagree on the truth-value of at least one formal wff. That's because all two type **II** models disagree on the truth-value of at least one formal wff. But that's not so for predicate logic. Consider these two type **I** models.

M_1 $\quad L(\neg, \rightarrow, \wedge, \vee, \forall, \exists;\ P_0^1)$

$\downarrow$

$L(\neg, \rightarrow, \wedge, \vee, \forall, \exists;$ 'taught in Brazil')
universe: all people living in the 300 or 400 block of Pinecone in Cedar City

$\downarrow$

<all people living in the 300 or 400 block of Pinecone in Cedar City; {R. L. Epstein}>

$\downarrow$

{T, F}

M_2 $\quad L(\neg, \rightarrow, \wedge, \vee, \forall, \exists;\ P_0^1)$

$\downarrow$

$L(\neg, \rightarrow, \wedge, \vee, \forall, \exists;$ 'hates dogs')
universe: all people living in the 300 or 400 block of Pinecone in Cedar City

$\downarrow$

<all people living in the 300 or 400 block of Pinecone in Cedar City;
{R. L. Epstein's male neighbor to the west}>

$\downarrow$

{T, F}

The models M_1 and M_2 differ at the level of realization and at the level of interpretation, since the predicates have different extensions. Yet they are the same in the sense that they validate exactly the same formal wffs. Compare:

$$\mathsf{M}_3 = \langle \{3, 6, 8, 10\};\ \{x : x \text{ is prime}\} \rangle$$

$$\mathsf{M}_4 = \langle \{3, 6, 9, 15\};\ \{x : x \text{ is even}\} \rangle$$

Here we have different universes and different predicates in extension, yet the models are the same in the sense that we can map the universe of M_3 to the universe of M_4 via the function φ given by $\varphi(3) = 6$, $\varphi(6) = 3$, $\varphi(8) = 9$, and $\varphi(10) = 15$. Then we can map assignments of references σ on M_3 to assignments of references σ^{φ} on M_4 via $\sigma^{\varphi}(x) = \varphi(\sigma(x))$, and that preserves truth in a model: $\sigma \models A$ iff $\sigma^{\varphi} \models A$.

Just as all that matters about a predicate is its extension, so all that matters about the universe of a model is that it is a collection and the objects in that collection satisfy or do not satisfy various predicates. Abstracted that far, two models whose universes can be put into a correspondence that preserves truth in the models are just notational variants of each other, at least as far as what we're paying attention to in our reasoning. We'll look at that idea in more detail in Chapter XII after we've seen more examples of mathematical models.

To apply mathematics to the study of classical predicate logic with the goal of applying classical predicate logic to classical mathematics, we can abstract further. But to do that we need to examine our idea of what it means to be a collection.

Exercises for Section A

1. What is the difference between a model of type **I** and one of type **II**?
2. Give two very different type **I** models that each realize two predicates and one name that result in the same model of type **II**. What properties of those predicates and name seem important to reasoning that have to be ignored to identify these models?
3. Define 'the extension of a predicate' for a type **I** model.
4. Give the condition for a formal atomic wff to be true in a type **II** model.
5. a. What is the relation of the extensional evaluation of predicates and the assumption that all predicates are extensional made in Chapter IV?
   b. What is the relation of the extensional evaluation of functions and the assumption of the extensionality of functions in Chapter VIII.C?
6. Give two very different type **II** models, each of which realizes two predicates and one name, for which there is a correspondence of their universes that preserves truth. What properties of those predicates and name seem important to reasoning that have to be ignored to identify these models?

B. Collections as Objects: Naive Set Theory

Mathematicians reify collections. That is, they consider a collection to be a thing, no less real than my desk or the number 7. That is a major assumption. Using it, mathematicians have analyzed collections using classical predicate logic, formulating various formal theories of what it is to be a collection. But such formalizations cannot help us use collections as things for abstracting and clarifying the semantics of predicate logic. We cannot pull ourselves up by our own bootstraps.

Rather, we must start with some naive, that is, informal notion of what a collection is and how collections relate to one another. We shall establish some agreements about this naive conception of collections and use only those assumptions in what follows. Such an informal analysis is currently deemed essential to the study of the semantics of predicate logic if we wish to apply classical predicate logic to classical mathematics. But I emphasize that this is not a trivial step.

It is not even clear that we have a notion of collection independent of our notion of predicate. That is, a collection is those objects that are picked out by (satisfy) a particular predicate. We can view a collection as an abstraction of a predicate, what we get when we ignore all aspects of a predicate other than its arity and what it applies to. A collection is the classical abstraction of a predicate. Mathematicians reify abstractions, and there is no harm in doing so now, so long as we recall where the abstraction came from.

Platonists, of course, disagree. Just as they take predicates to be abstract objects only some of which can be correlated to linguistic predicates, so they say collections are abstract objects, only some of which can be named. This difference in views will be important later when we wish to use infinitistic methods in studying predicate logic.

To avoid a misconception, note that whatever a collection is, it is not a heap. That is, it is not just many things in proximity. Consider, for example, the collection consisting of Ralph, all canaries, and the Eiffel Tower, or the collection of all natural numbers. Only in rare cases can we point to a collection, and then we must describe it in some way if we are to reason with it, for example, the collection of all objects on my desk, or the collection of all cars in that parking lot. The issue of naming and pointing in specifying or identifying collections is one step more problematic than the question of naming and pointing for objects.

Let's now turn to the *agreements about collections* that are generally accepted by mathematicians. Collections under this guise will be called *sets*.

The simple question of what metavariables we shall use for sets raises an important point. If we were to use lowercase letters for objects and capitals for sets, for example, a is in A, we would seem to be implying that objects and collections are different. But the point is that sets are things. Hence, a set can be an object in another set. In this section I shall use only lowercase letters for sets. We write '$\in$' for 'is an object (element) in the collection of', for example, 'Juney $\in$ all dogs'. We write '$\notin$' for 'is not an element of'.

Here, then, are the assumptions we will make about collections viewed as sets.

1. $\mathsf{a} = \mathsf{b}$ iff (for every x, $x \in \mathsf{a}$ iff $x \in \mathsf{b}$).

This criterion of *identity* for sets corresponds to the classical abstraction of predicates.

2. Given a set a and (informal) unary predicate P, there is a set consisting of those objects in a that satisfy P. We write $\{x \in \mathsf{a} : \mathsf{P}(x)\}$ for that set.

This assumption is what we have been doing all along in establishing the universe of a model. Here we further assume that given any predicate it makes sense to ask whether it applies or does not apply to the objects in any collection. In particular, there is an *empty set*, $\varnothing = \{x : x \neq x\}$.

3. Given two sets, there is a set consisting of exactly those objects that are in at least one of the sets. We write $\mathsf{a} \cup \mathsf{b}$ for $\{x : x \in \mathsf{a} \text{ or } x \in \mathsf{b}\}$, the *union* of a and b.

This corresponds to the use of inclusive 'or' in joining predicates: If $x \in \mathsf{a}$ iff $P(x)$ is true, and $x \in \mathsf{b}$ iff $Q(x)$ is true, then $x \in \mathsf{a} \cup \mathsf{b}$ iff $P(x)$ or $Q(x)$ is true. Using this and assumption 2, we also have $\mathsf{a} \cap \mathsf{b} = \{x : x \in \mathsf{a} \text{ and } x \in \mathsf{b}\}$, the *intersection* of a and b.

4. For any two sets there is a set they both belong to.

We use this assumption to assimilate pairs to sets, and hence to treat them as things. Let $\{a, b\}$ be the unordered pair consisting of a and b. Then (a, b), the *ordered pair* of first a and then b, can be identified with the unordered pair $\{\{a\}, \{a, b\}\}$.

In order to have that the extensions of predicates are sets, we need that subsets of ordered n-tuples exist for any $n \geq 1$. A set b is a *subset* of a set a, written $b \subseteq a$, means that for every x, if $x \in b$, then $x \in a$. And $b \subset a$ means $b \subseteq a$ and $b \neq a$.

5. Given any set a there is a set b consisting of all the subsets of a, which we call the *power set* of a or pow(a).

The collection of all ordered pairs from a is an element of pow(pow(pow(a))), as I'll let you show. So by assumption 5, it is a set. Hence, the extension of a binary predicate P relative to a is an element of pow(pow(pow(a))), which is guaranteed to exist via assumption 2. We write $a \times b$ for the collection of ordered pairs whose first element is from a and whose second element is from b, and that is a subset of pow(pow(pow($a \cup b$)))). We define inductively: $a^1 = a$, $a^{n+1} = a \times a^n$.

6. The natural numbers form a set, $\mathbb{N}$.

This assumption of the existence of an infinite set is deemed essential by classical mathematicians. Though accepted by most other mathematicians, it is not necessarily accepted in conjunction with the assumption about the relation between $\forall$ and $\exists$. Note that nowhere in our development of predicate logic so far have we used or needed the existence of an infinite set. We needed only that we could extend the counting sequence 1, 2, 3, . . . whenever we wanted. For example, when we took as a universe all primes it was enough that we had criteria for whether an object was a prime and for distinguishing primes, and a method for generating primes.

Functions, as discussed in Chapter VIII, can be viewed as relations. Given an (informal) n-ary function f from a to b, which we write as $f: a \to b$, there is a (informal) corresponding predicate contained in pow($a^n \times b$), namely, P_f where $P_f(a_1, \ldots, a_n, b)$ iff $f(a_1, \ldots, a_n) = b$. So by assumption 2, $\{(a_1, \ldots, a_n, b) : f(a_1, \ldots, a_n) = b\}$ is a set. An *infinite sequence* $a_1, a_2, \ldots, a_n, \ldots$ is just a function from the natural numbers to a set a, where $f(n) = a_n$, so by assumption 6 the corresponding predicate determines a set.

Using functions we can use any set for indexing elements. Given a function $f: a \to d$, write b_c for f(c), and $\bigcup_{y \in a} b_y$ for $\{x : x \in b_y \text{ for some } y \in a\}$.

Given an n-ary function f from a to b, the *range* of f is the collection of all $c \in b$ such that for some $a_1, \ldots, a_n \in a$, $f(a_1, \ldots, a_n) = c$.

7. Given any sets a and b, and a function $f: a \to b$, the range of f is a set.

These assumptions need not take us from the conception of sets as abstractions of linguistic predicates. Even in assumption 5 we may say that the only subsets that exist are those that correlate to predicates. However, the next and final assumption does take us from the conception of sets as correlated to linguistic predicates.

8. Given a set e such that for every $\mathsf{a} \in \mathsf{e}$, $\mathsf{a} \neq \varnothing$, and for every $\mathsf{a}, \mathsf{b} \in \mathsf{e}$, $\mathsf{a} \cap \mathsf{b} = \varnothing$, there is a set d and a *choice function* $\mathsf{f} : \mathsf{e} \to \mathsf{d}$ such that for every $\mathsf{a} \in \mathsf{e}$, $\mathsf{f}(\mathsf{a}) \in \mathsf{a}$.

That is, given a set of sets, there is a function that picks out ("chooses") one element from each of those sets. No method need be specified for how each element is picked out. This is the informal *axiom of choice*.

I shall not give a course on naive set theory here. *Hausdorff, 1957* does an excellent job for our purposes; *Lipschutz, 1964* is a good introduction, too. For a more advanced exposition see *Fraenkel, Bar-Hillel, and Levy, 1973*. My goal here is only to try to make clear on what basis we will abstract models, listing not only the crucial assumptions, but also a few consequences of those.

Mappings of sets Let φ be a function from a set a to a set b.

φ is *one-to-one* ('1-1', an *injection*) iff
for every $\mathsf{a}_1, \mathsf{a}_2 \in \mathsf{a}$, if $\varphi(\mathsf{a}_1) = \varphi(\mathsf{a}_2)$ then $\mathsf{a}_1 = \mathsf{a}_2$.

φ is *onto* iff for every $\mathsf{c} \in \mathsf{b}$, there is some $\mathsf{d} \in \mathsf{a}$ such that $\varphi(\mathsf{d}) = \mathsf{c}$.

φ is a *correspondence* (*bijection*) iff it is both 1-1 and onto.

φ is the *identity function* on b iff for every $\mathsf{c} \in \mathsf{b}$, $\varphi(\mathsf{c}) = \mathsf{c}$.

Two sets a and b have the *same cardinality*, written 'card(a) = card(b)' or '$\mathsf{a} \approx \mathsf{b}$', iff there is a correspondence from a to b. This formalizes the idea of two sets having the same number of elements. The cardinality of a is *less than* the cardinality of b, written card(a) < card(b), iff there is a 1-1 mapping from a to b, but there does not exist a 1-1 mapping from b to a. This formalizes the idea of one set having fewer elements than another.

We can have a 1-1 mapping from a set to itself that is not onto, for example, $\mathsf{f}(n) = 2n$ on the natural numbers. That is characteristic of what we mean by a set being infinite: a is *infinite* iff there is a 1-1 function from a to itself that is not onto.

We write card(a) = n, where n is a natural number and $n > 0$, to mean that $\mathsf{a} \approx \{1, 2, \ldots, n\}$, and we write card($\varnothing$) = 0. We say a set a is *countable* (or *enumerable* or *denumerable*) if a is finite or $\mathsf{a} \approx \mathbb{N}$. If $\mathsf{a} \approx \mathbb{N}$, we write card($\mathsf{a}$) = $\aleph_0$ (read *aleph-nought*) and say that a is *countably infinite.* If φ is a correspondence from $\mathbb{N}$ to a, we say that φ *enumerates* a.

I'll now show that there are *uncountable* (not countable) sets.

The real numbers

What are the real numbers? We'll consider that question in Chapters XVII–XIX. But for now consider the real numbers x such that $0 < x < 1$ as sequences $.x_1 x_2 \ldots x_n \ldots$ where each $x_n \in \{0, 1, \ldots, 9\}$ and the decimal (i) does not consist solely of 0's and (ii) does not end in a tail of 9's (we need (ii) to get unique representations, since, for example, $1.299999\ldots = 1.300000\ldots$). Thus, each is a function $\mathsf{f}: \mathbb{N} \to \{0, 1, \ldots 9\}$ that satisfies (i) and (ii), and the collection of all those forms a set, $[0, 1]$, contained in pow($\mathbb{N} \times \{0, 1, \ldots 9\}$). In Exercise 11 I suggest how to give a bijection from $[0, 1]$ to the collection of all real numbers. So by assumption 7, the collection of all real numbers, $\mathbb{R}$, is a set, too. We write card($\mathbb{R}$) = c. Now we can show that $\mathbb{R}$ is not countable by showing that $[0, 1]$ is not countable. If there were a correspondence φ from $\mathbb{N}$ to $[0, 1]$, we could define a real number by *diagonalizing* the set of $\varphi(n)$, setting $b = .b_1 b_2 \ldots b_n \ldots$, where $b_n = 1$ if the nth decimal place of $\varphi(n)$ is not 1, and 2 otherwise. Then for some n, $\varphi(n) = b$. But that cannot be, since b differs from $\varphi(n)$ in the nth decimal place. Hence, no such φ exists.

The following theorem puts these observations into a more general settting.

Theorem 1

a. If $\mathsf{b} \subseteq \mathsf{a}$, then card($\mathsf{b}$) $\leq$ card(a).

b. For every a, card(a) < card(pow(a)).

c. There is no set that contains all sets.

Proof (a) The identity function is a 1-1 mapping from b to a.

(b) Suppose there were a correspondence ψ from a to pow(a). Let $\mathsf{b} = \{\mathsf{c} : \mathsf{c} \in \mathsf{a}$ and $\mathsf{c} \notin \psi(\mathsf{c})\}$. By assumption 2, $\mathsf{b} \in$ pow(a). Hence, for some $\mathsf{d} \in \mathsf{a}$, $\psi(\mathsf{d}) = \mathsf{b}$. So we have $\mathsf{d} \in \psi(\mathsf{d})$ iff $\mathsf{d} \in \{\mathsf{c} : \mathsf{c} \in \mathsf{a}$ and $\mathsf{c} \notin \psi(\mathsf{c})\}$ iff $\mathsf{d} \notin \psi(\mathsf{d})$, a contradiction. Hence, there is no such ψ.

(c) Suppose there were a set U that contained all sets. Then in particular, since pow(U) is a set, pow(U) $\subseteq \mathsf{U}$. But then by (a), card(pow(U)) $\leq$ card(U), which is a contradiction. So there is no set of all things. ∎

If we assume (1)–(8) and take the universe U of a model to be a set, then we cannot take U to be *all things*. All things would have to include all sets, and by Theorem 1.c there is no set that contains all sets. Viewing collections as things according to assumptions (1)–(8) is incompatible with the view that logic is about all things in the sense that there is one model that will suffice for logic, the model of what there is.

In later chapters we'll need some other observations about the sizes of sets that are given in the exercises below.

Exercises for Section B

1. a. Give a mapping from the $\mathbb{N}$ to $\mathbb{N}$ that is 1-1 but not onto.
   b. Give a mapping from $\mathbb{N}$ to $\mathbb{N}$ that is onto but not 1-1.
   c. Give a mapping of $\mathbb{N}$ to $\mathbb{N}$ that is 1-1 and onto, but is not the identity map.
   d. Give a correspondence from the collection of all primes to $\mathbb{N}$.

2. Show that $\approx$ is an equivalence relation (see Chapter VI.D).

3. Show that the following are denumerable:
   a. The collection of all odd natural numbers.
   b. The integers.
   c. The collection of all numbers divisible by 7.

4. Given a set a, show that the collection of all ordered pairs from a is an element of pow(pow(pow(a))).

5. Show that the following are countable.
   a. Any subset of a countable set.
   b. The set of all pairs of natural numbers (including zero).
   (Hint: Consider the function $J(m, n) = {}^1\!/_2\,[(m+n)(m+n+1)] + m$.
   Show that J is a correspondence by showing that $J(m, n) =$ the number of pairs (x, y) such that $x + y < m + n$ or $(x + y = m + n$ and $x < m)$.)
   c. The set of all n-tuples of natural numbers for a fixed number n.
   (Hint: Use part (b) and induction on n.)
   d. The union of two countable sets.
   (Hint: Divide the natural numbers into evens and odds, then map the evens to one set, the odds to the other.)
   e. The countable union of countable sets, $\mathsf{a} = \bigcup_{n \in \mathbb{N}} \mathsf{a}_n$ (also written $\bigcup_n \mathsf{a}_n$) where each a_n is countable.
   (Hint: Make explicit a dovetailing procedure to enumerate a: take the first element of a_1, then the first element of a_2, then the second element of a_1, then the second element of a_2, then the first element of a_3, then the third element of $\mathsf{a}_1, \ldots$.)
   f. The set of all n-tuples of natural numbers for all n.

6. *The closure of a set under a set of functions*
   Let F be any set of functions from c to c. Let $\mathsf{a} \subseteq \mathsf{c}$. The *closure of* a *under* F is the smallest set $\mathsf{b} \subseteq \mathsf{c}$ such that $\mathsf{a} \subseteq \mathsf{b}$ and for every $\mathsf{g} \in \mathsf{F}$, if $\mathsf{a}_1, \ldots, \mathsf{a}_k \in \mathsf{b}$, then $\mathsf{g}(\mathsf{a}_1, \ldots, \mathsf{a}_k) \in \mathsf{b}$. That is:

   $$\mathsf{b} = \bigcup \{\mathsf{e} : \text{for every } \mathsf{g} \in \mathsf{F}, \text{ if } \mathsf{a}_1, \ldots, \mathsf{a}_k \in \mathsf{e}, \text{ then } \mathsf{g}(\mathsf{a}_1, \ldots, \mathsf{a}_k) \in \mathsf{e}\}$$

   a. What are the closures of the following sets under the given functions?
   i. In $\mathbb{N}$, $\{2, 4\}$ under $+$.
   ii. In $\mathbb{N}$, $\{1, 2\}$ under $+$.
   iii. In $\mathbb{N}$, $\{0, 2\}$ under $+$.
   iv. In $\mathbb{N}$, $\{0, 1\}$ under $+$.
   v. In $\mathbb{N}$, $\{1, 3, 9\}$ under $\cdot$.
   vi. In $\mathbb{N}$, $\{0, 1, 2, 3, 9\}$ under $\cdot$.
   vii. In $\mathbb{N}$, $\{2, 3\}$ under $+$ and $\cdot$.
   viii. In the integers, $\{2, 3\}$ under $+$ and $-$.
   ix. In the integers, $\{2, 4\}$ under $+$ and $-$.
   x. In $\mathbb{R}$, $\{1, 2, 3\}$ under $\cdot$ and $\sqrt{\ }$.

   b. Consider the inductive definition:

$$\mathsf{e}_0 = \mathsf{a}$$
$$\mathsf{e}_{n+1} = \mathsf{e}_n \cup \{\mathsf{g}(\mathsf{a}_1, \ldots, \mathsf{a}_k) \colon \mathsf{a}_1, \ldots, \mathsf{a}_k \in \mathsf{e}_n \text{ and } \mathsf{g} \in \mathsf{F}\}$$
$$\mathsf{e} = \bigcup_n \mathsf{e}_n$$

i. Show that e = the closure of a under F

ii. Show that if a and F are countable, then the closure of a under F is countable.

iii. Give a definition of the collection of all wffs of the propositional language as the smallest set closed under the connectives viewed as functions.

7. *The rationals,* $\mathbb{Q}$
The rationals are a subset of ordered pairs of natural numbers. Show directly that the set of all rational numbers, $\mathbb{Q}$, is countable.
(Hint: Consider this picture of *Cantor's tour of the rationals*:

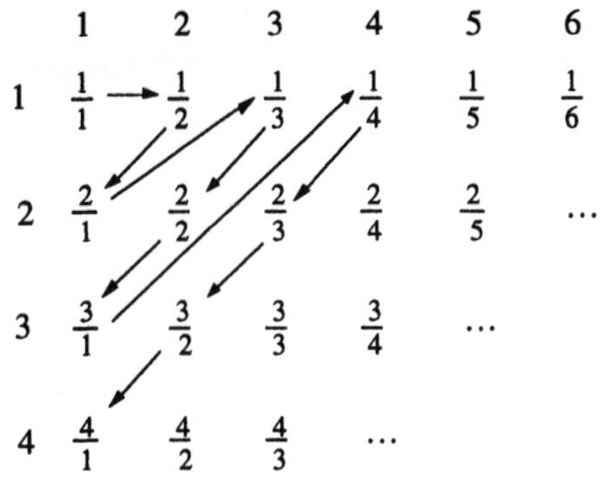

We can follow the path, skipping any fraction we've come to before, thus enumerating the positive rationals (compare the function in Exercise 5.b). Then we can do the same with the negative rationals. Show that we can combine the mappings by putting one on the evens bigger than 0, one on the odds, and mapping 0 to 0. See *Epstein and Carnielli, 2000* for a formal proof.)

8. *Algebraic numbers over the rationals*
A real number is algebraic over the rationals iff it is satisfies an equation $\mathsf{a}_0 + \mathsf{a}_1 x + \cdots + \mathsf{a}_n x^n = 0$ where each a_i is rational and some $i > 0$, $\mathsf{a}_i \neq 0$.
Each such equation has at most n solutions. Show that the set of numbers algebraic over the rationals is countable. (Hint: Use Exercise 5.e and 7.)

9. Show that $[0, 1) \approx \mathbb{R}$. (Hint: Consider the function φ defined by $\varphi(1/2) = 0$; $\varphi(x) = 2x/1{-}2x$ for $0 < x < 1/2$; and $\varphi(x) = 2x{-}2/{-}2x{-}1$ for $1/2 < x < 1$.)

10. Show how we can define $[0, 1)$ via binary expansions, that is, sequences all of whose terms come from $\{0, 1\}$.

11. a. $\mathbb{R}^2$ is the set of ordered pairs of real numbers. Show that $\mathbb{R}^2 \approx \mathbb{R}$.
(Hint: From any binary representation of a real number in $[0, 1)$ derive binary representations of two real numbers by taking every other digit in the sequence.)
b. Show that for every $n \geq 1$, $\mathbb{R}^n \approx \mathbb{R}$. (Hint: Use part (a) and induction on n.)
c. Given sets a_n such that for each n, $\mathsf{a}_n \approx \mathbb{R}$, show that $\bigcup_{n \geq 1} \mathsf{a}_n \approx \mathbb{R}$.
(Hint: Split $[0, 1)$ into countably many parts and proceed as in (a).)

12. Show that given any countable set a there is a set $\mathsf{b} \supseteq \mathsf{a}$ such that b has cardinality c.
(Hint: Take a correspondence from a to $\mathbb{N}$ and then a function from $\mathbb{N}$ to $\mathbb{R}$.
"Add" to a all the elements of $\mathbb{R}$ that are "left over" from that correspondence.)

<u>13</u>. Compare the proof of Theorem 1.b to the liar paradox (Chapter IV.D).

C. Classical Mathematical Models

Taking the extension of a predicate to be a set, we can further abstract the notion of a model of classical predicate logic. A *classical mathematical model* of the language $L(\neg, \rightarrow, \wedge, \vee, \forall, \exists, =;\ P_i^n, f_i^n, c_i)$ is:

(M) $$\mathsf{M} = \langle \mathsf{U};\ \mathsf{P}_0^1, \mathsf{P}_1^1, \ldots, \mathsf{P}_i^n, \ldots, \mathsf{f}_0^1, \mathsf{f}_1^1, \ldots, \mathsf{f}_i^n, \ldots, \mathsf{a}_0, \mathsf{a}_1, \ldots \rangle$$

1. U is a nonempty set.
2. For each $i \geq 0$, $n \geq 1$, P_i^n is a set and $\mathsf{P}_i^n \subseteq \mathsf{U}^n$.
3. For each $i \geq 0$, $\mathsf{a}_i \in \mathsf{U}$.
4. For each $i \geq 0$, $n \geq 1$, f_i^n is a function from U^n to U.

An *assignment of references* is a function σ from terms of the language to U satisfying:

1. For each i, $\sigma(c_i) = \mathsf{a}_i$.
2. For each i, for any terms $t_1, \ldots, t_n$,
   $\sigma(f_i^n(t_1, \ldots, t_n)) = \mathsf{f}_i^n(\sigma(t_1), \ldots, \sigma(t_n))$.

We define inductively 'σ *satisfies* A' as we did previously (Chapter IV.E.5) except that for atomic wffs we set:

$\sigma \vDash t_1 = t_2$ iff $\sigma(t_1) = \sigma(t_2)$

$\sigma \vDash P_i^n(t_1, \ldots, t_n)$ iff $(\sigma(t_1), \ldots, \sigma(t_n)) \in \mathsf{P}_i^n$

I'll often write $\mathsf{P}_i^n(\mathsf{b}_1, \ldots, \mathsf{b}_n)$ for $(\mathsf{b}_1, \ldots, \mathsf{b}_n) \in \mathsf{P}_i^n$.

All we've established about models in previous chapters continues to hold of these abstract versions.

What we've done so far can be seen as simplifying and abstracting from our original analysis of reasoning in ordinary language, for we could always trace back to type **I** models. The next step takes us into *mathematical logic.*

The fully general classical abstraction of predicate logic models
Any structure satisfying the conditions at (M) is a model of the formal language.

We no longer require that a classical mathematical model arises as an abstraction of a type **I** or even a type **II** model. Any structure will do to interpret the formal symbols. All connection between the formal symbols and reasoning in ordinary language can be cut. The motive for making this abstraction is a belief that a set can exist independently of any possibility of its being the extension of a linguistic predicate.

The fully general classical abstraction of models allows us to treat logic, including models and the notion of truth, as mathematical subjects, ones about which we can reason without regard to their original motivation, paying attention to structure only. It is a powerful abstraction and lies at the heart of what is now called 'mathematical logic'.

However, its introduction into logic must be done with some care, for it can lead us to sever logic from reasoning. It is not clear that a notion of set divorced from that of the extension of a linguistic predicate can be used in defining models and still yield a good formalization of the notion of validity. Perhaps we have too many models now, since we could have ones that correlate to no type **I** models. Perhaps we have too few models, since some collections of objects we might want to use as a universe are not sets. If too many, then informally valid inferences could be classified as invalid; if too few, then invalid inferences might be classified as valid. Both *Kreisel, 1967* and *Shapiro, 1991* discuss this point, which I'll return to in Chapter X.A.5.

In much of what follows the fully general abstraction will not be needed. I will try to point out key places where we seem to need it to establish a metalogical result. However, dealing with mathematical models will simplify much of what we wish to prove about the logic we have established, and I'll make the assumption from this point on that, unless I state the contrary, we'll be dealing with mathematical models. That we can no longer consider the collection of all things to be a universe of a model seems no great loss, for, as I discussed earlier (pp. 79 and 117), it is not clear what criterion of identity we could use for such a universe, and hence how we could reasonably talk of assignments of reference.

Exercises for Section C

1. Show that the collection of all assignments of references in a mathematical model is a set and is complete.
2. What do we mean by the notation $\mathsf{P}_i^n(\mathsf{b}_1, \ldots, \mathsf{b}_n)$?
3. Show that the applications of predicates and functions in a classical mathematical model are extensional.
4 Explain how using the fully general classical mathematical abstraction of models could sever the study of logic from reasoning in ordinary language.
5. *Shoenfield, 1967,* uses a different definition of truth in an (abstract) model, dispensing with the notion of a valuation relative to an assignment of references.

   Let $\mathsf{M} = \langle \mathsf{U}\,;\, \mathsf{P}_0^1, \mathsf{P}_1^1, \mathsf{P}_1^2, \ldots, \mathsf{P}_i^n, \ldots, \mathsf{a}_0, \mathsf{a}_1, \ldots \rangle$ be a structure as in the definition of classical mathematical model above (M). Shoenfield expands the formal language to include new name symbols $\{c_u : u \in \mathsf{U}\}$, regardless of whether U is countable. Then he sets:

   $m(t) = \mathsf{a}_i$ if $t = c_i$ and $m(t) = u$ if $t = c_u$

He then inductively defines truth in this model for *closed* formulas of the expanded language:

$\mathsf{M} \models P_i^n(t_1, \ldots, t_n)$ iff $(\mathsf{m}(t_1), \ldots, \mathsf{m}(t_n)) \in \mathsf{P}_i^n$.

The usual definitions for $\neg$, $\rightarrow$, $\wedge$, $\vee$.

$\mathsf{M} \models \exists x\, A$ iff for some name symbol c, $\mathsf{M} \models A(c/x)$.

$\mathsf{M} \models \forall x\, A$ iff $\mathsf{M} \models \neg \exists x \neg A$.

Then a closed wff A in the original language is true in this model iff viewed as part of the expanded language it is true in this model.

Show that for every closed A in the usual first-order language, and for every model M, $\mathsf{M} \models A$ in Shoenfield's sense iff $\mathsf{M} \models A$ by the definition for classical mathematical models given in this section.

6. *Mates, 1965,* uses a different definition of truth in an (abstract) model, also dispensing with valuations relative to assignments of reference. His definition is for any language with a countable infinity of name symbols. Given a structure M as at (M) above and c a name symbol, a structure N is a *c-variant* of M iff M and N differ at most in what object is assigned to c. For a name a, write $\mathsf{m}(a)$ for the object assigned to a by M. Mates defines truth in a model M inductively for closed formulas via:

$\mathsf{M} \models P_i^n(a_1, \ldots, a_n))$ iff $(\mathsf{m}(a_1), \ldots, \mathsf{m}(a_n)) \in \mathsf{P}_i^n$.

The usual definitions for $\neg$, $\rightarrow$, $\wedge$, $\vee$.

$\mathsf{M} \models \forall x\, A = \mathsf{T}$ iff $\mathsf{N} \models A(c/x)$ for every c-variant N of M where c is the first name symbol not appearing in A.

$\mathsf{M} \models \exists x\, A = \mathsf{T}$ iff $\mathsf{N} \models A(c/x)$ for some c-variant N of M where c is the first name symbol not appearing in A.

Show that for every closed A in the usual first-order language, and for every model M, $\mathsf{M} \models A$ in Mates' sense iff $\mathsf{M} \models A$ by the definition given for classical mathematical models in this section.

In Chapter IV.M of *Predicate Logic* I discuss why neither Shoenfield's nor Mates' definition of truth in a model is apt for a foundation for the formalization of reasoning.

X Axiomatizing Classical Predicate Logic

In this chapter we'll see how to axiomatize the valid wffs and the consequence relation of classical mathematical logic. We defined the general notion of syntactic consequence in Chapter II.C.2 and derived some properties of that. Here we will give a specific axiom system.

Again, we start with an idea of how to prove completeness and then add axioms that allow that proof. But rather than add as an axiom every wff that's needed, I'll establish syntactically that certain wffs are theorems (Section A.2). In some cases these will be only sketches of how we can give a derivation, relying on what we've proved earlier about the axiom system. You may wish to go first to Section A.3 to see the outline of the completeness proof.

A. An Axiomatization of Classical Predicate Logic

1. The axiom system

The axioms will be presented as schema in the language L which is $L(\neg, \rightarrow, \forall, \exists; P_0, P_1, \ldots, c_0, c_1, \ldots)$. That is, I will set out formal schema of wffs, each instance of which is meant to be taken as an axiom.

I. Propositional axioms

The axiom schema of **PC** in $L(\neg, \rightarrow)$, where A, B, C are replaced by wffs of L and the universal closure is taken:

$\forall \ldots \neg A \rightarrow (A \rightarrow B)$

$\forall \ldots B \rightarrow (A \rightarrow B)$

$\forall \ldots (A \rightarrow B) \rightarrow ((\neg A \rightarrow B) \rightarrow B)$

$\forall \ldots (A \rightarrow (B \rightarrow C)) \rightarrow ((A \rightarrow B) \rightarrow (A \rightarrow C))$

II. Axioms governing ∀

1. $\forall \ldots (\forall x\, (A \rightarrow B) \rightarrow (\forall x\, A \rightarrow \forall x\, B))$ — *distribution of* ∀

2. When x is not free in A,

a. $\forall \ldots (\forall x\, A \rightarrow A)$ — *superfluous quantification*

b. $\forall \ldots (A \rightarrow \forall x\, A)$

3. $\forall \ldots (\forall x \forall y\, A \rightarrow \forall y \forall x\, A)$ — *commutativity of* ∀

4. When term t is free for x in A,

$\forall \ldots (\forall x\, A(x) \rightarrow A(t/x))$ — *universal instantiation*

III. Axioms governing the relation between ∀ and ∃

5. a. $\forall \ldots (\exists x\, A \rightarrow \neg \forall x \neg A)$

b. $\forall \ldots (\neg \forall x \neg A \rightarrow \exists x\, A)$

Rule $\dfrac{A,\ A \rightarrow B}{B}$ A and B closed formulas — *modus ponens*

In this system only closed wffs are theorems. Hence, I'll use the Greek capitals Γ, Δ, Σ with or without subscripts for collections of closed wffs only, and if $\Gamma \vdash A$ then A is closed, too. I'll write $\Gamma, A \vdash B$ for $\Gamma \cup \{A\} \vdash B$. I'll write *Axiom* to mean 'axiom scheme'.

Note that axiom scheme 4 includes $\forall \ldots (\forall x\, A(x) \rightarrow A(x))$, since x is free for itself. It also includes $\forall \ldots (\forall x\, A(x) \rightarrow A(c/x))$ for every name symbol c.

As for propositional logic, we define for closed wffs:

Γ is *consistent* iff for every A, either $\Gamma \nvdash A$ or $\Gamma \nvdash \neg A$.

Γ is *complete* iff for every A, either $\Gamma \vdash A$ or $\Gamma \vdash \neg A$.

Γ is a *theory* iff for every A, if $\Gamma \vdash A$ then $A \in \Gamma$.

Theorem II.3 (p. 38) on syntactic consequence relations continues to hold, and Theorem II.6 (p. 41) about collections that are complete, consistent, and/or theories also applies. In particular, we have the following lemma.

Lemma 1

$\Gamma \cup \{A\}$ is consistent iff $\Gamma \nvdash \neg A$.

$\Gamma \cup \{\neg A\}$ is consistent iff $\Gamma \nvdash A$.

Γ is a complete and consistent theory iff for every closed A, exactly one of A, $\neg$A is in Γ.

Theorem 2 Soundness If $\Gamma \vdash A$, then $\Gamma \models A$.

Proof This will follow if every axiom is true in every model, since the rule preserves truth in a model.

The axioms in Group I are true by Theorem V.7.
The axioms in scheme 1 are true by Corollary V.9.a.
The axioms in schema 2.a and 2.b are true by Theorem V.1 and Theorem IV.4.
The axioms in scheme 3 are true by Theorem V.2 and Theorem IV.4.
The axioms in scheme 4 are true by Theorem V.13.
The axioms in schema 5.a and 5.b are true by Theorems IV.3 and IV.4. ■

2. Some syntactic observations

Lemma 3 The syntactic deduction theorem

a. $\Gamma, A \vdash B$ iff $\Gamma \vdash A \rightarrow B$.

b. $\Gamma \cup \{A_1, \ldots, A_n\} \vdash B$ iff $\Gamma \vdash A_1 \rightarrow (A_2 \rightarrow (\cdots \rightarrow (A_n \rightarrow B) \cdots))$.

Proof Since A, B, and all wffs in Γ are closed, the proof is as for **PC** (p. 41). ■

Lemma 4 If $\vdash A \rightarrow B$ and $\vdash B \rightarrow C$, then $\vdash A \rightarrow C$.

Proof We have $A \vdash B$ and $B \vdash C$. Hence, by Theorem II.4.f (the cut rule), $A \vdash C$. So by Lemma 3, $\vdash A \rightarrow C$. ■

Lemma 5 Let A be any wff and $y_1, \ldots, y_n$ a list of variables in alphabetic order that contains but is not necessarily limited to all variables free in A. Then:

$\vdash \forall y_1 \ldots \forall y_n\, A \rightarrow \forall \ldots A$

$\vdash \forall \ldots A \rightarrow \forall y_1 \ldots \forall y_n\, A$

Proof If each of $y_1, \ldots, y_n$ is free in A, we are done. So let $y_{i_1}, \ldots y_{i_m}$ be the list of the variables free in A in alphabetic order, where $n > m$, and we induct on n. The base is $n = 2$. Suppose y_1 is free in A, y_2 is not free in A. Then:

$\vdash \forall y_1\, A \rightarrow \forall y_2 \forall y_1\, A$	Axiom 2.b
$\vdash \forall y_2 \forall y_1\, A \rightarrow \forall y_1 \forall y_2\, A$	Axiom 3
$\vdash \forall y_1\, A \rightarrow \forall y_1 \forall y_2\, A$	Lemma 4

I'll leave the other two cases when $n = 2$ to you.

Now suppose the theorem is true for $< n$ variables, and let y_k be the first in the list of variables that is not free in A. We have:

1. $\vdash \forall y_1 \ldots \forall y_{k-1} (\forall y_k \forall y_{k+1} \ldots \forall y_n \mathrm{A} \rightarrow \forall y_{k+1} \ldots \forall y_n \mathrm{A})$ Axiom 2
2. $\vdash \forall y_1 \ldots \forall y_{k-1} \forall y_k \forall y_{k+1} \ldots \forall y_n \mathrm{A} \rightarrow \forall y_1 \ldots \forall y_{k-1} \forall y_{k+1} \ldots \forall y_n \mathrm{A}$ Axiom 1, (1), and Lemma 4
3. $\vdash \forall y_1 \ldots \forall y_{k-1} \forall y_{k+1} \ldots \forall y_n \mathrm{A} \rightarrow \forall \ldots \mathrm{A}$ induction
4. $\vdash \forall y_1 \ldots \forall y_{k-1} \forall y_k \forall y_{k+1} \ldots \forall y_n \mathrm{A} \rightarrow \forall \ldots \mathrm{A}$ (2), (3), and Lemma 4

I'll leave the other direction to you. ■

Lemma 6 Generalized modus ponens $\{\forall \ldots (\mathrm{A} \rightarrow \mathrm{B}), \forall \ldots \mathrm{A}\} \vdash \forall \ldots \mathrm{B}$.

Proof Let $(\forall \ldots)_1$ be the quantifications in the closure of $\mathrm{A} \rightarrow \mathrm{B}$. That is, $\forall \ldots (\mathrm{A} \rightarrow \mathrm{B})$ is $(\forall \ldots)_1 (\mathrm{A} \rightarrow \mathrm{B})$

1. $(\forall \ldots)_1 (\mathrm{A} \rightarrow \mathrm{B}) \vdash (\forall \ldots)_1 \mathrm{A} \rightarrow (\forall \ldots)_1 \mathrm{B}$ Axiom 1
2. $\forall \ldots \mathrm{A} \vdash (\forall \ldots)_1 \mathrm{A}$ Lemma 5
3. $\{(\forall \ldots)_1 (\mathrm{A} \rightarrow \mathrm{B}), \forall \ldots \mathrm{A}\} \vdash (\forall \ldots)_1 (\mathrm{B})$ (1), (2), and Theorem II.4.f
4. $(\forall \ldots)_1 \mathrm{B} \vdash \forall \ldots \mathrm{B}$ Lemma 5
5. $\{(\forall \ldots)_1 (\mathrm{A} \rightarrow \mathrm{B}), \forall \ldots \mathrm{A}\} \vdash \forall \ldots \mathrm{B}$ (3) and (4) ■

Recall from Chapter V.B (p. 104) that we can talk about a propositional form of a predicate logic wff.

Lemma 7 PC in predicate logic
If A has the form of a **PC**-tautology, then $\vdash \forall \ldots \mathrm{A}$.

Proof Suppose that one form of A is the **PC**-tautology B. By the completeness theorem for **PC**, there is a proof $\mathrm{B}_1, \ldots, \mathrm{B}_n = \mathrm{B}$ in **PC**. Inspecting the proof of that completeness theorem, you'll see that we can assume there is no propositional variable appearing in any of $\mathrm{B}_1, \ldots, \mathrm{B}_n$ that does not also appear in B. Let $\mathrm{B}_i{}^*$ be B_i with propositional variables replaced by predicate wffs just as they are replaced in B to obtain A. I'll show by induction on i that $\vdash \forall \ldots \mathrm{B}_i{}^*$. The lemma will then follow for $i = n$.

For $i = 0$, B_i is a **PC**-axiom, so it follows by the axioms of Group I. Suppose now that for all $j < i$, $\vdash \forall \ldots \mathrm{B}_j{}^*$. If B_i is a **PC**-axiom we are done. If not, then for some $j, k < i$, B_j is $\mathrm{B}_k \rightarrow \mathrm{B}_i$. Note that $(\mathrm{B}_k \rightarrow \mathrm{B}_i)^*$ is $\mathrm{B}_k{}^* \rightarrow \mathrm{B}_i{}^*$. So by induction we have $\vdash \forall \ldots \mathrm{B}_k{}^*$ and $\vdash \forall \ldots (\mathrm{B}_k{}^* \rightarrow \mathrm{B}_i{}^*)$. Hence by generalized *modus ponens*, $\vdash \forall \ldots \mathrm{B}_i{}^*$. ■

When invoking Lemma 6 or Lemma 7 I'll usually just say 'by **PC**', sometimes indicating the **PC**-wff. Note in particular we have:

Transitivity of $\rightarrow$ $\{\forall\ldots(A\rightarrow B),\ \forall\ldots(B\rightarrow C)\} \vdash \forall\ldots(A\rightarrow C)$

Lemma 8

a. $\vdash\forall\ldots(A(c/x) \rightarrow \exists x\, A(x))$ $\exists$-*introduction*

b. $\vdash\forall\ldots(\forall x\, A(x) \rightarrow \exists x\, A(x))$ $\forall$ *implies* $\exists$

Proof (a) By Axiom 4, $\vdash\forall\ldots(\forall x \neg A(x) \rightarrow \neg A(c/x))$. So by **PC** (transposition and double negation), $\vdash\forall\ldots(A(c/x) \rightarrow \neg\forall x \neg A(x))$. So by axiom 5.b and **PC**, $\vdash\forall\ldots(A(c/x) \rightarrow \exists x\, A(x))$.

(b) We have by Axiom 4, $\vdash\forall\ldots(\forall x\, A(x) \rightarrow A(c/x))$ for the first name symbol in the language. So by (a) and the transitivity of $\rightarrow$, we have (b). ■

It is difficult to show syntactically that if $\vdash\forall\ldots A$ then any permutation of the order of the variables in the universal closure also yields a theorem. Fortunately, all we need here is the following.

Lemma 9 If $y_1 \ldots y_n$ are the variables free in A in alphabetic order, then:
$\vdash\forall\ldots A \rightarrow \forall y_k\, \forall y_1 \ldots \forall y_{k-1}\, \forall y_{k+1} \ldots \forall y_n\, A$.

Proof We induct on n. The case $n = 2$ is Axiom 2. Suppose the lemma true for fewer than n variables. If $k \neq n$, then we are done by induction (replacing A by $\forall y_n$ A). If $k = n$, then by Axiom 2, Axiom 1, and Lemma 4:

$$\vdash\forall y_1 \ldots \forall y_{n-2}\,(\forall y_{n-1}\, \forall y_n\, A) \rightarrow \forall y_1 \ldots \forall y_{n-2}\,(\forall y_n\, \forall y_{n-1}\, A)$$

By induction we have:

$$\vdash\forall y_1 \ldots \forall y_{n-2}\, \forall y_n\,(\forall y_{n-1}\, A) \rightarrow \forall y_n\, \forall y_1 \ldots \forall y_{n-2}\,(\forall y_{n-1}\, A)$$

Hence, by Lemma 4, we are done. ■

The following lemma justifies that our logic does not distinguish any individual constant: What holds of a name symbol in all models must hold of any object.

Lemma 10 Let $B(x)$ be a formula with one free variable x.

a. If $\vdash B(c/x)$, then $\vdash\forall x\, B(x)$.

b. If $\Gamma\vdash B(c/x)$ and c does not appear in any wff in Γ, then $\Gamma\vdash\forall x\, B(x)$.

Proof (a) We proceed by induction on the length of a proof of $B(c/x)$. If the length of a proof is 1, then $B(c/x)$ is an axiom. I will show that (a) holds for any instance of Axiom 2.a; the other schema follow similarly.

If $B(c/x)$ is an instance of scheme 2.a, then it is:

$$\vdash(\forall\ldots)_1\,(\forall y\, A(c/x) \rightarrow A(c/x))$$

Another instance of Axiom 2.a is: $\vdash(\forall\ldots)_2\,(\forall y\, A(x) \rightarrow A(x))$. The only difference between $(\forall\ldots)_2$ and $(\forall\ldots)_1$ is that $\forall x$ appears in the former and not the latter. By Lemma 9, we have:

$$\vdash [(\forall \ldots)_2 (\forall y\, A(x) \rightarrow A(x))] \rightarrow [\forall x\, (\forall \ldots)_1 (\forall y\, A(x) \rightarrow A(x))]$$

Hence, by Lemma 4, we have $\forall x\, (\forall \ldots)_1 (\forall y\, A(x) \rightarrow A(x))$, as desired.

Suppose now that (a) is true for theorems with proofs of length m, $1 \leq m < n$, and the shortest proof of $B(c/x)$ has length n. Then for some closed A, $\vdash A$ and $\vdash A \rightarrow B(c/x)$, both of which have proofs shorter than length n. By induction, $\vdash \forall x\, (A \rightarrow B(x))$, so by Axiom 1, Axiom 2, and **PC**, $\vdash A \rightarrow \forall x\, B(x)$. Hence, $\vdash \forall x\, B(x)$.

(b) Suppose $\Gamma \vdash B(c/x)$. Then for some closed $D_1, \ldots, D_n \in \Gamma$, $\{D_1, \ldots, D_n\} \vdash B(c/x)$. Hence by the syntactic deduction theorem:

$$\vdash D_1 \rightarrow (D_2 \rightarrow \cdots \rightarrow (D_n \rightarrow B(c/x)) \ldots)$$

Since c does not appear in any of $D_1, \ldots, D_n$, we have by (a):

$$\vdash \forall x\, (D_1 \rightarrow (D_2 \rightarrow \cdots \rightarrow (D_n \rightarrow B(x)) \ldots))$$

So by Axiom 1 (repeated if necessary):

$$\vdash \forall x\, D_1 \rightarrow (\forall x\, D_2 \rightarrow \cdots \rightarrow (\forall x\, D_n \rightarrow \forall x\, B(x)) \ldots)$$

By Axiom 2, for each i, since D_i is closed, $\vdash D_i \rightarrow \forall x\, D_i$. By repeated use of *modus ponens*, we have $\{D_1, \ldots, D_n\} \vdash \forall x\, B(x)$. So $\Gamma \vdash \forall x\, B(x)$. ■

3. Completeness of the axiomatization

We'd like to proceed as in the proof for **PC**. Given a consistent Γ we can expand it to a complete and consistent $\Sigma \supseteq \Gamma$, as we did there. But to show that Γ has a model, we need a universe. If Γ is consistent, what objects will satisfy Γ?

The answer is in the language: the name symbols $c_0, c_1, \ldots$. But there's a problem. Suppose we let the universe of the model be $\{c_0, c_1, \ldots\}$. Consider:

$$\{\neg P_0(c_0, c_0), \neg P_0(c_1, c_0), \ldots, \neg P_0(c_n, c_0), \ldots, \exists x\, P_0(x, c_0)\}$$

This collection of wffs is consistent since the integers with P_0 interpreted as '<' is a model of it if we interpret c_i as the number i. But if we were to take as universe $\{c_0, c_1, \ldots\}$ we'd lack an element to make the last wff true.

So we add extra name symbols, $v_0, v_1, \ldots$ to our language to instantiate the existential formulas in Γ. That is, if $\exists x\, B(x) \in \Gamma$, we'll make sure that for some v_i, $B(v_i/x) \in \Sigma$. Then we can take the universe of a model for Σ to be the collection of name symbols $\{c_0, c_1, \ldots\} \cup \{v_0, v_1, \ldots\}$.

Formally, let $L(v_0, v_1, \ldots)$ be L augmented by the name symbols $v_0, v_1, \ldots$. That is, we add to the vocabulary of L name symbols $v_0, v_1, \ldots$ and define the collection of wffs inductively as before. We take our axiom schema to refer to this expanded language; all that we have proved up to this point continues to hold.

Theorem 11 Let Γ be a consistent set of closed wffs of L. Then there is a collection of closed wffs Σ in $L(v_0, v_1, \ldots)$ such that:

a. $\Gamma \subseteq \Sigma$.

b. Σ is a complete and consistent theory.

c. If $\exists x\, B \in \Sigma$ and x is free in B, then for some m, $B(v_m/x) \in \Sigma$.

d. For every wff $B(x)$ in $L(v_0, v_1, \ldots)$ with one free variable, if for each i, $B(v_i/x) \in \Sigma$, then $\forall x\, B(x) \in \Sigma$.

Proof Let $A_0, A_1, \ldots$ be a numbering of all the closed wffs of the expanded language $L(v_0, v_1, \ldots)$. Let $\vdash$ refer to derivations in this language.

Define Σ by stages:

$$\Sigma_0 = \Gamma$$

$$\Sigma_{n+1} = \begin{cases} \Sigma_n \cup \{\neg A_n\} & \text{if } \Sigma \vdash \neg A_n. \\ \Sigma_n \cup \{A_n\} & \text{if } \Sigma \nvdash \neg A_n \text{ and } A_n \text{ is not } \exists x\, B, \\ & \text{where } x \text{ is free in B.} \\ \Sigma_n \cup \{\exists x\, B,\ B(v_m/x)\} & \text{if } \Sigma \nvdash \neg A_n \text{ and } A_n \text{ is } \exists x\, B,\ x \text{ is free} \\ & \text{in B, and } v_m \text{ is the least of } v_0, v_1, \ldots \\ & \text{that does not appear in } \Sigma_n. \end{cases}$$

$$\Sigma = \bigcup_n \Sigma_n$$

We need to show that Σ satisfies (a)–(d). We have part (a) by construction.

For (b), I'll first show by induction that for each n, Σ_n is consistent. For $n = 0$ it's true by hypothesis. If it is true for n, then if Σ_{n+1} is defined by the first case it's immediate. If it is defined by the second case it follows by induction and Lemma 1. So suppose it is defined by the third case. Then $\Delta = \Sigma_n \cup \{\exists x\, B(x)\}$ is consistent by Lemma 1. Suppose now that Σ_{n+1} is not consistent. Then by Lemma 1, $\Delta \vdash \neg B(v_m/x)$. But then by Lemma 10, $\Delta \vdash \forall x \neg B(x)$. Hence, by Axiom 5 and **PC**, $\Delta \vdash \neg\exists x\, B(x)$. But this contradicts the consistency of Δ. So Σ_{n+1} is consistent.

It then follows that Σ is consistent, for if it were not, then some finite subset of it would be inconsistent, and hence some Σ_n would be inconsistent.

By construction, for every A, either $A \in \Sigma$ or $\neg A \in \Sigma$. So Σ is complete, and hence by Lemma 1, Σ is also a theory.

For (c), suppose $\exists x\, B(x) \in \Sigma$ and x is free in B. Then for some n and m, $\Sigma_{n+1} = \Sigma_n \cup \{\exists x\, B, B(v_m/x)\} \subseteq \Sigma$.

For (d), I'll show the contrapositive. Suppose $\forall x\, B(x) \notin \Sigma$. Then by (b), $\neg\forall x\, B(x) \in \Sigma$. Hence by Axiom 5 and **PC**, $\exists x \neg B(x) \in \Sigma$. So by (c), for some m, $\neg B(v_m/x) \in \Sigma$. So by the consistency of Σ, $B(v_m/x) \notin \Sigma$. ■

We say that a model is *countable* if its universe is countable.

Theorem 12 Löwenheim's Theorem
Every consistent collection of closed wffs in L has a countable model.

Proof Let Γ be a consistent collection of closed wffs of L. Let $\Sigma \supseteq \Gamma$ in $L(v_0, v_1, \ldots)$ be as in Theorem 11. We define an interpretation of $L(v_0, v_1, \ldots)$.

$$\mathsf{M} = < \{v_i, c_i : i \geq 0\}, \mathsf{P}_i^n, \ i \geq 0, n \geq 1, \ c_0, c_1, \ldots, v_0, v_1, \ldots >$$

where for $i \geq 0, n \geq 1$,

c_i interprets c_i

v_i interprets v_i

$(t_1, \ldots, t_n) \in \mathsf{P}_i^n$ iff $P_i^n(t_1, \ldots, t_n) \in \Sigma$

I'll let you show first by induction on the length of wffs that for any $d \in \{v_i, c_i : i \geq 0\}$ and for any assignment of references σ, for any wff B, if $\sigma(x) = d$, then $\sigma \models B(x)$ iff $\sigma \models B(d/x)$.

Now I'll show by induction on the length of wffs that for every closed wff A in $L(v_0, v_1, \ldots)$, $\mathsf{M} \models A$ iff $A \in \Sigma$. It is immediate if A is atomic. Suppose it true for all wffs shorter than A. I'll leave all the cases to you except when A is $\exists x$ B or $\forall x$ B.

Suppose that A is $\forall x$ B. If x is not free in B, then by axiom schema 2, $\forall x\, B \in \Sigma$ iff $B \in \Sigma$. So by induction, $\forall x\, B \in \Sigma$ iff $\mathsf{M} \models B$. By Theorem V.1, $\mathsf{M} \models B$ iff $\mathsf{M} \models \forall x\, B$. So $\forall x\, B \in \Sigma$ iff $\mathsf{M} \models \forall x\, B$

So suppose that x is free in B and $\forall x\, B(x) \in \Sigma$. Since Σ is a theory, by Axiom 4 for all $d \in \{v_i, c_i : i \geq 0\}$, $B(d/x) \in \Sigma$. Hence by induction, for every σ, $\sigma \models B(d/x)$, so for every σ, $\sigma \models B(x)$. Hence, $\mathsf{M} \models \forall x\, B(x)$.

In the other direction, if x is free in B and $\mathsf{M} \models \forall x\, B(x)$, then for every $i \geq 0$, if $\sigma(x) = v_i$, then $\sigma \models B(x)$. So for every i, for every σ, $\sigma \models B(v_i/x)$. Hence, for every i, $B(v_i/x) \in \Sigma$. So, since Σ satisfies the conditions of Theorem 11, $\forall x\, B(x) \in \Sigma$.

Now suppose A is $\exists x$ B. If x is not free in B, proceed as before, since we can derive $\vdash \exists x\, B \leftrightarrow B$ via Axiom 2 and Axiom 5 using **PC**. So suppose x is free in B. If $\exists x\, B(x) \in \Sigma$, then x must be the only variable free in $B(x)$. Since Σ satisfies the conditions in Theorem 11, for some m, $B(v_m/x) \in \Sigma$. Hence by induction, for any σ, $\sigma \models B(v_m/x)$, and so for any σ, if $\sigma(x) = v_m$, then $\sigma \models B(x)$. So $\mathsf{M} \models \exists x\, B(x)$. In the other direction, if $\mathsf{M} \models \exists x\, B(x)$, then for some σ, $\sigma \models B(x)$. Take one such σ, where $\sigma(x) = d$. Then $\sigma \models B(d/x)$. Since $B(d/x)$ is closed, $\mathsf{M} \models B(d/x)$, so $B(d/x) \in \Sigma$. By Lemma 8.a, $\vdash B(d/x) \rightarrow \exists x\, B(x)$, and so $\exists x\, B(x) \in \Sigma$.

To complete the proof of the theorem, define a structure N for L by taking M and deleting the interpretations of the v_i's (the universes are the same). Then for any closed wff A in L, $\mathsf{M} \models A$ iff $\mathsf{N} \models A$ by the partial interpretation theorem (Theorem V.15). So $\mathsf{N} \models \Gamma$. And N is countable. ■

Define the *theory of a model* M to be Th(M) = {A: M ⊨ A}.

Corollary 13

a. For any collection of closed wffs Σ in L, Σ is a complete and consistent theory iff there is a countable model M such that Σ = Th(M).

b. For any model M of L, there is a countable model M* such that Th(M) = Th(M*).

Proof I'll let you show that the set of wffs true in a model is a complete and consistent collection of wffs, from which you can show the corollary follows. ■

Corollary 14 ***Strong completeness*** for L(¬, →, ∀, ∃; $P_0, P_1, \ldots, c_0, c_1, \ldots$)
$\Gamma \vdash A$ iff $\Gamma \vDash A$.

Proof From left to right is Theorem 2. For the other direction, suppose $\Gamma \nvdash A$. Then $\Gamma \cup \{\neg A\}$ is consistent by Lemma 1. So by Theorem 12 it has a model M. Then M ⊨ Γ and M ⊭ A. ■

Corollary 15 ***Semantic compactness***

a. Γ has a model iff every finite subset of Γ has a model.

b. $\Gamma \vDash A$ iff for some $A_1, \ldots, A_n \in \Gamma$, $\{A_1, \ldots, A_n\} \vDash A$.

Proof (a) By Theorem 12, Γ has a model iff Γ is consistent. The result then follows as Γ is consistent iff every finite subset of Γ is consistent (Lemma II.7.c).
I'll leave (b) to you. ■

In the proof of Theorem 11 at each stage $n > 0$ we have to survey all consequences of Σ_n. That is the crucial nonconstructive part of the proof. *Bell and Slomson, 1969*, p. 103, show that the strong completeness theorem is equivalent to the axiom of choice for countable sets. As a result, the compactness theorem also does not have a constructive proof, so we do not have a constructive method for producing a model of Γ given models for each finite subset of Γ.[1] We have used the apparatus of an axiomatization to establish semantic compactness, but that is not necessary. *Bell and Slomson, 1969*, p. 102, give an entirely semantic proof.

4. Completeness for simpler languages

a. Languages with name symbols

Let L* be any language contained in L that has the same logical vocabulary and at least one name symbol. That is, L* may contain fewer predicate symbols than L, though it must have at least one or we would have no wffs. Then our axiomatization can be taken to refer to wffs of L*, the schema being instantiated only by wffs of that language. All that we've proved to this point applies equally to L* in place of L.

[1] Sometimes the semantic compactness theorem is stated as: Γ is satisfiable iff every finite subset of Γ is satisfiable, where *satisfiable* means 'has a model'.

b. Languages without name symbols

Let L* be any language contained in L that has the same logical vocabulary and at least one predicate symbol. The proof that $\vdash\forall\ldots(\forall x\,A(x) \rightarrow \exists x\,A(x))$ in Lemma 8.b required that there be names in our language. Without names, we add this as an axiom to our axioms governing the relation between ∀ and ∃ in order to prove the completeness theorem.

III. Axioms governing the relation between ∀ and ∃ Add to this group:

6. $\forall\ldots(\forall x\,A(x) \rightarrow \exists x\,A(x))$ *all implies exists*

---

c. Languages without ∃

We can delete '∃' as a primitive by setting: $\exists x\, A \equiv_{\mathrm{Def}} \neg\forall x\neg A$.
Then Axiom 5.a and Axiom 5.b of Group III can be deleted.

These observations apply to sublanguages of every language we study, and we can assume them whenever needed.

5. Validity and mathematical validity

In Chapter IX.C we asked whether the fully general classical abstraction of models results in the same semantic consequence relation as the informal notion of model in Chapter IV.

Theorem 16 If $\Gamma\vDash A$ is a valid inference in classical mathematical logic, then it is a valid inference according to the informal criteria of predicate logic without mathematical abstractions presented in Chapter IV.

Proof If $\Gamma\vDash A$ in classical mathematical logic, then by the completeness theorem, $\Gamma\vdash A$. Hence, there is a deduction of A from some $A_0, \ldots, A_n$ in Γ via the proof system of classical first-order logic. All the axioms and the rule of that system are valid in unabstracted (Type **I**) models, too. Hence, A is a valid consequence of Γ for Type **I** models, too. ■

The problem with this proof is that it presupposes some highly nonconstructive principles to establish the strong completeness theorem. To the extent that we accept those, we can claim that the notion of valid inference in classical mathematical logic does not go beyond the formalization of validity established in Chapter IV.

We are left, though, with the possibility that there are too many abstract set-theoretic models, invalidating some intuitively valid deductions. We have not established that our axiom system is strongly complete for informal validity. To a platonist classical mathematician, however, it would seem clear that every set-theoretic interpretation of the language does provide a model, since there simply "are" such predicates, abstract and divorced from language. On that assumption, classical mathematical validity could be claimed to be equivalent to informal validity for the first-order language of predicate logic.

Exercises for Section A

1. If A has just one variable x, derive the following syntactically.
   a. $\vdash \forall x \neg A(x) \rightarrow \neg \exists x\, A(x)$
   b. $\vdash \neg \forall x\, A(x) \rightarrow \exists x\, \neg A(x)$.
2. If x is not free in A, derive the following syntactically.
   a. $\vdash \forall \ldots (\exists x A \rightarrow A)$
   b. $\vdash \forall \ldots (A \rightarrow \exists x\, A)$.
3. a. Show syntactically that $\vdash \exists x_1\, (A(x_1) \rightarrow A(x_1))$. (Hint: Lemmas 7 and 8.)
   b. Explain why it is not bad that we can prove the existence of something in our logic. (Hint: Compare p. 110.)
4. a. Prove that for any model M, if Γ contains only wffs true in M, then: Γ is the set of all wffs true in M iff Γ is complete and consistent.
   b. Complete the proof of Corollary 13.
5. In the proof of Theorem 12, show by induction on the length of wff that for any $d \in \{v_i, c_i : i \geq 0\}$ and for any assignment of references σ, for any wff B, if $\sigma(x) = d$, then $\sigma \vDash B(x)$ iff $\sigma \vDash B(d/x)$.
6. *The rule of substitution in predicate logic*
   Let A(P) be a closed wff in which an n-ary predicate symbol P appears. Let B have exactly n variables free, where no variable y that is bound in B appears in A as $P(z_1, \ldots, y, \ldots, z_n)$. Let A(B/P) be the result of replacing every occurrence of P in A by B, retaining the same variables at each occurrence, i.e., for $P(z_1, \ldots, z_n)$ take $B(z_1, \ldots, z_n)$. The *rule of substitution* in predicate logic is: $\dfrac{\vdash A(P)}{\vdash A(B/P)}$
   a. Show that this is valid.
   b. Show how this allows easier proofs of the lemmas in Section A.2.
   c. Explain why it is a bad idea to use this rather than schema (see Section II.D.4).
7. *The substitutional interpretation of quantifiers*
   A different notion of model for predicate logic can be given by assuming that all objects in the universe are named in the formal language. If a is an object in the universe, let a be the first name of it in the (semi-) formal language. There is no notion of an assignment of references. Rather, the definition of a valuation on a model is given for closed wffs only:

   $\mathsf{M} \vDash P_i^n(a_1, \ldots, a_n)$ iff $\mathsf{P}_i^n(\mathsf{a}_1, \ldots, \mathsf{a}_n)$ for closed atomic wffs.

   The usual definitions for $\neg, \rightarrow, \wedge, \vee$.

   $\mathsf{M} \vDash \forall x\, A$ iff $\mathsf{M} \vDash A(c/x)$ for every name c of the language.

   $\mathsf{M} \vDash \exists x\, A$ iff $\mathsf{M} \vDash A(c/x)$ for some name c of the language.

a. Consider the realization in which the predicates are '— is a dog' and '— is a cat', the names are 'Princess', 'Juney', 'Anubis', 'Jagger', 'Buddy', and 'Birta', and the universe is all dogs that Richard L. Epstein has ever had. Show how to apply the definition above to give the truth-conditions for:
   $\forall x$ (x is a dog) $\rightarrow \neg\exists x$ (x is a cat).
   Compare that to the discussion in Chapter IV.E.6.
b. Consider the realization in which the predicates are '— is divisible by —' and '— is less than', the names are *1*, *2*, *3*, . . . and the universe is the set of natural numbers. Show how to apply the definition above to give the truth-conditions for $\exists x\, \forall y$ (x is less than y) $\rightarrow \exists z\, \forall w\, \neg$ (z is divisible by w).
c. Show that by repeatedly using the last two clauses, the definition above gives a truth-value to every closed wff of the language.
d. Show that for every closed A in the usual first-order language, A is valid for classical mathematical models iff it is true in all models in which the quantifiers are interpreted substitutionally.
e. Explain why the substitutional interpretation of quantifiers is not apt for a foundation for the formalization of reasoning in ordinary language (compare Chapter IV.B.1.)

B. Axiomatizations for Richer Languages

1. Adding '=' to the language

Let's call $L(\neg, \rightarrow, \forall, \exists, =; P_0, P_1, \ldots, c_0, c_1, \ldots)$ by the name 'L(=)'. For an axiomatization in L(=) we take the following.

Axioms

All schema in Groups I–III for wffs in L(=), with the same rule as before, plus the following schema.

IV. Axioms for equality

7. $\forall x\, (x = x)$ *identity*

8. For every $n+1$-ary atomic predicate P, *extensionality of atomic predications*

$$\forall \ldots \forall x\, \forall y\, (x = y \rightarrow (P(z_1, \ldots, z_k, x, z_{k+1}, \ldots, z_n) \rightarrow P(z_1, \ldots, z_k, y, z_{k+1}, \ldots, z_n))$$

Theorem 17 Soundness In L(=), if $\Gamma \vdash A$, then $\Gamma \vDash A$.

Proof This follows as in Theorem 2, noting that identity and the extensionality of atomic predications were shown to be valid in Chapter VI.B. ■

Theorem 18

a. $\vdash \forall x \forall y\, (x = y \rightarrow y = x)$

b. $\vdash \forall x\, \forall y\, \forall z\, (x = y \rightarrow (y = z \rightarrow x = z))$

Proof For (a) we have:

1. $\vdash \forall x\, (x = x)$	Axiom 7
2. $\vdash \forall x\, \forall y\, (x = y \rightarrow (x = x \rightarrow y = x))$	Axiom 8
3. $\vdash \forall x\, \forall y\, (x = x \rightarrow (x = y \rightarrow y = x))$	(2) and **PC**
4. $\vdash \forall x\, \forall y\, (x = y \rightarrow y = x)$	(1), (3), and Lemma 6

For (b):

1. $\vdash \forall z\, \forall y\, \forall x\, (y = x \rightarrow (y = z \rightarrow x = z))$	scheme 8
2. $\vdash \forall x\, \forall y\, (x = y \rightarrow y = x)$	part (a)
3. $\vdash \forall x\, \forall y\, \forall z\, (x = y \rightarrow (y = z \rightarrow x = z))$	(1), (2), **PC**, and Lemma 6 ∎

In using any instance of Axiom 7, Axiom 8, or Theorem 18, I'll say simply *by substitution of equals*.

Everything from Section A now carries over from L to L(=) except that we need to establish the following.

Theorem 19 Every consistent set of closed wffs of L(=) has a countable model in which '=' is interpreted as equality (the implicit identity of the universe).

Proof Let Γ be a consistent collection of closed wffs of L(=). Let $\Sigma \supseteq \Gamma$ be constructed as in the proof of Theorem 11. By the proof of Theorem 12, there is then a countable model M of Σ, and hence of Γ. In that model, write '$\simeq$' for the interpretation of '=', that is, $c \simeq d$ iff $(c = d) \in \Sigma$.

The relation $\simeq$ is an equivalence relation due to Axiom 7, Theorem 18, and Axiom 4. Due to Axiom 8, $\simeq$ is a congruence relation, too. So by Theorem VI.4 there is a model $\mathsf{M}/_{\simeq}$ of Σ in which '=' is interpreted as equality and $\mathsf{M}/_{\simeq}$ validates exactly the same wffs as M, and $\mathsf{M}/_{\simeq}$ is countable. ∎

Corollary 20 ***Strong completeness*** for L(=) $\quad \Gamma \vdash A$ iff $\Gamma \models A$.

If we do not stipulate that '=' has to be interpreted as identity, our axioms guarantee only that the interpretation of '=' is a congruence relation (see Chapter VI.D). Some authors allow for this. But it's a bad idea, because then '=' in our formalizations won't mean what we want. For example, $\exists !x\, A(x)$ would not correctly formalize 'There is a unique object that satisfies A'. We couldn't use '=' as a logical term in our formalizations as we did in Chapter VII.D and E. Logicians who allow the interpretation of '=' to be a congruence relation still wish to stipulate those models in which '=' is interpreted as equality, calling them *normal* models.

If we replace '=' with another symbol, say '$\approx$', we have a theory of congruence relations: $\mathsf{M} \models \Sigma$ iff the interpretation of '$\approx$' in M is a congruence relation on M. This is our first example of the formalization of a mathematical theory.

2. Adding function symbols to the language

Now let the language be L(¬, →, ∀, ∃, =; $P_0, P_1, \ldots, f_0, f_1, \ldots, c_0, c_1, \ldots$), which I'll call L(=; $f_0, f_1, \ldots$).

Axioms

All schema in Groups I–IV for wffs in L(=; $f_0, f_1, \ldots$) with the same rule as before.

V. Axioms for functions

9. For every $n + 1$-ary function symbol f, *extensionality of functions*

$$\forall \ldots \forall x \, \forall y \, (x = y \rightarrow (f(z_1, \ldots, z_k, x, z_{k+1}, \ldots, z_n) = f(z_1, \ldots, z_k, y, z_{k+1}, \ldots, z_n))$$

The soundness of this axiomatization follows from our requiring the evaluation of functions to be extensional in our models (Chapter VIII.C).

Theorem 21 Strong completeness for L(=; $f_0, f_1, \ldots$)

a. Every consistent set of closed wffs has a countable model in which '=' is interpreted as equality.

b. $\Gamma \vdash A$ iff $\Gamma \vDash A$.

Proof (a) The proof is just as for L(=), noting only that in the proof of Theorem VI.4, by Axiom 9 and Axiom 4 we can define:

$$\sigma/_{\approx}(f(u_1, \ldots, u_n)) = \sigma(f(t_1, \ldots, t_n))$$

where if u_i is x, then $u_i = x$, and if u_i is $[a_i]$, then $u_i = a_i$.

(b) This follows as for Corollary 14 above. ■

Exercises for Section B

1. For any wff A where t is free for x in A, derive syntactically:
   $\vdash \forall \ldots \forall x \, (x = t \rightarrow (A(x) \rightarrow A(t/x)))$.
   (Hint: Use Theorem 18.a and **PC** to get ↔ in place of → in Axiom 8. Then proceed by induction on the length of wffs.)
2. For any n-ary function symbol f, for any term t, derive syntactically:
   $\vdash \forall \ldots \forall x \, (x = t \rightarrow f(z_1, \ldots, z_k, x, z_{k+1}, \ldots, z_n) = f(z_1, \ldots, z_k, t, z_{k+1}, \ldots, z_n))$.
3. a. Show that for the sublanguage of L(=) in which there is only one binary predicate, axiom schema 7 and 8 can be replaced with just two axioms (not schema).
   b. Show that for any sublanguage of L(=) in which there are only a finite number of predicate symbols, axiom schema 7 and 8 can be replaced by a finite number of axioms (not schema).
4. Derive syntactically the extensionality of all predications with respect to '=', namely:

$\forall \ldots [x = t \rightarrow (A(x) \leftrightarrow A(t/x))]$ where t replaces some but not necessarily all occurrences of x in A for which it is free.

5. For the proof of Theorem 21.a, show that in the proof of Theorem VI.4, by using axiom scheme 9 and Axiom 4 we can define:

   $\sigma/_{\simeq}(f(u_1, \ldots, u_n)) = \sigma(f(t_1, \ldots, t_n))$
   where if u_i is x, then $t_i = x$, and if u_i is $[a_i]$, then $t_i = a_i$.

6. Show that for the completeness proof for L(=) we can modify the construction of the model in Theorem 11 to take the universe of M to be $\{v_i : i \geq 0\}$. (Hint: Consider the formulas $\exists x\,(x = c_i)$.)

7. a. Show that by adding to our axiomatizations propositional axioms governing $\wedge$ and $\vee$ we get an axiomatization for which $\wedge$ and $\vee$ can be taken as primitive.
   b. Derive syntactically from those axioms $\vdash \forall \ldots (A \wedge B) \rightarrow (\forall \ldots A \wedge \forall \ldots B)$.

8. Let M be a model for a semi-formal language L in which '=' is interpreted as the implicit identity of the universe. Show that there is a countable model N for a language $L^* \supseteq L$ in which '=' is interpreted as the predicate logic criterion of identity and for every A in L, $M \vDash A$ iff $N \vDash A$. (Hint: Add predicates '— is unique$_1$', '— is unique$_2$', ... and use Theorem 12.)

Taking open wffs as true or false

I noted before (p. 57) that some logicians view a wff such as $(x \geq y \vee y \geq x)$ as true in a model, reading the free variables as universally quantified. We saw in Theorem IV.2 that the inductive definition of truth in a model could be used for open wffs, too:

For every wff A with exactly n free variables $y_1, \ldots, y_n$,
$\upsilon(A) = T$ iff for every σ, $\sigma \vDash A$.

But we rejected that approach because reading open wffs as true or false is confusing. Still, it is not hard to axiomatize the resulting logic. We only need to add to our previous schema: $\forall \ldots A \rightarrow A$ and $A \rightarrow \forall \ldots A$.

But the main reason for adopting the convention that open wffs are read as true or false is to facilitate deductions. We needn't go through all the steps of proving that for every A that has **PC**-tautological form, $\vdash \forall \ldots A$. We can take as our propositional axiom schema in Group I the following:

(P) $\neg A \rightarrow (A \rightarrow B)$

$B \rightarrow (A \rightarrow B)$

$(A \rightarrow B) \rightarrow ((\neg A \rightarrow B) \rightarrow B)$

$(A \rightarrow (B \rightarrow C)) \rightarrow ((A \rightarrow B) \rightarrow (A \rightarrow C))$

Then if A has **PC**-tautological form, $\vdash A$, hence $\vdash \forall \ldots A$. A much simpler set of axiom schema overall suffices for a complete axiomatization.

Axioms

Propositional schema (P).

$\forall x\, A(x) \to A(t/x)$	when term t is free for x in A.
$\forall x\, A \to A$	(including when x is not free in A)
$\forall x\, (A \to B) \to (A \to \forall x\, B)$	where x is not free in A
$\forall x\, (x = x)$	
$x = y \to (A(x) \to A(y))$	where y replaces some but not necessarily all occurrences of x in A for which it is free

Rules: $\dfrac{A,\ A \to B}{B}$ *modus ponens* $\qquad \dfrac{A}{\forall x\, A}$ *generalization*

Mendelson, 1987 shows that this system is strongly complete for classical mathematical predicate logic when open wffs are taken as true or false in a model and ε is taken as a defined symbol (though he uses a different set of propositional axioms).

The degree to which this reading of open wffs facilitates the construction of proofs in predicate logic should not be discounted. Indeed, Carnielli and I used this system in our textbook on computability, *Epstein and Carnielli, 1989.*

XI The Number of Objects in the Universe of a Model

The *size* of a model is the number of elements in its universe. In particular, a model is finite if its universe is finite, infinite if its universe is infinite. In this chapter we'll see whether we can syntactically characterize how large a model of a theory must be. We'll look at models of $L(P_0, P_1, \ldots, f_0, f_1, \ldots, c_0, c_1, \ldots)$, which I'll call 'L', and models of L with '=', which I'll call 'L(=)'.

Characterizing the Size of the Universe

In Example 17 of Chapter VII (p. 131), we defined in the language L(=) formulas E_n for $n \geq 1$ such that in any model, for any σ, $\sigma \vDash E_n(y_1, \ldots, y_n)$ iff σ assigns distinct objects to $y_1, \ldots, y_n$. Now we define for $n \geq 1$ the following formulas that will allow us to characterize the size of a finite model.

$E_{\geq n} \equiv_{\text{Def}} \exists x_1 \ldots \exists x_n \, E_n(x_1, \ldots, x_n)$

$E_{<n+1} \equiv_{\text{Def}} \neg E_{\geq n}$

$E_{!n} \equiv_{\text{Def}} E_{\geq n} \wedge E_{\leq n+1}$

Theorem 1 In L(=), for each natural number $n \geq 1$, for every model M,

a. $M \vDash E_{\geq n}$ iff the universe of M has at least n elements.

b. $M \vDash E_{<n+1}$ iff the universe of M has at most n elements.

c. $M \vDash E_{!n}$ iff the universe of M has exactly n elements.

But, as we'll see next, there is no formula or even collection of formulas that hold in a model exactly if the model is finite.

Theorem 2 If a collection of wffs Σ in L or L(=) has arbitrarily large finite models, then Σ has an infinite model.

Proof First assume Σ is in L(=). Let $\Delta = \Sigma \cup \{E_{\geq n} : n \geq 1\}$. Given any finite set $\Gamma \subseteq \Delta$, there is a maximal m such that $E_{\geq m} \in \Gamma$. Hence, any model of Σ with $\geq m$ elements is a model of Γ. So every finite subset of Δ has a model. So by the compactness theorem (Corollary X.15), Δ has a model (the nonconstructive step). Any model of Δ is infinite, and a model of Δ is also a model of Σ.

If Σ is in L, view it as in L(=), proceed as above, and obtain an infinite model of Σ by deleting equality from the model of Δ. ■

Corollary 3 There is no collection of wffs Σ in L or L(=) such that $M \models \Sigma$ iff M is finite.

Proof If there were such a Σ, then it would have arbitrarily large finite models. But then by the theorem it would have an infinite model, too. ■

We can, however, characterize models with infinitely many elements using an infinite collection of wffs, though no finite collection of wffs will do.

Corollary 4

a. In L(=), $M \models \{E_{\geq n} : n \geq 1\}$ iff M is infinite.

b. There is no finite set of wffs Σ in L or L(=) such that $M \models \Sigma$ iff M is infinite.

Proof I'll leave (a) to you.

(b) Suppose there were such a collection Σ. Let A be a conjunction of all the wffs in Σ. Then $M \models A$ iff M is infinite. So $M \models \neg A$ iff M is finite, which is a contradiction to Corollary 3. Hence there is no such Σ. ■

Nor can we single out uncountable models.

Theorem 5 The Löwenheim-Skolem theorem If a collection of wffs Σ in L or L(=) has an infinite model, then Σ has a countably infinite model.

Proof By Löwenheim's theorem (Theorem X.12), every collection of wffs that has a model has a countable model. But that isn't quite good enough, for the model constructed in that proof is not guaranteed to be infinite.

So we expand the language to have additional name symbols $v_0, v_1, \ldots$ and, if not already in the language, '='. Then we define:

$$\Sigma^* = \Sigma \cup \{v_i \neq v_j : i, j \geq 0 \text{ and } i \neq j\}$$

Given any infinite model M of Σ with universe U, we can add the interpretation of '=' as identity and interpret the v_i as distinct elements of U, since U is infinite. Then we have a model M^* of Σ^*. Since Σ^* has a model, by Löwenheim's theorem, it has a countable model **N***. In this case **N*** must be infinite. By ignoring the interpretation of the v_i's, and possibly of '=', too, we have a countable infinite model of Σ by the partial interpretation theorem (Theorem V.15) ■

Corollary 6 If **M** is an infinite model for L or L(=), then there is a countably infinite model **N** such that Th(**M**) = Th(**N**).

Theorem 5 has consequences that may seem odd. Consider the theory T of the real numbers in, say, L(=; <, +, ·, 0, 1), that is, all wffs that are true of the real numbers on the usual interpretation of those symbols. Since T is consistent, T has a countable model. Though we can reason about the real numbers in classical predicate logic, what we say about them that's true is equally true of certain countable sets. Or consider a formal theory of sets, T in L(= ; ∈), formalizing the informal assumptions about sets of Chapter IX.B. Then T has a countable model, even though we know there must be uncountable sets. Here is what Andrzej Grzegorczyk says:

> [Theorem 5] is often believed to be paradoxical, as it shows that the ordinary methods of inference do not make it possible to single out non-denumerable domains. The theory of real numbers, *if we do not go beyond ordinary logical methods and ordinary symbolism*, always has a denumerable model, even though it may include very strong axioms. Likewise, set theory refers to sets of arbitrarily high powers, and yet has a denumerable model. Of course, those denumerable models of set theory or the theory of real numbers are very strange and do not comply with the intentions underlying those theories, though they satisfy the axioms. The paradox is called the *Skolem-Löwenheim paradox.*
>
> *Grzegorczyk, 1974,* p. 322 (italics added)

But there is no paradox: Classical predicate logic is based on assumptions about what we will pay attention to in the world and how we parse propositions. Those assumptions, as we have seen, limit the scope of reasoning we can do. But classical predicate logic hardly encompasses 'ordinary logical methods and ordinary symbolism'. To think it does would be as mistaken as for an Aristotelian to argue that it is paradoxical that we can't reason about relations such as '— is greater than —' with ordinary logical methods and ordinary symbolism, meaning in Aristotelian logic. In Chapter XIV we'll see natural extensions of what we've done in which we can formalize these theories and ensure that models of our theories are what we intend them to be.

Nowadays, the "paradox" is usually explained by saying that although a theory of sets says that there is a set that is uncountable, what it's really saying is that there is no function from the natural numbers to that set that is onto; that's true in a countable model because the model doesn't have "all" the functions there really are. For the history of the Skolem-Löwenheim paradox see *Moore, 1982*, and for a lively discussion of it, see *Shapiro, 1997*, especially p. 133, and *Shapiro, 1991.*

For the language without '=', however, we can't even characterize models with n elements. In order to show that, we will need to talk about models contained within other models.

Submodels A model N is a *submodel* of a model M iff:

1. $\mathsf{U_N} \subseteq \mathsf{U_M}$.
2. Every name symbol c realized in M is realized the same in N.
3. If a function symbol f is realized in M as f, then $\mathsf{U_N}$ is closed under f and f is realized as f restricted to $\mathsf{U_N}$ in N.
4. Given any atomic wff A and valuation σ on M, if for each x, $\sigma(x)$ is in $\mathsf{U_N}$, then $\sigma \vDash_{\mathsf{N}} \text{A}$ iff $\sigma \vDash_{\mathsf{M}} \text{A}$.

We say that the model N is M *restricted to* $\mathsf{U_N}$, and write $\mathsf{N} < \mathsf{M}$ or $\mathsf{N} = \mathsf{M}/_{\mathsf{U_N}}$.

Condition (4) requires, in particular, that all atomic predications in N are evaluated the same as they are in M.

Theorem 7 Let M be a model for the language L without '=' and V any set with $\text{card}(\mathsf{V}) > \text{card}(\mathsf{U_M})$. Then there is a model N with $\text{card}(\mathsf{U_N}) = \text{card}(\mathsf{V})$, such that $\mathsf{M} < \mathsf{N}$, and $\text{Th}(\mathsf{M}) = \text{Th}(\mathsf{N})$.

In particular, if M is a infinite countable model, then there is a model N of cardinality c such that $\mathsf{M} < \mathsf{N}$ and $\text{Th}(\mathsf{M}) = \text{Th}(\mathsf{N})$.

Proof The idea is simple: We take a set $\mathsf{W} \supset \mathsf{U_M}$ with $\text{card}(\mathsf{W}) = \text{card}(\mathsf{V})$, and identify all the elements in $\mathsf{W} - \mathsf{U_M}$ with one element of $\mathsf{U_M}$. The rest is just notation.

There is a 1-1 function φ from $\mathsf{U_M}$ to V, so we can take:

$$\mathsf{W} = \mathsf{U_M} \cup (\mathsf{V} - \{\mathsf{a}: \text{some } \mathsf{b} \in \mathsf{U_M},\ \varphi(\mathsf{b}) = \mathsf{a}\})$$

Choose one element u of $\mathsf{U_M}$ and define ψ from W to $\mathsf{U_M}$ via:

$$\psi(\mathsf{a}) = \begin{cases} \mathsf{a} & \text{if } \mathsf{a} \in \mathsf{U_M} \\ u & \text{if } \mathsf{a} \in \mathsf{W} - \mathsf{U_M} \end{cases}$$

Let P_i^n stand for the interpretation of P_i^n, and let f_i^n stand for the interpretation of f_i^n. Define the model N by taking W as universe and:

The interpretation of names is as in M.

$(\mathsf{d}_1, \ldots, \mathsf{d}_n) \in$ the interpretation of P_i^n iff $(\psi(\mathsf{d}_1), \ldots, \psi(\mathsf{d}_n)) \in \mathsf{P}_i^n$.

The value of the interpretation of f_i^n on $(\mathsf{d}_1, \ldots, \mathsf{d}_n)$ is $\mathsf{f}_i^n(\psi(\mathsf{d}_1), \ldots, \psi(\mathsf{d}_n))$.

Then $\mathsf{M} < \mathsf{N}$. We can map the assignments of references of N onto those of M via $\sigma^*(x) = \sigma(x)^*$. I'll let you show by induction on the length of wffs that for every wff A and all σ, $\sigma \vDash_{\mathsf{M}} \text{A}$ iff $\sigma^* \vDash_{\mathsf{N}} \text{A}$. Hence for closed A, $\mathsf{M} \vDash \text{A}$ iff $\mathsf{N} \vDash \text{A}$. ∎

This proof won't work for models of L(=) because for any a in $\mathsf{W} - \mathsf{U}_\mathsf{M}$, if $\sigma(x) = \mathsf{a}$ and $\sigma(y) = \mathsf{u}$, then $\sigma \vDash x = y$, so the interpretation of '=' would be only a congruence relation and not the implicit identity of the universe.

Corollary 8 The upward Löwenheim-Skolem theorem Suppose Σ in the language L without '=' has a model M, and V is any set with card(V) > card(U_M). Then Σ has a model N such that $\mathsf{M} < \mathsf{N}$ and card(U_N) = card(V).

Thus without '=' we cannot characterize a model with any particular number of elements, not even 47.

For L(=) we know that we cannot characterize only those models that have uncountably many elements. But is it possible to characterize models with countably many elements?

There is a standard argument for L(=) that claims to show Corollary 8 for infinite models. It proceeds by first taking $\mathsf{W} \supset \mathsf{U}_\mathsf{M}$ with card (W) = card(V), adding new name symbols to the language, $\{\mathrm{v}_\mathsf{a} : \mathsf{a} \in \mathsf{W}\}$, and then adding to Σ the collection of wffs $\{\mathrm{v}_\mathsf{a} \neq \mathrm{v}_\mathsf{b} : \mathsf{a}, \mathsf{b} \in \mathsf{W}$ and a is different from $\mathsf{b}\}$. Then it invokes the compactness theorem to show there is a model of this expanded collection of wffs. If you think it is legitimate to define an uncountable language and apply the compactness theorem to that, then such an argument will convince you that the upward Löwenheim-Skolem theorem holds in L(=) for infinite models. If you find such an extension of our methods questionable, then whether the upward Löwenheim-Skolem theorem holds in L(=) will remain an open question for you. I know of no proof that does not proceed in this way.

Exercises

1. Let T be the set of all wffs in L(<, +, ·, 0, 1) that are true of the natural numbers on the usual interpretation of the symbols. Show that T has a model of cardinality c. Can you show the same for L(=; <, +, ·, 0, 1)?
2. Let T be the set of all wffs in L(<, +, ·, 0, 1) that are true of the real numbers on the usual interpretation of the symbols. Show that T has a countable model. Can you show the same for L(=; <, +, ·, 0, 1)?
3. Let T be the set of all wffs in L(= ; $\in$) that are true of all sets.
   a. Show that T has a countable model.
   b. Explain why you think it is or is not legitimate to define such a theory.
4. Let T be the set of all wffs in L(<, +, ·, 0, 1) true of the natural numbers < 4,318.
   a. Show that T has a model of size 8,972,178.
   b. Can you show the same for L(=; <, +, ·, 0, 1)?
   c. Can you use the same proof as in (a) to show that T has a model of size 4,301?

Submodels and Skolem Functions

The Löwenheim-Skolem theorem was originally proved in a stronger form: Given an infinite model M of Σ, there is a countably infinite submodel of M that is also a model of Σ. The proof of that requires more work, but is more revealing.

Roughly the idea is that given any infinite model, to produce a countably infinite submodel we can start with any countable subset of the universe. Those elements will satisfy all wffs in Σ in which no existential quantifier appears. For every wff in Σ that contains existential quantifiers, we add elements that make the existential parts of it true. Since there are countably many wffs in Σ, and we need only add countably many elements for each of those, the submodel will, in the end, have only countably many elements. The problem is how to keep track of the elements we need to add.

As an example, let P be be a binary predicate and suppose we have a model M in which P is interpreted as P. Then:

> $M \models \forall x \exists y$ P(x, y) iff for every object a in the universe U of M, there is another object b in U such that P(a, b), that is (a, b) ∈ P.

If M is a model such that $M \models \forall x \exists y$ P(x, y), then given any a in U we can "choose" some b such that P(a, b). That is, we have (nonconstructively) a function $g: U \rightarrow U$ such that for every a, P(a, g(a)) is true. And if there is such a function, then $M \models \forall x \exists y$ P(x, y).

Similarly, I'll let you show: $M \models \forall x \exists y \forall z \forall w \exists u$ P(x, y, z, w, u) iff there is a function $f: U \rightarrow U$ and a function $g: U^4 \rightarrow U$ such that for every a, b, d in U, P(a, f(a), b, d, g(a, b, d, f(a))).

We can use functions to keep track of the elements we need. That's easier if we consider only wffs in normal form. Recall from Chapter V.C (p. 107) that a formula is in prenex normal form iff it has a variable free matrix, every variable that appears in the prefix of B appears also in the matrix of B, and no variable appears twice in the prefix of B. In particular, every closed quantifier-free wff is in prenex normal form. In the Normal Form Theorem (p. 108) we showed that for every wff there is a wff in prenex normal form that is semantically equivalent to it.

A *constant function* f is a function such that there is just one value b such that for all $a_1, \ldots, a_n$, $f(a_1, \ldots, a_n) = b$.

Skolem functions Let A be a closed wff of L or L(=) in prenex normal form. Let M be a model with universe U. A *complete set of Skolem functions for* A *in* M is any set of functions satisfying:

> Case 1: If A is quantifier free or A has only universal quantifiers in its prefix, the function is the identity on U and $M \models$ A.
>
> Case 2: If A has the form $\exists w_1 \ldots \exists w_s$ B where B is quantifier-free or B has the form $\forall y_1 \ldots \forall y_n$ C where C is quantifier-free, then the functions are constant functions $h_i : U \rightarrow U$ for $1 \leq i \leq s$, such that for any σ, for any a ∈ U, if for all i, $\sigma(w_i) = h_i(a)$, then $\sigma \models$ B.

If A is not of a form described in Case 1 or Case 2, then A has an alternation of quantifiers ∀∃ in its prefix. Let $\forall y_1, \ldots, \forall y_n$ be the universal quantifiers appearing in its prefix (reading from left to right), and $\exists z_1, \ldots, \exists z_m$ the existential quantifiers (reading from the left) that are preceded (to the left) by at least one universal quantifier. Let $a(j)$ be the number of quantifiers to the left of $\exists z_j$ for $1 \le j \le m$. We have two cases.

Case 3: The prefix of A begins with a universal quantifier. Then the functions are $g_j : U^{a(j)} \rightarrow U$ such that, for all σ, $\sigma \vDash B$ whenever:

$$\sigma(y_1) = a_1, \ldots, \sigma(y_n) = a_n$$
$$\sigma(z_1) = g_1(a_1, \ldots, a_{a(1)})$$
$$\sigma(z_j) = g_j(a_1, \ldots, a_{a(j)}, \sigma(z_1), \ldots, \sigma(z_{j-1})) \text{ for } j > 1.$$

Case 4: The prefix begins with $\exists w_1 \ldots \exists w_s$. The functions are constant functions $h_i : U \rightarrow U$ for $1 \le i \le s$ and $g_j : U^{a(j)+s} \rightarrow U$ such that for all σ, for all $a \in U$, $\sigma \vDash B$ whenever:

$$h_i(a) = b_i, \text{ and } \sigma(w_i) = b_i, \text{ and } \sigma(y_i) = a_i \text{ for all } i,$$
$$\sigma(z_1) = g_1(b_1, \ldots, b_s, a_1, \ldots, a_{a(1)}),$$
$$\sigma(z_j) = g_j(b_1, \ldots, b_s, a_1, \ldots, a_{a(j)}, \sigma(z_1), \ldots, \sigma(z_{j-1})) \text{ for } j > 1.$$

Lemma 9 For every closed wff A of L or L(=) that is in prenex normal form, and any model M, M$\vDash$A iff there is a complete set of Skolem functions for A in M.

Proof If A has the form described in Case 1, the lemma is immediate.

If A has the form in Case 2, suppose M$\vDash$A. Then for some σ, $\sigma \vDash$B. Define the constant functions $h_i : U \rightarrow U$ by setting for all $a \in U$, $h_i(a) = \sigma(w_i)$. This is a complete set of Skolem functions. On the other hand, given a complete set of Skolem functions, let σ be an assignment and $a \in U$ such that $\sigma(w_i) = h_i(a)$. Then $\sigma \vDash$B, so M$\vDash$A.

If A has the form for Case 3, suppose first that it has the form $\forall y \exists z$ B, where B is quantifier-free. Then M$\vDash$A iff for all σ there is some τ that differs from σ only in what it assigns z such that $\tau \vDash$B. Now if M$\vDash$A, to define a Skolem function g, given any $a \in U$, choose a σ such that $\sigma(y) = a$, and set $g(a) = \tau(z)$. On the other hand, if there is a Skolem function g, then given any σ, choose a τ such that $\tau(z) = g(\sigma(y))$, and we will have that $\sigma \vDash$A. So M$\vDash$A.

The other parts of this case are messy, so I'll leave them to you. If A is of the form described in Case 4, proceed by amalgamating the proofs for Cases 2 and 3. ■

Theorem 10 The downward Löwenheim-Skolem theorem
If a collection of wffs Σ in L or L(=) has an infinite model M, then there is a countably infinite submodel N of M such that N$\vDash\Sigma$.

Proof We only need to show the case where M is uncountable. Let Σ be a collection of wffs with model M, where U_M is uncountable. Let $f_0, f_1, \ldots$ be the interpretations in M of the function symbols (if there are any), and $F = \{f_0, f_1, \ldots\}$.

By the normal form theorem (p. 108), we may assume that each of the wffs in Σ is in prenex normal form. So for each $A \in \Sigma$, "choose" by Lemma 9 a complete set of Skolem functions, S(A) (this is the major nonconstructive step in this proof). For each A, S(A) is finite. Let $S(\Sigma) = \bigcup\{S(A): A \in \Sigma\}$. Since there are countably many wffs in Σ, the set $S(\Sigma)$ is countable (the countable union of countable sets is countable, Exercise 5.e of Chapter IX.B). Hence, $S(\Sigma) \cup F$ is countable.

"Choose" a set $V \subseteq U$ that is infinite and countable. Define W to be the closure of V under the functions in $S(\Sigma) \cup F$ (Exercise 6 of Chapter IX.B). That is,

$$V_0 = V$$

$$V_{n+1} = V_n \cup \{g(a_1, \ldots, a_k): a_1, \ldots, a_k \in V_n \text{ and } g \in S(\Sigma) \cup F\}$$

$$W = \bigcup_n V_n$$

I'll let you show that for all $g \in S(\Sigma) \cup F$ and $a_1, \ldots, a_k \in W$, $g(a_1, \ldots, a_k) \in W$. Hence $M/_W$ is a model. And $M/_W$ has a complete set of Skolem functions for each $A \in \Sigma$. Hence by Lemma 9, $M/_W \models A$. Since V and $S(\Sigma) \cup F$ are countable, so is W. Since $V \subseteq W$, W is infinite. ∎

Corollary 11 In L or L(=), if M is an infinite model, then there is a countably infinite submodel N of M such that $\text{Th}(N) = \text{Th}(M)$.

XII Formalizing Group Theory

We now turn to testing how suitable classical mathematical predicate logic is for formalizing mathematics. In this and succeeding chapters we'll look at formalizations of particular mathematical theories. In doing so we'll also develop new methods of analyzing reasoning.

A. A Formal Theory of Groups

What exactly is it we wish to formalize? What is a group?

Rotations and flips of a square

Consider a square. If we rotate the square any multiple of 90° in either direction it lands exactly on the place where it was before. If we flip the square over its horizontal or vertical or diagonal axis, it ends up where it was before. Any one of these operations followed by another leaves the square in the same place, and so is the same result as one of the original operations. To better visualize what we're doing, imagine the square to have labels at the vertices and track which vertex goes to which vertex in these operations:

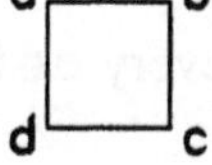

There is one operation that leaves the square exactly as it is: Do nothing. For each of these operations there is one that undoes it. For example: the diagonal flip that takes vertex c to vertex a is undone by doing that same flip again; rotating the square 90° takes vertex a to vertex b, and then rotating 270° returns the square to the original configuration.

Already we have a substantial abstraction. I don't believe that rotations and flips leave the square drawn above in the same place, because I can't draw a square with such exact precision (nor can a machine), and were I (or a machine) to cut it out and move it, it wouldn't be exactly where it was before. We are either imagining that the square is so perfectly drawn, abstracting from what we have in hand, choosing to ignore the imperfections in the drawing and the movement, or we are dealing with abstract, "perfect" objects, platonic squares, of which the picture above is only a suggestion.

Permutations
A *permutation* of the set {1, 2, 3} is a 1-1, onto function from the set to itself. For example, $\varphi(1) = 2$, $\varphi(2) = 3$, and $\varphi(3) = 1$.

Any permutation followed by another permutation yields a permutation. There is one permutation that leaves everything unchanged, the identity function: $\varepsilon(1) = 1$, $\varepsilon(2) = 2$, $\varepsilon(3) = 3$. Given any permutation there is one that undoes it. For example, the function defined by $\psi(1) = 3$, $\psi(2) = 1$, $\psi(3) = 2$ undoes φ above: φ followed by ψ is ε, and similarly, ψ followed by φ is ε.

We can also consider permutations of the set of natural numbers $\leq n$ for any n. We also have the permutations of the set of all natural numbers. Any permutation followed by another is a permutation, there is one permuation that leaves everything unchanged (the identity), and given any permutation there is one that undoes it.

The integers and addition
We can add any two integers. There is exactly one integer which, when we add it to anything, leaves that thing unchanged, namely, 0. And given any integer there is another which on addition yields back 0, namely, the negative of that integer.

Mathematicians noted similarities among these and many more examples. Abstracting from them, in the sense of paying attention to only some aspects of the examples and ignoring all others, they arrived at the definition of 'group'. A typical definition is:

A *group* is a non-empty set G with a binary operation $\circ$ such that:

i. For all $\mathsf{a}, \mathsf{b} \in \mathsf{G}$, $\mathsf{a} \circ \mathsf{b} \in \mathsf{G}$.
ii. There is an $\mathsf{e} \in \mathsf{G}$, called the *identity*, such that for every $\mathsf{a} \in \mathsf{G}$, $\mathsf{a} \circ \mathsf{e} = \mathsf{e} \circ \mathsf{a} = \mathsf{a}$.
iii. For every $\mathsf{a} \in \mathsf{G}$, there is an $\mathsf{a}^{-1} \in \mathsf{G}$, called the *inverse* of a, such that $\mathsf{a} \circ \mathsf{a}^{-1} = \mathsf{a}^{-1} \circ \mathsf{a} = \mathsf{e}$.
iv. The operation is *associative*, that is, for every $\mathsf{a}, \mathsf{b}, \mathsf{c} \in \mathsf{G}$, $(\mathsf{a} \circ \mathsf{b}) \circ \mathsf{c} = \mathsf{a} \circ (\mathsf{b} \circ \mathsf{c})$.

Let's check that all the examples above are groups.

For the first example, what is the collection? We must imagine a rotation or a flip to be a thing and view all those together as a set. The operation is then the composition of following one of the "objects" by another. The identity is the process of leaving all unchanged; and inverses exist as you were to show. I'll also leave to you to show that the operation is associative.

For the second example, the set is the collection of permutations, viewing a function as a thing; the operation is composition of functions; the identity is the identity function; and inverses exist as you can prove. I'll let you establish associativity.

For the third example, the set is the integers, the operation is addition, the identity is 0, and the inverse of n is $-n$. The operation is associative because $(n+m)+j = n+(m+j)$ for every n, m, j that are integers.

How can we formalize this informal theory? First, we need a formal language with two function symbols, one for the operation and one for the inverse, and a name symbol for the identity. Let's take $L(\neg, \rightarrow, \wedge, \vee, \forall, \exists, =;\ f_1^2, f_1^1, c_0)$. In this language the only predicate is equality, and all claims are assertions of equality.

In order to make our work easier for us to read, I'll use the symbols '$\circ$' for 'f_1^2', '$^{-1}$' for 'f_1^1', and 'e' for 'c_0' and write $L(=;\ \circ, ^{-1}, e)$ for this language. Then we can take the following axioms for a formal theory of groups.

G, the *theory of groups* in $L(=;\ \circ, ^{-1}, e)$

G1 $\forall x_1\, (e \circ x_1 = x_1 \;\wedge\; x_1 \circ e = x_1)$

G2 $\forall x_1\, (x_1 \circ x_1^{-1} = e \;\wedge\; x_1^{-1} \circ x_1 = e)$

G3 $\forall x_1\, \forall x_2\, \forall x_3\, ((x_1 \circ x_2) \circ x_3 = x_1 \circ (x_2 \circ x_3))$

G = Th(*G1*, *G2*, *G3*)

We say that *G1*, *G2*, *G3* are the *axioms* of the theory, the *nonlogical* axioms. From now on, whenever I present a theory by listing (nonlogical) axioms, it's meant that the theory is the collection of syntactic consequences of those axioms. Note that *G1*, *G2*, *G3* are particular wffs, not schema. We call a structure a *group* iff it is a model of **G**.

But how can we use the symbols $\circ, ^{-1}, e$ in the formal theory? All our work so far has been directed to formalizing informal notions, yet here we've put back the same symbols we were trying to formalize. Compare: We don't use '— is a dog' in the formal language. We don't use the phrase 'is the same as' in the formal language, but its formalization, '='. We take an informal notion and ignore all aspects of it save those we choose to formalize.

But here there is no informal notion. There are many examples, but we're not formalizing any specific one of those. Rather, we have an informal theory, not an informal notion, and we are making that theory more precise, more amenable to logical investigations. Mathematicians have already made the abstraction, choosing which properties of the examples to pay attention to and which to ignore. There is, in this case, no semi-formal language; there is only the formal language, with symbols chosen for ease of reading, and models for that.

Note some differences that arise in the formal version of the theory of groups compared to the informal theory:

1. There is no need to say formally that a group is a non-empty set of objects. That's built into our semantics: the universe of a model is a non-empty set.

2. There is no need to say that the set is closed under $\circ$. For a model to satisfy the axioms it must interpret '$\circ$' as an everywhere-defined binary function on the universe. Nor need we say that the set is closed under $^{-1}$.

3. There is no need to say that an object e exists. Names refer in our first-order predicate logic, so in any model 'e' must have a referent.

Let's look at how a mathematician proves a claim about groups and compare that to a formal version.

Theorem 1 (*Informal*) In every group the identity is unique.

Proof Suppose there were two identities in a group, e_1 and e_2. We have:

$e_1 \circ e_2 = e_1$	since e_2 is a right-hand identity
$e_1 \circ e_2 = e_2$	since e_1 is a left-hand identity
$e_1 = e_2$	substitution of equals ∎

For the formal version of Theorem 1, recall that in Example 19 of Chapter VII we showed that for any wff $A(x)$, the wff $\exists! x\, A(x)$ is true in a model iff there is exactly one object in the model that satisfies $A(x)$, where:

$$\exists! x\, A(x) \equiv_{\text{Def}} \exists x\, A(x) \wedge \forall x\, \forall y\, ((A(x) \wedge A(y)) \rightarrow x = y)$$

y is the least variable not appearing in $A(x)$

Theorem 2 (*Formal*) $G \vdash \exists! x_2\, \forall x_1\, (x_2 \circ x_1 = x_1 \;\wedge\; x_1 \circ x_2 = x_1)$

Proof By $\exists$-introduction (Lemma X.8.a),

$$G \vdash \exists x_2\, \forall x_1\, (x_2 \circ x_1 = x_1 \wedge x_1 \circ x_2 = x_1).$$

Let $A(x, y)$ be $(y \circ x = x \wedge x \circ y = x)$. Using the definition of '$\exists! x$' and **PC**, we need only show:

$$G \vdash \forall x_2\, \forall x_3\, (\forall x_1\, A(x_1, x_2) \wedge \forall x_1\, A(x_1, x_3) \rightarrow x_2 = x_3)$$

First, we may assume the Tarski-Kuratowski equivalences as theorems of our logic (from Chapter V.C, but here we need syntactic derivations, which are given in Theorem 2 of the Appendix). Using that and **PC** we have:

$\vdash \forall x_2\, \forall x_3\, (\forall x_1\, A(x_1, x_2) \wedge \forall x_1\, A(x_1, x_3) \rightarrow$
$(\forall x_1\, (x_1 \circ x_2 = x_1) \wedge \forall x_1\, (x_3 \circ x_1 = x_1))$

$\vdash \forall x_2\, \forall x_3\, (\forall x_1\, (x_1 \circ x_2 = x_1) \wedge \forall x_1\, (x_3 \circ x_1 = x_1) \rightarrow$
$(x_3 \circ x_2 = x_3 \wedge x_3 \circ x_2 = x_2))$ by substitution twice

$\vdash \forall x_2\, \forall x_3\, ((x_3 \circ x_2 = x_3 \wedge x_3 \circ x_2 = x_2) \rightarrow x_3 = x_3 \circ x_2 \wedge x_3 \circ x_2 = x_2))$
by substitution of equals and **PC**

$\vdash \forall x_2\, \forall x_3\, ((x_3 = x_3 \circ x_2 \wedge x_3 \circ x_2 = x_2) \rightarrow x_3 = x_2)$
by substitution of equals and **PC**

$\vdash \forall x_2\, \forall x_3\, (\forall x_1\, A(x_1, x_2) \wedge \forall x_1\, A(x_1, x_3) \rightarrow x_3 = x_2)$
by **PC** and generalized *modus ponens*. ■

Some observations: First, I did not give a full derivation of the theorem. I only showed that there is such a derivation using what we established earlier in the text; with this proof we could produce a derivation.

Second, the formal derivation follows the same lines as the informal proof, but is longer and no more convincing. But creating convincing, easy to read proofs was not the point of formalizing deduction. Rather, we are trying to ensure precision and analysis of the tools of reasoning we use. If we really want a good model of the *process* of reasoning, of the methods of proof, then we would need to consider methods of proof considerably more, which is a subject in itself.

The formal version shows we need the group axioms to establish the existence of an element that acts as identity, but uniqueness is a consequence of just the logical axioms. So this proof of uniqueness will apply to other theories with an identity, too.

Alternatively, we could have used Theorem 1 to prove Theorem 2 directly: $\exists! x_2\, \forall x_1\, (x_2 \circ x_1 = x_1 \wedge x_1 \circ x_2 = x_1)$ is true in every model, so by strong completeness $\mathbf{G} \vdash \exists! x_2\, \forall x_1\, (x_2 \circ x_1 = x_1 \wedge x_1 \circ x_2 = x_1)$. That way of proceeding highlights the nonconstructiveness of the proof of the strong completeness theorem, for it gives us no clue how to produce a derivation from **G**.

Here is another example. I'll show that for every object in a group there is only one element in the group that acts as its inverse.

Theorem 3 (*Informal*) Given any group and element a of it, if $\mathsf{b} \circ \mathsf{a} = \mathsf{a} \circ \mathsf{b} = \mathsf{e}$ and $\mathsf{c} \circ \mathsf{a} = \mathsf{a} \circ \mathsf{c} = \mathsf{e}$, then $\mathsf{b} = \mathsf{c}$.

Proof We have: $\mathsf{b} \circ (\mathsf{a} \circ \mathsf{c}) = \mathsf{b} \circ \mathsf{e} = \mathsf{b}$. We also have: $\mathsf{b} \circ (\mathsf{a} \circ \mathsf{c}) = (\mathsf{b} \circ \mathsf{a}) \circ \mathsf{c} = \mathsf{e} \circ \mathsf{c} = \mathsf{c}$. So $\mathsf{b} = \mathsf{c}$. ■

Theorem 4 (*Formal*) $\mathbf{G} \vdash \forall x_1\, \exists! x_2\, (x_1 \circ x_2 = \mathrm{e} \wedge x_2 \circ x_1 = \mathrm{e})$

Proof Abbreviate $(x_1 \circ x_2 = \mathrm{e} \wedge x_2 \circ x_1 = \mathrm{e}) \wedge (x_1 \circ x_3 = \mathrm{e} \wedge x_3 \circ x_1 = \mathrm{e})$ by $\mathrm{B}(x_1, x_2, x_3)$. I'll let you fill in the justification for each step below.

$$\forall x_1 \, \forall x_2 \, \forall x_3 \, (\mathrm{B}(x_1, x_2, x_3) \to x_2 \circ (x_1 \circ x_3) = x_2 \circ \mathrm{e})$$
$$\forall x_1 \, \forall x_2 \, \forall x_3 \, (\mathrm{B}(x_1, x_2, x_3) \to x_2 \circ (x_1 \circ x_3) = x_2)$$
$$\forall x_1 \, \forall x_2 \, \forall x_3 \, (\mathrm{B}(x_1, x_2, x_3) \to x_2 \circ (x_1 \circ x_3) = (x_2 \circ x_1) \circ x_3)$$
$$\forall x_1 \, \forall x_2 \, \forall x_3 \, (\mathrm{B}(x_1, x_2, x_3) \to (x_2 \circ x_1) \circ x_3 = \mathrm{e} \circ x_3)$$
$$\forall x_1 \, \forall x_2 \, \forall x_3 \, (\mathrm{B}(x_1, x_2, x_3) \to (x_2 \circ x_1) \circ x_3 = x_3)$$
$$\forall x_1 \, \forall x_2 \, \forall x_3 \, (\mathrm{B}(x_1, x_2, x_3) \to (x_2 = x_3) \qquad \blacksquare$$

Mathematicians consider various kinds of groups, paying attention to more in particular examples. For example, a group is *abelian* if for every a, b in the group, $\mathsf{a} \circ \mathsf{b} = \mathsf{b} \circ \mathsf{a}$. We can formalize reasoning about this kind of group.

Abelian G, the *theory of abelian groups*

G1, *G2*, *G3*, and

Abelian $\equiv_{\mathrm{Def}} \forall x_1 \, \forall x_2 \, (x_1 \circ x_2 = x_2 \circ x_1)$

Theorem 5 (*Informal*) There are abelian groups. There are groups that are not abelian.

Proof The integers with addition form an abelian group. But the group of rotations and flips of a square is not abelian, as you can show. ■

Theorem 6 (*Formal*)

$\mathbf{G} \cup \{Abelian\}$ has a model.

$\mathbf{G} \cup \{\neg \, Abelian\}$ has a model.

$\mathbf{G} \nvdash Abelian$

$\mathbf{G} \nvdash \neg Abelian$

Proof The first two are established in Theorem 5. The second two follow by the strong completeness theorem. ■

In Theorem 6 we converted a theorem about models to one about syntactic consequences. You won't see that kind of proof in a mathematics text. Once mathematicians have given the axioms of a theory, they normally work only semantically. To a mathematician, a claim is a theorem of group theory just means that it is a semantic consequence of the axioms. Mathematicians are rarely concerned about how they prove that a claim is a semantic consequence of the axioms: any and all ways to prove are acceptable, so long as no further "mathematical" axioms are invoked beyond those of the original theory.

So why should mathematicians be interested in classical mathematical logic? The history of mathematics is replete with examples of proofs from an informal theory that assumed mathematical claims beyond those of the axioms. There were many proofs of Euclid's parallel postulate from his other axioms that were later shown to have used other geometric assumptions; indeed, we now know that there can be no proof of Euclid's parallel postulate from his other axioms without assuming a geometrical claim that is equivalent to the parallel postulate.

Further, many informal methods of proof used by mathematicians were either considered suspicious or led to anomalies. In such cases the methods of proof were simply discarded, like the use of infinitesimals in the theory of real numbers, or were no longer considered just a method of proof but rather mathematical axioms, ones to be kept track of in the assumptions of a theory, like the axiom of choice.

In our work here, we are careful to distinguish semantic from syntactic consequences. The relationship between the syntax and semantics often clarifies mathematical problems, isolating assumptions, as we'll see in many instances below.

Mathematicians prove various theorems about abelian groups. For example, if for every a, b ∘ a = a, then b = e. We can formalize that as:

$$\mathbf{AbelianG} \vdash \forall x_1 \, (\forall x_2 \, (x_2 \circ x_1 = x_2) \rightarrow x_1 = \mathrm{e}))$$

Mathematicians are particularly interested in the theory of finite groups, that is, ones whose universe is finite. But we cannot give a formal theory of those.

Theorem 7 There is no collection of wffs $\Sigma \supseteq \mathbf{G}$ in $L(= ; \circ, ^{-1}, \mathrm{e})$ such that M is a model of Σ iff M is a finite group.

Proof There are arbitrarily large finite groups, namely, for every n, the permutations on $\{1, 2, \ldots, n\}$. Hence, the theorem follows by Theorem XI.2. ∎

On the other hand, with Corollary XI.4 we can formulate a theory of infinite groups using an infinite number of axioms that we can describe in a clear and constructive way.

Inf G, the *theory of infinite groups*

G1, *G2*, *G3*, and $\{E_{\geq n} : n \geq 1\}$

Mathematicians prove theorems not only about groups in general and about various kinds of groups, but also about subgroups of a group. A *subgroup* of a group is a subset of the group that is itself a group under the original group operation. For instance, the rotations are a subgroup of the rotations and flips of a square; the even integers with addition are a subgroup of the integers with addition. In the study of nonabelian groups the commutator subgroup of a group is important: $\{a^{-1} \circ b^{-1} \circ a \circ b$: a, b are elements of the group$\}$. In the exercises I ask you to prove that G_1 is a subgroup of G_2 iff G_1 is a submodel of G_2 for the theory **G**. But we can't formalize theorems about subgroups in a theory that extends **G** because that would involve quantifying over subcollections of the group, and we're prevented from doing that by the way we've chosen to avoid self-referential paradoxes (Example 15 of Chapter VII).

Exercises for Section A

1. a. Show that the collection of flips and rotations of a square is closed under composition.
   b. How does your proof in (a) relate to the formal theory? Why isn't there an axiom for this?
   c. Show that each "element" of the collection of flips and rotations of a square has an inverse, and show that the operation of composition on this set is associative.
   d. Show that the group of rotations and flips of a square is not abelian.
2. a. For the example of permutations of $\{1, 2, 3\}$, show that each permutation has an inverse and that the operation of composition is associative.
   b. Show the same for permutations of the entire set of natural numbers.
3. Fill in the justification for each step in the derivation in the proof of Theorem 4.
4. a. Give an informal proof that for any group, for any element b, if for some a, $\mathsf{b} \circ \mathsf{a} = \mathsf{a}$, then $\mathsf{b} = \mathsf{e}$.
   b. Convert the proof in (a) to a syntactic derivation for:
   $\mathbf{G} \vdash \forall x_1 (\exists x_2 (x_2 \circ x_1 = x_2) \rightarrow x_1 = \mathrm{e}))$.
5. Prove: G_1 is a subgroup of G_2 iff G_1 is a submodel of G_2 for the theory $\mathbf{G}$.
6. Show that the following are models of **G**, specifying the inverse operation and identity.
   a. The real numbers except 0 with multiplication.
   b. The *natural numbers modulo* 7. That is, $\{0, 1, 2, 3, 4, 5, 6\}$ with the operation of $+$ defined by: $x + y =$ the remainder of $(x + y)$ when divided by 7.
   c. The *natural numbers modulo* $n + 1$, where $n \geq 1$. That is, $\{0, 1, \ldots, n\}$ with the operation defined by: $x + y =$ the remainder of $(x + y)$ when divided by $n + 1$.
   d. The positive rational numbers with the operation of multiplication.
7. List all submodels of the following models of **G**:
   a. The natural numbers modulo 3.
   b. The natural numbers modulo 4.
   c. The natural numbers modulo p, where p is a prime.
   d. The permutations of $\{1, 2, 3\}$.
   e. The rotations and flips of a square.
8. *The collection of all groups is not a set*
   Show that the collection of all models of **G** cannot be a set. (Hint: Given any set, show that the collection of permutations of it is a group.)

B. On Definitions

1. Eliminating 'e'

There are other formulations of group theory. Here is one from *Kleene, 1967.*

GK in $L(=\,;\circ, ^{-1}, e)$

GK1 $\forall x_1\,(x_1 \circ e = x_1)$

GK2 $\forall x_1\,(x_1 \circ x_1{}^{-1} = e)$

G3 $\forall x_1\,\forall x_2\,\forall x_3\,((x_1 \circ x_2) \circ x_3 = x_1 \circ (x_2 \circ x_3))$

But don't we need $\forall x_1\,(e \circ x_1 = x_1)$? In Kleene's theory that's derivable.

Theorem 8 **GK = G**

Proof I'll let you show that **G**⊢GK1 and **G**⊢GK2. So by the cut rule (Theorem II.4.f) we have that **GK** ⊆ **G**, since **G** is a theory. Now we'll show that **GK**⊢*G1* and **GK**⊢ *G2* to obtain similarly that **G** ⊆ **GK**.

To show that **GK**⊢ *G2*, we need only show (via the Tarski-Kuratowski algorithm and **PC**) that **GK**$\vdash \forall x_1\,(x_1{}^{-1} \circ x_1 = e)$. Informally, we have:

$$
\begin{aligned}
x^{-1} \circ x &= (x^{-1} \circ x) \circ e && \text{by GK1}\\
&= (x^{-1} \circ x) \circ (x^{-1} \circ (x^{-1})^{-1}) && \text{by GK2}\\
&= (x^{-1} \circ (x \circ (x^{-1} \circ (x^{-1})^{-1})) && \text{by G3}\\
&= (x^{-1} \circ ((x \circ x^{-1})) \circ (x^{-1})^{-1}) && \text{by G3}\\
&= ((x^{-1} \circ (e \circ (x^{-1})^{-1}) && \text{by GK2}\\
&= x^{-1} \circ (x^{-1})^{-1} && \text{by GK1}\\
&= e && \text{by GK2}
\end{aligned}
$$

Here we see how easy a proof looks if open wffs are taken as implicitly universally quantified propositions, using without comment substitutions of equals. I'll let you convert this to a formal proof.

To show that **GK**⊢*G1* we need only show that **GK**$\vdash \forall x_1\,(e \circ x_1 = x_1)$. Informally, and I'll let you convert this to a formal proof, we have:

$$
\begin{aligned}
e \circ x &= (x \circ x^{-1}) \circ x && \text{by GK2}\\
&= x \circ (x^{-1} \circ x) && \text{by GK3}\\
&= x \circ e && \text{by GK2}\\
&= x && \text{by GK1 and substitution of equals}
\end{aligned}
$$

■

Kleene's theory is a simplification of **G**. We can simplify even further. In Theorem 1 we showed that the identity **e** in a group is the unique object such that for every **a**, **a** ∘ **e** = **e** ∘ **a** = **a**. So we don't need a name for the identity. We can take e as an abbreviation for 'the unique element y such that for every x, $y \circ x = x \circ y = x$'. This is an example of a more general procedure for defining terms that we'll need.

Definability of names A name symbol c is *definable* in a theory Γ iff there is a wff $\mathrm{C}(x)$ with one free variable x in which c does not appear such that:

$$\Gamma \vdash \exists ! x\, \mathrm{C}(x) \quad \text{and} \quad \Gamma \vdash \mathrm{C}(\mathrm{c}).$$

In that case, $\mathrm{C}(x)$ is a *definition* of c.

I'll let you show that if there is one definition of a name symbol in a theory, then there are infinitely many definitions of it.

If a theory has a name symbol that can be defined, then the theory with that symbol eliminated should be equivalent to the old one. But equivalent how? Suppose that c is definable by $\mathrm{C}(y)$ in a theory Γ. We can define a mapping $*$ from the language of Γ to the language of Γ without c by induction on the length of wffs.

If A is atomic and c does not appear in A, then $\mathrm{A}^* = \mathrm{A}$.
If A is atomic and c appears in A, then $\mathrm{A}^* = \forall x\, (\mathrm{C}(x) \to \mathrm{A}(x/\mathrm{c}))$
where x is the least variable not appearing in A or C.

Extend $*$ to all wffs *homophonically*:

$(\neg \mathrm{A})^*$ is $\neg \mathrm{A}^*$	$(\mathrm{A} \to \mathrm{B})^*$ is $\mathrm{A}^* \to \mathrm{B}^*$
$(\mathrm{A} \vee \mathrm{B})^*$ is $\mathrm{A}^* \vee \mathrm{B}^*$	$(\forall x\, \mathrm{A})^*$ is $\forall x\, \mathrm{A}^*$
$(\mathrm{A} \wedge \mathrm{B})^*$ is $\mathrm{A}^* \wedge \mathrm{B}^*$	$(\exists x\, \mathrm{A})^*$ is $\exists x\, \mathrm{A}^*$

For any collection of wffs Σ in the language L, define $\Sigma^* = \{\mathrm{A}^* : \mathrm{A} \in \Sigma\}$.

Now we define an associated mapping of models. Since c does not appear in $\mathrm{C}(x)$, we have that $\exists ! x\, \mathrm{C}(x) \in \Gamma^*$. So if M is a model of Γ^*, we can modify M to be a model of the language with c by interpreting c as the unique element satisfying the interpretation of C; call the result M^*. For each assignment of references σ on M, we have an assignment of references σ^* on M^* by taking $\sigma(\mathrm{c})$ to be the interpretation of c. Every assignment on M^* is σ^* for some σ.

$$\text{L} \xrightarrow{\ *\ } \text{L with c deleted}$$

$$\text{models of } \Gamma \xleftarrow{\ *\ } \text{models of } \Gamma^*$$

Lemma 9

a. For every model M of Γ^*, for every wff A of $\mathrm{L}(\Gamma)$, $\mathsf{M} \vDash \mathrm{A}^*$ iff $\mathsf{M}^* \vDash \mathrm{A}$.

b. For every model M of Γ^*, M^* is a model of Γ.

c. For every model N of Γ there is a model M of Γ^* such that $\mathsf{N} = \mathsf{M}^*$.

Proof (a) I'll show that for every atomic wff A, $\sigma \vDash \mathrm{A}^*$ iff $\sigma^* \vDash \mathrm{A}$. It will then follow that $\sigma \vDash \mathrm{A}^*$ iff $\sigma^* \vDash \mathrm{A}$ for every A, since $*$ is extended to all wffs homophonically. Since the mapping of assignments of references is onto, part (a) will then follow.

So suppose A is atomic. It's immediate if c does not appear in A. So suppose c appears in A. Then:

$$\begin{aligned}\sigma\vDash A^* \text{ iff } & \sigma\vDash\forall x\,(C(x)\to A(x/c))\\ \text{iff } & \text{for every } \tau \text{ that differs from } \sigma \text{ only in what it assigns } x,\\ & \quad \text{if } \tau\vDash C(x), \text{ then } \tau\vDash A(x/c)\\ \text{iff } & \text{for every } \tau \text{ that differs from } \sigma \text{ only in what it assigns } x,\\ & \quad \text{if } \tau(x)=\sigma^*(c), \text{ then } \tau\vDash A(x/c)\\ \text{iff } & \sigma^*\vDash A(c)\end{aligned}$$

Part (b) then follows from (a). For part (c), suppose $\mathsf{N}\vDash\Gamma$. Let M be N with the interpretation of c deleted. Then $\mathsf{M}^* = \mathsf{N}$, and $\mathsf{M}\vDash\Gamma^*$ by (a). ∎

Theorem 10 Suppose that $C(x)$ is a definition of c in theory Γ.

a. For all Δ, A in L, $\Delta\vDash_\Gamma A$ iff $\Delta^*\vDash_{\Gamma^*} A^*$.

b. For all Δ, A in L, $\Delta\vdash_\Gamma A$ iff $\Delta^*\vdash_{\Gamma^*} A^*$.

c. $\Gamma^* = \{A\colon A\in\Gamma$ and c does not appear in A$\}$.

d. Γ^* is a theory.

Proof (a) This follows by Lemma 9. Then (b) follows by strong completeness.

(c) Let $\Delta = \{A\colon A\in\Gamma$ and c does not appear in A$\}$. Since for $A\in\Delta$, $A^* = A$, it follows that $\Delta\subseteq\Gamma^*$. To show that $\Gamma^*\subseteq\Delta$, suppose $A^*\in\Gamma^*$. Then $A^{**} = A^*$ since c does not appear in A^*. Yet by (a), $A^*\in\Gamma$, since $A^{**}\in\Gamma^*$. So $A^*\in\Delta$.

(d) If $\Gamma^*\vdash B$, then by (c), $\{A\colon A\in\Gamma$ and c does not appear in A$\}\vdash B$. Hence, $\Gamma\vdash B$. Hence, $B^*\in\Gamma^*$. But B^* is B. ∎

We can now apply this general method of eliminating names to eliminate 'e' from the theory of groups.

Lemma 11 $E(x_2) \equiv_{\text{Def}} \forall x_1\,(x_2\circ x_1 = x_1 \ \wedge\ x_1\circ x_2 = x_1)$ is a definition of 'e' in **G**.

Proof By Theorem 2, $\mathbf{G}\vdash\exists!x_2\,E(x_2)$, and *G1* is $E(e/x_2)$. ∎

We can take the theory of groups without 'e' to be **G***. But the translations of the axioms of **G** won't serve as an axiomatization of **G***.

*G1** is $\forall x_2\,(E(x_2)\to E(x_2))$.

*G2** is $\forall x_2\,(E(x_2)\to\forall x_1\,(x_1\circ x_1^{-1} = x_2))$
$\qquad\wedge\ \forall x_2\,(E(x_2)\to\forall x_1\,(x_1^{-1}\circ x_1 = x_2))$.

*G3** is *G3*.

Here *G1** is a tautology. So a model of *G2** and *G3** may have no element that satisfies $E(x_2)$; consider, for example, the non-zero integers with addition (the conjuncts in *G2** are conditionals). We need to add the assumption that there is a unique element satisfying E. In the general case, we have the following.

Theorem 12 If C(x) is a definition of c in theory Γ, and $\Gamma = \mathrm{Th}(\Delta)$, then $\Gamma^* = \mathrm{Th}(\Delta^* \cup \{\exists! x\, \mathrm{C}(x)\})$.

Proof Let $\Sigma = \mathrm{Th}(\Delta^* \cup \{\exists! x\, \mathrm{C}(x)\})$. We have that $\Sigma \subseteq \Gamma^*$, as $\exists! x\, \mathrm{C}(x) \in \Gamma^*$. It remains to show that $\Gamma^* \subseteq \Sigma$.

Let $\mathrm{E} \in \Gamma^*$. Then for some $\mathrm{D} \in \Gamma$, D^* is E. Let $\mathrm{D}_1, \ldots, \mathrm{D}_n$ be a proof of D from Γ. Consider $\mathrm{D}_1^*, \ldots, \mathrm{D}_n^*$. For each i, D_i^* is either D_i itself, or is $\forall x\, (\mathrm{C}(x) \rightarrow \mathrm{D}_i(x/\mathrm{c}))$. I'll show by induction that for each i,

(2) $\forall x\, (\mathrm{C}(x) \rightarrow \mathrm{D}_i(x/\mathrm{c})) \in \Sigma$

We have that $\mathrm{D}_1 \in \Delta$, so $\mathrm{D}_1^* \in \Delta^* \subseteq \Sigma$. If D_1^* is $\forall x\, (\mathrm{C}(x) \rightarrow \mathrm{D}_1(x/\mathrm{c}))$ we are done. If not, then $\mathrm{D}_1(x/\mathrm{c})$ is just D_1, and $\mathrm{D}_1 \vDash \forall x\, (\mathrm{C}(x) \rightarrow \mathrm{D}_1)$, so by strong completeness we are done. Now suppose that (2) is true for all $j < i$. If $\mathrm{D}_i \in \Delta$, we are done as for $i = 1$. If not, for some $j, k < i$, D_j is $\mathrm{D}_k \rightarrow \mathrm{D}_i$. We have by induction $\forall x\, (\mathrm{C}(x) \rightarrow \mathrm{D}_j(x/\mathrm{c})) \in \Sigma$ and $\forall x\, (\mathrm{C}(x) \rightarrow \mathrm{D}_k(x/\mathrm{c})) \in \Sigma$. So by **PC**, $\forall x\, (\mathrm{C}(x) \rightarrow \mathrm{D}_i(x/\mathrm{c})) \in \Sigma$.

So $\forall x\, (\mathrm{C}(x) \rightarrow \mathrm{D}) \in \Sigma$. If D^* is $\forall x\, (\mathrm{C}(x) \rightarrow \mathrm{D})$, we are done. If not, D^* is D. But $\{\exists! x\, \mathrm{C}(x), \forall x\, (\mathrm{C}(x) \rightarrow \mathrm{D})\} \vDash \mathrm{D}$, as you can show. Hence, by strong completeness, $\mathrm{D} \in \Sigma$. ■

So $\mathbf{G}^* = \mathrm{Th}(G1^*, G2^*, G3^*, \exists! x_2\, \mathrm{E}(x_2))$. But since *G1** is a tautology, and *G3** is *G3*, we can obtain a simpler axiomatization.

G* in $\mathrm{L}(= ; \circ, {}^{-1})$

*G*1* $\exists! x_2\, \mathrm{E}(x_2)$

*G*2* $\forall x_2\, (\mathrm{E}(x_2) \rightarrow \forall x_1\, (x_1 \circ x_1^{-1} = x_2))$
$\wedge\ \forall x_2\, (\mathrm{E}(x_2) \rightarrow \forall x_1\, (x_1^{-1} \circ x_1 = x_2))$

G3 $\forall x_1\, \forall x_2\, \forall x_3\, ((x_1 \circ x_2) \circ x_3 = x_1 \circ (x_2 \circ x_3))$

The theory **G*** is a simpler theory than **G** in the sense that it uses fewer primitives. But it is notably harder to read. In any use of it, we would want to add 'e' as a defined symbol.

We can simplify this theory further by eliminating the function symbol '${}^{-1}$'. It's actually easier to see this in a general settting. I'll deal with eliminating predicates first, because that's easiest.

2. Eliminating '-1'

Definability of predicates An n-ary predicate symbol P is *definable* in a theory Γ iff there is some wff $C(y_1, \ldots, y_n)$ in which P does not appear and in which exactly $y_1, \ldots, y_n$ are free and are all distinct such that:

$$\Gamma \vdash \forall y_1 \ldots \forall y_n (P(y_1, \ldots, y_n) \leftrightarrow C(y_1, \ldots, y_n))$$

In that case C is a *definition* of P.

Suppose that C is a definition of P in theory Γ. For any B, define:

$B^o = B(C/P)$ that is, B with every occurrence of $P(z_1, \ldots, z_n)$ replaced by $C(z_1, \ldots, z_n)$ if there is any such occurrence.

For any collection of wffs Σ in the language L, define $\Sigma^o = \{A^o\colon A \in \Sigma\}$.

We can define an associated mapping of models of Γ^o to models of Γ: Given a model M of Γ^o, let M^o be M with the interpretation of P given by C. That is, for any assignment of references σ, $\sigma \vDash P(z_1, \ldots, z_n)$ iff $\sigma \vDash C(z_1, \ldots, z_n)$.

Lemma 13

a. For every model M of Γ^o, M^o is a model of Γ.

b. For every model N of Γ there is a model M of Γ^o such that $\mathsf{N} = \mathsf{M}^o$.

Proof Part (a) follows from the theorem on the division of form and content (Theorem V.16), as you can show. Part (b) I'll leave to you. ■

Theorem 14 Suppose that C is a definition of P in theory Γ in language L.

a. For all Δ, A in L, $\Delta \vDash_\Gamma A$ iff $\Delta^o \vDash_{\Gamma^o} A^o$.

b. For all Δ, A in L, $\Delta \vdash_\Gamma A$ iff $\Delta^o \vdash_{\Gamma^o} A^o$.

c. $\Gamma^o = \{A\colon A \in \Gamma$ and P does not appear in A$\}$.

d. Γ^o is a theory.

I'll let you establish the following, using the proof of Theorem 12 as a guide.

Theorem 15 If A is a definition of P in a theory $\Gamma = Th(\Delta)$, then $\Gamma^o = Th(\Delta^o)$.

Definability of functions An n-ary function symbol f is *definable* in a theory Γ that contains '=' iff there is some wff $C(y_1, \ldots, y_n, z)$ in which f does not appear and in which exactly $y_1, \ldots, y_n, z$ are free and are all distinct such that:

i. $\Gamma \vdash \forall y_1 \ldots \forall y_n \forall z\, (f(y_1, \ldots, y_n) = z \rightarrow C(y_1, \ldots, y_n, z))$

ii. $\Gamma \vdash \forall y_1 \ldots \forall y_n \exists! z\, C(y_1, \ldots, y_n, z)$

In that case F is a *definition* of f.

I'll let you show the following.

Lemma 16 The function f is definable by F in theory Γ iff
$\Gamma \vdash \forall y_1 \ldots \forall y_n \forall z\, (\, f(y_1, \ldots, y_n) = z \leftrightarrow C(y_1, \ldots, y_n, z))$.

Suppose that F is a definition of f in theory Γ in language L. Let L′ be L without f. The translation of L to L′ is very complicated, but we've already seen how to do it in Chapter VIII.E. Take the translation given on pp. 148–149 there which we used to translate a function into a predicate and simply replace the new predicate symbol 'P' there by 'F'; call that new mapping #. The existential assumption α_f is just condition (ii) for the definability of f. In that section we showed how to give a mapping # of models M of Γ# (which includes (ii)) to models M# of Γ such that for every A, M$\models$A# iff M#$\models$A. We then have the following as for Theorem 10.

Theorem 17 Suppose that F is a definition of f in theory Γ in language L.

a. For all Δ, A in L, $\Delta \models_\Gamma A$ iff $\Delta^{\#} \models_{\Gamma^{\#}} A^{\#}$.

b. For all Δ, A in L, $\Delta \vdash_\Gamma A$ iff $\Delta^{\#} \vdash_{\Gamma^{\#}} A^{\#}$.

c. $\Gamma^{\#} = \{A\colon A \in \Gamma$ and f does not appear in A$\}$.

d. $\Gamma^{\#}$ is a theory.

Lemma 18 $I(x_1, x_2) \equiv_{\text{Def}} (x_1 \circ x_2 = e \ \wedge\ x_2 \circ x_1 = e)$ is a definition of '$^{-1}$' in **G**.

Proof This follows by Theorem 2, *G2*, and Lemma 16. ∎

So **G#** is a theory of groups in the language L(= ; ∘, e). For an axiomatization of **G#**, I'll first let you establish the following, much as for Theorem 12.

Theorem 19 If $F(y_1, \ldots, y_n, z)$ is a definition of f in theory Γ, and $\Gamma = Th(\Delta)$, then $\Gamma^{\#} = Th(\Delta^{\#} \cup \{\forall y_1 \ldots \forall y_n \exists! z\, F(y_1, \ldots, y_n, z)\})$.

I'll now let you show that the following is an axiomatization of **G#**.

G# in L(= ; ∘, e)

G1, *G3*

$\forall x_1\, \exists!\, x_2\, I(x_1, x_2)$

We can now eliminate both '$^{-1}$' and 'e' from the theory of groups by combining these translations. Define a mapping $^{+}$ from the language L(= ; ∘, $^{-1}$, e) to L(= ; e) by first using the mapping # to eliminate '$^{-1}$', and then the mapping * to eliminate 'e'. I'll let you prove the following.

Theorem 20

a. For all Δ, A, $\Delta \vDash_{\mathbf{G}} A$ iff $\mathbf{G}^{\dagger} \vDash_{\mathbf{G}^{\dagger}} A^{\dagger}$.

b. For all Δ, A, $\Delta \vdash_{\mathbf{G}} A$ iff $\mathbf{G}^{\dagger} \vdash_{\mathbf{G}^{\dagger}} A^{\dagger}$.

c. $\mathbf{G}^{\dagger} = \{A: A \in \mathbf{G}$ and neither 'e' nor '$^{-1}$' appears in A$\}$.

d. $\mathbf{G}^{\dagger}$ is a theory.

I'll let you show that the following is an axiomatization of $\mathbf{G}^{\dagger}$.

$\mathbf{G}^{\dagger}$ in L(= ; ∘)

G† *1* $\exists ! x_2\, E(x_2)$

G† 2 $\forall x_1 \exists ! x_2 (\forall x_3\, E(x_3) \rightarrow (x_1 \circ x_2 = x_3 \ \wedge\ x_2 \circ x_1 = x_3))$

G3 $\forall x_1 \forall x_2 \forall x_3 ((x_1 \circ x_2) \circ x_3 = x_1 \circ (x_2 \circ x_3)$

Some would like to say that **G**, **G***, **G#**, and **G†** are all the same theory, the theory of groups. That suggests some abstract nonlinguistic theory of groups that these somehow represent. I would rather say that we can identify these theories through the mappings we have established.

3. Extensions by definition

Exactly when we can identify two theories as the same and when we can interpret one theory in another is the subject of Chapter XX. Here I'll summarize what we've done for theories related by definitions.

Extensions of theories Let Σ and Δ be theories with $L(\Sigma) \subseteq L(\Delta)$.

Δ is an *extension* of Σ if $\Sigma \subseteq \Delta$.

Δ is a *conservative extension* of Σ if Δ is an extension of Σ, and for each A in $L(\Sigma)$, $\Sigma \vdash A$ iff $\Delta \vdash A$. (The only new theorems use symbols in $L(\Delta)$ that are not in $L(\Sigma)$.)

Δ is an *extension by definitions* of Σ if Δ is a conservative extension of Σ and every name, predicate, and function symbol in $L(\Delta)$ that is not in $L(\Sigma)$ has a definition in Σ.

Theorem 21 Suppose that Σ and Γ are theories, and Γ is an extension by definitions of Σ. Then there is a mapping * of $L(\Gamma)$ to $L(\Sigma)$ and a corresponding mapping * of models of Σ to models of Γ

$$\begin{array}{ccc} L(\Gamma) & \xrightarrow{\ *\ } & L(\Sigma) \\ \text{models of } \Gamma & \xleftarrow{\ *\ } & \text{models of } \Sigma \end{array}$$

which satisfy the following:

a. For every model M of Σ, and every A in $L(\Delta)$, $\mathsf{M} \models A^*$ iff $\mathsf{M}^* \models A$.

b. If $\mathsf{N} \models \Delta$, then there is a unique model M of Σ such that $\mathsf{M}^* = \mathsf{N}$.

c. For all Δ, A in $L(\Gamma)$, $\Delta \models_\Gamma A$ iff $\Delta^* \models_\Sigma A^*$.

d. For all Δ, A in $L(\Gamma)$, $\Delta \vdash_\Gamma A$ iff $\Delta^* \vdash_\Sigma A^*$.

e. $\Gamma^* = \Sigma$.

Exercises for Section B

1. Prove $\mathbf{G} \vdash GK1$ and $\mathbf{G} \vdash GK2$.

2. a. Prove that if there is one definition of a name symbol c in a theory then there are infinitely many definitions of c.
   b. Prove that if there is one definition of a predicate symbol P in a theory then there are infinitely many definitions of P.

3. Let Γ be the set of all wffs in $L(= ; +, \cdot, 0, 1)$ that are true of the natural numbers.
   a. Show that 0 and 1 are definable in Γ.
   b. Show that the collection of all wffs true of the natural numbers in $L(= ;$ '—is even', $+, \cdot, 0, 1)$ is an extension by definition of Γ.
   c. Show that the collection of all wffs true of the natural numbers in $L(=; <, +, \cdot, 0, 1)$ is an extension by definition of Γ.

4. Show that for atomic wffs the translation that eliminates 'e' from **G** is:

$$(t = u)^* \text{ is } \begin{cases} t = u & \text{if e does not appear in either of these terms} \\ \forall x\,(\mathrm{E}(x) \rightarrow \forall x\,(t(x) = u(x))) & \end{cases}$$

   if e appears in one of the terms, where x is the first variable that does not appear in either t or u, and $t(x) = u(x)$ is $t = u$ with every occurrence of e replaced by x

5. We can give an alternate global translation to eliminate 'e' from **G**. Show that the following satisfies Lemma 9 and Theorem 10:
   For all A, $A^* =$ A if 'e' does not appear in A; otherwise $\forall x\,(\mathrm{C}(x) \rightarrow \mathrm{A}(x/\mathrm{c}))$ where x is the least variable not appearing in A or C

6. Give an inductive definition of the map eliminating a predicate symbol.

7. Complete the proof of Lemma 13.

8. Prove Theorem 15.

9. Prove Lemma 16: A function symbol f is definable by F in theory Γ iff $\Gamma \vdash \forall y_1 \ldots \forall y_n \forall z\,(\mathrm{f}(y_1, \ldots, y_n) = z \leftrightarrow \mathrm{F}(y_1, \ldots, y_n, z))$

10. Show that the axiomatization of $\mathbf{G}\#$ is correct.

11. Show that the axiomatization of $\mathbf{G}^\dagger$ is correct.

12. Define explicitly the mapping † from $L(= ; \circ, ^{-1}, e)$ to $L(= ; \circ)$.

13. Show that **G** is a conservative extension of $\mathbf{G}^*$, $\mathbf{G}\#$, and $\mathbf{G}^\dagger$.

XIII Linear Orderings

In this chapter we'll look at formalizations of the theory of linear orderings. This will lead us to consider when we should identify two models as the same.

A. Formal Theories of Orderings

What is an ordering? Consider some examples.

1. The natural numbers

The usual ordering of the natural numbers corresponds to the way we learn to count. We can define this ordering in terms of addition: $n < m \equiv_{\text{Def}}$ there is some $k \neq 0$ such that $m + k = n$. There is a least element in this ordering, 0, but no greatest element.

2. The integers, $\mathbb{Z}$

The usual ordering has neither a least nor a greatest element.

3. The set $\{0, 1, 2, 3, 4, 5, 6, 7\}$

The usual ordering gives a set with both a first and last element.

4. The natural numbers $\mathbb{N}$ with ordering $n < m$ iff n divides m.

Here $3 < 9$ and $2 < 4$, but neither $2 < 9$ nor $9 < 2$. This is a *partial* order.

5. The set $\{0, 1, 2, 3, 4, 5, 6, 7\}$ *with the ordering* $0 < 1 < 2 < 3 < 0$ and $5 < 6 < 7$.

This is a partial order with one *cycle*, that is, a sequence in the ordering that returns to where it started.

6. The rationals, $\mathbb{Q}$

Taking the usual order, there is no first or last element. This ordering is *dense*: Given any two elements $r < q$, there is another element p such that $r < p < q$. Compare this to the ordering of the natural numbers, which is *discrete*: Given any m there is an n such that $m < n$ and for no p, do we have $m < p < n$.

8. The real numbers, $\mathbb{R}$

This ordering is dense, too, and has no first or last element.

Except for (5), the orderings above are *strict* orderings: for every a, it's not the case that $a < a$. From each we can define a non-strict ordering, $\leq$, by setting $a \leq b$ iff $a < b$ or $a = b$. But mathematicians usually reserve the word *ordering* for relations on sets that contain no cycles. As mathematicians we say a (*strict*) *order* on a set S is a relation, which we denote by '<' (and its negation by $\nless$), such that:

1. For every a, $a \nless a$. *anti-reflexivity*
2. For every a, b, if $a < b$ then $b \nless a$. *anti-symmetry*
3. For every a, b, c, if $a < b$ and $b < c$, then $a < c$. *transitivity*

The *size* of the ordering is the size of the set S.

We can formalize this informal theory of orders with the following axioms in the language $L(=; P_1^2)$, where I'll write '<' for P_1^2 and x, y, z for x_1, x_2, x_3.

Ord, the *theory of orderings* in L(= ; <)

O1 $\forall x \, \neg(x < x)$ *anti-reflexivity*

O2 $\forall x \forall y \, (x < y \rightarrow \neg(y < x))$ *anti-symmetry*

O3 $\forall x \forall y \forall z \, ((x < y \wedge y < z) \rightarrow x < z)$ *transitivity*

Note that *O1*, *O2*, *O3* are specific wffs, not schema.

Just as for the theory of groups (Theorem XII.7), I'll let you show that it is not possible to characterize the theory of finite orderings.

Theorem 1 There is no collection of wffs $\Sigma \supseteq$ **Ord** in L(= ; <) such that M is a model of Σ iff M is a finite ordering.

In distinction to partial orderings, we say an ordering is *linear* if it satisfies the *trichotomy* law: For any a, b, $a < b$ or $b < a$ or $a = b$.

LO, the *theory of linear orderings* in L(= ; <)

Axioms *O1*, *O2*, *O3* plus

O4 $\forall x \forall y \, (x < y \vee y < x \vee x = y)$ *trichotomy*

We've seen examples of linear orderings with endpoints and linear orderings without endpoints. A linear ordering has a *last endpoint* or last element if it satisfies $\neg\forall x\, \exists y\, (x < y)$; it has a *first endpoint* if it satisfies $\neg\forall x\, \exists y\, (y < x)$. In Exercise 4 I ask you to show that any model of **LO** that lacks an endpoint is infinite.

We've also seen examples of dense orderings and discrete orderings. Of particular interest are dense linear orderings without endpoints, such as $\mathbb{Q}$ and $\mathbb{R}$.

DLO, the *theory of dense linear orderings without endpoints*, in L(= ; <)

LO (that is, *O1–O4*) plus:

O5	$\forall x\, \forall y\, (x < y \rightarrow \exists z\, (x < z \wedge z < y))$	*density*
O6	$\forall x\, \exists y\, (x < y)$	*no last endpoint*
O7	$\forall x\, \exists y\, (y < x)$	*no first endpoint*

Since any model of **DLO** lacks endpoints, it is infinite, as you can show. Before we study **DLO** further, we need to consider when we can identify two models of a theory as being the same for all our purposes in classical predicate logic.

Exercises for Section A

1. Why do we write '<' for P_1^2? (Compare the discussion for group theory, pp. 193–194.)
2. Consider the ordering on the natural numbers given by $n \prec m$ iff (n and m are both even and $n < m$, or n and m are both odd and $n < m$, or n is even and m is odd).
   a. Is this ordering linear? (Hint: Draw a picture.)
   b. Does this ordering have a first element? A last element?
   c. Is this ordering dense?
   d. Is this ordering discrete?
3. Consider the usual ordering on $\mathbb{R} - \{0\}$.
   a. Is this ordering linear?
   b. Does this ordering have a first element? A last element?
   c. Is this ordering dense?
4. Show that any model of **LO** that lacks at least one endpoint is infinite. (Hint: Prove that finite models have endpoints.)
5. Show that $\mathbb{Q}$ with its usual ordering is a model of **DLO**.
6. a. Give an example of a dense ordering that is not linear. (Hint: Two copies of the rationals.)
   b. Give an ordering of the complex numbers that is dense and not linear.
7. a. Give an example of an ordering that is neither discrete nor dense.
   b. Give a formal theory of orderings that are neither discrete nor dense.
8. a. Give a model of Th(*O1–O6*, $\neg$*O7*), the theory of dense linear orderings with a first element.

b. Give a model of Th(*O1–O5*, ¬*O6*, *O7*), the theory of dense linear orderings with a last element.
c. Give a model of Th(*O1–O5*, ¬*O6*, ¬*O7*), the theory of dense linear orderings with endpoints.

9. Give an example of an infinite non-dense linear ordering with both first and last element.

B. Isomorphisms

When should we classify two models as the same? Perhaps two models that validate exactly the same formal wffs should qualify.

Elementary equivalence Given models M and N for a language, M is *elementarily equivalent* to N iff for every A, $M \models A$ iff $N \models A$. That is, Th(M) = Th(N).

But elementarily equivalent models, as we have seen, can have different size universes. To identify models we need a 1-1 correspondence between their universes that allows us to see how formal wffs are validated the same in both structures.

First we define what it means for a mapping to preserve structure.

Homomorphisms Let M and N be models, and φ a map $\varphi: U_M \to U_N$.

φ is a *homomorphism with respect to the* (realization of the) *predicate symbol* P if:
P is realized as P_M in M and as P_N in N, and for every $a_1, \ldots, a_n \in U_M$, $P_M(a_1, \ldots, a_n)$ iff $P_N(\varphi(a_1), \ldots, \varphi(a_n))$.

φ is a *homomorphism with respect to the* (realization of the) *function symbol* f if:
f is realized as f_M in M and as f_N in N, and for every $a_1, \ldots, a_n \in U_M$, $\varphi(f_M(a_1, \ldots, a_n)) = f_N(\varphi(a_1), \ldots, \varphi(a_n))$.

φ is a *homomorphism with respect to the* (realization of the) *name symbol* c if:
c is realized in M as a and $\varphi(a)$ realizes c in N.

φ is a *homomorphism* if it is a homomorphism with respect to every predicate, name, and function symbol realized in M.

Here are some examples to help understand these definitions.

1. For the language L(= ; +, e), consider the groups:

M = <ℤ; +, 0>, where ℤ is the set of integers.
N = The group of rotations and flips of the square of Chapter XII.A.

Define γ from M to N by $\gamma(z)$ = the rotation *r* through 90° done *z* times. Then γ is a homomorphism, for $\gamma(z) \circ \gamma(y) = r$ done *z* times followed by *r* done *y* times = *r* done $z + y$ times = $\gamma(z + y)$. But γ is not 1-1, since $\gamma(0) = \gamma(4)$.

2. As in Example 1, but define ψ from M to N by setting for all z, $\psi(z) =$ the identity (which leaves the square unmoved). This is a homomorphism.

3. For the language $L(= ; +, e)$, consider the groups:

$\mathsf{M} = <\mathbb{Z}; +, 0>$

$\mathsf{N} = <\mathbb{R}; \cdot, 1>$, where $\mathbb{R}$ is the set of real numbers and '+' is interpreted as $\cdot$.

The mapping δ from M to N via $\delta(z) = z$ is 1-1, but it is not a homomorphism: $\delta(3 + -3) = \delta(0) = 0$, but $\delta(3) \cdot \delta(-3) = 3 \cdot -3 = -9$. Also, $\delta(0) \neq 1$.

4. As in Example 3, but define β from M to N by $\beta(z) = z$ for $z > 0$, $\beta(z) = {}^1/_z$ for $z < 0$, and $\beta(0) = 1$. Then β is 1-1 and a homomorphism, as you can check. But β is not onto.

5. For the language $L(= ; +, e)$, consider the groups:

$\mathsf{M} = <\mathbb{Z}; +, 0>$

$\mathsf{N} = <\mathsf{E}; +, 0>$, where E is the set of even integers

The map ρ from M to N via $\rho(z) = 2z$ is 1-1 and onto and is a homomorphism. I'll let you show that for every wff A in $L(= ; +, e)$, $<\mathbb{Z}; +, 0> \models A$ iff $<\mathsf{E}; +, 0> \models A$.

6. For the language $L(= ; <)$, consider the orderings:

$\mathsf{M} = <\{0, 1, 2, 3, 4, 5, 6, 7\}; <>$

$\mathsf{N} = <\mathbb{N}; <>$

The identity map from M to N is a homomorphism and is 1-1, but it is not onto.

7. For the language $L(= ; <)$, the map φ from $<\mathbb{N}; <>$ to $<\mathbb{Z}; <>$ via $\varphi(x) = -x$ is not a homomorphism, since $3 < 7$, but $-3 \nless -7$.

8. For the language $L(= ; <)$, the identity map from $<\mathbb{Q}; <>$ to $<\mathbb{R}; <>$ is a homomorphism, but is not onto.

9. For the language $L(= ; +, e)$, consider $\mathsf{Z} = <\mathbb{Z}; +, 0>$. For the language $L(= ; +, \cdot, e, c>$, consider $\mathsf{C} = <\mathbb{C}; +, \cdot, 0+0i, i>$, where $\mathbb{C}$ is the set of complex numbers.

Let φ be the map $\varphi(z) = z+0i$. Then φ is 1-1 and is a homomorphism with respect to + and 0, as you can check. Let $\varphi(\mathbb{Z})$ be the range of φ, that is $\varphi(\mathbb{Z}) = \{\varphi(z) : z \in \mathbb{Z}\}$. Then $\mathsf{C}/_{\varphi(\mathbb{Z})}$ is a submodel of C.

The assignments of references of Z can be mapped 1-1 onto the assignments of references of $\mathsf{C}/_{\varphi(\mathbb{Z})}$ via $\sigma^{\varphi}(x) = \varphi(\sigma(x))$. I'll let you show that for every A in $L(= ; +, e)$, $\mathsf{Z} \models A$ iff $\mathsf{C}/_{\varphi(\mathbb{Z})} \models A$.

With these examples in mind, we can make the following definitions.

Isomorphisms and the image of a map Given models M and N, a map $\varphi: U_M \to U_N$ is an *injection* if it is a 1-1 homomorphism. In particular, each predicate, function, and name symbol realized in M must be realized in N. The map φ is an *isomorphism* if it is a 1-1 and onto homomorphism and N realizes exactly the same predicate, function, and name symbols as M. In that case M and N are *isomorphic*, and we write $M \simeq N$. An isomorphism from a model to itself is called an *automorphism*.

When $M = \langle U; P_1, P_2, \ldots, f_1, f_2, \ldots, a_1, a_2, \ldots \rangle$, and φ is a homomorphism from M to another model, the *image of* M *under* φ is:

$\varphi(M) = \langle \varphi(U); \varphi(P_1), \varphi(P_2), \ldots, \varphi(f_1), \varphi(f_2), \ldots, \varphi(a_1), \varphi(a_2), \ldots \rangle$, where

$$\varphi(U) = \{\varphi(b) : b \in U\}$$

$$\varphi(P) = \{(\varphi(b_1), \ldots, \varphi(b_n)) : (b_1, \ldots, b_n) \in P\}$$

$$\varphi(f) = \{ (\varphi(b_1), \ldots, \varphi(b_n), \varphi(f(b_1, \ldots, b_n)) : b_1, \ldots, b_n \in U\}.$$

The mapping in Example 5 is an isomorphism; the mappings in Examples 4, 6, 8, and 9 are injections. Now we'll see how wffs are validated in the same way in models that are isomorphic.

Lemma 2 Let M and N be models that realize the same name symbols (if any). Let $\varphi: U_M \to U_N$ be a 1-1 onto mapping that is a homomorphism of names. Then φ induces a 1-1 onto mapping of assignments of references for M to those for N via $\sigma^{\varphi}(x) = \varphi(\sigma(x))$.

Proof Since φ is an isomorphism of names, σ^{φ} is an assignment of references. That mapping of assignments is 1-1:

$$\begin{aligned} \sigma^{\varphi} = \tau^{\varphi} \quad &\text{iff for every variable } x,\ \sigma^{\varphi}(x) = \tau^{\varphi}(x) \\ &\text{iff for every variable } x,\ \varphi(\sigma(x)) = \varphi(\tau(x)) \\ &\text{iff for every variable } x,\ \sigma(x) = \tau(x) \qquad \text{since } \varphi \text{ is 1-1} \\ &\text{iff } \sigma = \tau \end{aligned}$$

The mapping of assignments is onto: Given any τ for N, for each i, let b_i be the unique element in U_M such that $\varphi(b_i) = \tau(x_i)$. This is possible since φ is 1-1 and onto. Define σ on U_M via $\sigma(x_i) = b_i$. Then $\sigma^{\varphi} = \tau$. ∎

Theorem 3 The isomorphism theorem

a. If φ is an injection from M into N, then $\varphi(M)$ is a submodel of N.

b. If φ is an isomorphism from M to N, then for every closed wff A, for every σ on M, $\sigma \models A$ iff $\sigma^{\varphi} \models A$.

c. If there is an isomorphism from M to N, then for every closed wff A, $M \models A$ iff $N \models A$.

Proof (a) Since φ is a homomorphism of names, every name realized in $\varphi(M)$ is realized the same as in N. Because φ is a homomorphism with respect to all the predicates realized in M, the submodel condition on predications is satisfied. Finally, suppose that f is a function symbol that is realized as f_N in N and as f_M in M. Let U be the universe of M. If $c_1, \ldots, c_n \in \varphi(U)$, then there are $b_1, \ldots, b_n \in U$ such that $\varphi(b_i) = c_i$, and $f_N(c_1, \ldots, c_n) = \varphi(f_M(b_1, \ldots, b_n))$. So $\varphi(U)$ is closed under f_N. Hence, $\varphi(M)$ is a submodel of N.

(b) We have by Lemma 2 that φ induces a mapping of assignments of references of M to assignments of references of N. Note that exactly the same formal wffs are realized in M and N.

We proceed by induction on the complexity of A to show that for every σ, $\sigma \models A$ iff $\sigma^\varphi \models A$. Since the mapping of assignments of references is 1-1 and onto, we will then have $M \models A$ iff $N \models A$. We begin with atomic wffs.

$$\begin{aligned} \sigma \models P(t_1, \ldots, t_n) \quad &\text{iff} \quad (\sigma(t_1), \ldots, \sigma(t_n)) \in P_M \\ &\text{iff} \quad (\sigma^\varphi(t_1), \ldots, \sigma^\varphi(t_n)) \in P_N \qquad \text{as } \varphi \text{ is a homomorphism} \\ &\text{iff} \quad \sigma^\varphi \models P(t_1, \ldots, t_n) \end{aligned}$$

I'll leave to you the cases when A is of the form $B \wedge C$, $B \vee C$, $\neg B$, or $B \rightarrow C$. If A is of the form $\exists x\, B$, then:

$$\begin{aligned} \sigma \models \exists x\, B \quad &\text{iff} \quad \text{for some } \tau \text{ on } M \text{ that disagrees with } \sigma \text{ at most} \\ &\qquad \text{in what it assigns } x,\ \tau \models B \\ &\text{iff} \quad \text{for some } \tau^\varphi \text{ on } N \text{ that disagrees with } \sigma^\varphi \text{ at most} \\ &\qquad \text{in what it assigns } x,\ \tau^\varphi \models B \quad \text{(by induction, as } \varphi \text{ is 1-1 and onto)} \end{aligned}$$

I'll leave to you the case when A is of the form $\forall x\, B$. Part (c) then follows. ∎

We can interpret Theorem 3 as saying that if M is isomorphic to N, then N is just a notational variant of M: *we can identify* M *and* N *as the same for all our purposes in classical predicate logic.*

I'll leave the proof of the following to you.

Theorem 4 Let Γ be a theory and α a name, predicate, or function symbol definable in Γ. Then for any models M and N, if φ an isomorphism from M to N with respect to the realization of every name, predicate, and function symbol in $L(\Gamma)$ other than α, then φ is a homomorphism with respect to the realization of α, too.[1]

Theorem 4 provides a way, called *Padoa's method*, to show a categorematic symbol α is *not* definable in Γ: Exhibit two models of the theory and a mapping that is a homomorphism of the realization of every categorematic symbol in $L(\Gamma)$, but which is not a homomorphism of the realization of α.

[1] The converse of Theorem 4 is called *Beth's theorem*, and it also holds:
Suppose that for any models M and N, if φ is an isomorphism from M to N with respect to every name, predicate, and function symbol in $L(\Gamma)$ other than α, then φ is a homomorphism with respect to α. Then α is definable in Γ. (See, for example, p. 81 of *Shoenfield.*)

Exercises for Section B

1. Are the following pairs of models isomorphic?

   a. $\langle \{x : n \in \mathbb{N} \text{ and } x \text{ is prime}\};\ \{(x, y) : y \text{ is the first prime larger than } x\}\rangle$
      $\langle \mathbb{N};\ \{(n, n+1) : n \in \mathbb{N}\}\rangle$

   b. $\langle \mathbb{R};\ \{(x, y, z) : x + y = z\}\rangle$
      $\langle \mathbb{R} - \{0\};\ \{(x, y, z) : x \cdot y = z\}\rangle$

   c. $\langle \mathbb{R};\ \{x : x > 0\},\ \{(x, y, z) : x + y = z\}\rangle$
      $\langle \mathbb{N};\ \{x : x > 0\},\ \{(x, y, z) : x + y = z\}\rangle$

   d. $\langle \mathbb{N};\ \{(x, y, z) : x + y = z\}$
      $\langle \mathbb{Z};\ \{(x, y, z) : x - y = z\}$

   e. $\langle \{2, 4, 6, 8, 14, 32, 117, 412, 1018\};\ \{x : x \text{ is even}\}\rangle$
      $\langle \{3, 27, 288, 318, 354, 1182, 2445, 6125\};\ \{x : x \text{ is divisible by } 3\}\rangle$.

2. Give two examples of automorphisms that are not the identity map. (Hint: Consider the integers as a model for the theory of groups and reverse the positive with the negative.)

3. Show that **M** is a submodel of **N** iff the identity map is an injection of **M** into **N**.

4. a. For the language L(=), show that all models of size 7 are isomorphic.
   b. Show that there are non-isomorphic models of L(=).
   c. Show that there are models of L(=) that are not elementarily equivalent.

5. a. For the language L(= ; <), show that all models of **LO** of size 7 are isomorphic.
   b. Is the same true if '=' is not in the language? (Hint: Count with variables.)
   c. Show that there are non-isomorphic models of L(<) without '='.
   d. Show that there are models of L(<) without '=' that are not elementarily equivalent.

6. Show that if Σ is a theory in a language that contains '=' and has both finite and infinite models, then it has models that are not elementarily equivalent.

7. Show that $\langle \mathbb{Z}; +, 0\rangle$ is not elementarily equivalent to $\langle \mathbb{R}; \cdot, 1\rangle$.

8. Show that if **M** is a model of **DLO** and **N** is a model of linear orderings with endpoints, then **M** is not isomorphic to **N**.

9. Derive the partial interpretation theorem (Theorem V.15) as a corollary to the isomorphism theorem.

10. a. Show that the isomorphism theorem applies to models of type **I**.
    b. Give an example of two very different models that are isomorphic. What properties of the predicates, functions, and/or names in these models seem important to reasoning that have to be ignored to identify these models?

11. For the language L(= ; <), consider the model with universe $\mathbb{N}$ and ordering given by $n \prec m$ iff (n and m are both even and $n < m$, or n and m are both odd and $n < m$, or n is even and m is odd). Is this elementarily equivalent to $\mathbb{N}$ with its usual ordering?

12. Give an example of two orderings such that there is a 1-1 homomorphism from each to the other, but they are not isomorphic.

C. Categoricity and Completeness

Theorem 5 Any countable model of **DLO** is isomorphic to the rational numbers with their usual ordering.

Proof I'll let you show that the rationals with their usual ordering, $<\mathbb{Q}; <>$, is a model of **DLO**.

Suppose $<S; \prec>$ is a model of **DLO**. Since $\mathbb{Q}$ is countable (Exercise 7, Section IX.B), we can enumerate its elements, $\{q_0, q_1, \ldots\}$. We can also enumerate the elements of S as $\{s_0, s_1, \ldots\}$. We don't assume that these enumerations respect the orderings. We'll define inductively pairs (a_i, b_i) satisfying:

a. $a_i \in \mathbb{Q}$ and $b_i \in S$.

b. For every $q \in \mathbb{Q}$ there is some i such that $q = a_i$.

c. For every $s \in S$ there is some i such that $s = b_i$.

d. $\{a_0, \ldots, a_n\}$ is isomorphic to $\{b_0, \ldots, b_n\}$ via $\varphi(a_i) = b_i$.

The theorem will then follow, as you can show.

Stage 0: $a_0 = q_0$ and $b_0 = s_0$.

Stage $n + 1$, $(a_0, b_0), \ldots, (a_n, b_n)$ have been defined.

If n is even, let $a_{n+1} = q_i$ for the least i such that $q_i \notin \{a_0, \ldots, a_n\}$.

There are three possibilities:

(i) $a_{n+1} < a_j$ for all $j \leq n$.
Let $b_{n+1} = s_i$ for the least i such that $s_i < b_j$ for all $j \leq n$.
Such exists because S has no endpoints.

(ii) $a_{n+1} > a_j$ for all $j \leq n$.
Let $b_{n+1} = s_i$ for the least i such that $s_i > b_j$ for all $j \leq n$.
Such exists because S has no endpoints.

(iii) $a_j < q_{n+1} < a_{j+1}$ for some $j < n$.
Let $b_{n+1} = s_i$ for the least i such that $b_j < s_i < b_{j+1}$.
Such exists because S is dense.

If n is odd, let $b_{n+1} = s_i$ for the least i such that $s_i \notin \{b_0, \ldots, b_n\}$.
Proceed as in the case for n even, reversing the roles of $\mathbb{Q}$ and S.

I'll leave to you to show that (a)–(d) are satisfied. ■

Categorical theories A theory is *categorical* if it has a model, and all models of it are isomorphic. A theory is $\aleph_0$-*categorical* or *categorical in* $\aleph_0$ or *countably categorical* if it has a countably infinite model and all countably infinite models of it are isomorphic. A theory is c-*categorical* or *categorical in* c if it has a model M such that $\text{card}(U_M) = \text{card}(\mathbb{R})$, and all such models are isomorphic.

We've shown that **DLO** is $\aleph_0$-categorical. But we can't show it's categorical because of the following theorem.

Theorem 6 If Γ is a categorical theory, then there is a natural number *n* such that every model of Γ has size *n*.

Proof If Γ has finite models of different size, then it is not categorical. If Γ has both a finite model and an infinite model, it is not categorical. If Γ has only infinite models, then by the Löwenheim-Skolem theorem (Theorem XI.5), it has a countable model. So by Theorem XI.7, if Γ is in a language without '=', it has an uncountable model. If Γ is a model for the language with '=', then, assuming the construction described after Corollary XI.8 is legitimate, Γ has an uncountable model. ■

There are categorical theories, for example, the theory **LO** with the additional axiom $E_{!493}$ guaranteeing that every model has exactly 493 elements.

A theory that is $\aleph_0$-categorical is the closest we can come in classical predicate logic to ensuring that we have just one countably infinite model we intend for our formalization. Having only one countable model, up to isomorphism, will ensure that the theory is complete. Recall that a theory Γ is complete iff for every closed A, $A \in \Gamma$ or $\neg A \in \Gamma$. First we'll relate completeness to elementary equivalence.

Theorem 7 Γ is complete iff all models of Γ are elementarily equivalent.

Proof Suppose that all models of Γ are elementarily equivalent and that for some A, $\Gamma \nvdash A$ and $\Gamma \nvdash \neg A$. Then both $\Gamma \cup \{\neg A\}$ and $\Gamma \cup \{A\}$ are consistent, and hence have models. But models for those cannot be elementarily equivalent, since they validate different wffs, which is a contradiction. Hence, Γ is complete.

I'll let you establish the other direction. ■

Theorem 8 If a theory Γ is $\aleph_0$-categorical and every model of Γ is infinite, then Γ is complete.

Proof We'll show that all models of Γ are elementarily equivalent, and then by Theorem 7 we'll be done. Let M and N be models of Γ. Since they are infinite, by Corollary XI.6 there are countable models M′ and N′ with Th(M′) = Th(M), and Th(N′) = Th(N). By hypothesis $M' \simeq N'$, so Th(M′) = Th(N′). So Th(M) = Th(N). ■

Corollary 9 **DLO** is complete.

So for every model M of **DLO**, Th(M) = **DLO**. There aren't any wffs that can distinguish Th(M) from **DLO**. Any two models of **DLO** are elementarily equivalent. In particular, $\langle \mathbb{R}; < \rangle$ is elementarily equivalent to $\langle \mathbb{Q}; < \rangle$.

Every time we discover elementarily equivalent models that aren't isomorphic, we see a limitation on our methods of reasoning imposed by our use of classical predicate logic. With our limited vocabularly and with the abstractions we have made, we cannot distinguish formally what we clearly know is different informally. It might seem that this is acceptable, since we agreed that all we're going to pay attention to in our reasoning is what we've formalized in classical predicate logic. But such an example shows we do, and should, pay attention to more.

But actually for **DLO** the situation is worse. Though it is $\aleph_0$-categorical, it is not **c**-categorical: $\langle \mathbb{R}; =, < \rangle$ is not isomorphic to $\langle \mathbb{R} - \{0\}; =, < \rangle$, as I ask you to show in Exercise 9. In Chapter XV.A we'll see a theory that is **c**-categorical but is not $\aleph_0$-categorical.

The completeness of **DLO** has an important consequence.

Corollary 10 There is a constructive procedure for determining for every A in L(= ; <) whether **DLO**$\vDash$A.

Proof I'll present the procedure informally. I'll show how to make it more precise in Chapter XV.B.2.

Proving theorems is an effective (constructive) procedure: We can enumerate the theorems of **DLO** by starting with the axioms and constructing proofs, making sure we go through all possible proofs. The theorems are the last wffs in the proofs.

To decide whether **DLO**$\vdash$A, generate out the theorems of **DLO**. Eventually, because **DLO** is complete, either A or $\neg$A will show up on the list. If it's A, then **DLO**$\vdash$A; if it's $\neg$A, then **DLO**$\nvdash$A. ■

The method given here for deciding whether a wff is a consequence of **DLO** has no relation to the form of the wff. In an appendix to this chapter I present another method that, though more complicated, clearly illustrates a semantic procedure.

It is essential in Theorem 8 that Γ has only infinite models. Let **Eq** be the theory of equality, that is, predicate logic with just the one predicate symbol '='. Any two models of **Eq** that have the same number of elements are isomorphic, since '=' just distinguishes distinct elements. But there are models of both **Eq** $\wedge \forall x \, \forall y \, (x = y)$ and **Eq** $\wedge \neg \forall x \, \forall y \, (x = y)$, so **Eq** is not complete. However, we can generalize Theorem 8 to theories that have only infinite models.

Theorem 11 The Łos-Vaught theorem
Suppose Γ is a theory all of whose models are infinite. Suppose also that Γ has a model M such that all models N of Γ with card(N) = card(M) are isomorphic. Then Γ is complete.

Proof Let N_1 and N_2 be any models of Γ. As they are infinite, by Corollary XI.6, there are countable models $N_1{}'$ and $N_2{}'$ such that Th($N_1{}'$) = Th(N_1) and Th($N_2{}'$) = Th(N_2). By the upward Löwenheim-Skolem theorem (for languages with '=', too, if you believe that, p. 187), there are models $N_1{}^*$ and $N_2{}^*$ such that card($N_1{}^*$) = card($N_2{}^*$) = card(M) and both Th($N_1{}^*$) = Th(N_1) and Th($N_2{}^*$) = Th(N_2). By hypothesis, N_1* is isomorphic to N_2*. SoTh(N_1*) = Th(N_2*) and hence Th(N_1) = Th(N_2). Thus, every two models of Γ are elementarily equivalent. Hence by Theorem 7, Γ is complete. ■

Corollary 12 If a theory Γ is c-categorical and every model of Γ is infinite, then Γ is complete.

Exercises for Section C

1. Give an example of a theory all of whose models have size 6 which is not categorical. (Hint: Find another group with the same size as the group of permutations of {1, 2, 3}.)
2. Show that the theory of **LO** with the added axiom $E_{!47}$ is categorical.
3. Prove: If Γ is a complete and consistent theory, then for every model M of Γ, Th(M) = Γ.
4. Prove that if Γ is complete, then all models of Γ are elementarily equivalent.
5. Show that the following are not complete:
   a. The theory of linear orderings.
   b. The theory of dense linear orderings.
   c. The theory of linear orderings with no first element and no last element.
   d. The theory of abelian groups.
   e. The theory of infinite groups.
6. For the language L(=), show that any complete theory that has a finite model is categorical.
7. *Least upper bounds*
   In a linearly ordered set S define the *least upper bound* (l.u.b.) of a set of elements V ⊆ S to be the unique element b, if there is one, such that for all c ∈ V, c ≤ b, and any d ∈ S such that for all c ∈ V, c ≤ d, satisfies b ≤ d. Prove that if φ is an isomorphism from a linearly ordered set S to a linearly ordered set U and b is the l.u.b. of V ⊆ S, then φ(b) is the l.u.b. of φ(V).
8. a. Show that ‹R – {0} ; <› is a model of **DLO**.
   b. Show that ‹R; =, <› is not isomorphic to ‹R – {0} ; =, <›.
   (Hint: Use Exercise 7 and $\{-1/n : n \geq 1\}$.)
9. Let Q* be a copy of the rationals (relable them). Let M = ‹R ∪ Q*; =, <›, where a < b iff (i) a, b ∈ R and a < b, or (ii) a, b ∈ Q* and a < b in the ordering inherited from Q, or (iii) a ∈ R and b ∈ Q*.
   a. Show that neither ‹R; =, <› nor ‹R – {0} ; =, <› is isomorphic to M.
   b. Show that there are uncountably many models of **DLO** of cardinality c no two of which are isomorphic.
10. a. Show that the theory of dense linear orderings with a first element but no last element is not c-categorical.
    (Hint: Consider $\{x : 0 \leq x < 1\}$ ⊆ R and Exercise 9.a.)
    b. Show that the theory is $\aleph_0$-categorical.
    c. Show that the theory is complete.

11. a. Do Exercise 10 for the theory of dense linear orderings with a last element but no first element.
    b. Do Exercise 10 for the theory of dense linear orderings with neither a last element nor a first element.

12. Axiomatize the following theories of intervals of real numbers.
    a. Th($\langle \{x : 0 \leq x \leq 1\}; =, < \rangle$)
    b. Th($\langle \{x : 0 \leq x < 1\}; =, < \rangle$)
    c. Th($\langle \{x : 0 < x < 1\}; =, < \rangle$)

D. Set Theory as a Foundation of Mathematics?

Classical predicate logic comes with the reputation that it is useful for formalizing mathematics. Indeed, that is supposed to be its greatest value.

We've just seen that classical predicate logic is apt for formalizing the theory of dense linear orderings with no first or last elements. We can also accurately formalize our informal axioms of group theory in predicate logic. But we've seen some substantial limitations. We cannot formalize the theory of finite groups, nor prove theorems in group theory that involve subgroups, nor formalize the theory of finite orderings.

Before we draw any conclusions about the utility and limitations of predicate logic in formalizing mathematics we should consider further examples. Chapters XV–XIX are devoted to formalizing reasoning about the natural numbers, the integers, the rationals, the real numbers, and Euclidean geometry. But one route to rectifying these deficiencies has been proposed by some logicians which we can look at here: Formalize set theory and do all mathematics within that.

It would be far too big a task to include a formalization of set theory in this book. In any case, there are many good presentations of set theory already available, for example, *Fraenkel, Bar-Hillel, and Levy.* But to give you an idea, we take a language L($=$; P_1^2) and formulate axioms corresponding to our informal ones (Chapter IX.B). For example:

1. $\forall x_1 \forall x_2 (x_1 = x_2 \leftrightarrow \forall x_3 (P_1^2(x_3, x_1) \leftrightarrow P_1^2(x_3, x_2))$

3. $\forall x_1 \forall x_2 \exists x_3 \forall x_4 (P_1^2(x_4, x_3) \leftrightarrow (P_1^2(x_4, x_1) \vee P_1^2(x_4, x_2)))$

Already we have an issue about this formalization: Should we write '$\in$' for P_1^2, as we used '$<$' for P_1^2 in the formal theory of orderings or '${}^{-1}$' for 'f_1^1' in the theory of groups? That very much depends on what we think we're formalizing.

We can take the view that the informal theory of sets, just like the informal theory of groups or of orderings, is a theory based on many examples: collections of whole numbers, collections of groups, collections of wffs, collections of dogs, collections of students, the collection consisting of Ralph, all canaries, and the Eiffel Tower. Mathematicians have already made the abstraction, choosing what to pay attention to in the examples and what to ignore. There is no semi-formal language;

there is only the formal language, with symbols chosen for ease of reading, and models for that.

Alternatively, one could argue that the informal theory of sets is an abstraction of one single notion, our understanding of the words 'is an element of', for which the mathematician uses the symbol '∈'. Just as we would use P_1^1 for '— is a dog', we should use P_1^2 for 'is an element of'. It's not good enough to say that we're interested only in collections of abstract objects, the kind that mathematicians reason about. Mathematicians often reason about things such as all the pigs in Denmark or all airline flights from Albuquerque, New Mexico to Las Vegas, Nevada on June 29, 2001 when doing statistical analyses. If there is just one notion of collection that we are formalizing, it must be ample enough to encompass all of these.

Allowing for one of these views, the argument for formalizing all of mathematics within set theory is roughly that the theory of infinite groups and theorems about subgroups of a group really involve a notion of set or collection. We can't talk about these without considering arbitrary collections and reasoning about them. So the natural place to formalize group theory is within set theory. And ditto for linear orderings. Further, we have to use sets as universes of models in our metatheory, and that, too, can and should be formalized within set theory. With some (considerable) effort we can formalize within set theory all the work we've done in this book: a language is a certain kind of set, a model is a certain kind of ordered set, truth in a model is a relation between a language and a model, etc. Then within that formalization we can further formalize group theory. Set theory, it's proposed, is the natural foundation of mathematics.

And of logic, it would seem. But then we approach a vicious circle. Wasn't logic used to establish set theory? Well, that's no worse than our usual symbiosis between mathematics and logic. We use mathematics to investigate our abstractions of reasoning, and then use the formal logic to investigate mathematics, which we then use to investigate the formal logic, which . . . . Except there's a difference.

If we formalize all of mathematics and all of logic within set theory, as it's claimed we can and should do, what will serve as models of our formal theory of collections? Sets?

If we think that set theory is formalizing an informal theory of many different examples, we could say that the semantics are those sets that are appropriate to the particular model. But then we don't need the set theory, since those collections are already adequate. We've just interposed one more level of syntax between the formal language and what in the world we're talking about.

If we think that set theory is formalizing just one notion, the informal idea of 'is an element of', then it might seem that by using set theory for the semantics we have made more precise our assumptions. But it still looks like we've just interposed one more level of syntax between the formal language and what in the world we're talking about.

Worse, set theory involves its own limitations. Within set theory we cannot reason about all models of, say, group theory, because that is not a set (Exercise 8 of Chapter XII.A). Yet that is exactly the kind of reasoning we did in Chapter XII, which we believe involves no contradiction.

Perhaps set theory should be only one among many theories, logic being independent of it as we use an informal notion of set and collection in our semantics, an informal notion that need be no more precise and extensive than the mathematics we are formalizing. And some mathematics should be formalized within set theory, say mathematics that depends on the nature of collections.

In any case, a formalization of group theory in set theory looks quite odd and far from our original notion of group. Anyone familiar with the material in this text through Chapter III could write down a theory of groups given the informal theory. But even if you know set theory and the theory of sets pretty well, it's hard to write a formal theory of groups within that, for we have to define '$\circ$', '$^{-1}$', and 'e'. The utility and naturalness of first-order logic as a tool in formalizing mathematics is hardly obvious if we are to do group theory within first-order set theory. What's left is a theoretical foundation for mathematics. We could do all of mathematics within set theory if we wanted.

It seems to me that the limitations in using classical predicate logic to formalize mathematics are real and substantial, and replacing all theories with just one does not overcome them. Rather, we should look for extensions of the classical predicate logic we've developed—which we'll begin in the next chapter.

For a much fuller treatment of the relationship of set theory to the foundations of logic and mathematics, with many references to other views, see *Shapiro, 1991.*

Decidability by Elimination of Quantifiers

In the proof of Corollary 10 I described a constructive procedure for deciding for any wff whether it is a theorem of **DLO**, that is, to show that **DLO** is *decidable*. But that procedure has no relation to the form of the wff that we're deciding.

In this section I'll present a more direct way to decide of any given wff whether it or its negation is a theorem of **DLO**, using the work on prenex normal forms of Chapter V (pp. 107–108)). This will also give another proof that **DLO** is complete. The method will be useful for several theories in later chapters, so I'll do it in detail here. In an exercise you can see an application of it to give a decision procedure for a theory that is not complete.

Suppose we are given a wff A of L(= ; <). We're going to produce in a constructive manner wffs that are semantically equivalent to A until we reach one that is true in all models or false in all models. Then by the completeness theorem, in the first case, **DLO**$\vdash$A, and in the second case, **DLO**$\vdash\neg$A.

***Elimination of quantifiers for* DLO** There is a constructive procedure using elimination of quantifiers for determining for every A in L(= ; <) whether **DLO**$\vdash$A. The procedure also establishes that **DLO** is complete.

Proof Let A be a wff in L(= ; <). By the normal form theorem (p. 108), we can find a wff that is semantically equivalent to A and is in prenex normal form. So assume now that A is in prenex normal form:

$Q_1 z_1 \ldots Q_k z_k$ B where each Q_i is either ∀ or ∃, and B has no quantifiers.

What we'll do now is eliminate Q_k. That is, we'll produce, in a constructive manner, a wff C such that A is equivalent to $Q_1 z_1 \ldots Q_{k-1} z_{k-1}$ C. In doing so we'll find parts of wffs that are true in every model or false in every model, and I'll write T or F for those, ignoring the variables in them. (If you like, you could write '$x = x$' or '$\neg(x = x)$' for those for some variable x that does not appear in A, and then we'd have at the end of our process a wff $\forall x$ E, where E has formulas only of that form and is clearly decidable.)

First, let's suppose that Q_k is ∃. I'll write 'z' for 'z_k'.

We can assume that B is in propositional disjunctive normal form. We can eliminate all occurrences of negation by the following equivalences, based on the trichotomy law:

$$\neg(x < y) \leftrightarrow (y < x) \vee (y = x)$$

$$\neg(x = y) \leftrightarrow (y < x) \vee (x < y)$$

Using De Morgan's laws (Exercise 4, p. 35), we can then again assume that we have B in propositional disjunctive normal form. That is, B is $(B_1 \vee \cdots \vee B_r)$ where each B_i is a conjunction of atomic wffs. Since $\exists y\,(D \vee E)$ is semantically equivalent to $\exists y\, D \vee \exists y\, E$, the problem reduces to eliminating the quantifier in formulas of the form

$C_1 \wedge \cdots \wedge C_m$ where each C_i is an atomic wff.

Suppose we have such a wff, C.

We proceed by induction on m. For $m = 1$ we have one of the following cases:

(i)	$\exists z\,(z < z)$	(v)	$\exists z\,(z = z)$
(ii)	$\exists z\,(z < y)$	(vi)	$\exists z\,(x = z)$
(iii)	$\exists z\,(y < z)$	(vii)	$\exists z\,(z = x)$
(iv)	$\exists z\,(x < y)$	(viii)	$\exists z\,(x = y)$

Then we can eliminate the quantifier:

(i) Replace this with F.
(ii) Replace this with T. (There are no endpoints in a model.)
This is regardless of whether y is bound by ∀ or ∃.
(iii) Replace this with T. (There are no endpoints in a model.)
(iv) The quantifier is superfluous.
(v) Replace this by T.
(vi) Replace this by T.
(vii) Replace this by T.
(viii) The quantifier is superfluous.

Now suppose that we have eliminated the quantifier whenever we have m conjuncts, and C is $(C_1 \wedge \cdots \wedge C_m) \wedge C_{m+1}$.

If for some i, z does not appear in C_i, we may assume it is C_{m+1}, and since $\exists z\,[(C_1 \wedge \cdots \wedge C_m) \wedge C_{m+1}]$ is equivalent to $(\exists z\,(C_1 \wedge \cdots \wedge C_m)) \wedge C_{m+1}$, we are done by induction and the Tarski-Kuratowski algorithm:

(‡) $\exists z\,(D \wedge E) \equiv D \wedge \exists z E$ if z does not appear in D

$\exists z\,(D \wedge E) \equiv \exists z D \wedge E$ if z does not appear in E

So assume that z appears in each C_i. By **PC**, using '$\bigwedge$' to indicate a conjunction of the indexed wffs, we must eliminate the quantifer from:

$$\exists z\,(\bigwedge_i (z < y_i) \wedge \bigwedge_j (t_j < z) \wedge \bigwedge_k (z = w_k))$$

If for some i or j, y_i or t_j is z then we can replace the entire wff by F.

If for some k, z is w_k, then we can delete that conjunct, since it is always true, and we are done by induction.

So assume that for every i, j, k, the variable z is distinct from each of y_i, t_j, w_k.

If $\bigwedge_i (z < y_i)$ has more than one conjunct, we have:

$$\begin{aligned}(z < y_1 \wedge \cdots \wedge z < y_n) \equiv\\ &[(y_1 < y_2 \wedge (z < y_1 \wedge z < y_3 \wedge \cdots \wedge z < y_n)]\\ &\vee [(y_2 < y_1 \wedge (z < y_2 \wedge z < y_3 \wedge \cdots \wedge z < y_n)]\\ &\vee [(y_1 = y_2 \wedge (z < y_1 \wedge z < y_3 \wedge \cdots \wedge z < y_n)]\end{aligned}$$

Since z does not appear in the first conjunct, we are done by (‡) and induction.

If $\bigwedge_j (t_j < z)$ has more than one conjunct, we can proceed similarly.

If $\bigwedge_k (z = w_k)$ has more than one conjunct, then we can replace it with $\bigwedge_k (w_1 = w_k)$, and we're done as before.

If C is $\exists z\,(z < y \wedge t < z \wedge (z = w))$, it is equivalent to $\exists z\,(w < y \wedge t < w \wedge (z = w))$, which we can finish using (‡).

If C is $\exists z\,(z < y \wedge t < z)$, we can replace it with $(t < y)$.

If C is $\exists z\,(z < y \wedge (z = w))$, it is equivalent to $\exists z\,(w < y \wedge (z = w))$, and we're done as before.

If C is $\exists z\,(t < z \wedge (z = w))$, proceed as in the previous case.

If C has only one conjunct, we're done by induction.

Hence, we've finished the case where Q_k is $\exists$. If Q_k is $\forall$, then A is equivalent to:

$$Q_1 z_1 \ldots \neg\exists z_k \neg B \quad \text{where each } Q_i \text{ is either } \forall \text{ or } \exists, \text{ and B has no quantifiers}$$

Put $\neg$B in disjunctive normal form, eliminate negations, and proceed as before, noting that when a negation appears before T, we replace it with F, and vice-versa.

Thus, we have eliminated the first quantifier. Now eliminate each succeeding quantifier in the same manner. We will finally reach a formula that is a disjunction of conjunctions of (wffs that are equivalent to) T or F. For each conjunction, if it has one conjunct that is F, replace it by F. Otherwise replace it by T. Then we have a disjunction of (wffs that are equivalent to) T or F. If any one of them is T, A is true in every model, and **DLO**$\vdash$A. If all of them are F, A is false in every model, and hence **DLO**$\vdash\neg$A.

Thus, **DLO** is decidable and complete. ■

Exercises

Let **DLO$^-$** = Th(*O1–O5*) be the theory of dense orderings.

1. Show that **DLO$^-$** is not complete.

2. Apply the method of elimination of quantifiers to **DLO⁻** to show that it is decidable. (Hint: When you come to a case with a wff for which there is a model in which it is true and a model in which it is false, replace it with U ('undecided'.) When applying a negation to U, replace it with U. When evaluating the final wff that is a disjunction of conjunctions of T, F, or U, if the wff is true in all models it's a theorem; if false in all models its negation is a theorem; if true in some models and false in others, neither it nor its negation is a theorem.)

XIV Second-Order Classical Predicate Logic

A. Quantifying Over Predicates?

Consider the semi-formal proposition:

(1) $\forall x_1\,(x_1 \text{ is a dog} \vee \neg(x_1 \text{ is a dog}))$

In any model this is true or false.

Consider now a formal version of (1):

(2) $\forall x_1\,(P_1(x_1) \vee \neg P_1(x_1))$

We do not say that this is a proposition. But given a model of the formal language we do speak of (2) as true or false, understanding it as true or false for that interpretation of the predicate symbol 'P_1'.

But we say more: (2) is true in every model. That is,

For every interpretation of 'P_1', '$\forall x_1 (P_1(x_1) \vee \neg P_1(x_1))$' is true.

The self-reference exclusion principle (Chapter IV.D) excludes the predicate '— is true' from the semi-formal language. But instead of talking of interpretations and truth, we could just assert:

For every predicate P_1, $\forall x_1 (P_1(x_1) \vee \neg P_1(x_1))$.

We still have a semantic predicate, '— is a predicate', in the proposition. We can eliminate that by noting that just by the form of the following expression we know that 'P_1' is meant to stand for a predicate:

(3) For every P_1, $\forall x_1 (P_1(x_1) \vee \neg P_1(x_1))$

The self-reference exclusion principle is no longer violated. If we can devise an explanation of quantification over predicates, then we will be able to bring into the semi-formal language a significant part of our metalogical reasoning. We would like to do that as part of our general goal of making explicit what has been implicit, formalizing our assumptions within a semi-formal language. If we further make the mathematical abstraction that a predicate is to be identified with a subset of n-tuples of the universe of a model, then we will be able to overcome some of the limitations on formalizing mathematics that we met in the last chapter.

B. Predicate Variables and Their Interpretation: Avoiding Self-Reference

1. Predicate variables

We would like to formalize and give a semantic analysis of (3). To begin, we shall need new predicate variables, for despite what I wrote there, we cannot use the predicate symbols as variables in a semi-formal language. For suppose we did, and realized P_0^1 as '— is a dog'. Then consider $\forall P_0^1 \forall x_1 (P_0^1(x_1) \vee \neg P_0^1(x_1))$. That would have to be realized as $\forall P_0^1 \forall x_1 (x_1 \text{ is a dog} \vee \neg (x_1 \text{ is a dog}))$, so that P_0^1 would not be acting as a variable.

Every predicate has an arity, as we've noted in indexing each predicate symbol. Shall we allow predicate variables to stand for any predicate whatsoever, regardless of arity? If so, what would we mean by $\forall x\, X(x)$ when we let 'X' stand for '— is the brother of —'? We could say that the form of the formula indicates that X can stand for a unary predicate only, but that would mean we would have to have X stand for two different predicates in '$\forall x\, X(x) \vee \exists y \exists x \neg X(x,y)$'. Our assumption that words are types excludes that possibility. So a predicate variable should stand for a predicate of a fixed arity. We'll take as predicate variables: X_0^1, X_0^2, X_0^3, . . . , X_1^1, X_1^2, X_1^3, ... I'll use $X, Y, Z, Y_0, Y_1, \ldots, Z_0, Z_1, \ldots$ as metavariables ranging over these formal variables, trusting that in informal contexts we can see what arity is required. I'll continue to use $P, Q, Q_0, Q_1, \ldots$ as informal metavariables for predicates.

As before we have name symbols to be realized as names of objects in the universe, and variables $x_0, x_1, \ldots$ that *range over*—that is, are allowed to take as temporary reference—objects of the universe. These latter we call *individual variables* or *first-order variables*. We still have predicate symbols to be realized as particular predicates. And now we have predicate variables that will range over predicates (in an as yet undetermined sense). The latter are called *second-order variables*. The name *first-order (predicate) logic* or *elementary (predicate) logic* is used for what we have up to now called simply 'predicate logic'. *Second-order (predicate) logic* is our previous predicate logic with allowance made for using variables for predicates and quantifications involving those variables, as presented in this chapter.

If predicates are linguistic items, then they are objects, individual things. If you consider predicates to be abstract, then that means you countenance abstract things, and predicates again are things. In either case, since predicates are things, we can carry over our understanding of the use of variables from the previous chapters, particularly Chapter IV. A predicate variable is like a blank for a predicate: 'X(Ralph)' is not a proposition, neither true nor false, until we fill in the blank with a predicate. If we let 'X' stand for '— is square', then 'X(Ralph)' is false, under the usual interpretation of those words.

To form propositions using predicate variables we can use quantifications. As before, let's restrict ourselves to the quantifications 'for all' and 'for some'. The use of the same symbols '$\forall$' for 'for all' and '$\exists$' for 'for some' will emphasize that quantification over predicates is, after all, quantification over objects, though objects of a special kind.

So now we can form a formal version of (3):

$$\forall X_1^1\ \forall x_0\,(X_1^1(x_0) \vee \neg X_1^1(x_0))$$

or using metavariables:

(4) $$\forall X\, \forall x\,(X(x) \vee \neg X(x))$$

To make the discussion easier to follow in the rest of Section B, I'm going to concentrate almost exclusively on unary predicate variables, though what I will say carries over straightforwardly to predicate variables of every arity. And I'll use metavariables throughout.

2. The interpretation of predicate variables

Let's apply our previous understanding of variables and quantifiers to (4). We'll let a realization of the language be as before: predicate symbols are realized as predicates, function symbols as functions, name symbols as names, and we have a universe of objects. Then we take assignments of references just as before, only now each assignment will have to assign a predicate to each predicate variable.

(5) $\upsilon(\forall X\, \forall x\, (X(x) \vee \neg X(x))) = \mathsf{T}$

iff for every assignment of references γ, $\upsilon_\gamma \models \forall X\, \forall x\, (X(x) \vee \neg X(x))$

iff for every assignment of references γ, for every assignment of references τ that differs from γ at most in what it assigns X, $\upsilon_\tau \models \forall x\, (X(x) \vee \neg X(x))$

iff for every assignment of references γ, for every assignment of references τ that differs from γ at most in what it assigns X, for every assignment of references σ that differs from τ at most in what it assigns x, $\upsilon_\sigma \models X(x) \vee \neg X(x)$

iff for every assignment of references γ, for every assignment of references τ that differs from γ at most in what it assigns X, for every assignment of references σ that differs from τ at most in what it assigns x, $\upsilon_\sigma \models X(x)$ or not $\upsilon_\sigma \models X(x)$.

As before, in any semantic analysis we finally arrive at a place where atomic predications determine the truth-value of the whole.

But there are problems in ensuring that this process yields an inductive or even coherent definition of truth in a model. And they all have to do with self-reference.

We know that self-reference can lead to contradictions, as discussed in Chapter IV.D and Example 15 of Chapter VII.C. The self-reference exclusion principle excludes ' '— is a dog' is a dog' from the semi-formal realization. But since, as we've agreed, '— is a dog' is a thing (whether you think that inscription is the predicate or is standing for an abstract predicate), it could be in the universe of our realization. So we could set $\sigma(X)$ to be '— is a dog', and $\sigma(x)$ to be '— is a dog'. And then in the evaluation of $\upsilon_\sigma(X(x))$ we have the problems we were so careful to avoid. The solution is to require:

No predicate over which we are quantifying is an object in the universe.

So $\sigma(x)$ cannot be '— is a dog'.

This same problem appears in the semi-formal language. If we allow X(Y) to be a wff, then in a semantic analysis we could have $\sigma(X) = \sigma(Y)$, and we have ' '— is a dog' is a dog' all over again. So we exclude such concatenations of symbols as wffs. That is, we will allow as wffs $X(t_1, \ldots, t_n)$ if the t_i are terms (individual variables, or names, or compound names), but we will not allow

$X(Y_1, \ldots, Y_n)$, nor $X(Y, y)$, nor $X(Y, y, c)$. Thus, the atomic predications we'll consider will be those that are predicated of objects in the universe, just as before.

We make these restrictions because we wish to avoid contradictions engendered by self-reference. To the extent that the self-reference exclusion principle is only a temporary simplification in formalizing reasoning that we later hope to lift, so, too, will these restrictions be only simplifications.

Returning now to the evaluation at (5), we know that we have to determine $\nu_\sigma(X(x))$ for all σ before we can complete the analysis. And we know, further, that $\sigma(x)$ is not a predicate.

Since $X(x)$ is an atomic wff, it seems the assignment of a truth-value to it under the assignment σ ought to be primitive, just as we take $\nu_\sigma(P_1^1(x))$ as primitive when '— is a dog' realizes P_1^1. If so, we must impose consistency conditions on the predications. In addition to the conditions on consistency of predications we already have for our models (Chapter IV.E.2), we now need the following.

Cons 1 *Consistency of predications with respect to variables and terms.*
For any σ and τ, for all terms t and u, and all variables X and Y,
if $\sigma(t) = \tau(u)$ and $\sigma(X) = \tau(Y)$, then $\nu_\sigma(X(t)) = \nu_\tau(Y(u))$.

In addition, we'll also require extensionality of predications with respect to how we refer to the predicate in the predication.

Cons 2 *Extensionality with respect to predicate variables and atomic predicates.*
For any σ and τ, for any predicate symbol P_i^1, and all terms
t and u, if $\sigma(t) = \tau(u)$ and $\sigma(X) =$ the realization of P_i^1,
then $\nu_\sigma(X(t)) = \nu_\tau(($the realization of $P_i^1)(u))$.

For example, if '— is a dog' is both $\sigma(X)$ and the realization of P_0^1, and $\sigma(x)$ is the lamp on my table, and 'Bud' is a name in the semi-formal language for that lamp, then Cons 2 requires us to have for every τ, $\nu_\sigma(X(x)) = \nu_\tau($Bud is a dog$)$.

Given these consistency conditions, we'd like to take $\nu_\sigma(X(x))$ as primitive for every σ. This would be regardless of whether you believe 'Bud is a dog' comes equipped with a truth-value or we have the liberty of making one model in which it is true and another in which it is false. How you determine the truth-values of these predications is not for logic to decide.

But some atomic predications cannot be taken as primitive. If $\sigma(X)$ is a compound formula of the first-order part of the language (or what the compound represents), say '$\forall y$ (y is a cat $\rightarrow x$ is bigger than y)', then $\nu_\sigma(X(x))$ should be the same as $\nu_\sigma(\forall y$ (y is a cat $\rightarrow x$ is bigger than y)). And that is determined already by the primitive predications ν_τ(y is a cat) and ν_τ(x is bigger than y) for all τ, using our previous definition of truth in a model. In order to ensure a consistent inductive definition of truth in a model we have to proceed in two stages: we define truth-conditions for the first-order part of the language as before, then, using the values so determined, we impose the following condition.

Cons 3 *Consistency of predications with respect to predicate variables and compound predicates.*
For any σ and τ, for any formula A of the first-order part of the semi-formal language that has exactly one free variable x, for all terms t and u, if $\sigma(t) = \tau(u)$ and $\sigma(X) = A$ (or what A represents), then $\upsilon_\sigma(X(t)) = \upsilon_\tau(A(u/x))$.

But a problem remains: σ may assign a second-order formula to X. For example, we could have:

(6) $\sigma(X) = \forall X\,(X(x) \vee x \text{ is a dog})$

We have self-reference, vicious circularity: The evaluation of $\upsilon_\sigma(X(x))$ involves $\upsilon_\sigma(\forall X\,(X(x) \vee x \text{ is a dog}))$, which itself involves $\upsilon_\sigma(X(x))$. We cannot establish an inductive definition of truth; it's not even clear we could have any coherent definition of truth.

Opinion now splits on what restriction should be made. All agree that we cannot let X be free in the assignment of σ(X). However, as we shall see in Section F.4, some logicians accept apparently circular evaluations such as (6). For now, let's adopt the following to avoid possible self-referential problems.

Cons 4 *The reference assigned to* X *cannot be any formula of the semi-formal language that contains predicate variables.*

Actually, we should require that σ(X) cannot be any ordinary language predicate that we'd readily formalize as a wff of the semi-formal language that involves a predicate variable, but that must remain an informal restriction.

With conditions Cons 1–4 we can give an inductive definition of truth: We start with assignments of references to the variables, we ascribe truth-values to the first-order atomic predications, and then establish, simultaneously, truth-values for each first-order formula relative to all assignments of references. Then we extend to formulas involving predicate variables the truth-assignments relative to assignments of references by the same method of interpreting quantifiers we used for individual variables, as illustrated at (5). I'll let you formulate Cons 1–4 for predicate variables of all arities.

To make this more precise, we need to set out exactly the language of second-order logic.

Higher-order logics

Our consistency conditions and the restrictions on what predicates can be taken as (temporary) reference of a predicate variable are all of one piece:

> *We distinguish between predicates and the objects of which they can be predicated; we do not predicate a predicate of a predicate.*

By distinguishing between predicates that can be predicated of predicates, what are called *third-order* objects, and predicates that can be predicated of objects only, we could allow for a predication such as: ''— is the brother of —' is symmetric'.

A third-order predicate could not be predicated of another third-order predicate, but only of individuals and predicates as we previously understood those. Allowing for third-order variables we could develop a third-order logic. And then a fourth-order logic in which we could make predications about third-order predicates; and a fifth-order logic, and—the sky's the limit—even a four hundred and seventy-seventh-order logic. But for this book I'll stop with second-order logic.

For an introduction to higher-order logics, see *Copi, 1971*.

C. A Formal Language for Second-Order Logic, L_2

The formal language of second-order logic, L_2

We add to the vocabulary of the formal language of first-order logic with equality:

predicate variables: $X_0^1, X_0^2, \ldots, X_1^1, X_1^2, \ldots, X_i^n, \ldots$ for $i \geq 0, n \geq 1$

We now call $x_0, x_1, \ldots$ *individual variables.* A *term* is defined as before.

The definition of a *well-formed-formula* (*wff*) is given by:

1. *Atomic wffs*: For every $i \geq 0, n \geq 1$, and terms t_j, $1 \leq j \leq n$,
   $(P_i^n(t_1, \ldots, t_n))$ is an atomic wff.
   $(X_i^n(t_1, \ldots, t_n))$ is an atomic wff.
   $t_1 = t_2$ is an atomic wff.
   All atomic wffs have *length* 1.
2. If A is a wff to which the number n is assigned, then each of $(\neg A)$, and, for any $i \geq 0, m \geq 1$, $(\forall x_i\, A)$, $(\exists x_i\, A)$, $(\forall X_i^m\, A)$ and $(\exists X_i^m\, A)$ is a *compound* wff with *length* $n + 1$.
3. If A and B are wffs and the maximum of the lengths assigned to A and to B is n, then each of $(A \rightarrow B)$, $(A \wedge B)$, and $(A \vee B)$ is a *compound* wff with *length* $n + 1$.
4. A concatenation of symbols is a *wff* iff it is assigned some length $n \geq 0$.

The unique readability of wffs can be proved much as for wffs of our previous formal languages (pp. 63–64). I'll continue to use the conventions on deleting parentheses in informal presentations of wffs (p. 63) along with a new convention that we may delete the outer parentheses in '($X_i^n(t_1, \ldots, t_n)$)'.

The full name of this language, listing all its primitives is:

$$L(\neg, \rightarrow, \wedge, \vee, \forall, \exists, =; x_0, x_1, \ldots, X_0^1, X_0^2, X_0^3, \ldots, X_1^1, X_1^2, X_1^3, \ldots, P_0^1, P_1^1, \ldots, P_0^2, P_1^2, \ldots, f_0^1, f_1^1, \ldots, f_0^2, f_1^2, \ldots, c_0, c_1, \ldots)$$

That is a very long name, so I'll simply use the subscript '2' to indicate that second-order variables are among the primitives, calling the language L_2, or

$$L_2(\neg, \rightarrow, \wedge, \vee, \forall, \exists, =; P_0, P_1, \ldots, f_0, f_1, \ldots, c_0, c_1, \ldots)$$

The language we used previously is now called the *first-order part of* L_2.

The definitions concerning the complexity of formal wffs and free and bound variables from Chapter III (p. 65) carry over with little or no modification. I'll discuss here the ones we'll need in this chapter. I'll leave to you the definition of what it means to say a *symbol appears* in a wff.

The definition of the *scope* of an occurrence of a quantifier is extended:

In ($\forall X_i^n$ A) the scope of the initial $\forall X_i^n$ is A.

In ($\exists X_i^n$ A) the scope of the initial $\exists X_i^n$ is A.

The definition of a *free variable* is extended:

An occurrence of a variable X in A is *bound in* A if it immediately follows an occurrence of a symbol '$\forall$' or '$\exists$' or lies within the scope of a quantifier '$\forall$X' or '$\exists$X'. If it is not bound in A, it is *free in* A.

In a wff ($\forall$X A), respectively ($\exists$X A), every free occurrence of X in A is *bound by* the initial $\forall$X, respectively $\exists$X.

As before, the wffs that will be symbolic versions of propositions are those in which no occurrence of a variable is free; we call those *closed wffs.* A wff that is not closed is *open.* If $Y_1, \ldots, Y_m, y_1, \ldots y_n$ appear free in A, we write $A(Y_1, \ldots, Y_m, y_1, \ldots y_n)$, though other variables may appear free in A, too. The *universal closure* of a wff A is:

$$\forall \ldots A \equiv_{\text{Def}} \forall X_{j_1}^{k_1} \ldots \forall X_{j_m}^{k_m} \forall x_{i_1} \ldots \forall x_{i_n} A$$

where $x_{i_1}, \ldots, x_{i_n}$ is the list of all individual variables that occur free in A, with $i_1 < \cdots < i_n$, and $X_{j_1}^{k_1}, \ldots, X_{j_m}^{k_m}$ is a list of all predicate variables that occur free in A, such that for all i, $j_i < j_{i+1}$, or $j_i = j_{i+1}$ and $k_i < k_{i+1}$

The formal language may have fewer or no name symbols, fewer or no predicate and function symbols, and may or may not include the equality predicate; I'll assume the modifications necessary for those cases when we need them. Unlike the first-order language, however, we may take the language to have no predicate symbols or function symbols—not even the equality predicate—and no name symbols. We call that the *language of pure second-order logic*. In that language we still have closed formulas that we can construe as propositions, for example, $\forall X_1^1\ \forall x_0\ (X_1^1(x_0) \vee \neg X_1^1(x_0))$.

Exercises for Sections A–C

1. Give at least two motives for extending classical predicate logic to allow for quantification over predicates.
2. For each of Cons 1–4:
   a. Give an informal statement of it.
   b. Say whether it is a restriction on the semantics or the syntax.
   c. Explain why the restriction is needed.
3. a. State each of Cons 1–4 for binary predicates.
   b. State each of Cons 1–4 for predicates of all arities.
4. Prove the unique readability of wffs of L_2.
5. Give a definition of 'symbol α appears in wff A' for L_2.
6. Which of the following are wffs of L_2, according to the conventions on deleting parentheses?
   a. $\forall X_1^1\ (P_1^1(x_1))$
   b. $\forall X_1^1\ (X_1^1(P_1^1(x_1)))$
   c. $\exists x_1\ (X_0^2(x_1, X_1^1))$
   d. $\exists x_1\ \forall x_2\ (X_0^2(x_1,x_2) \to \neg X_0^1(x_1))$
   e. $\forall x_2\ \exists X_1^1\ \exists x_1\ (X_1^1 \to \neg(c_0 = x_2 \vee x_1 = x_2))$
   f. $\exists x_1\ X_4^2(x_1, x_2, x_3, x_4)$
   g. $\exists x_{16}\ X_2^3(x_{16}, x_2, x_4)$
   h. $\forall X_1^1\ (X_1^2(c_1, X_1^1(c_0))$
7. For each of the following (abbreviations of) wffs:
   i. Identify the scope of each quantifier.
   ii. State which occurrences of variables are free and which are bound.
   iii. State whether the wff is a sentence or an open expression, as well as whether it is atomic.
   iv. Give the universal closure of the wff.
   a. $X_1^1(c_0) \to \exists x_1\ X_1^1(x_1)$
   b. $\neg\forall X_{47}^2\ (X_{47}^2(x_1, x_2) \to \exists X_{47}^2\ (X_{47}^2(c_0, c_1)))$
   c. $P_1^1(x_1) \vee \neg X_1^1(x_1)$
   d. $\forall X_1^1\ (X_1^1(c_0) \wedge X_1^2(c_0, x_1)) \to \exists x_1\ \exists X_1^2\ (X_1^2(x_1, c_0))$

D. Realizations and Models

For a realization of L_2 the semi-formal language should be just as before: predicate symbols (if any) are realized as particular predicates, function symbols as functions, and name symbols as names. The formal symbol '=' is allowed as a predicate in the semi-formal language. If we are using the language of pure second-order logic or pure second-order logic with equality, the semi-formal language and the formal language will be the same.

As before we shall have to specify a collection of objects for the universe of the realization, those objects that we allow as temporary reference for the individual variables. But now we must specify the collection of predicates which the predicate variables may be assigned as temporary reference, the *range of the predicate variables.*

We have the restriction from Section B that we shall allow as reference for a predicate variable only a predicate of objects and not a predicate of predicates. That is, among the predicates that the predicate variables can take as reference there is none that is an object of the universe, nor a formula of the semi-formal language that contains a predicate variable. Essentially, the predicates a predicate variable can take as temporary reference are those we previously considered were *predicates for this universe*, that is, expressions (or what the expressions represent) such that whenever the blanks are filled with names of objects (are given as temporary reference objects of the universe), a proposition results.

For example, for the universe of all natural numbers, if we cannot agree that '— loves —' has a truth-value no matter what names appear in the blanks, then we shall not allow it to be taken as reference of a predicate variable. If '$\exists y$ (x is prime $\wedge\ x > y$)' is a formula of the semi-formal language to which we will assign semantic properties, then it's a predicate on the universe of natural numbers.

For '$\forall$' and '$\exists$' to continue to be acceptable formalizations of 'for all' and 'there exists', we must have that all predicates on the universe are in the range of the predicate variables. Though the collection of "all things" is too vague—if it makes sense at all—to be used as the universe of a realization, we may accept that the collection of all predicates on a specified universe is not so vague as to be ill-defined. What exactly 'all predicates' means will depend on particular conceptions of predicates, as we'll see in Section F below. In any case, the range of the predicate variables will not be empty.

So by specifying the universe of a realization we have specified the range of the predicate variables (relative to any particular conception of predicates).

Now we need to specify what we mean by an assignment of references and ensure that we have enough of them for '$\forall$' to formalize 'for all'. Then we can define what it means for an assignment to satisfy a formula.

Assignments of references An *assignment of references to the variables* is a method to assign:

To each x_i an object of the universe, for $i \geq 0$.

To each X_i^n in the semi-formal language an n-ary predicate in the range of the predicate variables, for $i \geq 0$ and $n \geq 1$.

We require also that each assignment assigns to each name of the realization an object of the universe such that all assignments assign that name the same object.

Completeness of the collection of assignments of references
There is at least one assignment of references.

For every assignment of references σ and any object of the universe, for any $i \geq 0$ there is an assignment τ that differs from σ at most in that it assigns that object to x_i.

For every assignment of references σ and any n-ary predicate in the range of the predicate variables, for any $i \geq 0$, there is an assignment τ that differs from σ at most in that it assigns that predicate to X_i^n.

Valuations
We take as primitive for every assignment of references σ the *valuation based on* σ for the first-order atomic formulas, υ_σ: First-order atomic wffs $\rightarrow \{T, F\}$. This must satisfy the first-order requirement on consistency of predications (p. 85), using the implicit identity of the universe to interpret '='. Then, as before, all υ_σ are simultaneously extended to all first-order wffs (p. 91).

We then extend all υ_σ to the atomic wffs containing predicate variables:

1. If $\sigma(X_i^n)$ is not a predicate of the semi-formal language, then the value of $\upsilon_\sigma(X_i^n(t_1, \ldots, t_n))$ is taken as primitive, subject to the requirement of consistency of predications (Cons 1 above).
2. If $\sigma(X_i^n)$ is a predicate (represented by) $A(y_1, \ldots, y_n)$ of the semi-formal language, then $\upsilon_\sigma(X_i^n(t_1, \ldots, t_n)) = \upsilon_\sigma(A(t_1/y_1, \ldots, t_n/y_n))$. (Cons 2 and Cons 3 above).

Then all υ_σ are extended simultaneously to all formulas of the realization using the previous clauses and:

$\upsilon_\sigma(\forall X\, A) = T$ iff for every assignment of references τ that differs from σ at most in what it assigns as reference to X, $\upsilon_\tau(A) = T$

$\upsilon_\sigma(\exists X\, A) = T$ iff for some assignment of references τ that differs from σ at most in what it assigns as reference to X, $\upsilon_\tau(A) = T$

If $\upsilon_\sigma(A) = T$, we say σ *satisfies* or *validates* A and write '$\upsilon_\sigma \vDash_2 A$' or '$\sigma \vDash_2 A$'.

Just as in the first-order case, at each stage of an analysis of the truth-value of a compound proposition we will have to account for assigning reference to at most a finite number of occurrences of a single variable not assigned reference before. I'll let you show by induction on the length of wffs, using the unique readability of wffs, that for every σ and every wff A a unique truth-value is determined for $\upsilon_\sigma(\mathrm{A})$.

Note that if A is in the first-order part of the language, then for any σ, $\sigma \models_2 \mathrm{A}$ iff $\sigma \models \mathrm{A}$ according to the usual first-order definition of truth in a model. We have extended, not altered, our previous definition of truth in a model. I'll let you prove the following, which is the second-order version of Theorem IV.1.

Theorem 1 Given any wff A and assignments of references σ and τ, if for every variable x free in A, $\sigma(x) = \tau(x)$, and for every variable X free in A, $\sigma(\mathrm{X}) = \tau(\mathrm{X})$, then $\upsilon_\sigma(\mathrm{A}) = \upsilon_\tau(\mathrm{A})$. And for closed A, $\upsilon_\sigma(\mathrm{A}) = \mathsf{T}$ for every assignment of references σ iff $\upsilon_\sigma(\mathrm{A}) = \mathsf{T}$ for some assignment of references σ.

Truth in a model
A valuation υ on all closed wffs of the semi-formal language is given by:

$\upsilon(\mathrm{A}) = \mathsf{T}$ iff for every assignment of references σ, $\upsilon_\sigma(\mathrm{A}) = \mathsf{T}$.

$\upsilon(\mathrm{A}) = \mathsf{F}$ iff $\upsilon(\mathrm{A}) \neq \mathsf{T}$.

A *second-order model* is then the realization, the collection of assignments of references, primitive evaluations of atomic predications, the method of extension of those to all wffs, and the valuation υ. A proposition of the realization is *true in the model* if $\upsilon(\mathrm{A}) = \mathsf{T}$, *false* if $\upsilon(\mathrm{A}) = \mathsf{F}$.

Validity and semantic consequence for second-order models are defined in the usual way, writing '$\models_2$' in place of '$\models$'. *Second-order classical predicate logic* is then this notion of model and semantic consequence for the formal language.

Exercises for Section D

1. For each part of Exercise 7 of Sections A–C give a semi-formal wff that could realize it.
2. Show that the range of the predicate variables is not empty.
3. Give the truth-conditions (see p. 93) for the following propositions:
   a. $\forall x_1\ (x_1$ is a dog $\vee \exists \mathrm{X}_0^1 \neg \mathrm{X}_0^1(x_1))$
   b. $\exists x_1 \forall \mathrm{X}_1^2\ (\mathrm{X}_1^2(\text{Ralph, Ralph}) \vee \neg \mathrm{X}_1^2(x_1, x_1))$
4. Show by induction on the length of wffs, using the unique readability of wffs, that for every σ and every wff A a unique truth-value is determined for $\upsilon_\sigma(\mathrm{A})$.
5. Prove Theorem 1.

E. Examples of Formalization

1. Every predicate is true of Ralph or is not true of Ralph.

$\forall X\, (X(\text{Ralph}) \vee \neg X(\text{Ralph}))$

predicates: none
names: 'Ralph'
functions: none

$\forall X_1^1\, (X_1^1(c_0) \vee \neg X_1^1(c_0))$

Analysis The proposition appears to be excluded from formalization by the self-reference exclusion principle because of the phrase 'is true of'. But in our models ''— is a dog' is true of Ralph' iff 'Ralph is a dog' is true.

2. For every predicate X, X(c) ∨ ¬X(c).

Not formalizable.

Analysis A fully formal version of Example 2 would be $\forall X_0^1\, (X_0^1(c_0) \vee \neg X_0^1(c_0))$. To view that as a proposition instead of the form of a proposition confuses the metalogic with the logic, for either a name must be substituted for 'c_0' or a universal quantifier over names added in order to convert the wff into a proposition rather than the form of a proposition. Example 2 can serve as a scheme of propositions formalizable in second-order logic, but not as a single proposition that can be formalized.

3. Everything true of Ralph is true of Juney.
***Therefore,* if Ralph is a dog, then Juney is a dog.**

$\forall X\, (X(\text{Ralph}) \rightarrow X(\text{Juney}))$
Therefore, Ralph is a dog → Juney is a dog.

predicates: 'is a dog'
names: 'Ralph', 'Juney'
functions: none

$$\frac{\forall X_1^1\, (X_1^1(c_0) \rightarrow X_1^1(c_1))}{P_1^1(c_0) \rightarrow P_1^1(c_1)}$$

Analysis We say a predicate is true of an object; quantifying over those things that are true of Ralph can be viewed as quantifying over predicates. Hence the argument can be put into a semi-formal language.

The deduction is valid because every first-order predicate of the language is in the range of the predicate variables, or if you prefer, for every first-order predicate of the semi-formal language there is a predicate corresponding to it in the range of the predicate variables. *First-order predicates are to predicate variables as the names in the semi-formal language are to individual variables.*

4. No predicate is green.

Not formalizable.

Analysis I take this example to be a proposition, as I expect a platonist would (though I take it to be false, and a platonist would say it is true). But it is not formalizable for two reasons.

First, consider the assertion: ''— is a dog' is green'. The predicate '— is green' is being applied to a predicate, not to an object of the universe. We cannot use '— is green' in that way in second-order logic.

Second, we have no way to quantify over all predicates, but only over all unary predicates, or all binary predicates, or . . . . In Example 1 it was implicit that unary predicates were being spoken of because the predicates were applied to a single object. But in Example 4 we mean that no predicate of any arity is green.

It is not our goal to devise a logic of predicates comparable to a logic of things generally. If it were, then our inability to formalize Example 4 would be fatal: We could not compare various conceptions of predicates within the semi-formal language. Predicate logic, whether we allow quantification over predicates or not, is a logic of things. We began with the assumption that the world is made up of things and that propositions are about things. Predicates came to our attention via that assumption, and our interest in them is solely in that context.

5. Predicate (3) of Chapter 3 applies to Ralph.

Not formalizable.

Analysis We do not allow for names of predicates in our semi-formal languages.

6. Every two things are related by some relationship.

$$\forall x\, \forall y\, (x \neq y \rightarrow \exists \mathrm{X}\ \mathrm{X}(x,y))$$

predicates: '='
names: none
functions: none

$$\forall x_1\, \forall x_2\, (\neg(x_1 = x_2) \rightarrow \exists\, \mathrm{X}_1^2\, (\mathrm{X}_1^2(x_1,x_2)))$$

Analysis I've understood 'two things' in its ordinary way, and not in the mathematician's irregular use of it as meaning 'things referred to by two variables'. The semi-formal wff contains no categorematic words and satisfies the self-reference exclusion principle. The semantic predicate 'x and y are related by X' is replaced by $\mathrm{X}(x,y)$.

The example is valid, because we can take $\mathrm{X}(x,y)$ to refer to the formula $\neg(x = y)$, since $\forall x\, \forall y\, (x \neq y \rightarrow x \neq y)$ is valid. But that seems cheap. What about a stronger version of Example 6?

(a) Every two things are related by some relationship other than the negation of the identity relation.

We might argue that (a) is informally valid, but we quantify over predicates as units and have no way to recognize their internal structure, nor can we say that one is the same as another, for that would require third-order predications. We'll return to this in Example 8.

7. Ralph has some quality that distinguishes him from all other things.

$\exists X\, (X(\text{Ralph}) \wedge \forall y\, (y \neq \text{Ralph} \rightarrow \neg X(y)))$

predicates: '='
names: 'Ralph'
functions: none

$\exists X_1^1\, (X_1^1(c_1) \wedge \forall x_0\, (\neg(x_0 = c_1) \rightarrow \neg X_1^1(x_0)))$

Analysis To formalize this example we need to identify qualities with unary predicates. But what do we mean by 'a quality that distinguishes him from all other things'? I take that to mean that the quality (predicate) is true of Ralph and of no other things. So the proposition is valid, for we can take X to be '$y = \text{Ralph}$'.

8. *The identity of indiscernibles*
If every property applies to both *x* and *y* or to neither *x* nor *y*, then *x* is identical to *y*.

$\forall x\, \forall y\, (\forall X\, [(X(x) \wedge X(y)) \vee (\neg X(x) \wedge \neg X(y))] \rightarrow x = y)$

predicates: '='
names: none
functions: none

$\forall x_0\, \forall x_1\, (\forall X_1^1\, [(X_1^1(x_0) \wedge X_1^1(x_1)) \vee (\neg X_1^1(x_0) \wedge \neg X_1^1(x_1))] \rightarrow x_0 = x_1)$

Analysis This principle, discussed in Chapter VI.C, can now be formalized.

Does this mean that in second-order logic we have captured the "true" notion of identity? The answer depends on what you think predicates are. The discussion in Chapter VI.C has not been advanced just because we can now give a formal version of the principle of identity of indiscernibles. What we see, instead, is that disagreements about the nature of predicates can no longer be resolved by pointing out that in any formalized version of reasoning the only "properties" we consider are those which are "expressible" by predicates of the semi-formal language.

Some say that the informal example here characterizes identity. They give *a second-order definition of identity* in pure second-order logic:

$x = y \equiv_{\text{Def}} \forall X\, (X(x) \leftrightarrow X(y))$

They say that this captures "true" identity in the sense that $\sigma \models x = y$ iff $\sigma(x)$ is identical to $\sigma(y)$. In that case, we could formalize 'Every two things are related by some relationship other than the negation of the identity relation' as:

$$\forall x\, \forall y\, (x \neq y \rightarrow \exists X\ X(x,y) \wedge \neg \forall x\, \forall y\, (\forall Y\, (Y(x) \leftrightarrow Y(y)) \leftrightarrow \neg X(x, y))$$

9. If the extensions of the predicates *P* and *Q* are the same, then *P* is the same as *Q*.

Not formalizable.

Analysis It's easy enough to write down a formula that would seem to work for unary predicates: $\forall X\, \forall Y\, (\forall x\, (X(x) \leftrightarrow Y(x)) \rightarrow X = Y)$. But '=' is a relation on objects of the universe, not on predicates; we don't apply predicates to predicates. We would need a second-order version of equality to formalize this.

The example, however, suggests a way to define equality on predicates in second-order logic:

$$X = Y \equiv_{\text{Def}} \forall x\, (X(x) \leftrightarrow Y(x))$$

This is a good definition in the sense that it reflects our informal idea of what '=' for predicates should mean in classical mathematical models.

10. The extension of '— is a dog' is a predicate.

$\exists X\, \forall x\, (X(x) \leftrightarrow x$ is a dog)

predicates: '— is a dog'
names: none
functions: none

$$\exists\, X_0^1\, (\forall x_1\, (X_0^1(x_1) \leftrightarrow P_1^1(x_1)))$$

Analysis Regardless of what conception of predicates we have, this is true in any model in which '— is a dog' is in the semi-formal language. To accept '— is a dog' in the semi-formal language requires that we accept that it is (or represents) a predicate on the universe, and hence its extension is in the range of the predicate variables.

Exercises for Section E

Formalize and discuss each exercise below or explain why it is not formalizable.

1. Some relationship relates every two things
2. Everything is related to something.
3. The only relation between Ralph and Marilyn Monroe is that he's smaller than she was.

4. There are seven differences between Marilyn Monroe and Ralph, one of which is that she is dead.
5. The only relation between Ralph and Marilyn Monroe is in Ralph's mind.
6. Everything true of Ralph is true of Juney and vice-versa, though they are not the same.
7. There is no quality that everything has.
8. There is no quality that nothing has.
9. No predicate that can't be expressed in English is true of Ralph.
10. Every predicate is either linguistic or abstract.

F. Classical Mathematical Second-Order Predicate Logic

1. The abstraction of models

We begin as we did for first-order logic. We first identify a predicate with its extension (Chapter IX.A). Then we abstract models to identify the extension of a predicate with a subset of the collection of n-tuples of the universe of the model (Chapter IX.C). For the first-order part of the language, a model is:

$$\mathsf{M} = < \mathsf{U} ; \mathsf{P}_0^1, \mathsf{P}_1^1, \ldots, \mathsf{P}_i^n, \ldots, \mathsf{f}_0^1, \mathsf{f}_1^1, \ldots, \mathsf{f}_i^n, \ldots, \mathsf{a}_0, \mathsf{a}_1, \ldots > \quad i \geq 0, n \geq 1$$

1. U is a non-empty set.
2. $\mathsf{P}_i^n \subseteq \mathsf{U}^n$.
3. f_i^n is a function from U^n to U.
4. $\mathsf{a}_i \in \mathsf{U}$.

An assignment of references is any function from terms to elements of U satisfying:

For each i, for each σ, $\sigma(c_i) = \mathsf{a}_i$.

For each i, for each σ, for any terms $t_1, \ldots, t_n$,

$$\sigma(f_i^n(t_1, \ldots, t_n)) = \mathsf{f}_i^n(\sigma(t_1), \ldots, \sigma(t_n)).$$

For first-order logic we make a further abstraction, or at least it is a further abstaction for those who hold that there are sets that aren't expressible as extensions of predicates. We say that any structure satisfying conditions (1)–(4) is a model of the formal language.

We've already agreed that the range of the predicate variables in a second-order model is the collection of all predicates on the universe. So for the classical mathematical abstraction, the range of the predicate variables is the set of all subsets of n-tuples of the universe for all $n \geq 1$. That set is guaranteed to exist by the assumptions of our naive set theory (p. 159). So no further abstraction is needed.

The fully general classical abstraction of second-order predicate logic models
Any structure $< \mathsf{U}\,;\, \mathsf{P}_0^1, \mathsf{P}_1^1, \ldots, \mathsf{P}_i^n, \ldots, \mathsf{f}_0^1, \mathsf{f}_1^1, \ldots, \mathsf{f}_i^n, \ldots\; \mathsf{a}_0, \mathsf{a}_1, \ldots >$ satisfying conditions (1)–(4) is a model of the formal language, where the range of the predicate variables is the set of all subsets of n-tuples of the universe for $n \geq 1$.

Assignments of references are defined as for first-order classical mathematical models adding:

$$\sigma(\mathrm{X}_i^n) \subseteq \mathsf{U}^n$$

Truth in a model is defined as before, except that at the atomic stage we set:

$$v_\sigma(\mathrm{P}_i^n(t_1, \ldots, t_n)) \text{ iff } (\sigma(t_1), \ldots, \sigma(t_n)) \in \mathsf{P}_i^n$$

$$v_\sigma(\mathrm{X}_i^n(t_1, \ldots, t_n)) \text{ iff } (\sigma(t_1), \ldots, \sigma(t_n)) \in \sigma(\mathrm{X}_i^n)$$

When we need to distinguish this definition from the first-order definition, I'll write '$\sigma \vDash_2 \mathrm{A}$' for $v_\sigma(\mathrm{A}) = \mathsf{T}$.

These are the *classical mathematical second-order models* or *classical mathematical models of second-order logic.* I will leave to you to show that each of Cons 1–3 are satisfied by these models. And if A is in the first-order part of the language, then for any assignment of references σ, $\sigma \vDash_2 \mathrm{A}$ iff $\sigma \vDash \mathrm{A}$ by the usual definition in the first-order part of the model.

Validity and semantic consequence are defined in the usual way: $\vDash_2 \mathrm{A}$ iff A is true in every classical mathematical second-order model; $\Gamma \vDash_2 \mathrm{A}$ iff A is true in every classical mathematical second-order model in which all the wffs of Γ are true.

2. All things and all predicates

With the classical mathematical abstraction, second-order logic becomes more than just tied to the study of sets. For any assignment of references σ, $\sigma \vDash \mathrm{X}(x)$ iff $\sigma(x) \in \sigma(\mathrm{X})$, so second-order logic becomes one way to do set theory. Indeed, *Takeuti, 1975*, pp. 129–133, suggests that this is a better way to do the set theory needed for mathematics. But the classical mathematical abstraction is not compatible with many conceptions of predicates, nor with reasoning that pays attention to more about predicates than just their form and their extension, as I've discussed in Chapter X of *Predicate Logic*.

Though the classical mathematical abstraction of second-order models looks clear, there is a problem. Mathematicians and logicians do not agree about what, in the simplest case corresponding to unary predicates, 'all subsets of the universe' comprises. Different conceptions of predicates yield different understandings of that phrase.

We did not have this problem with first-order logic. We have some general notion of what it means to be a thing, some core upon which we can agree: trees, dogs, people, ... Our agreement about that core gives us confidence to develop a logic that formalizes how we should reason about things. A large periphery of this notion provokes disagreements: electrons, quarks, numbers, uncountable sets, abstract propositions, ... Nonetheless, we can reason together using the logic we have developed on the basis of our core understandings so long as we agree that whatever kind of things we wish to reason about satisfy the semantic assumptions we made in establishing classical predicate logic. Differences about what things exist can be accommodated as different first-order theories.

For any first-order model the only predicates that concern us are those that are (represented) in the semi-formal language. It may be that there are inexpressible abstract predicates, but that won't affect our reasoning. By making the classical mathematical abstraction we do rule out some conceptions of predicates. But that abstraction does not affect our reasoning within a semi-formal language, though it perhaps affects our notion of validity (Chapter X.A.5).

In contrast, different conceptions of predicates do affect reasoning with specific models of the second-order language, even given the classical mathematical abstraction. The range of the second-order variables, which seemed fixed by the phrase 'all subsets of', depends on how we understand the nature of predicates and, correspondingly, our understanding of sets. We'll return to this issue after we look at some examples of formalizations.

3. Examples of formalization

11. There are countably many things.

$$\exists \mathrm{Y}\, (\forall x\, \exists! z\, \mathrm{Y}(x,z) \wedge$$
$$\exists y\, \forall \mathrm{X}\, ([\mathrm{X}(y) \wedge \forall x\, \forall z\, ((\mathrm{X}(x) \wedge \mathrm{Y}(x,z)) \rightarrow \mathrm{X}(z))] \rightarrow \forall x\, \mathrm{X}(x)))$$

predicates: '='
functions: none
names: none

$$\exists\, \mathrm{X}_1^2\, (\forall x_1\, \exists! x_3\;\, \mathrm{X}_1^2(x_1, x_3) \wedge \exists x_2\, \forall\, \mathrm{X}_0^1\, ((\mathrm{X}_0^1(x_2) \wedge$$
$$\forall x_1\, \forall x_3\, ((\mathrm{X}_0^1(x_1) \wedge \mathrm{X}_1^2(x_1, x_3)) \rightarrow \mathrm{X}_0^1(x_3))) \rightarrow \forall x_1\, \mathrm{X}_0^1(x_1)))$$

Analysis I understand the proposition to mean 'There are exactly countably many things', not 'There are at least countably many things.' I'll show that the formalization is true in a model iff the universe is countable.

Suppose the universe is countable. Since the universe is not empty, let either $\mathsf{a}_1, \ldots, \mathsf{a}_n$ or $\mathsf{a}_1, \ldots, \mathsf{a}_n, \ldots$ be an enumeration of the universe without repetitions, depending on whether the universe is finite or infinite. If the universe is finite, the relation $\{(\mathsf{a}_1, \mathsf{a}_2), \ldots, (\mathsf{a}_{n-1}, \mathsf{a}_n), (\mathsf{a}_n, \mathsf{a}_n)\}$ is in the range of the predicate variables. If the universe is infinite, the relation $\{(\mathsf{a}_i, \mathsf{a}_{i+1}) : i \geq 1\}$ is in the range of the predicate variables. Given any subset P of the universe such that

$\mathsf{a}_1 \in \mathsf{P}$ and for every a_i, if $\mathsf{a}_i \in \mathsf{P}$ then $\mathsf{a}_{i+1} \in \mathsf{P}$, we have by induction that every object of the universe is in P. I'll let you show that P satisfies the formal wff following the initial $\exists$Y, and hence that the formalization is true in the model.

Now suppose the formalization is true in a model. Then there is a binary relation R on the universe satisfying the wff following the initial quantifier in the formal wff. There is also some object of the universe, call it a_1, such that R and a_1 satisfy the second conjunct of that wff. So define by induction for all $i \geq 1$:

$$\mathsf{a}_{i+1} \text{ is the unique object such that } (\mathsf{a}_i, \mathsf{a}_{i+1}) \in \mathsf{R}$$

We do not rule out that $\mathsf{a}_i = \mathsf{a}_{i+1}$. If 'Y' is interpreted as R, 'y' as a_1, and 'X' as $\{\mathsf{a}_i : i \geq 1\}$, we have that '$\mathrm{X}(y) \wedge \forall x \forall z\, (\mathrm{X}(x) \wedge \mathrm{Y}(x,z) \rightarrow \mathrm{X}(z))$' is true, and hence '$\forall x\, \mathrm{X}(x)$' is true in the model. So every object in the universe is in $\{\mathsf{a}_i : i \geq 1\}$.

Note that the formalization is in pure second-order logic with equality.

12. There are infinitely many things.

$$\exists \mathrm{Y}\, (\forall x\, \exists ! z\, \mathrm{Y}(x,z) \wedge \forall x\, \forall y\, \forall z\, (\mathrm{Y}(x,z) \wedge \mathrm{Y}(y,z) \rightarrow x = y) \wedge \exists z\, \forall x\, \neg \mathrm{Y}(x,z))$$

predicates: '='
functions: none
names: none

$$\exists \mathrm{X}_1^2\, (\forall x_1\, \exists ! x_3\; \mathrm{X}_1^2(x_1,x_3) \wedge \forall x_1\, \forall x_2\, \forall x_3\, (\mathrm{X}_1^2(x_1,x_3) \wedge \mathrm{X}_1^2(x_2,x_3) \rightarrow x_1 = x_2) \wedge \exists x_3\, \forall x_1\, \neg\, \mathrm{X}_1^2(x_1,x_3))$$

Analysis Our definition of what it means to be infinite (p. 160) is 'There is a 1-1 function that is not onto'. Thus, the formal wff is true in exactly those models in which the universe is infinite.

In first-order logic we can formalize this example, but only by using an infinite collection of wffs (Corollary XI.4).

13. There is exactly a countable infinity of things.

Example 11 $\wedge$ Example 12

Analysis The formalization is true in a model iff the universe of the model is countable and infinite. The upward Löwenheim-Skolem theorem (Corollary XI.8) fails in second-order logic. Note that this formalization is in pure second-order logic with equality. By using the definition of '=' from Example 8, we can formalize the example in pure second-order logic.

14. There are uncountably many things.

The negation of Example 11.

Analysis This is true in a model iff the model has an uncountable universe. The downward Löwenheim-Skolem theorem (Theorem XI.5) also fails for second-order logic. This formalization, too, is in pure second-order logic with equality.

15. There are infinitely many stars.

Inf ('— is a star')

predicates: 'is a star', '='
functions: none
names: none

Analysis We can use Example 12 to define a new quantifier, 'There are infinitely many'.

$$\text{Inf}(\text{W}) \equiv_{\text{Def}} \exists \text{Y}\,(\forall x\,(\text{W}(x) \rightarrow \exists! z\,(\text{W}(z) \wedge \text{Y}(x,z))) \wedge \forall x\, \forall y\, \forall z\,(\text{W}(x) \wedge \text{W}(y) \rightarrow (\text{Y}(x,z) \wedge \text{Y}(y,z) \rightarrow x = y)) \wedge \exists z\,(\text{W}(z) \wedge \forall x\, \neg \text{Y}(x,z))$$

In any model, a subset $\mathfrak{A}$ of the universe satisfies Inf (W) iff $\mathfrak{A}$ is infinite. I'll leave to you to write out a fully formal equivalent of Example 15.

16. There is a countable infinity of stars.

Count ('— is a star') $\wedge$ Inf ('— is a star')

predicates: 'is a star', '='
functions: none
names: none

Analysis Using Example 11 as a guide we can define a new quantifier, 'There are countably many'.

$$\text{Count}(\text{W}) \equiv_{\text{Def}} \exists \text{Y}\,(\forall x\,(\text{W}(x) \rightarrow \exists! z\,(\text{W}(z) \wedge \text{Y}(x,z))) \wedge \exists y\,(\text{W}(y) \wedge \forall \text{X}\,([\text{X}(y) \wedge \forall x\, \forall z\,(\text{W}(x) \wedge \text{W}(z) \wedge \text{X}(x) \wedge \text{Y}(x,z)) \rightarrow \text{X}(z)] \rightarrow \forall x\,(\text{W}(x) \rightarrow \text{X}(x)))$$

In any model, a subset $\mathfrak{A}$ of the universe satisfies Count(W) iff $\mathfrak{A}$ is countable.

17. There are only finitely many things.

The negation of Example 12.

Analysis In first-order logic we cannot formalize this example, even with an infinite collection of wffs. I'll leave to you as an exercise to define a quantifier 'There are only finitely many'.

18. Some relation is symmetric.

$\exists X\, \forall x\, \forall y\, (X(x,y) \rightarrow X(y,x))$

predicates: none
functions: none
names: none

$\exists X_1^2\, (\forall x_1\, \forall x_2\, (X_1^2(x_1,x_2) \rightarrow X_1^2(x_2,x_1))$

Analysis The predicate '— is symmetric' is third-order. But we can characterize what it means to be symmetric in terms of how a predicate relates elements of the universe. That is, we can replace the third-order predicate '— is symmetric' with a compound second-order predicate.

19. There is an ordering on the universe.

$\exists X\, (\forall x\, \neg X(x,x) \wedge \forall x\, \forall y\, (X(x,y) \rightarrow \neg X(y,x))$
$\wedge\, \forall x\, \forall y\, \forall z\, (X(x,y) \wedge X(y,z) \rightarrow X(x,z)))$

predicates: none
functions: none
names: none

$\exists X_1^2\, (\forall x_1\, \neg X_1^2(x_1,x_1) \wedge \forall x_1\, \forall x_2\, (X_1^2(x_1,x_2) \rightarrow \neg X_1^2(x_2,x_1))$
$\wedge\, \forall x_1\, \forall x_2\, \forall x_3\, (X_1^2(x_1,x_2) \wedge X_1^2(x_2,x_3) \rightarrow X_1^2(x_1,x_3)))$

Analysis The formalization is true in a model iff there is an ordering on the universe (p. 208), where that ordering may be partial (indeed, it need not relate any elements at all). Generally, we define:

$$\text{Order}(X) \equiv_{\text{Def}} \forall x\, \neg X(x,x) \wedge \forall x\, \forall y\, (X(x,y) \rightarrow \neg X(y,x))$$
$$\wedge\, \forall x\, \forall y\, \forall z\, (X(x,y) \wedge X(y,z) \rightarrow X(x,z))$$

So Example 19 is '∃X Order(X)'. For any model and any binary relation **R** on the universe of that model, **R** satisfies Order(X) iff **R** is a (possibly partial) order.

20. There is a linear ordering on the universe.

$\exists X\, (\text{Order}\,(X) \wedge \forall x\, \forall y\, (X(x, y) \vee X(y, x) \vee x = y))$

predicates: none
functions: none
names: none

$\exists X_1^2\, (\text{Order}(X_1^2) \wedge \forall x_1\, \forall x_2\, (X_1^2(x_1,x_2) \vee X_1^2(x_2,x_1) \vee x_1 = x_2))$

Analysis This is true in a model iff there is a relation on the universe that is a linear ordering (p. 208)

The formalization is in pure second-order logic with equality. So given any model with any relations and operations, we can assert that there is a linear ordering on it, which may or may not be one of the relations of the model. For example, we can take the formalization to appear in $L_2(P_0^2, P_1^2, f_0^2, c_0, c_1)$, and it is true in the model $\langle \mathbb{N};=,$ — is divisible by —, — is the square of —, $+, 0, 1\rangle$, even though neither of the relations that are interpreted in the model are linear orderings.

21. A theory of countable dense linear orderings with no endpoints.

DLO $\wedge$ Example 11

predicates: '=', '<'
functions: none
names: none

Analysis By '**DLO**' I mean axioms *O1–O7* of the theory of dense linear orderings (p. 209). A *theory* here has to be understood as the collection of semantic consequences of a collection of wffs, since we have not discussed axiomatizing second-order logic.

The formalization is true in a model iff the interpretation of '<' on the universe is a countable dense linear ordering. Since any countable dense linear ordering is isomorphic to the rationals with their usual ordering (Theorem XIII.5), every model of this formalization in $L_2(=;<)$ is isomorphic to the rationals. We have a categorical theory of dense linear orderings. In second-order logic, unlike first-order logic, we can have a categorical theory with an infinite model.

22. A theory of finite groups.

G $\wedge$ Example 17

predicates: '='
functions: '$\circ$', '$^{-1}$'
names: 'e'

Analysis By '**G**' I mean axioms *G1*, *G2*, *G3* (p. 193) for the theory of groups in $L(=;\circ,{}^{-1},e)$.

An interpretation of $L_2(=;\circ,{}^{-1},e)$ is a model of this formalization iff it is a finite group. So the semantic consequences of this formalization are exactly those wffs true of finite groups. This contrasts with first-order logic where we can have no theory of finite groups (Theorem XII.7).

23. Every group has a proper subgroup.

Formalizable?

Analysis With the assumptions we have made in establishing second-order classical mathematical predicate logic we cannot quantify over all groups because we cannot have a universe that contains all groups (see Exercise 8 of Chapter XII.A and the discussion in Chapter IV on p. 79 as well as Theorem IX.1).

But we can talk about all groups by formulating a theory of groups. A claim formulated in that language is true of all groups iff it is a consequence of that theory. For example, we can formalize 'In every group the identity is unique' as the metalogical assertion:

$$\mathbf{G} \models \exists! y\, \forall x\, (y \circ x = x \ \wedge \ x \circ y = x).$$

For Example 23 we can formulate a claim in $L_2(= ; \circ, e, {}^{-1})$ that is true in a model of **G** just in case the model has a proper subgroup, where a subgroup is proper just in case it is not the identity or the entire group:

$$\exists X\, ([X(e) \wedge \forall x\, \forall y\, (X(x) \wedge X(y) \rightarrow X(x \circ y)) \wedge \forall x\, (X(x) \rightarrow X(x^{-1}))] \wedge \neg \forall x\, X(x) \wedge \neg \forall x\, (X(x) \rightarrow x = e))$$

Then Example 23 becomes a metalogical assertion: This wff is a semantic consequence of **G** in $L_2(= ; \circ, e, {}^{-1})$. Since there are models in which the wff is false (for example, the natural numbers modulo as in Exercise 7 of Chapter XII.A), Example 23 is false.

24. There is a group that has no proper subgroup.

Not formalizable?

Analysis We can say that there exists a group satisfying a condition by making a metalogical assertion that the theory of groups plus a wff formalizing that condition has a model: 'For some group' is translated into 'There is a model of **G** that'. To formalize the true claim 'There is a group with no proper subgroup' we say 'There is a model of $\mathbf{G} \cup \{\neg A(x)\}$', where $A(x)$ is the formal wff of Example 23.

25. Every nonempty collection of natural numbers has a least element.

$$\forall X\, (\exists x\, X(x) \rightarrow \exists y\, (X(y) \wedge \forall z\, (X(z) \rightarrow y \leq z)))$$

predicates: '≤', '='
functions: none
names: none

$$\forall X_1^1\, (\exists x_1\, X_1^1(x_1) \rightarrow \exists x_2\, (X_1^1(x_2) \wedge \forall x_3\, (X_1^1(x_3) \rightarrow P_1^2(x_2, x_3))))$$

Analysis I've assumed we want to formalize this example in the context of a formal theory of the natural numbers (as we'll see in the next chapter), including a (possibly defined) predicate '$x \leq y$'. The example is a claim true about the natural numbers that we could not formalize in a first-order theory of the natural numbers, since we could not quantify over subsets of the universe.

26. There is a least set X such that $0 \in X$ and for all x, if $x \in X$, then $x + 1 \in X$.

$\exists X\,(X(0) \wedge \forall x\,(X(x) \rightarrow X(x+1)) \wedge$
$\quad \forall Y\,((Y(0) \wedge \forall y\,(Y(y) \rightarrow Y(y+1))) \rightarrow \forall z\,(X(z) \rightarrow Y(z))))$

predicates: none
functions: '+'
names: 0, 1

$\exists X_0^1\,(X_0^1(c_0) \wedge \forall x_1\,(X_0^1(x_1) \rightarrow X_0^1(f_0^2(x_1, c_1))) \wedge$
$\forall X_1^1\,((X_1^1(c_0) \wedge \forall x_2\,(X_1^1(x_2) \rightarrow X_1^1(f_0^2(x_2, c_1)))) \rightarrow \forall x_3\,(X_0^1(x_3) \rightarrow X_1^1(x_3))))$

Analysis Again I've assumed we want to formalize this example in the context of a formal theory of the natural numbers.

By 'the least set' satisfying a condition, mathematicians mean a set that is contained in all sets satisfying that condition (the intersection of those sets, if that satisfies the condition).

We can formalize the example by using '$X(x)$' for '$x \in X$', since for any assignment of references σ, $\sigma \vDash X(x)$ iff $\sigma(x) \in \sigma(X)$. In any model whose universe contains the natural numbers and 0, 1, and + are interpreted in the obvious way, this example is true.

Exercises for Sections F.1–F.3

1. Show that Cons 1–3 are satisfied in every classical mathematical second-order model. In particular, explain why we do not need to add as a further requirement to our classical second-order models that if $\sigma(X_1^n)$ is a predicate $A(y_1, \ldots, y_n)$ of the semi-formal language, then $\nu_\sigma(X_1^n(t_1, \ldots, t_n)) = \nu_\sigma(A(t_1/y_1, \ldots, t_n/y_n))$.
2. Show that the collection of assignments of references in a classical mathematical second-order model is a set and is complete.
3. Why does the classical mathematical abstraction make second-order logic a way to do set theory?
4. Give a formalization of Example 12 in pure second-order logic without equality using the definition of '=' from Example 8.
5. Formalize Example 15 in first-order logic using infinitely many wffs.
6. a. Define a quantifier 'There are only finitely many'.
   b. Formalize 'There are only finitely many dogs'.
7. Formalize in first-order logic 'Any group with 5 elements has no proper subgroup.' (Hint: Talk about any subcollection with 2, 3, or 4 elements by using distinct variables and equality.)
8. Show that Example 21 is equivalent to:
   $\forall x\,(X(x) \leftrightarrow \forall Y\,([Y(0) \wedge \forall y(Y(y) \rightarrow Y(y+1))] \rightarrow Y(x))$.

9 Assume we are working within a second-order theory of the real numbers. Formalize:

a. Example 8 of Chapter I.F, namely:
Let A and B be sets of real numbers such that
(a) every real number is either in A or in B;
(b) no real number is in A and in B;
(c) neither A nor B is empty;
(d) if $\alpha \in$ A, and $\beta \in$ B, then $\alpha < \beta$.
Then there is one (and only one) real number γ such that $\alpha \leq \gamma$ for all $\alpha \in$ A, and $\gamma \leq \beta$ for all $\beta \in$ B.

b. Given any set, there is a least upper bound for it.
(See Exercise 7 of Chapter XIII.C.)

c. Every nonempty set of real numbers that is bounded below has a least element.

d. Example 7 of Chapter I.F, namely:
Suppose $\{s_n\}$ is monotonic. Then $\{s_n\}$ converges if and only if it is bounded.
(A sequence $\{s_n : n \geq 0\}$ is monotonic iff for each n, $s_n < s_{n+1}$.)
(Hint: Talk about sequences of objects by using a 1-1 onto relation between the set of natural numbers and the set of elements in a set.)

10. a. Formalize 'There are more stars than planets'.
b. Show how to define in second-order logic a two-place quantifier: 'There are more A than B' where A and B are wffs with a single variable free.

11. In a language of +, $\cdot$, 0, and 1, formalize: 'There are as many prime numbers as there are natural numbers'. (Hint: Define the set of prime numbers.)

12. Formalize and/or discuss: 'No symmetric relation is reflexive'.

13. Formalize and/or discuss: 'There is only one group with 7 elements.'

14. Formalize and/or discuss: 'There is an infinite group.'

15. Formalize and/or discuss: 'Every infinite group has a proper subgroup.'

16. Formalize and/or discuss: 'Every group with a prime number of elements has no proper subgroup.'

17. Devise a normal form theorem for second-order logic comparable to the one for first-order logic (p. 108).

4. The comprehension axioms

Consider the following scheme of wffs:

(7) $$\exists X\, \forall y_1 \ldots \forall y_n\, (X(y_1, \ldots, y_n) \leftrightarrow A(y_1, \ldots, y_n))$$
where exactly $y_1, \ldots, y_n$ are free in A.

It is not controversial to say that every instance of this scheme is valid when A is a formula in the first-order part of the language. When we established what it meant to be a model for the language of predicate logic, we said that each such formula A is a (representative of a) predicate on the universe. In the classical mathematical abstraction of first-order models, no matter the wff we put in for A, it has an extension that is a subset of the n-tuples of the universe.

For example, in the language L(=; <), we have:

(8) $\exists X\ \forall x\ \forall y\ (X(x, y) \leftrightarrow x < y)$

A single predicate is asserted to exist. And for every model there is such a predicate.

But consider also:

(9) $\forall x\ \exists X\ \forall y\ (X(x, y) \leftrightarrow x < y)$

For every object in the universe, there is a predicate determined by that object corresponding to the objects that are greater than it. For example, for the interpretation $\prec$ of '<' as the usual ordering on the natural numbers, (9) asserts that there are infinitely many predicates, corresponding to the sets: $\{y: 0 \prec y\}$, $\{y: 1 \prec y\}$, $\{y: 2 \prec y\}$, $\{y: 3 \prec y\}$, ... That we are committed to the existence of these can be seen in how we evaluate a wff such as:

$$\exists x\ \forall y\ (x < y)$$

Given an assignment of references σ, $\sigma \vDash \exists x\ \forall y\ (x < y)$ iff for every τ that differs from σ at most in what it assigns x, $\tau \vDash \forall y\ (x < y)$. For each assignment $\tau(x)$, a different evaluation is given, which amounts to viewing $(x < y)$ as a unary predicate when x is assigned a reference. As a consequence of adopting the semantics for first-order logic, for any first-order wff A in which $z_1, \ldots, z_m, y_1, \ldots, y_n$ are all the variables free in A, we should also accept:

(10) $\forall z_1 \ldots \forall z_m\ \exists X\ \forall y_1 \ldots \forall y_n\ (X(y_1, \ldots, y_n) \leftrightarrow A(y_1, \ldots, y_n))$

Each z_i is a parameter in (10): for each choice of assignments for $z_1, \ldots, z_m$ a predicate is asserted to exist.

Many mathematicians, however, argue that it is not only useful but essential to allow A to be a second-order formula in (7) or (10), as we did in Example 26. But not all instances of (7) using second-order formulas for A are true. Consider:

$$\exists X\ \forall x\ (X(x) \leftrightarrow \neg X(x))$$

If this were true, we'd have a contradiction.

The mathematical community is split. There are those who say that we should accept all instances of (10) for any wff A of second-order logic, so long as X itself is not free in A. Indeed, predicate variables should be allowed as parameters, too. There simply are such predicates, regardless of how we may describe them, so the full scheme is valid.

There are others who say that to accept the full scheme involves us unavoidably in circular definitions and evaluations in which an object (a predicate) is defined or evaluated in terms of a collection of objects of which it is a part. Such definitions and evaluations, they say, should be shunned because they lead to contradictions. Even were contradictions not threatened, it's argued, such definitions and evaluations are incoherent. The set of natural numbers does not satisfy the formalization in Example 26 because satisfaction doesn't make sense when it's circular. Those who argue this way develop mathematics without using such definitions and evaluations. See *Hazen, 1983* and *Simpson, 1999* for discussions of this view.

Thus, a difference in conceptions of predicates, even accepting the classical mathematical abstraction of models, leads to different second-order logics. Anyone who accepts the semantics of first-order logic should accept a *minimal* scope for the second-order quantifiers.

The comprehension$_1$ axioms For any *first-order* wff A in which $y_1, \ldots, y_n$ are free: $\forall \ldots \exists X\ \forall y_1 \ldots \forall y_n\ (X(y_1, \ldots, y_n) \leftrightarrow A(y_1, \ldots, y_n))$

But what is acceptable to only some in the mathematical community is that the following scheme is valid, where predicate variables, too, are allowed as parameters.

The comprehension$_2$ axioms For any wff A of the second-order language in which $y_1, \ldots, y_n$ are free and X is not free:

$$\forall \ldots \exists X\ \forall y_1 \ldots \forall y_n\ (X(y_1, \ldots, y_n) \leftrightarrow A(y_1, \ldots, y_n))$$

Requiring the comprehension axioms (either the first- or second-order version) to be true in all our models is not the same as requiring that the only sets of n-tuples in the range of the predicate variables are those that are extensions of open formulas. Even those who take predicates to be linguistic do not assume that regardless of the universe, each predicate on the universe corresponds to a first-order formula, for language is not fixed, while the formal language is. In any case, we have no way to characterize syntactically models in which the only predicates that exist are those that satisfy the comprehension axioms, for we don't have variables to range over wffs of the language to assert: $\forall X\ \exists A\ (X(y_1, \ldots, y_n) \leftrightarrow A(y_1, \ldots, y_n))$.

Thus, *requiring the comprehension$_1$ axioms to be true in our models ensures that we have a minimally acceptable formalization of* 'for all' *and* 'there exists' *as applied to predicate variables.* If X is an n-ary predicate symbol free in B(X), and A is a first-order formula of the semi-formal language that has exactly n free variables, we can write B(A/X) for the wff that results by replacing each free occurrence of X as $X(t_1, \ldots, t_n)$ in B with $A(t_1, \ldots, t_n)$. In models in which the comprehension axioms are true, the following schema are valid, where we have two versions corresponding to the two versions of the comprehension axioms.

Universal instantiation for predicates

$\models \forall \ldots (\forall X\, B(X) \rightarrow B(A/X))$

for any A such that: (*comprehension*$_1$) A is a first-order formula.
(*comprehension*$_2$) X is not free in A

Existential generalization for predicates

$\models \forall \ldots (B(A/X) \rightarrow \exists X\, B(X))$

for any A such that: (*comprehension*$_1$) A is a first-order formula.
(*comprehension*$_2$) X is not free in A

Example 3 (p. 237) is an instance of universal instantiation for first-order predicates. The following is an example of existential generalization for first-order predicates (twice):

Ralph is a dog $\rightarrow \neg$ Ralph meows.
Therefore, $\exists X\, \exists Y\, (X(\text{Ralph}) \rightarrow \neg Y(\text{Ralph}))$

We have seen that differences in views about the nature of predicates can lead to different second-order logics. And we have seen that we can formulate those differences syntactically. This is as far as we will go now in relating the syntax and semantics of second-order logic. We'll see in Chapter XV.E that there are significant limitations on axiomatizing second-order logic.

Exercises for Section F.4

1. Consider the language of equality with one binary function and two names, which we can write as $L_2(=; +, 0, 1)$. Let $\langle \mathbb{N}; \text{plus, zero, one} \rangle$ be a model of this language. For each of the following describe colloquially what predicate(s) are asserted to exist and whether the wff is true in the model:
   a. $\exists X\, \forall x\, \forall y\, \forall z\, (X(x, y, z) \leftrightarrow x + y = z)$
   b. $\forall x\, \exists X\, \forall y\, \forall z\, (X(y, z) \leftrightarrow x + y = z)$
   c. $\forall x\, \forall y\, \exists X\, \forall z\, (X(z) \leftrightarrow x + y = z)$
   d. $\exists X\, \forall x\, \forall y\, (X(x, y) \leftrightarrow \exists z\, (x + z = y))$
   e. $\exists X\, \forall x\, \forall y\, (X(x, y) \leftrightarrow (x + 1 = y))$
   f. $\exists X\, \forall x\, \forall y\, (X(x, y) \leftrightarrow ((x + 1) + 1 = y))$
   g. $\forall z\, \exists X\, \forall x\, \forall y\, (X(x, y) \leftrightarrow (x + z = y))$
   h. $\exists X\, \forall x\, (X(x) \leftrightarrow \forall Y\, ([Y(0) \wedge \forall z\, (Y(z) \rightarrow Y((z + 1) + 1)) \rightarrow \forall z\, Y(z)] \rightarrow Y(x))$
   i. $\exists X\, \forall x\, (X(x) \leftrightarrow \forall Y\, (Y((x + 1) + 1)))$
   j. $\forall Y\, \exists X\, \forall x\, (X(x) \leftrightarrow Y((x + 1) + 1))$

k. $\exists X\, \forall x\, (X(x) \leftrightarrow X((x+1)+1)))$

l. $\exists X\, \forall x\, (X(x) \leftrightarrow \forall Y\, \exists y\, Y(y, x+1))$

m. $\forall Y\, \exists X\, \forall x\, (X(x) \leftrightarrow \exists y\, Y((x+y)+1))$

n. $\forall Y\, \exists X\, \forall x\, \forall y\, (X(x, y) \leftrightarrow (Y(0) \wedge Y(1) \rightarrow Y(x+y))$

2. How does Cons 4 relate to the debate about what comprehension axioms are valid in second-order logic?

3. a. Explain why the comprehension$_1$ axioms are valid.
   b. Does that justification depend on our making the classical abstraction of models?

4. Why do some mathematicians and logicians reject the comprehension$_2$ scheme as valid?

5. a. Why is '$\exists X\, \forall x\, (X(x) \leftrightarrow (x \neq x))$' not a paradox that comes from accepting the comprehension$_1$ axioms?
   b. Why is '$\exists X\, \forall x\, (X(x) \leftrightarrow \neg X(x))$' not a paradox that comes from accepting the comprehension$_2$ axioms?

6. Show that *all implies exists* is a semantic consequence in second-order logic:

   $\forall \ldots \forall X\, B(X) \vDash \forall \ldots \exists X\, B(X).$

   Does it matter whether we assume the comprehension axioms? Does it matter whether we use mathematical models?

7. The *closure of a set under a function* is defined in Exercise 6 of Chapter IX.B (p. 162).
   a. Restate Example 26 in terms of the least set closed under a function.
   b. Formalize in L(= ; +, ·, 0, 1) 'There is a least set that is closed under the function *the largest prime* $\leq x$'.
   c. In a model of L(=, +, ·, 0, 1) whose universe is the natural numbers with the usual interpretation of those symbols, what set satisfies the formalization of (b)? Give a definition of that set that does not use a comprehension$_2$ axiom.
   d. Formalize the principle behind induction: Any set that contains 0 and is closed under '+ 1' is the set of all natural numbers.

8. Compare the issue of which comprehension axioms are valid with how we should formalize assumption 2 of our naive set theory (Chapter IX.B, p. 158).

9. Compare the substitutional interpretation of quantifiers in first-order logic (Exercise 7, p. 177) to restricting the range of quantifiers in second-order logic to contain only (predicates corresponding to) first-order formulas of the semi-formal language.

10. a. Second-order logic has predicate variables for every arity $n \geq 1$. What would a 0-ary predicate variable stand for? (Hint: What in our reasoning does not require a variable to be true or false?)
    b. Suppose we expand the language L_2 to include 0-ary predicates. Formulate comprehension axioms for those variables.
    c. Suppose we expand the language L_2 to include 0-ary predicates and assume universal instantiation. Show that we could then axiomatize classical propositional logic with four axioms (not schema).

G. Quantifying over Functions

In mathematics it is common to quantify over functions, as in 'There is a continuous nowhere differentiable function' or 'There is no onto function from the natural numbers to the real numbers'. We can already formalize such claims in second-order logic by converting talk of functions into talk of predicates (Chapter VIII.E). But just as we introduced function symbols into the language of first-order logic to simplify formalizations in mathematics, we can introduce quantification over functions into second-order logic.

For a language L_{2F} of second-order logic with quantification over functions, add to the vocabulary of L_2 the following.

function variables: $F_0^1, F_0^2, \ldots, F_1^1, F_1^2, \ldots, F_i^n, \ldots$ for $i \geq 0, n \geq 1$

I'll use $F, G, G_0, G_1, \ldots$ as metavariables for function variables. We define the *terms* of the language inductively by adding to the definition on p. 141:

If $t_1, \ldots, t_m$ are terms of depth $\leq n$, and one of them has depth n, then for every $i \geq 0, m \geq 1$, $F_i^m(t_1, \ldots, t_n)$ is a term and has *depth* $n + 1$.

A *first-order term* is a term in which no function variable appears.

In the definition of *wff* we now understand 'term' in this broader sense, and we add the following clause:

If A is a wff of length n, then for every $i \geq 0, m \geq 1$,
each of $(\forall F_i^m A)$, $(\exists F_i^m A)$ is a wff of length $n + 1$.

The definition of the *scope of a quantifier* is extended by adding:

In $(\forall F\, A)$ the scope of the initial $\forall F$ is A.
In $(\exists F\, A)$ the scope of the initial $\exists F$ is A.

To the definition of *free variable* add:

An occurrence of a variable F in A is *bound in* A if it immediately follows an occurrence of a symbol '$\forall$' or '$\exists$' or lies within the scope of a quantifier '$\forall F$' or '$\exists F$'. If it is not bound in A, it is *free in* A.
In a wff $(\forall F\, A)$, respectively, $(\exists F\, A)$, every free occurrence of F in A is *bound by* the initial $\forall F$, respectively, $\exists F$.

A *closed wff* is, again, one in which no occurrence of a variable is free.

Since we will use this language only for formalizing mathematics, I'll assume the classical mathematical abstraction of models (for a more general discussion, see Chapter X.K of *Predicate Logic*). A classical mathematical model of L_{2F} is a classical mathematical model of L_2 where:

The range of the function variables, that is, those functions that can be assigned as reference, is the set of all functions from U^n to U for all $n \geq 1$.

The issues about what 'all' means in this are exactly the same as discussed in Section F.4, since functions can be translated into predicates (Chapter VIII.E).

We extend each assignment of references to all terms by induction:

> If $y_1, \ldots, y_n$ do not appear in the (not necessarily first-order) terms $t_1, \ldots, t_n$, and for all i, $1 \leq i \leq n$, $\sigma(t_i) = \sigma(y_i)$, then $\sigma(\mathrm{F}_i^n(t_1, \ldots, t_n)) = \sigma(\mathrm{F}_i^n(y_1, \ldots, y_n))$.

We then simultaneously extend ν_σ for all σ to all wffs of the language as before by adding at the last step:

$\nu_\sigma(\forall \mathrm{F}\ \mathrm{A}) = \mathsf{T}$ iff for every assignment of references τ that differs from σ at most in what it assigns as reference to F, $\nu_\tau(\mathrm{A}) = \mathsf{T}$

$\nu_\sigma(\exists \mathrm{F}\ \mathrm{A}) = \mathsf{T}$ iff for some assignment of references τ that differs from σ at most in what it assigns as reference to F, $\nu_\tau(\mathrm{A}) = \mathsf{T}$

Truth, validity, and semantic consequence are defined as usual, and when necessary I'll denote those with the subscript '2F', that is, as $\vDash_{2F}$.

Let's look at some examples.

27. There are countably many things.

$\exists \mathrm{F}\, \exists x\, \forall \mathrm{X}\, (\mathrm{X}(x) \wedge \forall y\, ((\mathrm{X}(y) \rightarrow \mathrm{X}(\mathrm{F}(y))) \rightarrow \forall z\, \mathrm{X}(z)))$

predicates: none
functions: none
names: none

$\exists \mathrm{F}_0^1\, \exists x_1\, \forall \mathrm{X}_0^1\, (\mathrm{X}_0^1(x_1) \wedge \forall x_2\, ((\mathrm{X}_0^1(x_2) \rightarrow \mathrm{X}_0^1(\mathrm{F}_0^1(x_2))) \rightarrow \forall x_3\, \mathrm{X}_0^1(x_3)))$

Analysis I've replaced the use of a binary predicate variable in Example 11 with a unary function variable. We no longer need equality in the formalization.

28. There are infinitely many things.

$\exists \mathrm{F}\, (\forall x\, \forall y\, (x \neq y \rightarrow \mathrm{F}(x) \neq \mathrm{F}(y)) \wedge \exists x\, \forall y\, (\mathrm{F}(y) \neq x))$

predicates: '='
functions: none
names: none

$\exists \mathrm{F}_0^1\, (\forall x_1\, \forall x_2\, (x_1 \neq x_2 \rightarrow \mathrm{F}_0^1(x_1) \neq \mathrm{F}_0^1(x_2)) \wedge \exists x_1\, \forall x_2\, (\mathrm{F}_0^1(x_2) = x_1))$

Analysis This is Example 12 with the use of a binary predicate variable replaced by a unary function variable.

29. There are more stars than planets.

$\exists F\, [\forall x\, \forall y\, (x \neq y \rightarrow F(x) \neq F(y)) \wedge \forall x\, (x \text{ is a planet} \leftrightarrow F(x) \text{ is a star})]$
$\quad \wedge \neg \exists G\, [\forall x\, \forall y\, (x \neq y \rightarrow G(x) \neq G(y)) \wedge \forall x\, (x \text{ is a star} \leftrightarrow$
$\qquad G(x) \text{ is a planet})]$

predicates: '=', '— is a star', '— is a planet'
functions: none
names: none

$\exists F_0^1\, [\forall x_1\, \forall x_2\, (x_1 \neq x_2 \rightarrow F_0^1(x_1) \neq F_0^1(x_2)) \wedge \forall x_1\, (P_0^1(x_1) \leftrightarrow P_1^1(F_0^1(x_1))]$
$\quad \wedge \neg \exists F_1^1\, [\forall x_1\, \forall x_2\, (x_1 \neq x_2 \rightarrow F_1^1(x_1) \neq F_1^1(x_2))$
$\quad \wedge \forall x_1\, (P_1^1(x_1) \leftrightarrow P_0^1(F_1^1(x_1)))]$

Analysis The formalization is true in a model iff there is a 1-1 function from the extension of '— is a planet' to the extension of '— is a star', but none from the extension of '— is a star' to that of '— is a planet'. If, as we suspect, there are only finitely many stars and planets, then the formalization is certainly correct. If there are infinitely many stars or planets, then the formalization is correct because that's what we took 'more than' to mean for infinite sets (Chapter IX.B, p. 160).

30. 'The largest number whose square is less than or equal to a given number' is a function.

$\exists F\, \forall x\, \forall y\, (F(x) = y \leftrightarrow (y \cdot y \leq x \wedge \forall z\, (z \cdot z \leq x \rightarrow z \leq y)))$

predicates: '=', '≤'
functions: none
names: none

$\exists F_0^1\, \forall x_1\, \forall x_2\, (F_0^1(x_1) = x_2 \leftrightarrow$
$\quad (P_0^2(f_0^2(x_2, x_2), x_1) \wedge \forall x_3\, P_0^2(f_0^2(x_3, x_3), x_1) \rightarrow P_0^2(x_3, x_2)))$

Analysis The formalization is true in every model in which the wff $\forall x\, \exists! y\, (y \cdot y \leq x \wedge \forall z\, (z \cdot z \leq x \rightarrow z \leq y))$ is true, for example, the model whose universe is the natural numbers with multiplication and the usual ordering. Example 30 is an instance of a comprehension scheme for L_{2F}.

Alternatively, we could formalize the example as:

$\exists F\, \forall x\, (F(x) \cdot F(x) \leq x \wedge \forall z\, (z \cdot z \leq x \rightarrow z \leq F(x))$

31. There are bounded monotonically increasing functions on the real numbers that have no greatest value.

$\exists F\, (\forall x\, \forall y\, (x < y \rightarrow F(x) < F(y)) \wedge \exists z\, \forall x\, (F(x) \leq z) \wedge$
$\quad \neg \exists x\, \forall y\, (F(y) \leq F(x)))$

predicates: '=', '<'
names: none
functions: none

$$\exists F_0^1 \, (\forall x_1 \, \forall x_2 \, ((P_0^2(x_1, x_2) \rightarrow P_0^2(F_0^1(x_1), F_0^1(x_2))) \wedge$$
$$\exists x_3 \, \forall x_1 \, (P_0^2(F_0^1(x_1), x_3) \vee F_0^1(x_1) = x_3) \wedge$$
$$\neg \exists x_1 \, \forall x_2 \, (P_0^2(F_0^1(x_2), F_0^1(x_1)) \vee F_0^1(x_2) = F_0^1(x_1)))$$

Analysis I've formalized this example assuming the context of a theory of the real numbers, using '≤' as an abbreviation. In any model in which the universe is the real numbers with their usual ordering it is true: arctan(x) is a function on the real numbers, and hence is in the range of the variables, regardless of whether it corresponds to a formula in the formal language.

32. There are uncountably many functions on the real numbers.

Not formalizable.

Analysis This example requires third-order predications.

33. A theory of groups in pure second-order logic with equality.

$$\exists F \, \exists G \, \exists H \, (\forall x \, \forall y \, (H(x) = H(y)) \wedge$$
$$\forall x \, (F(x, H(x)) = x) \wedge$$
$$\forall x \, \forall y \, (F(x, G(x)) = H(x)) \wedge$$
$$\forall x \, \forall y \, \forall z \, (F(F(x, y), z) = F(x, F(y, z))))$$

predicates: '='
functions: none
names: none

$$\exists F_0^2 \, \exists F_0^1 \, \exists F_1^1 \, (\forall x_1 \, \forall x_2 \, (F_1^1(x_1) = F_1^1(x_2)) \wedge$$
$$\forall x_1 \, (F_0^2(x_1, F_1^1(x_1)) = x_1) \wedge$$
$$\forall x_1 \, \forall x_2 \, (F_0^2(x_1, F_0^1(x_1)) = F_1^1(x_1)) \wedge$$
$$\forall x_1 \, \forall x_2 \, \forall x_3 \, (F_0^2(F_0^2(x_1, x_2), x_3) = F_0^2(x_1, F_0^2(x_2, x_3))))$$

Analysis I've formalized Kleene's axiomatization **GK** (p. 199). The name 'e' is replaced by the requirement that there be a constant function satisfying the conditions, and functions are asserted to exist that play the roles of '∘' and '$^{-1}$'.

Exercises for Section G

1. Why is it not necessary in second-order classical mathematical models to impose conditions on consistency and extensionality of applications of functions?

2. Formalize the quantifier 'There are countably many' using function variables.
3. Formalize 'The extension of '— is a dog' has the same number of elements as the universe.' (Hint: Compare Example 29.)
4. Formalize in $L_2(+, \cdot, 0, 1)$ 'There are exactly as many prime numbers as there are natural numbers.'
5. Use Example 29 to define a quantifier in second-order logic 'There are exactly as many (objects satisfying) A as B'.
6. Formalize and/or discuss 'Every continuous function on a closed interval has a greatest value in that interval.'
7. Formalize and/or discuss: 'There are two groups with different group operations'.
8. Formulate versions of the comprehension$_1$ axioms and of the comprehension$_2$ axioms for L_{2F}.
9. Formulate a theory of groups in pure second-order logic (without equality).

H. Other Kinds of Variables and Second-Order Logic

1. Many-sorted logic

In mathematics and science we often distinguish the kinds of things that we are quantifying over. For example, in this text I have informally used different kinds of variables for different kinds of objects: p, q, r for propositions; *i, j, k, n, m* for natural numbers; A, B, C, . . . for wffs. We can model that practice in first-order logic.

Suppose, for example, we are formalizing reasoning in plane geometry. We say that $P, Q, R, S, P_0, P_1, \ldots$ will stand for points and $l, m, n, l_0, l_1, \ldots$ will stand for lines. We write, for instance, $l \parallel m$ for 'l is parallel to m', and $P*Q*R$ for 'Q lies between P and R'. We'd never write $P \parallel l$, or $P \parallel Q$, or $1*Q*R$, for they seem ill-formed. But we could write those; they'd just be false for any assignment of references to l, P, Q, R. To formalize that kind of reasoning we have two choices.

Many-sorted logic

We can add to our first-order language new variables $w_0, w_1, \ldots$, retaining the same formation rules for the language. Then for the semantics we adopt two universes, $\mathsf{U}_1, \mathsf{U}_2$ for every interpretation, requiring that for every assignment of references σ, for every i, $\sigma(x_i)$ is in U_1 and $\sigma(w_i)$ is in U_2. The rest of the definition of a model is as before. Here we allow as a wff $P_1^2(x_1, w_1)$, for example, even if we are interpreting P_1^2 as a predicate that holds only of objects from U_1; in that case for every σ, $\sigma \nvDash P_1^2(x_1, w_1)$. This simplifies the formation rules by not requiring us to say in advance which predicates are to apply to which kind of objects;

syntax can be independent if we allow falsity as the default truth-value. This is a *2-sorted* language in *2-sorted first-order logic.*

Relative quantification in first-order logic
We can take a proposition such as 'For every l and m, if $l \parallel m$, then $m \parallel l$' and rewrite it as 'For every line l and every line m, if $l \parallel m$, then $m \parallel l$'. This is a relative quantification (Chapter VII.A) that we can formalize as:

$$\forall x\, \forall y\, (x \text{ is a line} \wedge y \text{ is a line} \rightarrow (x \parallel y \rightarrow y \parallel x))$$

So for the formal language we can take two unary predicate symbols, say P_1^1 and P_2^1, and realize P_1^1 as '— is a point' and P_2^1 as '— is a line'. We could then formalize 'For every $Q \cdots$' as '$\forall x\, (P_1^1(x) \rightarrow \cdots)$', and 'For every $l \cdots$' as '$\forall x\, (P_2^1(x) \rightarrow \cdots)$'.

Which of these approaches to use seems to be a matter of taste. We can justify that observation by showing how to translate from the 2-sorted language to our usual first-order language:

$$(P_i^n)^* = \begin{cases} P_{i+2}^1 & \text{if } n = 1 \\ P_i^n & \text{if } n > 1 \end{cases}$$

$$(c_i)^* = c_i$$

$$(x_i)^* = x_{2i}$$

$$(w_i)^* = x_{2i+1}$$

$$(P_i^n(t_1, \ldots, t_n))^* \text{ is } (P_i^n)^*(t_1^*, \ldots, t_n^*)$$

$$(t = u)^* \text{ is } t^* = u^*$$

$\neg$, $\rightarrow$, $\wedge$, $\vee$ are translated homophonically

$$(\forall x_i A)^* = \forall x_i^*\, (P_1^1(x_i^*) \rightarrow A^*)$$

$$(\exists x_i A)^* = \exists x_i^*\, (P_1^1(x_i^*) \wedge A^*)$$

$$(\forall w_i A)^* = \forall w_i^*\, (P_2^1(w_i^*) \rightarrow A^*)$$

$$(\exists w_i A)^* = \exists w_i^*\, (P_2^1(w_i^*) \wedge A^*)$$

We supplement the translation by adding as an axiom in the usual first-order language:

(11) $\forall x_1\, (P_1^1(x_1) \leftrightarrow \neg P_2^1(x_1))$

As an exercise I ask you to show that this translation induces a translation of the 2-sorted models to first-order models that satisfy (11) such that for every model M and every closed wff A, $M \vDash A$ iff $M^* \vDash A^*$.

I'll leave to you to formulate a definition of a language with n different kinds or *sorts* of variables and corresponding models using n different sorts of objects. We call such a language a (*first-order*) *many-sorted language*. The translation of such a language to the usual first-order language is just a modification of the case for $n = 2$ above.

We could add function symbols to a many-sorted language. But then there is the issue of whether some functions are meant to be defined on just one sort of variable. We'll look at what to do with functions that are undefined on some inputs in Chapter XXI.

2. General models for second-order logic

The semantics of second-order logic are based on viewing predicates as things. We speak of all predicates or assert the existence of a predicate, and those quantifications are evaluated just as they were when we talked about things in general. The difference is that once we have chosen a universe for a model, we don't also choose a universe for the range of the predicate variables. The range of the predicate variables is fixed.

But, as in some of our discussions in this chapter, we might consider using smaller subsets of the sets of n-tuples of the universe to serve as the range of the predicate variables. Just as we allowed any set of things to be the universe of a model, we could allow any set of predicates on that universe to be the range of the predicate variables.

A *general* classical mathematical model of the language L_2 of second-order logic is defined just as before except that we select a set B as the range of the predicate variables such that:

$$\mathsf{B} \neq \varnothing \text{ and } \mathsf{B} \subseteq \bigcup_n \{\mathsf{U}^n : n \geq 1\}$$

We also have to make a proviso that for every n for which there is an n-ary predicate variable in the language, there is at least one element from U^n in B, because quantification as we have analyzed it requires a non-empty range for variables. Truth in a model, validity, and semantic consequence relative to these models are defined in the usual way. This is *general second-order logic.*

Now consider, '$\exists X\, \forall x\, X(x)$'. This is valid in second-order logic, but it is not valid with respect to general models. There is certainly a subset of the universe that makes the formula true, namely, the entire universe, but that may not be in the range of the predicate variables in a general model. The quantifier '$\exists$' in general second-order logic no longer serves as a good formalization of 'exists', nor '$\forall$' of 'for all'.

On the other hand, any formula that is true in all general second-order models is also true in all standard models of second-order logic, since a *standard* model—that is, one in which all predicates are in the range of the predicate variables—is also a general model.

In general second-order logic we treat predicates as just things that bear some structural relation to the other things in the universe. We can allow any collection of things to serve as the "universe" of the predicate variables, so long as they bear the right structural relation to other things. So general second-order logic is really a two-sorted first-order theory. In Chapter X of *Predicate Logic* I give a detailed translation of general second-order logic to first-order logic.

Exercises for Section H

1. Give two examples from a mathematics or science text that use different sorts of (apparently first-order) variables in an argument. How many different sorts of variables are used in this book? (See the Index of Notation.)
2. Formalize in a 2-sorted language: Given a line and a point not on that line, there is one and only one line through that point that is parallel to the first line.
   (Hint: Use a predicate for '— lies on —'.)
3. Formalize the following claims of real number theory in a two-sorted language.
   a. Suppose $\{s_n\}$ is monotonic. Then $\{s_n\}$ converges if and only if it is bounded.
      (A sequence $\{s_n: n \geq 0\}$ is monotonic iff for each n, $s_n < s_{n+1}$.)
      Compare your formalization to the formalization you gave for Exercise 9.d of Sections F.1–F.3.
   b. $\lim_{n \to 0} {}^1/_n = 0$.
   c. $\lim_{n \to 0} \sin(nx)$ does not exist.
4. Let L_1 be the usual language of first-order logic.
   a. Define a formal first-order language L_{2S} of two sorts of variables, $x_0, x_1, \ldots,$ $w_0, w_1, \ldots$.
   b. Define the notion of model for L_{2S}.
   c. Let * be the translation of the language L_{2S} to L_1 sketched above. Show that * induces a 1-1, onto mapping of models of L_{2S} to models of L_1 that satisfy (11) such that for every model M of L_{2S} and every closed wff A of L_{2S}, M⊨A iff M*⊨A*.
5. What is a general mathematical model of second-order logic?
6. Discuss which of the formalizations in Examples 1–26 of this chapter are good formalizations in general second-order logic.

XV The Natural Numbers

Our goal in developing predicate logic has been to have a logic suitable for formalizing mathematics. We've seen some examples of formalization in the previous chapters, but a crucial test of our logic is how well we can formalize the most fundamental part of mathematics we all reckon we understand: arithmetic.

We'll start with a simple theory of counting and progress to theories that encompass more of the arithmetic of the natural numbers. In trying to answer questions about decidability and completeness, we'll have to consider what functions we can compute within the theory, which will lead us to consider what we mean by 'compute'. A twist on the liar paradox will lead us to see some major limitations on how much of arithmetic we can formalize, and indeed how much of any mathematics we can formalize in classical predicate logic.

Much of the work in this chapter was originally devised to try to answer questions about the role of the infinite in mathematics, as you can read in Epstein and Carnielli, *Computability*.

A. The Theory of Successor

As children we learn the counting numbers 1, 2, 3, 4, . . . as a sequence we know how to extend as far as we want. We learn how to add and multiply counting numbers, and we learn that it's useful to start the list with 0.

So let's begin by formalizing just this counting, that is, the sequence that starts with 0 and progresses by adding one: 0, 0 + 1, (0 + 1) + 1, ((0 + 1) + 1) +1, ... That is, we intend to formalize a particular example. Hence, unlike group theory, which was already an informal theory, we will have a semi-formal language, let's say L(= ; S, o), where 'o' is a name for 0 and 'S' is a name for the function of adding one, called the *successor function*. We can take this as a realization of L(= ; f_0^1, c_0). But our goal is to axiomatize what is true of this single example.

To make our work easier to read, we can informally delete parentheses between successive uses of 'S' and use an inductive abbreviation for *numerals*:

$$S^0 x \equiv_{\text{Def}} x$$

$$S^{n+1} x \equiv_{\text{Def}} SS^n x$$

For example, 'S^3x' is an abreviation for 'SSSx'. This abbreviation is in the metalogic: the numeral '3' is part of a short-hand way to write a formal wff and is not part of the formal language, just as n is not part of the formal language.

Our goal is to characterize counting on the natural numbers. Here are some axioms that mathematicians have found suitable.

S, the *theory of successor* in L(= ; S, o)

S1 $\forall x \forall y (Sx = Sy \to x = y)$

S2 $\forall x (o \neq Sx)$

S3 $\forall y (y \neq o \to \exists x (y = Sx))$

T_1 $\forall x (Sx \neq x)$

T_2 $\forall x (S^2x \neq x)$

T_3 $\forall x (S^3x \neq x)$

⋮

T_n $\forall x (S^n x \neq x)$ for $n \geq 1$

Each of these is indeed true of the natural numbers, $\mathbb{N}$, with 0 interpreting 'o' and the function '+1' interpreting 'S'. That is, $\langle \mathbb{N} ; +1, 0 \rangle$ is a model of **S**.

Theorem 1

a. Every model of **S** is infinite.

b. There are infinitely many non-isomorphic countable models of **S**.

c. **S** is $\mathfrak{c}$-categorical.

d. **S** is complete and decidable.

e. **S** = Th(<ℕ; +1, 0>).

Proof (a) Suppose we have a model of *S1* and *S2*, where a interprets 'o' and g interprets 'S'. Consider the sequence: $a, g(a), \ldots, g^{n+1}(a), \ldots$, where $g^{n+1}(a) = g(g^n(a))$. By *S2* there is no i such that $g^i(a) = a$. Suppose that the model is finite. Then there is a minimal n and $i < n$ such that $g^n(a) = g^i(a)$. If $i = 1$, then $g^{n-1}(a) = a$, a contradiction. But if $i > 1$, then $g^{n-1}(a) = g^{i-1}(a)$, which contradicts the minimality of n. So the model cannot be finite.

(b) and (c) Consider a set $B = \{\ldots b_{-2}, b_{-1}, b_0, b_1, b_2, \ldots\}$ (for example, the integers relabeled). Consider:

$N = <\mathbb{N} \cup B;\ f,\ 0>$

$f(n) = n + 1$, for n a natural number; $f(b_j) = b_{j+1}$ for j an integer.

Then N is countable and, as you can show, N is a model of **S**, and N is not isomorphic to <ℕ; +1, 0>.

Now given any model $M = <U;\ g,\ e>$ of **S**, there is one and only one injection of <ℕ; +1, 0> into M (it takes 0 to e). Call the image of ℕ in M the *standard part* of M. If $c \in U$ is not in the standard part of M, then by *S3* there is some d such that $g(d) = c$. Since *S3* and for all $n \geq 1$, T_n are in **S**, c must be part of an infinite chain $\ldots c_{-2}, c_{-1}, c_0 = c, c_1, c_2, \ldots$, where for each integer i, $g(c_i) = c_{i+1}$, and if $i \neq j$, then $c_i \neq c_j$. We call such a chain a *Z-chain*.

So M is a model of **S** iff M consists of a standard part plus some number (finite or infinite, possibly none) of Z-chains. Given two models each with a Z-chain, there is an injection of the one Z-chain onto the other. So (using the axiom of choice) any two models of **S** are isomorphic iff they have the same number of Z-chains.

Thus, there are infinitely many non-isomorphic countable models of **S**, depending on whether there are 0, 1, 2, . . . , or countably many Z-chains. But each Z-chain is countable, so the only way we can have a model of cardinality c is to have c-many Z-chains (Exercise 5.d of Chapter IX.B), and any two such models are isomorphic.

(d) and (e) **S** is complete by Corollary XIII.12. So we can decide if a given wff is a theorem of **S** as in the proof of Corollary XIII.10.

(f) As <ℕ; +1, 0> is a model of **S**, and **S** is complete, **S** = Th(<ℕ; +1, 0>). ∎

So our axiom system does indeed characterize counting on the natural numbers in the sense that the theory it gives is exactly the wffs true of counting in the natural numbers. But it doesn't characterize it in the sense that it is not categorical, even for countable models.

We used the axiom of choice to show that **S** is complete. But it is possible to give a constructive proof using the method of elimination of quantifiers from Chapter XIII (see *Enderton, 1972*).

Do we really need an infinite list of axioms for Th(<ℕ; +1, 0>)? A theory Γ is *finitely axiomatizable* if there is a finite collection of wffs Σ such that Γ = Th(Σ).

Lemma 2 Th(Δ) is finitely axiomatizable iff for some finite $\Sigma \subseteq \Delta$, Th(Δ) = Th(Σ).

Proof Suppose Th(Δ) = Th($A_1, \ldots, A_n$). Let B = $A_1 \wedge \cdots \wedge A_n$. Then Th(Δ) = Th(B). So $\Delta \vdash B$. So there are $D_1, \ldots, D_m$ in Δ with $\{D_1, \ldots, D_m\} \vdash B$. So Th(Δ) = Th($\{D_1, \ldots, D_m\}$). The other direction is immediate. ■

Theorem 3 **S** is not finitely axiomatizable.

Proof Suppose **S** were finitely axiomatizable. Then for some finite $\Sigma \subseteq$ **S**, Th(Σ) = **S**. There must be a maximal n such that T_n is in Σ. Taking any $b \notin \mathbb{N}$, we then have a model M = <ℕ ∪ B; g, 0> with B = $\{b, g^1(b), \ldots, g^n(b)\}$ where $g^{n+1}(b) = b$ and $g(i) = i + 1$ for $i \in \mathbb{N}$. But M is not a model of **S** because $M \nvDash T_{n+1}$. So **S** is not finitely axiomatizable. ■

Thus, **S** is the simplest theory we can have for proving all there is to prove in a first-order language about the natural number sequence as just a sequence.

But we're interested in the arithmetic of the natural numbers. Are addition and multiplication already implicit in **S** = Th(<ℕ; +1, 0>)? What do we mean by 'implicit'? We already have a notion of a function being definable in a theory (Chapter XII.B.2), but here we are talking about a function on the universe of a model being definable in the theory of that model.

Definability in a model An n-ary predicate P in a model M is *definable in* Th(M) in language L iff there is a wff $C(y_1, \ldots, y_n)$ of L in which exactly $y_1, \ldots, y_n$ are free and distinct such that for every assignment of references σ on M,

$$\sigma \vDash C(y_1, \ldots, y_n) \text{ iff } P(\sigma(y_1), \ldots, \sigma(y_n)).$$

An n-ary function f in a model M is *definable* in Th(M) in language L iff there is a wff $C(y_1, \ldots, y_n, y_{n+1})$ of L in which exactly $y_1, \ldots, y_n, y_{n+1}$ are free and distinct such that for every assignment of references σ on M,

$$\sigma \vDash C(y_1, \ldots, y_n, y_{n+1}) \text{ iff } f(\sigma(y_1), \ldots, \sigma(y_n)) = \sigma(y_{n+1}).$$

It's not hard to relate this new notion of definability to our old one. Given Th(M) in L and a predicate P on $(U_M)^n$, let L_P be L expanded to include a new predicate symbol 'P' and M_P be M expanded with the interpretation of 'P' as P. Similarly, if f is defined on $(U_M)^n$, define L_f and M_f. I'll let you prove the following.

Lemma 4

a. P is definable in Th(M) in L iff Th(M_P) is a conservative extension of Th(M) and P is definable in Th(M_P).

b. f is definable in Th(M) in L iff Th(M_f) is a conservative extension of Th(M) and f is definable in Th(M_f).

Theorem 5 Neither addition nor multiplication is definable in $\mathbf{S} = \mathrm{Th}(\langle \mathbb{N}; +1, 0\rangle)$.

Proof Suppose to the contrary that addition is definable in **S**, say by the formula C. Then each of the following are consequences of **S**, since they are true of $\langle \mathbb{N}; +1, 0\rangle$:

B_1 $\forall x\, \forall y\, \exists! z\, C(x, y, z)$

B_2 $\forall x\, \forall y\, (C(x, y, o) \rightarrow x = o \wedge y = o)$

B_3 $\forall x\, \forall y\, \forall z\, (C(Sx, y, Sz) \rightarrow C(x, y, z))$

B_4 $\forall x\, \forall y\, \forall z\, \forall w\, (C(x, y, z) \wedge C(y, x, w) \rightarrow z = w)$

B_5 $\forall x\, \forall y\, (\neg C(x, Sy, x))$

By compactness, let $\Sigma \subseteq \mathbf{S}$ be finite such that $\Sigma \vdash B_1 \wedge B_2 \wedge B_3 \wedge B_4 \wedge B_5$ and $\Sigma \supseteq \{S1, S2, S3\}$. There must be a largest n such that $T_n \in \Sigma$, which we can assume is ≥ 2.

So as in the proof of Theorem 3 we have a model M of Σ where $\mathsf{b} \notin \mathbb{N}$ and $M = \langle \mathbb{N} \cup \mathsf{B};\ \mathsf{g},\ 0\rangle$, with $\mathsf{B} = \{\mathsf{b}, \mathsf{g}^1(\mathsf{b}), \ldots, \mathsf{g}^n(\mathsf{b})\}$ where $\mathsf{g}^{n+1}(\mathsf{b}) = \mathsf{b}$ and $\mathsf{g}(i) = i + 1$ for $i \in \mathbb{N}$. Since *S2* is true in M, $\mathsf{g}^i(\mathsf{b}) \neq 0$ for all i. As B_1 is true in M, C defines a function in this model, say $+$. Consider now $\mathsf{b} + \mathsf{g}^n(\mathsf{b})$.

Suppose $\mathsf{b} + \mathsf{g}^n(\mathsf{b}) = m$ for a natural number m. Then $\mathsf{g}^{k(n+1)}(\mathsf{b}) + \mathsf{g}^n(\mathsf{b}) = \mathsf{g}^m(0)$, where k is the least number such that $k(n + 1) > m$. But then since B_3 is true in M, we would have $\mathsf{g}^{k(n+1)-m}(\mathsf{b}) + \mathsf{g}^n(\mathsf{b}) = 0$, which contradicts B_2 being true in M. Nor can we have $\mathsf{b} + \mathsf{g}^n(\mathsf{b}) = \mathsf{b}$, since B5 is true in M.

So for some $j \geq 1$, $\mathsf{b} + \mathsf{g}^n(\mathsf{b}) = \mathsf{g}^j(\mathsf{b})$. We cannot have $j = n$ since B_4 and B_5 are true in M and $\mathsf{b} = \mathsf{g}^{n+1}(\mathsf{b})$. So $j < n$. But then by B_3 and B_4, we get that $\mathsf{b} + \mathsf{g}^{n-j}(\mathsf{b}) = \mathsf{b}$, which is a contradiction on B_5 being true in M. Hence, no such model can exist and there is no such formula C. So addition is not definable in **S**.

But then multiplication is not definable in **S**, because if it were, then addition would be definable, too (see Theorem 18.c below). ∎

Exercises for Section A

1. Write out T4 in the fully formal language using no abbreviations.
2. In the proof of Theorem 1, verify that **N** is a model of **S** and show it is not isomorphic to $\langle \mathbb{N}; +1, 0\rangle$.
3. Let $M = \langle U; \mathsf{s}, \mathsf{e}\rangle$ be a model of **S**. Show that there is one and only one injection of $\langle \mathbb{N}; +1, 0\rangle$ into M.
4. a. Show that $\langle \mathbb{R}; +1, 0\rangle$ is a model of **S** and consists of a standard part plus uncountably many Z-chains.

   b. Show two models of **S** are isomorphic iff they have the same number of Z-chains.

5. Prove: For all natural numbers m, n, $\mathbf{S} \vdash S^m o \neq S^n o$ iff $m \neq n$.
   (Hint: Show the wff is true in $\langle \mathbb{N}; +1, 0 \rangle$.)
6. a. Show that Γ is a conservative extension of **S** iff $\Gamma \supseteq \mathbf{S}$ and Γ is consistent.
   b. Prove Lemma 4.
7. Show that both 0 and 1 are definable in **S**.
8. Show that the set $\{n \colon n \leq 47\}$ is definable in **S**.
9. Show that the relation 'is less than' on the natural numbers is not definable in **S**.
10. Prove the decidability and completeness of **S** via elimination of quantifiers.
11. a. In L(= ; <) axiomatize the theory **T** of discrete linear orderings with first but no last element in which every element other than the first has an immediate predecessor and an immediate successor. (An immediate precedessor of a is an element $\mathsf{b} < \mathsf{a}$ such that there is no c with $\mathsf{b} < \mathsf{c} < \mathsf{a}$, and similarly for immediate successor).
   b. In L(= ; <, S), show that $\mathbf{T} \cup \mathbf{S}$ is an extension by definition of **T**.
   c. Show that $\mathbf{T} \cup \mathbf{S}$ is not an extension by definition of **S**.
   d. Show that **T** is complete and decidable by elimination of quantifiers.
   (Compare *Kreisel and Krivine, 1967*.)
   e. Show that $\mathbf{T} = \text{Th}(\langle \mathbb{N}; \text{—is less than—} \rangle)$.

B. The Theory Q

1. Axiomatizing addition and multiplication

We want a theory of arithmetic that includes at the very least successor, addition and multiplication on the natural numbers. I'll use $\oplus$ and $*$ as names for the latter two. So our semi-formal language will be L(= ; S, $\oplus$, $*$, o), realizing L(= ; f_0^1, f_1^2, f_2^2, c_0). To make our work easier to read, I'll write '$t \oplus u$' instead of $\oplus(t, u)$, and '$t * u$' for $*(t, u)$, using the same conventions we already adopted for the theory of successor. For example, $o * (S^2 o \oplus So)$ abbreviates $*(o, \oplus(S(S(o)), S(o)))$.

In trying to characterize addition, consider how we can calculate $4 + 3$ inductively:

$$4 + 0 = 4$$
$$4 + 1 = (4 + 0) + 1 = 4 + 1 = 5$$
$$4 + 2 = (4 + 1) + 1 = 5 + 1 = 6$$
$$4 + 3 = (4 + 2) + 1 = 6 + 1 = 7$$

The value of $n + (m + 1)$ is obtained from the value of $(n + m)$ by adding 1.

Similarly, we can calculate multiplication inductively using addition. The value of $n \cdot (m + 1)$ is obtained from the value of $n \cdot m$ by adding m.

We can formalize these definitions, along with the basic definition of successor, in a very simple theory. I'll write x, y for x_1 and x_2.

Q *in* L(= ; S, ⊕, ∗, o)

Q1 $\forall x\, \forall y\, (Sx = Sy \rightarrow x = y)$

Q2 $\forall x\, (0 \neq Sx)$

Q3 $\forall y\, (y \neq o \rightarrow \exists x\, (y = Sx))$

Q4 $\forall x\, (x \oplus o = x)$

Q5 $\forall x\, \forall y\, (x \oplus Sy = S(x \oplus y))$

Q6 $\forall x\, (x * o = o)$

Q7 $\forall x\, \forall y\, (x * Sy = (x * y) \oplus x)$

Note that these are axioms, not schema of axioms. Thus, **Q** is the set of consequences of just the one wff:

$$Q \equiv_{\text{Def}} Q1 \wedge Q2 \wedge Q3 \wedge Q4 \wedge Q5 \wedge Q6 \wedge Q7$$

Note also that *Q1*, *Q2*, *Q3* are the axioms *S1*, *S2*, *S3* of the theory of successor.

The *standard model* of **Q** is <ℕ ; +1, +, · , 0>, where + and · are addition and multiplication on the natural numbers as we learned them in grade school. Does **Q** capture all we want about addition and multiplication, as **S** does for successor?

Theorem 6 Every model of **Q** is infinite, but **Q** is not $\aleph_0$-categorical, **Q** is not complete, and **S** ⊄ **Q**.

Proof That every model of **Q** is infinite follows as in the proof of Theorem 1.a.

Now consider the interpretation $\mathfrak{M}$ with universe the natural numbers plus two other objects. Call those two 'α' and 'β', and interpret 'S' as $\mathfrak{s}$, '⊕' as $\mathfrak{f}$, and ∗ as $\mathfrak{g}$ according to the following tables:

$\mathfrak{s}$	
n	$n+1$
α	α
β	β

$\mathfrak{f}$	n	α	β
m	$m + n$	β	α
α	α	β	α
β	β	β	α

$\mathfrak{g}$	0	$n \neq 0$	α	β
m	0	$m \cdot n$	α	β
α	0	β	β	β
β	0	α	α	α

Then $\mathfrak{M}$ is countable and I'll let you show that $\mathfrak{M}$ is a model of **Q**.

Since $\mathfrak{s}(\beta) = \beta$, $\mathfrak{M} \models \neg\forall x\, (Sx \neq x)$. So **S** ⊄ **Q**.

As you can check, $\mathfrak{M} \models \neg\forall x\, \forall y\, (x \oplus y = y \oplus x)$. So $\mathfrak{M}$ is not elementarily equivalent to <ℕ ; +1, +, · , 0>, and hence not isomorphic to it, either. So **Q** is not complete, and **Q** is not $\aleph_0$-categorical. ■

There are lots of other simple claims about the natural numbers that we can't prove in **Q** (Exercise 6). Nonetheless, we've formalized enough in **Q** to get the right answers for S, ⊕ and ∗ when we use numerals.

Theorem 7 For all natural numbers m, n, k,

a. $\vdash_Q S^m o = S^n o$ iff $m = n$.

b. $\vdash_Q S^m o \oplus S^n o = S^k o$ iff $m + n = k$.

c. $\vdash_Q S^m o * S^n o = S^k o$ iff $m \cdot n = k$.

Proof (a) First suppose that $m \neq n$. We'll show $\vdash_Q S^m o \neq S^n o$ by induction on n, first assuming that $m < n$. Our basis is $m = 0$, $n = 1$, and by axiom Q2, $\vdash_Q o \neq So$. So suppose $\vdash_Q S^m o \neq S^n o$ and $m + 1 < n + 1$. By axiom Q1, $\vdash_Q SS^m o = SS^n o \rightarrow S^m o = S^n o$. So by **PC**, $\vdash_Q S^m o \neq S^n o \rightarrow SS^m o \neq SS^n o$. Hence by *modus ponens*, $\vdash_Q SS^m o \neq SS^n o$, and we've proved (a) for $m < n$. We then have (a) for $n < m$ since for any terms t, u $\vdash t \neq u \rightarrow u \neq t$.

Now suppose $\vdash_Q S^m o \neq S^n o$. By substitution of '=', $\vdash_Q S^m o = S^m o$. Since **Q** is consistent (it has a model), we can't have $\vdash_Q \neg(S^m o = S^m o)$. Hence, $m \neq n$.

(b) First I'll show by induction on n if $m + n = k$, then $\vdash_Q S^m o \oplus S^n o = S^k o$. If $n = 0$, $\vdash_Q S^m o \oplus o = S^m o$ by axiom Q4. So suppose it's true for r, and $n = r + 1$. Then (here we're using what we know about the natural numbers) for some s, $k = s + 1$, and $m + r = s$. By induction, $\vdash_Q S^m o \oplus S^r o = S^s o$. By substitution of equals, $\vdash_Q S(S^{m+r} o) = S^{s+1} o$. By Q5, $\vdash_Q S^m o \oplus SS^r o = S(S^{m+r} o)$. So by substitution of equals, $\vdash_Q S^m o \oplus S^n o = S^k o$.

Suppose now that $\vdash_Q S^m o \oplus S^n o = S^k o$ and $\vdash_Q S^m o \oplus S^n o = S^r o$. Then by substitution of equals, we have $\vdash_Q S^k o = S^r o$, and hence by (a), $k = r$.

(c) The proof for multiplication is similar, and I'll leave it to you. ■

Exercises for Section B.1

1. Write each of the following wffs without any abbreviations:
   a. $\forall x\, (S^3 o \neq (S^2 o \oplus (x * x))$
   b. $\forall x\, \forall y\, (S^4 y = (S^2 o * (x * y))$
   c. $\forall x\, \forall y\, (S^2 y \oplus So = S^2 o \oplus Sy)$
   d. $S^{47} o = S^{283} o$
2. Determine whether each wff in Exercise 1 is true of the natural numbers.
3. Do all the steps in the inductive evaluation of multiplication to calculate $4 \cdot 3$.
4. Give a proof in **Q** that $\vdash S^3 o \oplus S^2 o = S^5 o$.
5. Show that **M** in the proof of Theorem 6 is a model of **Q**.
6. Using **M** in the proof of Theorem 6, show that the following are not theorems of **Q**.
   a. $\forall x\, \forall y\, (x \oplus y = y \oplus x)$
   b. $\forall x\, \forall y\, \forall z\, (x \oplus (y \oplus z) = (x \oplus y) \oplus z$
   c. $\forall x\, (0 \oplus x = x)$
   d. $\forall y\, \forall z\, \neg(\exists x\, (Sx \oplus y = z) \wedge \exists x\, (Sx \oplus z = y)$
   e. $\forall x\, \exists y\, \forall z\, (x * (y * z) = (x * y) * z)$
   f. $\forall x\, \forall y\, (x * y = y * x)$
   g. $\forall x\, \forall y\, \forall z\, (x * (y \oplus z) = (x * y) \oplus (x * z))$

2. Proving is a computable procedure

Our axioms are sufficient for us to prove the correct values for successor, addition and multiplication for the numerals. We can use **Q** to compute those values, since proving is a constructive procedure in **Q**, or in any other axiom system. To make this precise, though, we need to be clear about what we mean by 'axiom system'.

Axiom system An *axiom system* for a collection of closed wffs Γ is a (possibly infinite) list of closed wffs $A_1, A_2, \ldots$ such that we can constructively decide of any wff whether it is on the list, and $\Gamma = \text{Th}(\{A_1, A_2, \ldots\})$. If there is an axiom system for Γ, we say that Γ is *axiomatizable.*

Note that this requires that if an axiom system is presented by schema, we must be able to constructively recognize whether a wff is an instance of one of the schema.

Each of **Q**, **DLO**, and **G** is axiomatizable, since they are given by finite lists of axioms. So is **S**, since we can tell of any wff whether it is on the list of axioms of **S**. So, too, is $\text{Th}(\langle \mathbb{N}; +1, 0\rangle)$, since $\text{Th}(\langle \mathbb{N}; +1, 0\rangle) = \mathbf{S}$, which is axiomatizable.

Given any axiom system, *proving from the axioms is a computable procedure*:

1. Given any wff, we can decide whether it is an axiom of our system.
2. Given wffs A, B, C, we can constructively decide whether C is the result of applying *modus ponens* (that is, whether B is A→C).
3. Given any sequence of wffs $B_1, B_2, \ldots, B_n$, we can constructively decide whether it is a proof; that is, for each i, B_i is an axiom or for some $j, k < i$, B_k is $B_j \rightarrow B_i$.

Further, for any axiom system we can *constructively enumerate the theorems*:

a. We have a constructive numbering of all wffs, $C_1, C_2, \ldots$.
b. The fundamental theorem of arithmetic says that every natural number $n \geq 2$ can be uniquely written as $n = p_1^{a_1} \cdot \ldots \cdot p_k^{a_k}$, where $p_1 < \ldots < p_k$ are primes, and for all i, $0 < a_i$.
c. We can constructively enumerate all finite sequences of wffs: Let the nth sequence, for $n \geq 2$, be $C_{a_1}, \ldots, C_{a_k}$, where $n = p_1^{a_1} \cdot \ldots \cdot p_k^{a_k}$.
d. We can constructively enumerate all sequences of wffs that are proofs: The nth proof sequence is the one corresponding to the nth sequence of wffs that is a proof.
e. We can constructively enumerate all theorems: B_n is the final wff in the nth proof sequence.

Now we can set out generally the kind of decision procedure we used before.

Theorem 8 Any complete axiomatizable theory is decidable.

Proof Let Γ be a complete axiomatizable theory. If Γ is inconsistent, we're done. If Γ is consistent, to decide whether $\Gamma \vdash A$, generate out the theorems of Γ. Because Γ is complete, either A or $\neg A$ will show up on the list. If it's A, then $\Gamma \vdash A$; if it's $\neg A$, then $\Gamma \nvdash A$. ∎

3. The computable functions and Q

The theory **Q** is strong enough to allow us to compute addition and multiplication correctly. But what about other functions such as exponentiation? Is there already a formula A such that $\vdash_{\mathbf{Q}} A(S^m o, S^n o, S^k o)$ iff $m^n = k$? Or do we need to add a new function symbol 'exp' and axioms for an inductive definition such as:

$$\forall x\,(x \text{ exp } o = 1)$$
$$\forall x\,\forall y\,(x \text{ exp } Sy = (x * y) * x)$$

Representability of functions and predicates In languages that contain S and =,

A k-ary function f on the natural numbers is *represented by* A *in theory* Γ iff for every $n_1, \ldots, n_k, m$, if $f(n_1, \ldots, n_k) = m$, then both:

$$\Gamma \vdash A(S^{n_1}o, \ldots, S^{n_k}o, S^m o)$$
$$\Gamma \vdash \forall x\,(x \neq S^m o \rightarrow \neg A(S^{n_1}o, \ldots, S^{n_k}o, x))$$

A k-ary predicate P on the natural numbers is *represented by* A *in theory* Γ iff for every $n_1, \ldots, n_k$:

If $P(n_1, \ldots, n_k)$, then $\Gamma \vdash A(S^{n_1}o, \ldots, S^{n_k}o)$.

If not-$P(n_1, \ldots, n_k)$, then $\Gamma \vdash \neg A(S^{n_1}o, \ldots, S^{n_k}o)$.

I'll let you prove the following, using the proof of Theorem 7 as a guide.

Lemma 9 In any language that contains $L(= ; S, \oplus, *, o)$,

a. If Γ is an inconsistent theory, then all functions on the natural numbers are representable in Γ.

b. If Γ is a consistent theory such that $\Gamma \supseteq \mathbf{Q}$, and the function f on the natural numbers is representable by A in Γ, then for all $n_1, \ldots, n_k, m$, $f(n_1, \ldots, n_k) = m$ iff $\Gamma \vdash A(S^{n_1}o, \ldots, S^{n_k}o, S^m o)$.

c. If Γ is a consistent theory such that $\Gamma \supseteq \mathbf{Q}$, and the predicate P on the natural numbers is representable by A in Γ, then for all $n_1, \ldots, n_k$, $P(n_1, \ldots, n_k)$ iff $\Gamma \vdash_{\mathbf{Q}} A(S^{n_1}o, \ldots, S^{n_k}o)$.

If a function f is representable by A in **Q**, then we can compute its values: To calculate the value $f(n_1, \ldots, n_k)$ start proving theorems, which is a computable procedure. Since **Q** is consistent, there has to be just one formula of the form $\vdash_{\mathbf{Q}} A(S^{n_1}o, \ldots, S^{n_k}o, S^m o)$ on the list. When we reach it, we have that $f(n_1, \ldots, n_k) = m$. So every function that is representable in **Q** is computable.

I've been using the words 'computable' and 'constructive' informally here. But, as you can read in Epstein and Carnielli, *Computability*, these notions have been formalized. The set of *recursive functions* on the natural numbers is the smallest class that (i) contains the constant function with output 0, the successor function, and for each $k, i \geq 1$, the function that for input $n_1, \ldots, n_k$ outputs n_i, and (ii) is closed under composition of functions, definition of functions by induction, and searching for the least value that satisfies a condition already established as recursive. The class of recursive functions can be defined inductively. *Church's thesis* says that the formalization is good: The class of recursive functions is exactly the class of computable functions.

To say that a set of objects that aren't numbers or sequences of numbers (like the wffs of a formal language) is computable means that we can number all the objects constructively (we can't escape the informal notion when numbering things that aren't numbers). To say that a subset of those objects, say $\mathcal{A}$, is computable means that the function f given by $f(n) = 1$ if n is the number of an object that is in $\mathcal{A}$, and $f(n) = 0$ otherwise, is computable (recursive).

What is remarkable about the simple, weak system **Q** is that we have formalized enough about arithmetic in it to be able to represent every computable function. That is, if f is a computable (recursive) function on the natural numbers, there is some formula A that represents f in **Q**. We've also shown that every function that's representable in **Q** is computable.

***Computability and* Q** The computable/recursive functions are exactly those that can be represented in **Q**.

It would take many pages to present enough study of the recursive functions to prove this and give a motivation for accepting Church's Thesis. That's why Walter Carnielli and I wrote the textbook *Computability*. Even to show that exponentiation is representable in **Q** is hard. So I'll continue to use the terms 'computable' and 'constructive' informally here, referring you to that book for the formal details.

4. The undecidability of Q

We've seen that we can constructively enumerate the theorems of **Q**. Can we go further and decide of every wff whether it is a theorem of **Q**? The answer is no, because using a constructive numbering of all wffs we can understand talk about theorems and proving as assertions about natural numbers, and that would allow for a self-referential paradox to be constructed within **Q** itself.

In order to show that, for any wff A let $[\![A]\!]$, the *Gödel number* of A, be the number assigned to A in the constructive numbering of wffs. To say a theory **T** is *decidable* now means that the set $\{n: n = [\![A]\!] \text{ and } \vdash_{\mathbf{T}} A\}$ is computable.

Theorem 10 **Q** is not decidable.

Proof Suppose that **Q** is decidable. Then the following relation on the natural numbers is computable:

$$\begin{aligned} \mathsf{R}(m, n) \text{ iff } & \text{ for some A with exactly one free variable } x_1, \\ & m = [\![A]\!] \text{ and } \vdash_{\mathbf{Q}} A(S^n o / x_1) \end{aligned}$$

So R is representable in **Q**, say by $B(x, y)$. Consider the *diagonalization* of R :

$$\mathsf{D} = \{n: \text{not } \mathsf{R}(n, n)\}$$

D is representable in **Q** by (relabeling variables) $\neg B(x_1, x_1)$. Since **Q** is consistent:

$$\vdash_{\mathbf{Q}} \neg B(S^n o, S^n o) \text{ iff } S^n o \in \mathsf{D}$$

Let $k = [\![\neg B(x_1, x_1)]\!]$. Then we have a contradiction:

$$\begin{aligned} k \in \mathsf{D} & \text{ iff } \vdash_{\mathbf{Q}} \neg B(S^k o, S^k o) && \text{since } \neg B(x_1, x_1) \text{ represents } \mathsf{D} \\ & \text{ iff } \mathsf{R}(k, k) && \text{by the definition of } \mathsf{R} \\ & \text{ iff } k \notin \mathsf{D} && \text{by the definition of } \mathsf{D} \end{aligned}$$

■

Exercises for Sections B.2–B.4

1. What does it mean to say that a collection of wffs is axiomatizable?
2. Every theory is countable and, hence, has an enumeration. Why does this not entail that every theory is axiomatizable?
3. Show that every inconsistent theory is decidable.
4. a. Is every decidable collection of closed wffs axiomatizable?
   b. Is every axiomatizable collection of closed wffs decidable?
5. Why shouldn't we allow for noncomputable proofs?
6. Prove Lemma 9.
7. Show that 'x is less than y' is representable in **Q** by $\exists z\, (z \neq o \wedge x + z = y)$.
8. Compare the proof of Theorem 10 to the proofs that there is no set of all sets and that the real numbers are not countable (p. 161).

C. Theories of Arithmetic

1. Peano Arithmetic and Arithmetic

The theory **Q** is strong enough to allow us to calculate the values of all computable functions. But we saw that there are many simple truths of arithmetic we can't prove in **Q**. Before we look for further axioms to add to **Q**, what do we know generally about theories Γ that *extend* **Q**, that is, $\Gamma \supseteq \mathbf{Q}$ in $L(= ; S, \oplus, *, o)$?

Theorem 11 For any consistent theory Γ that extends **Q**,

a. All the computable functions are representable in Γ.

b. If Γ is axiomatizable, then a function is representable in Γ iff it is computable.

c. Γ is undecidable.

Proof (a) Since Γ extends **Q**, if a function is representable in **Q**, it is representable in Γ. So all the computable functions are representable in Γ.

(b) If Γ is axiomatizable, then proving is a computable procedure in Γ, just as it was in **Q**. So, since Γ is consistent, any function representable in **T** is computable.

(c) All we needed in the proof of Theorem 10 is that the theory is consistent and all computable functions are representable in it. ■

Among all the claims true about the natural numbers, what should we add to **Q**? The characteristic method of proof for the natural numbers, which we've relied on extensively in our metalogic, is induction (Appendix to Chapter I): Given any predicate P, if P is true for 0, and if for any n, if P is true for n then P is true for $n + 1$, then P is true for all natural numbers. Can we add this to **Q**?

In the first-order language of arithmetic the only predicates we can talk about are those that correspond to first-order wffs. We have no way to quantify over the numerals of the language, but only over arbitrary elements of the universe. So the best we can do in formalizing the principle of induction in an extension of **Q** is the following scheme.

The principle of first-order induction, *Induction*$_1$

$$\forall \ldots ([A(o) \wedge \forall x\,(A(x) \rightarrow A(Sx))] \rightarrow \forall x\, A(x))$$

for every wff A in which x appears free.

Given any wff of the formal language, we can tell if it is an instance of the scheme of first-order induction. So the following theory is axiomatizable.

Peano Arithmetic, **PA**, in L(= ; S, ⊕, ∗, o)

Q1–Q7

All instances of *Induction*$_1$

The first-order scheme of induction adds tremendous power to **Q**. We can now prove all the claims in Exercise 6 of Section B.1 that we couldn't prove in **Q**. Here's an example, where x, y, z stand for x_1, x_2, x_3, and '$\vdash$' stands for $\vdash_{\mathbf{PA}}$. I'll use the commutativity of quantifiers (Theorem V.2), as justified by the completeness theorem.

$$\vdash_{\mathbf{PA}} \forall z\, \forall x\, \forall y\, (x \oplus (y \oplus z) = (x \oplus y) \oplus z)$$

Let A(z) be $\forall x\, \forall y\, (x \oplus (y \oplus z) = (x \oplus y) \oplus z)$.

1.	$\vdash \forall y\, (y \oplus o = y)$	Q4
2.	$\vdash \forall x\, \forall y\, (x \oplus (y \oplus o) = x \oplus y)$	substitution of equals
3.	$\vdash \forall x\, \forall y\, ((x \oplus y) \oplus o = x \oplus y)$	Q4
4.	$\vdash$ A(o)	substitution of equals
5.	$\vdash \forall y\, \forall z\, (y \oplus Sz = S(y \oplus z))$	Q5
6.	$\vdash \forall z\, \forall x\, \forall y\, (x \oplus (y \oplus Sz) = x \oplus S(y \oplus z))$	substitution of equals
7.	$\vdash \forall z\, \forall x\, \forall y\, (x \oplus S(y \oplus z) = S(x \oplus (y \oplus z)))$	Q5
8.	$\vdash \forall z\, \forall x\, \forall y\, ((x \oplus y) \oplus Sz = S((x \oplus y) \oplus z))$	Q5
9.	$\vdash \forall z\, (A(z) \rightarrow A(Sz))$	(6), (7), (8) and substitution of equals
10.	$\vdash (A(o) \wedge \forall z\, (A(z) \rightarrow A(Sz))) \rightarrow \forall z\, A(z)$	*Induction*$_1$
11.	$\vdash \forall z\, A(z)$	

We can also prove in **PA** each of the axioms T_n of **S**, for $n \geq 1$.

$\vdash_{\mathbf{PA}} \forall x\, (S^n x \neq x)$

1. $\vdash S^n o \neq o$ — Q2 and substitution
2. $\vdash S^{n+1}x = Sx \rightarrow S^n x = x$ — Q1
3. $\vdash \forall x\, (S^n x \neq x \rightarrow S^{n+1}x \neq Sx)$ — **PC**
4. $\vdash (S^n o \neq o \wedge \forall x\, (S^n x \neq x \rightarrow S^{n+1}x \neq Sx)) \rightarrow \forall x\, (S^n x \neq x)$ — an instance of *Induction*$_1$
5. $\vdash \forall x\, (S^n x \neq x)$

Thus, **S** $\subset$ **PA**. Can we now prove all first-order claims true of the natural numbers? The collection of all those wffs is a theory.

Arithmetic *in* L(= ; S, $\oplus$, $*$, o)

Arithmetic = Th(<ℕ ; +1, +, · , 0>)

Theorem 12

a. **Arithmetic** is not decidable.

b. **Arithmetic** is not axiomatizable.

c. **PA** $\subsetneq$ **Arithmetic**.

d. There is a countable model of **Arithmetic** not isomorphic to <ℕ ; +1, +, · , 0>.

Proof (a) This follows from Theorem 11.c.

(b) If **Arithmetic** were axiomatizable, it would be decidable (Theorem 8). But that contradicts (a). So **Arithmetic** is not axiomatizable.

(c) **PA** is axiomatizable, so **PA** $\neq$ **Arithmetic**. But, as we've tacitly assumed above, all instances of the first-order theory of induction are true of the natural numbers. So **PA** $\subset$ **Arithmetic**.

(d) I suggest how to prove this as Exercise 13 below. ■

Corollary 13 If Γ is a consistent axiomatizable theory in $L(= ; S, \oplus, *, o)$, then there is some wff true of the natural numbers that cannot be proved in Γ.

Corollary 14 If Γ is a consistent axiomatizable extension of **Q** all of whose theorems are true of the natural numbers, then there is some wff A true of the natural numbers such that $\Gamma \nvdash A$ and $\Gamma \nvdash \neg A$.

Corollary 15 Church's theorem The collection of valid wffs in any formal language that contains $L(= ; f_0^1, f_1^2, f_2^2, c_0)$ is undecidable.

Proof Recall that **Q** is the theory of the single axiom Q. Suppose there were a decision procedure for valid wffs. Then we could decide for any wff whether $\models Q \rightarrow A$. By the semantic deduction theorem, $\models Q \rightarrow A$ iff $Q \models A$. By the completeness theorem, $Q \models A$ iff $\vdash_Q A$. So we would have a decision procedure for **Q**, which is a contradiction. Hence, there is no decision procedure for validity. ■

Corollary 16 In $L(= ; S, \oplus, *, o)$, every theory $\Gamma \subseteq$ **Arithmetic** is undecidable.

Proof Consider $Th(\Gamma \cup \mathbf{Q})$. Since $\mathbf{Q} \subseteq$ **Arithmetic**, $Th(\Gamma \cup \mathbf{Q}) \subseteq$ **Arithmetic**. So $Th(\Gamma \cup \mathbf{Q})$ is consistent, and hence by Theorem 11.c, it is undecidable. Now $\Gamma \cup \mathbf{Q} \vdash A$ iff $\Gamma \vdash Q \rightarrow A$. So if Γ were decidable, $\Gamma \cup \mathbf{Q}$ would be too. Hence, Γ is undecidable. ■

From Corollary 16 we can conclude that **S** in $L(= ; S, \oplus, *, o)$ is undecidable. Yet **S** in $L(= ; S, o)$ is decidable. *Whether a theory is decidable can depend on the language in which it is formulated.*

Exercises for Section C.1

1. Prove in **PA** all the claims in Exercise 6 of Section B.1.
2. Prove in **PA** $\forall x \forall y (x \oplus y = y \oplus x)$.
   (Hint: Prove $\forall x \forall y (x \oplus Sy = Sx \oplus y)$.)
3. Prove Corollary 13.
4. Prove Corollary 14.
5. Show that the set of valid wffs in the language without functions is undecidable.
   (Hint: Compare Chapter VIII.E.)
6. Prove that $Induction_1 \vdash Q3$, so that Q3 can be deleted from the axioms of **PA**.
7. Prove that there is a function on $\mathbb{N}$ that is not representable in **Arithmetic**.
   (Hint: How many functions can be represented in **Arithmetic**?)

8. Let **Exp** be the theory in L(= ; S, ⊕, ∗, o) of **Q** plus the two axioms:
   $\forall x_1 \, (x_1 \text{ exp o} = 1)$
   $\forall x_1 \, \forall x_2 \, (x_1 \text{ exp } Sx_2 = (x_1 * x_2) * x_1)$
   Show that **Exp** is undecidable and is not complete.

9. Show that a function on the natural numbers is representable in **Arithmetic** iff it is definable in **Arithmetic**. (Hint: To get that representability implies definability, show that if **Arithmetic**$\vdash \exists x \, A(x)$, then for some n, **Arithmetic**$\vdash A(S^n o)$.)

10. Explain why Church's theorem shows that in first-order logic we cannot replace proving with a method like the truth-table method for propositional logic.

11. What reason do you have to believe that all the instances of *Induction*$_1$ are true?

12. *There is a countable model of* **Arithmetic** *not isomorphic to the natural numbers*
    a. Let **T** = Th(**Arithmetic** ∪ $\{c_1 \neq o, c_1 \neq So, \ldots, c_1 \neq S^n o, \ldots\}$). Show that <ℕ ; +1, +, · , 0> is a model of every finite subset of **T**.
    b. Show that **T** has a countable model, M′. (Hint: Compactness.)
    c. Show that M′ is infinite.
    d. Show that M $\equiv_{\text{Def}}$ (M′ with the reference of 'c_1' deleted) is a model of **Arithmetic**.
    e. Show that M is not isomorphic to <ℕ ; +1, +, · , 0>. (Hint: Every element of <ℕ ; +1, +, · , 0> has a name.)

13. *Characterizing nonstandard models of* **Arithmetic**
    Let M be a *nonstandard* model of **Arithmetic**, that is, M is countable and is not isomorphic to <ℕ ; +1, +, · , 0>. Prove:
    a. M consists of a standard part plus some number (finite or infinite, possibly none) of Z-chains. (Hint: Compare the proof of Theorem 1.)
    b. The standard part of M is isomorphic to <ℕ ; +1, +, · , 0>. (Hint: Use the names of the elements of the standard part.)

    So we can assume the standard part of M is <ℕ ; +1, +, · , 0> . Use n, m, k to stand for elements of ℕ. Call any element of M that is not in the standard part a *non-standard number.* Let s, +, ⁎ be the intepretations in M of S, ⊕, ∗.

    c. If a, b are in the same Z-chain, then for some n, $s^n a = b$ or $s^n b = a$.
    d. If a, b are in the same Z-chain, then for some n, either $a + n = b$ or $b + n = a$. (Hint: $\forall x \, \forall y \, (S^n x = y \rightarrow y = x + S^n o)$ is true of the natural numbers.)
    e. If a is non-standard, then a + a is not in a's Z-chain. (Hint: $\forall x \, \forall y \, \forall z \, (x + y = x + z \rightarrow y = z)$ is true of the natural numbers.)
    f. The predicate 'less than' on the natural numbers is representable, and its interpretation < is a linear order on M. (Hint: Use the hint for (e) for the conditions for a linear order.)
    g. If a and b are from different Z-chains and a < b, then c < d for every c in the Z-chain of a , every d in the Z-chain of b. (Hint: sa ≤ b, but sa ≠ b, since they're in different Z-chains. So sa < b, and similarly for $s^n a$. Now work backwards in a's Z-chain.)
    h. The following relation is a linear ordering of Z-chains: Z-chain A ≠ Z-chain B and there is some a in A, some b in B, such that a < b.
    i. Every natural number is less than every non-standard number. (Hint: 0 and < .)

j. If a is non-standard, there is some non-standard b with b + b = a or b + b + 1 = a. (Hint: Every natural number is even or odd.)
k. There is no least non-standard Z-chain.
l. If Z-chain A is less than Z-chain B, then there is a Z-chain C such that A is less than C and C is less than B.
(Hint: Suppose a < b, where a∈A and b∈B. Then there is a c such that c + c = a + b or c + c + 1 = a + b. That c can't be standard, nor in a's Z-chain nor b's Z-chain.)
m. The ordering of Z-chains is a dense linear order with first but no last element.
n. The non-standard Z-chains can be put into 1-1 correspondence with the rational numbers ≥ 0. (Hint: Compare Theorem XIII.5.)
o. All countable non-standard models of **Arithmetic** are isomorphic.

14. Formalize and prove in **PA** the following versions of the principle of induction.
a. *The least number principle*
If a claim is true of some number, then there is a least number of which it is true. (Hint: Prove it informally and convert that to a formal one, using Theorem 7.a.)
b. *Alternate form of induction*
If a claim is true for 0, and if whenever it is true for all numbers less than x it is true for x, then it is true for all x.
c. *The method of infinite descent*
Given a claim, if whenever it is true for a number x it is true for some number strictly less than x, then the claim is false for all x.

2. The languages of arithmetic

We have formulated our theories of arithmetic in the language L(= ; S, ⊕, ∗, o). But we always understood the successor function as 'add one'. It's not hard to show that we can formulate our theories of arithmetic in the language L(= ; ⊕, ∗, o, ł). First, we have by Q5 and Q4:

$$\mathbf{Q} \vdash \forall x\, (Sx = x \oplus So)$$

Consider the translation from the language L(= ; S, ⊕, ∗, o) to the language L(= ; ⊕, ∗, o, ł) given by letting A* be A with every occurrence of 'Sx' replaced by '$x \oplus$ ł'. For any Γ, let Γ* be {A*: A∈Γ}. Given any model M of **Q***, we can convert it to a model M* of **Q** by deleting the interpretation of 'ł' and interpreting 'S' by the function sa = a + 1, where 1 is the interpretation of 'ł'. Further, as you can show, every model of **Q** is M* for some model M⊨**Q***. I'll let you prove the following.

Theorem 17

a. Given any model N of **Q**, there is a model M of **Q*** such that M* = N.

b. M⊨A* iff M*⊨A

c. For any consistent theory Γ ⊇ **Q**, Γ⊨A iff Γ*⊨A*.

d. *Q1*–Q7** is an axiomatization of **Q***.

We can devise translations from consistent theories extending **Q** to consistent theories in other languages, too. I'll let you prove the following.

Theorem 18

a. The predicate 'is less than' on ℕ is definable in **Arithmetic** by:
$\exists z\,(z \neq \mathrm{o} \wedge x \oplus z = y)$

b. 0 is definable in **Arithmetic** by: $x \oplus x = x$.
1 is definable in **Arithmetic** by: $x * x = x$.

c. In L(= ; S, $*$), addition on ℕ is definable in Th(<ℕ; +1, ·>) by:
$(x = 0 \wedge y = 0 \wedge z = 0) \vee (z \neq 0 \wedge \mathrm{S}(x * z) * \mathrm{S}(y * z) = \mathrm{S}((z * z) * \mathrm{S}(x * y)))$.

d. In L(= ; <), the successor function is definable in Th(<ℕ; is less than>) by:
$(x < y) \wedge \neg\exists z\,(x < z \wedge z < y)$.

e. In L(= ; <, $*$), addition on ℕ is definable in Th(<ℕ; is less than, ·>).

So we can talk of **Q** and extensions of **Q** as formulated in any of:

L(= ; S, ⊕, $*$, o) L(= ; ⊕, $*$, o, ł) L(= ; ⊕, $*$) L(= ; S, $*$) L(= ; <, $*$)

Some say this is because there is a body of abstract truths that comprise the arithmetic of the natural numbers. But what those "truths" are, I cannot say.

But *multiplication is not definable in* Th(<ℕ; +1, +>): If it were, the theory would be undecidable, but Mojzesz Presburger devised a variation on the method of elimination of quantifiers to show that Th(<ℕ; +1, +>) is complete and decidable (see, for example, *Enderton, 1972*). Similarly, *multiplication is not definable in* Th(<ℕ; is less than>), as then that theory would be undecidable; yet the method elimination of quantifiers can be used to show it is decidable. So neither L(= ; S, ⊕) nor L(= ; <) suffice for a language of arithmetic.

Exercises for Section C.2

1. Prove Theorem 17.
2. Write out an axiomatization of **Q***.
3. Show that Th(<ℕ; is less than>) is decidable by the method of elimination of quantifiers.

D. The Consistency of Theories of Arithmetic

We've said that **Q** is consistent because it has a model, <ℕ; +1, +, ·, 0>. But in showing that <ℕ; +1, +, ·, 0> is a model of **Q**, we've assumed we know arithmetic. To show that our simple formal theory of a fragment of arithmetic is consistent, we've assumed our informal theory of arithmetic.

In our informal theory of arithmetic we're not explicit about how much of the induction scheme we accept. We normally don't use very complicated instances of it. Yet we assumed that all instances of the induction scheme are true of the natural numbers when we claimed that **PA** has a model and hence is consistent. We would like to prove that **Q** and **PA** are consistent using much weaker assumptions.

Just what do we need in order to prove the consistency of these theories? The answer requires several steps, all involving variations on the diagonal method used to prove Theorem 10. First, we know that we can't decide which claims in the first-order language of arithmetic are in **Arithmetic**. That is, the set of Gödel numbers of wffs true of $\langle \mathbb{N}; +1, +, \cdot, 0\rangle$ is not computable. But we can show more.

Theorem 19 Arithmetical truth is not representable in arithmetic
The set of Gödel numbers of sentences in the first-order theory of arithmetic is not representable in **Arithmetic**.

Proof Let $\mathsf{T} = \{n$: for some A, $n = [\![A]\!]$ and A is true of the natural numbers$\}$.

Suppose that T is representable in **Arithmetic** by the formula T. In that case, the set of wffs that are false of their own Gödel number is also representable:

$\mathsf{F} = \{m$: for some A, $m = [\![A]\!]$ and A has exactly one free variable x_1, and $A(S^m o)$ is false of the natural numbers$\}$

That's because we can constructively recognize whether a wff has exactly one free variable x_1, and whether a term is of the form '$S^m o$'. Relabeling variables, we can assume that a wff that represents F is $D(x_1)$ with x_1 its one free variable. Let $k = [\![D(x_1)]\!]$. We then have:

$D(S^k o)$ is true of the natural numbers iff $k \in \mathsf{F}$ iff $D(S^k o)$ is false.

That's a contradiction. So T is not representable in **Arithmetic**. ∎

Even more so, then, arithmetical truth is not definable in any consistent theory weaker than **Arithmetic**, such as **Q** or **PA**.

In 1931, Kurt Gödel observed that we could replace 'true' by 'provable' in this method to get a sentence that, instead of being a version of the liar paradox, truthfully asserts (via the numbering) its own unprovability. We define a predicate on the natural numbers:

$\mathsf{W}(n, m) \equiv_{\text{Def}}$ for some A with exactly one free variable x_1, $n = [\![A]\!]$ and m is the number of a proof sequence in **Q** that ends with $A(S^n o)$.
(That is, m is the number of a proof in **Q** of $A(S^n o)$.)

From Section B.2 we know that W is computable. So W is representable in **Q**, say by $W(x_1, x_2)$. Let $k = [\![\neg\exists x_2\, W(x_1, x_2)]\!]$. This has just x_1 free. Define:

$$k = [\![\neg\exists x_2\, W(x_1, x_2)]\!]$$

$$U_Q \equiv_{\text{Def}} \neg\exists x_2\, W(S^k o, x_2)$$

In terms of the numbering of wffs and proofs, U_Q says that it itself is unprovable.

Theorem 20

a. If **Q** is consistent, then $\nvdash_{\mathbf{Q}} \mathrm{U}_{\mathbf{Q}}$ and $\mathrm{U}_{\mathbf{Q}}$ is true of the natural numbers.

b. If all the theorems of **Q** are true of the natural numbers, then $\nvdash_{\mathbf{Q}} \neg \mathrm{U}_{\mathbf{Q}}$.

Proof (a) Suppose that $\vdash_{\mathbf{Q}} \mathrm{U}_{\mathbf{Q}}$. That is, there is a proof sequence in **Q** that ends with $\mathrm{U}_{\mathbf{Q}}$. Let m be the least number that codes such a proof. Then $\mathsf{W}(k, m)$. Hence, $\vdash_{\mathbf{Q}} \mathrm{W}(\mathrm{S}^k\mathrm{o}, \mathrm{S}^m\mathrm{o})$. So by existential generalization, $\vdash_{\mathbf{Q}} \exists x_2\, \mathrm{W}(\mathrm{S}^k\mathrm{o}, x_2)$. So by **PC**, $\vdash_{\mathbf{Q}} \neg \mathrm{U}_{\mathbf{Q}}$, which is a contradiction on the consistency of **Q**. So $\mathrm{U}_{\mathbf{Q}}$ is not provable in **Q**. So there is no m such that $\mathrm{W}(\mathrm{S}^k\mathrm{o}, \mathrm{S}^m\mathrm{o})$. So $\mathrm{U}_{\mathbf{Q}}$ is true of the natural numbers.

(b) If all the theorems of **Q** are true of the natural numbers, then **Q** is consistent. Hence, $\mathrm{U}_{\mathbf{Q}}$ is true of the natural numbers, and so $\nvdash_{\mathbf{Q}} \neg \mathrm{U}_{\mathbf{Q}}$. ∎

The only place we use anything peculiar to **Q** in the proofs of Theorem 20 is in the numbering of the proofs of **Q**. The same procedure can be used to produce a wff U_Γ for any other theory Γ, so long as Γ is consistent and all the computable functions are representable in it.

Corollary 21 ***Gödel's first incompleteness theorem*** Suppose $\Gamma \supseteq \mathbf{Q}$.

a. If Γ is a consistent axiomatizable theory, then there is a wff U_Γ such that $\Gamma \nvdash \mathrm{U}_\Gamma$, and U_Γ is true of the natural numbers.

b. If all the theorems of Γ are true of the natural numbers, then $\Gamma \nvdash \neg \mathrm{U}_\Gamma$.

For any theory Γ, a closed wff A such that $\Gamma \nvdash \mathrm{A}$ and $\Gamma \nvdash \neg \mathrm{A}$ is called *formally undecidable relative to* Γ. That word 'formally' is important, because in Theorem 20 we can see from outside the system that $\mathrm{U}_{\mathbf{Q}}$ is actually true.

Though the formulas U_Γ show that we can find formally undecidable wffs, these are not wffs that seem "natural" in arithmetic. *Paris and Harrington, 1977* have produced an example of a combinatorial principle true of the natural numbers that is formally undecidable relative to **PA**; *Isaacson, 1996* has a survey and discussion of other formally undecidable sentences that seem mathematically natural.

Now we can investigate whether we can prove the consistency of **PA**. By definition, **PA** is consistent iff there is no A such that $\vdash_{\mathbf{PA}} \mathrm{A}$ and $\vdash_{\mathbf{PA}} \neg \mathrm{A}$. So consider the relation $\mathsf{R}(n, m, p, q)$ on the natural numbers:

$n = [\![\mathrm{A}]\!]$ and m is the number of a proof sequence in **PA** that ends with A

$p = [\![\neg \mathrm{A}]\!]$ and q is the number of a proof sequence in **PA** that ends with $\neg$A

This is computable. Hence, there is some formula $\mathrm{C}(x_1, x_2, x_3, x_4)$ that represents it in **PA**. Define the closed formula:

$$\mathrm{Consis}_{\mathbf{PA}} \equiv_{\mathrm{Def}} \forall x_1 \forall x_2 \forall x_3 \forall x_4 \neg \mathrm{C}(x_1, x_2, x_3, x_4)$$

Then $\mathrm{Consis}_{\mathbf{PA}}$ is true of the natural numbers iff **PA** is consistent.

Theorem 22 Gödel's second incompleteness theorem
If **PA** is consistent, then $\nvdash_{\mathbf{PA}} \text{Consis}_{\mathbf{PA}}$.

Proof (sketch) The proof of Corollary 21.a for **PA** is just the proof of Theorem 20.a with the computable predicate for numbers of proofs the one for **PA** rather than **Q**. We may convert that entire proof into talk about natural numbers instead of our formal system via the numberings of wffs and sequences of wffs. Nowhere do we use anything about wffs being true of the natural numbers in proving:

If **PA** is consistent, then $\nvdash_{\mathbf{PA}} U_{\mathbf{PA}}$.

We also know that $U_{\mathbf{PA}}$ formalizes the proposition that $U_{\mathbf{PA}}$ is not provable in **PA**. So we can establish in our informal theory of arithmetic:

(Coded as talk about the natural numbers) If $\text{Consis}_{\mathbf{PA}}$, then $U_{\mathbf{PA}}$.

The entire proof of this claim about the natural numbers can be formalized in **PA**. Formalizing that informal proof is very long and tedious. You can find it proved in detail in a slightly different version in *Shoenfield, 1967*, pp. 211–213, (his theory *N* is **Q**, and his theory *P* is **PA**). So we have:

$\vdash_{\mathbf{PA}} \text{Consis}_{\mathbf{PA}} \rightarrow U_{\mathbf{PA}}$

Since we also have $\nvdash_{\mathbf{PA}} U_{\mathbf{PA}}$, it follows that $\nvdash_{\mathbf{PA}} \text{Consis}_{\mathbf{PA}}$. ■

To prove that **PA** *is consistent, we need more of arithmetic than is formalized within* **PA**. The same reasoning applies to any consistent theory Γ that extends **PA**. That is, we can formulate a predicate 'Consis_{Γ}' that is true of the natural numbers iff Γ is consistent, and $\Gamma \nvdash \text{Consis}_{\Gamma}$, though we have to be careful exactly how we formulate the predicate Consis_{Γ} (see *Feferman, 1960*).

Nor can we prove the consistency of **Q** within **Q** (see *Bezboruah and Shepherdson, 1976*). It is possible, however, to give a proof within **PA** that **Q** is consistent (in *Shoenfield, 1967* an informal proof is on p. 51, and the description of how to convert it to one in **PA** is on p. 214). But that doesn't really help us, since the consistency of **PA** is more in question than the consistency of **Q**.

Nor can we prove that **Arithmetic** is consistent within **Arithmetic**, because we can't prove anything within **Arithmetic**, since it's not axiomatizable.

To prove the consistency of one of our theories of arithmetic, we need a stronger theory whose consistency is more in question. In Epstein and Carnielli, *Computability* you can read about the significance of this for whether there is any finitary way to justify that arithmetic is consistent.

Exercises for Section D

1. Show that the claims in Exercise 6 of Section B.1 are formally undecidable relative to **Q**.
2. Fermat's last theorem is: For all $n, m, p > 0$ and any $k \geq 3$, $n^k + m^k \neq p^k$.

Show that there is a formula F such that Fermat's last theorem is false iff for some n, m, $p > 0$ and $k \geq 3$, $\mathbf{Q} \vdash \mathrm{F}(\mathrm{S}^m\mathrm{o}, \mathrm{S}^m\mathrm{o}, \mathrm{S}^p\mathrm{o}, \mathrm{S}^k\mathrm{o})$.
(Hint: Exponentiation is representable in **Q**.)

3. Let A be any formally undecidable wff relative to **Q** such that there is some computable predicate $\mathcal{A}$ such that A is false of the natural numbers iff there are $n_1, \ldots, n_k$ and $\mathcal{A}(n_1, \ldots, n_k)$. Show that if A is false of the natural numbers, then a counterexample to it can be proved in **Q**. Why doesn't this show that A is not formally undecidable relative to **Q**?

4. *ω-consistent theories*
A theory **T** in the first-order language of arithmetic is *ω-consistent* means that for every wff B, if for all n, $\vdash_{\mathbf{T}} \mathrm{B}(\mathrm{S}^n\mathrm{o})$, then $\nvdash_{\mathbf{T}} \exists x \neg \mathrm{B}(x)$. This is a syntactic criterion that can take the place of the semantic requirement in Corollary 21.b. Prove:
   a. If **T** is ω-consistent, then **T** is consistent.
   b. If **T** is ω-consistent and $\mathbf{T} \supseteq \mathbf{Q}$, then $\nvdash_{\mathbf{T}} \neg \mathrm{U}_{\mathbf{T}}$.
   c. Show that there is a theory $\mathbf{T} \supseteq \mathbf{Q}$ that is consistent and is not ω-consistent.
   (Hint: Let **T** be $\mathbf{Q} \cup \{\exists x \neg(\mathrm{o} \oplus x = x)\}$and use Exercise 6.d of Section B.1.)

5. *Rosser's theorem*
Set $\mathcal{W}^*(n, m) \equiv_{\mathrm{Def}}$ for some A with exactly one free variable x_1, $n = [\![\neg \mathrm{A}]\!]$ and m is the number of a proof sequence in **Q** that ends with $\neg \mathrm{A}(\mathrm{S}^n\mathrm{o})$.
This is computable and representable in **Q**, say by W*. With W defined as in this section, and '<' defined as in Theorem18.a, set:

$$\mathrm{B}(x_1) \equiv_{\mathrm{Def}} \forall x_2\,(\mathrm{W}(x_1, x_2) \rightarrow \exists x_3\,(x_3 < x_2 \wedge \mathrm{W}^*(x_1, x_3)))$$

Let $m = [\![\mathrm{B}(x_1)]\!]$. Set:

$$\mathrm{V} \equiv_{\mathrm{Def}} \forall x_2\,(\mathrm{W}(\mathrm{S}^m\mathrm{o}, x_2) \rightarrow \exists x_3\,(x_3 < x_2 \wedge \mathrm{W}^*(\mathrm{S}^m\mathrm{o}, x_3)))$$

   a. Show that $\mathcal{W}(m, n)$ is true iff n is the number of a proof in **Q** of V.
   b. Show that $\mathcal{W}^*(m, n)$ is true iff n is the number of a proof in **Q** of $\neg$V.
   c. Prove that if **Q** is consistent, then $\nvdash_{\mathbf{Q}} \mathrm{V}$ and $\nvdash_{\mathbf{Q}} \neg \mathrm{V}$.
   (Hint: Prove in **Q** the wff $(x < \mathrm{S}^{n+1}\mathrm{o} \rightarrow (x = \mathrm{o} \vee \cdots \vee x = \mathrm{S}^n\mathrm{o})$ and the trichotomy law for '<'.)

E. Second-Order Arithmetic

We've seen serious limitations in formalizing arithmetic in first-order classical predicate logic. Can we accomplish more with second-order logic?

Any second-order theory of arithmetic should formalize addition and multiplication. So we'll consider extensions of **Q** in second-order languages that contain $\mathrm{L}_2(=; \mathrm{S}, \oplus, *, \mathrm{o})$. Recall that $\mathcal{Q}$ is the conjunction of the axioms of **Q**.

The first-order version of proof by induction, the scheme *Induction*$_1$, is not enough to characterize the arithmetic of the natural numbers: **PA ≠ Arithmetic.** But in second-order logic we can formalize induction with a single axiom, letting X stand for X^1_1, x for x_1, and y for x_2.

The second-order axiom of induction, *Induction*

$\forall X\,((X(o) \wedge \forall y\,(X(y) \rightarrow X(Sy))) \rightarrow \forall x\, X(x))$

We previously said that the comprehension$_1$ axioms are valid in second-order logic (Chapter XIV.F.4). So every instance of the first-order scheme of induction is a semantic consequence of the second-order axiom of induction. That is, for every A that is an instance of *Induction*$_1$, we have *Induction* $\vDash_2$ A.

$\mathbf{PA_2}$, *second-order Peano arithmetic* in $L_2(=; S, \oplus, *, o)$

$\mathbf{PA_2} = \{A: PA_2 \vDash_2 A\}$ where $PA_2 \equiv_{\text{Def}} Q \wedge \textit{Induction}$

Our big assumption in this section is: $\langle \mathbb{N}; +1, +, \cdot, 0\rangle$ is a model of $\mathbf{PA_2}$.

Theorem 23 Every model of $\mathbf{PA_2}$ is isomorphic to $\langle \mathbb{N}; +1, +, \cdot, 0\rangle$.

Proof Let M be a model of $\mathbf{PA_2}$. Let $\mathsf{X} = \{\mathbf{e}, \mathbf{se}, \mathbf{sse}, \ldots\} = \{\mathbf{s}^n\mathbf{e} : n \geq 0\}$. Then X satisfies the antecedent of *Induction*, and *Induction* is true in M, so every element of the universe is in X. I'll let you show that the mapping φ from $\langle \mathbb{N}; +1, +, \cdot, 0\rangle$ to M via $\varphi(n) = \mathbf{s}^n\mathbf{e}$ is an isomorphism. ■

Corollary 24 In $L_2(=, S, \oplus, *, o)$, $\langle \mathbb{N}; +1, +, \cdot, 0\rangle \vDash A$ iff $\mathbf{PA_2} \vDash A$.

Proof We've assumed that every consequence of $\mathbf{PA_2}$ is true of $\langle \mathbb{N}; +1, +, \cdot, 0\rangle$. If $\langle \mathbb{N}; +1, +, \cdot, 0\rangle \vDash A$, then by Theorem 23 it is true in every model of $\mathbf{PA_2}$. ■

We can number the wffs of L_2, and then the notion of *decidability* for second-order logic is the same as for first-order logic.

Corollary 25 $\mathbf{PA_2}$ is not decidable.

Proof By Corollary 24, **Arithmetic** = {A: A is a first-order wff and $A \in \mathbf{PA_2}$}. So if $\mathbf{PA_2}$ were decidable, **Arithmetic** would be decidable, which is a contradiction. ■

Corollary 26 The set of valid wffs of $L_2(=; S, \oplus, *, o)$ is not decidable.

Proof In $L_2(=, S, \oplus, *, o)$, $A \in \mathbf{PA_2}$ iff $PA_2 \vDash A$. But $PA_2 \vDash A$ iff $\vDash PA_2 \rightarrow A$. So if we could decide validity in L_2, we would have a decision procedure for $\mathbf{PA_2}$, which is a contradiction. ■

Corollary 27 The set of valid wffs of the language of pure second-order logic with equality is not decidable.

Proof We can devise a translation from $\mathbf{PA_2}$ to the valid wffs of $L_2(=)$. Write H for the function variable F_1^1, F for F_1^2, G for F_2^2, and X for X_1^1, Yfor X_2^1. For any x,

y, y_1, y_2, define the following open formula of $L_2(=)$:

$$\gamma \equiv_{Def} X(x) \wedge \forall Y\,(Y(x) \wedge \forall y\,(Y(y) \rightarrow X(y)) \wedge \forall y\,(Y(y) \rightarrow Y(H(y)))$$
$$\rightarrow \forall y\,(X(y) \rightarrow Y(y))) \qquad \textit{(Induction)}$$
$$\wedge \forall y_1 \forall y_2\ [X(y_1) \wedge X(y_2) \rightarrow (H(y_1) = H(y_2) \rightarrow y_1 = y_2) \qquad (Q1)$$
$$\wedge\ x \neq H(y_1) \qquad (Q2)$$
$$\wedge\ y_2 \neq x \rightarrow \exists y_1\,(y_2 = H(y_1)) \qquad (Q3)$$
$$\wedge\ F(y_1, x) = y_1 \qquad (Q4)$$
$$\wedge\ F(y_1, H(y_2)) = H(F(y_1, y_2)) \qquad (Q5)$$
$$\wedge\ G(y_1, x) = x \qquad (Q6)$$
$$\wedge\ G(y_1, H(y_2)) = F(G(y_1, y_2), y_1)] \qquad (Q7)$$

In any second-order model M with universe U, if U is infinite, then there is an assignment of references σ such that $\sigma(X) = \mathsf{A}$, $\sigma(H) = \mathsf{h}$, $\sigma(F) = \mathsf{f}$, $\sigma(G) = \mathsf{g}$, and $\sigma(x) = \mathsf{o}$ such that $\sigma \models_2 \gamma$. This is because we can take a countably infinite subset A of U and map the natural numbers onto it; there are then functions corresponding to successor, addition, and multiplication on A via that mapping. For such a σ, by Theorem 23, $\langle \mathsf{U}; \mathsf{h}, \mathsf{f}, \mathsf{g}, \mathsf{o}\rangle \simeq \langle \mathsf{N}; +1, +, \cdot, 0\rangle$.

Now we define a translation φ from $L_2(=; S, \oplus, *, o)$ to $L_2(=)$ inductively:

If A is atomic, $\varphi(A) = \exists X\, \exists x\, \exists H\, \exists F\, \exists G\, (\gamma \rightarrow A(H/S, F/\oplus, G/*, x/o))$
where $A(H/S, F/\oplus, G/*, x/o)$ is A with every occurrence of S replaced by H, $\oplus$ by F, $*$ by G, and o by x, and x, y, y_1, y_2 are in order the least variables not appearing in A

$\neg, \rightarrow, \wedge, \vee$ are translated homophonically

If A is $\forall z$ B, then $\varphi(A) = \exists X\, \exists x\, \exists H\, \exists F\, \exists G\, (\gamma \rightarrow \forall z\, (X(z) \rightarrow \varphi(B))$
where x, y, y_1, y_2 are in order the least variables not appearing in A

If A is $\exists z$ B, then $\varphi(A) = \exists X\, \exists x\, \exists H\, \exists F\, \exists G\, (\gamma \rightarrow \exists z\, (X(z) \wedge \varphi(B))$
where x, y, y_1, y_2 are in order the least variables not appearing in A

Then $\mathbf{PA_2} \models_2 A$ iff $\models_2 \varphi(A)$. ■

Corollary 28 Second-order logic is not decidable
The set of valid wffs of the language of pure second-order logic is undecidable.

Proof We can use the definition of '=' in second-order logic (in Example 8 of Chapter XIV) to eliminate '=' in the translation in the proof of Corollary 27. ■

Corollary 29 If Γ is a second-order theory in L_2 with at least one countably infinite model, then Γ is undecidable.

Proof The same translation as in the proof of Corollary 28 serves to show that $\mathbf{PA_2} \models_2 A$ iff $\Gamma \models_2 \varphi(A)$. ■

In second-order logic we can use the same notion of syntactic consequence we already have: axioms plus rules of proof (Chapter II.C.2). Here, however, we want to be explicit in requiring that for any collection of rules of proof we can decide if a particular sequence of wffs is an instance of one of the rules. Then, as before, a collection of wffs is *axiomatizable* means that it is exactly the collection of syntactic consequences of an axiom system.

Theorem 30 Second-order logic is not axiomatizable The collection of valid wffs of second-order logic is not axiomatizable.

Proof Suppose there were an axiom system for second-order logic. Then proving with that system is computable. But for any first-order wff B, either $\vdash PA_2 \to B$ or $\vdash PA_2 \to \neg B$, since $\mathbf{PA_2}$ is complete. So we'd have a decision procedure for **Arithmetic**, which is impossible. ∎

In our earlier terminology (p. 39), Theorem 30 says that there is no complete axiomatization of second-order logic. That is, *second-order logic is incomplete.*

Theorem 31 Second-order logic is not compact
Let $\Sigma = \{Q \wedge \mathit{Induction},\ c_1 \neq o,\ c_1 \neq So,\ c_1 \neq S^2o, \ldots,\ c_1 \neq S^n o, \ldots\}$.
Every finite subset of Σ has a second-order model, but Σ has no second-order model.

Proof For any finite subset $\Gamma \subset \Sigma$, there is a largest n such that '$c_1 \neq S^n o$' is in Γ. So $\langle \mathbb{N}; +1, +, \cdot, 0 \rangle$ with c_1 interpreted as $n + 1$ is a model of Γ. But by Theorem 23 there is no model of Σ. ∎

In first-order logic we have an axiom system that proves all valid wffs and characterizes first-order semantic consequence. But we cannot find a set of wffs that axiomatizes the truths of arithmetic, nor, because of the completeness theorem, any decidable set of wffs whose semantic consequences are the truths of arithmetic. We can only characterize the truths of arithmetic in terms of what is true in a specific model. Even then the set of arithmetic truths has a countable model that isn't isomorphic to the natural numbers (Exercises 12 and 13 of Section C.1 above).

In second-order logic we cannot find an axiom system that proves all valid wffs. But we do have a finite set of wffs that semantically characterizes the truths of arithmetic, and its only model is the natural numbers.

We cannot prove the consistency of any interesting first-order fragment of arithmetic by finitary means; we need a stronger theory whose consistency is more doubtful. The consistency of second-order arithmetic is more doubtful still, relying on assumptions about the set theory of the semantics.

The incompleteness of second-order logic does not seem to be a serious drawback to most mathematicians. As discussed on in Chapter XII (pp. 196–197), mathematicians are normally interested only in the semantic consequences of axioms. Lack of a proof theory does not worry or interest them. And there are plenty of good candidates for further axioms of second-order logic. In particular, various versions of the comprehension scheme arise naturally and reflect assumptions mathematicians make in proving theorems in mathematics. See, for example, *Simpson, 1999.*

Exercises for Section E

1. Show that the mapping in the proof of Theorem 23 is an isomorphism.
2. a. Using Theorem 7 show that every model of $\mathbf{PA_2}$ is infinite.
   b. Use Example 12 of Chapter XIV to show that every model of $\mathbf{PA_2}$ is countable.
   c. Why was the axiom of choice not invoked in proving Theorem 23? What big assumption did we make in proving that theorem?
3. Why won't a proof like the one for Theorem 30 show that first-order logic is not axiomatizable?
4. Why won't a proof as for Theorem 31 show that first-order logic is not compact?

F. Quantifying over Names

No first-order theory of arithmetic in which we can compute addition and multiplication can ensure we're speaking of our intended model $\langle \mathbb{N} ; +1, +, \cdot, 0 \rangle$. In second-order logic, on the other hand, we can characterize the natural numbers categorically, but we have to make assumptions concerning the nature of collections and predicates that are not required for first-order logic and that seem, on the face of it, to be extraneous to the study of the natural numbers. In the first-order language $L(= ; S, \oplus, *, o)$ we have names for all the natural numbers: o, So, S^2o, S^3o, ... The problem is that we cannot say, 'Every object has a name.' Here we'll look at what we get if we add that kind of quantification.

The language L_{qn} of quantification over names

We start with the usual first-order language (including function symbols) and add:

name variables a_i for $i \geq 0$.

The collection of terms of the language is defined as before, except name variables are now atomic terms, too. The definition of *wffs* now includes:

If A is a wff, then $\forall a_i$ (A) is a wff.
If A is a wff, then $\exists a_i$ (A) is a wff.

We can make all the usual definitions for the formal language, including the definition of the *universal closure* of a wff, which requires that all name variables be universally quantified, with the name variables appearing after the individual variables in increasing order. The definition of *closed term* is as before: no variable, including any name variable, appears in a closed term. For any closed term t, let $A(t/a_i)$ be A with every occurrence of a_i replaced by t.

The logic $\mathbf{L_{qn}}$ A *model* for L_{qn} is the same as a first-order model with the following clauses added to the definition of satisfaction:

For any assignment of references to variables, σ :

$$\sigma \models \forall a_i\,(A) \text{ iff } \sigma \models A(t/a_i) \text{ for every closed term } t.$$

$$\sigma \models \exists a_i\,(A) \text{ iff } \sigma \models A(t/a_i) \text{ for some closed term } t.$$

$\mathbf{L_{qn}}$, the *first-order logic of quantifying over names* is:

The semantic consequence relation of these models, denoted $\models_{\mathbf{qn}}$.

For any collection of wffs Σ in L_{qn} we'll call the *theory* of Σ the set of semantic consequences of Σ, that is $\text{Th}_{\mathbf{qn}}(\Sigma) = \{A : \Sigma \models_{\mathbf{qn}} A\}$.

A big difference between the first-order logic of quantifying over names and second-order logic is that the Löwenheim-Skolem theorem holds here. The proof is just as for first-order logic.

Theorem 32 In $\mathbf{L_{qn}}$ if a set of sentences Σ has an infinite model M, then M has a countable submodel of Σ.

I'll let you show that the following wff is true in a model iff every object in the model has a name.

$$AN \equiv_{\text{Def}} \forall x_1 \, \exists a_0 \, (x_1 = a_0)$$

QN, the theory of *named arithmetic* in $L_{qn}(=; S, \oplus, *, o)$

$$\mathbf{QN} = \text{Th}_{\mathbf{qn}}(Q \wedge AN)$$

I'll let you now show the following.

Theorem 33 Every model of **QN** is isomorphic to $<\mathbb{N}\,; +1, +, \cdot\,, 0>$.

Thus, the upward Löwenheim-Skolem theorem fails because, as we intended, we have a categorical theory of the natural numbers.

Corollary 33

a. For any A in $L_{qn}(=; S, \oplus, *, o)$, A is true of $<\mathbb{N}\,; +1, +, \cdot\,, 0>$ iff $\mathbf{QN} \models A$.

b. For any wff A in $L_{qn}(=; S, \oplus, *, o)$ that is an instance of the first-order induction scheme *Induction*$_1$, $\mathbf{QN} \models A$.

c. **QN** is not decidable.

d. $\mathbf{L_{qn}}$ is not decidable.

e. $\mathbf{L_{qn}}$ is not axiomatizable.

f. $\mathbf{L_{qn}}$ is not compact: Every finite subset of $\Sigma = \{Q \wedge AN\} \cup \{c_1 \neq S^n o: n \geq 0\}$ has a model, but Σ has no model.

Proof Each part follows as for second-order arithmetic, replacing *Induction* by *AN* in the proofs, except for (b), which follows from (a). ■

The virtues and flaws of the logic of quantifying over names compared to first-order logic are the same as for second-order logic, with one big exception: To prove that $\mathbf{PA_2}$ has only one model we made the big assumption that $\langle \mathbb{N}; +1, +, \cdot, 0\rangle$ is a model of $\mathbf{PA_2}$, which assumes a lot about infinite sets. But to prove that **QN** is categorical we need assume only what we assumed in formalizing arithmetic in first-order logic, namely, $\langle \mathbb{N}; +1, +, \cdot, 0\rangle$ is a model of **Q**.[1]

No new entities are demanded nor are further assumptions about the nature of language and reasoning required in order to employ quantification over names: the usual language of first-order logic already allows for names, and the process of naming is an essential part of the semantics of first-order logic. Names are things, too, and hence objects that can be quantified over. Quantifying over names does violate the self-reference exclusion principle by allowing talk of the syntax of the language within the language. But no contradiction arises, since **LN** is consistent for the same reason **Q** is: It has a model.

Thus, it is possible to give a categorical characterization of the arithmetic of the natural numbers without making any assumptions about language, the world, or reasoning beyond what we already do for first-order logic.[2]

Exercises

1. Write a formula in $\mathbf{L}_{qn}$ that is true in a model iff there is some object that does not have a name in the formal language.

2. Show that the set of wffs valid on the substitutional interpretation of the quantifiers (Exercise 7, p. 177) is not the same as the set of first-order wffs in $\mathbf{L_{qn}}$.

3. Compare quantification over names to quantifying over predicates conceived as linguistic objects.

4̲. Define a second-order version of the logic of quantifying over names and formalize Example 2 of Chapter XIV.E.

---

[1] After devising the system here I found that Saul Kripke, *1976*, pp. 354–355, described in a footnote what appears to be the same semantics, though I have not been able to find work by him or anyone else that followed up on that. Kripke suggested formulating the logic as a 2-sorted one, but that replaces quantification over names with quantification over named individuals.

[2] These remarks contrast with the long-held view that assumptions about the nature of sets or infinitary assumptions are needed to formalize more of arithmetic than can be done in first-order logic. See, for example, Daniel Isaacson, *1996*, pp. 202–203 and p. 210.

XVI The Integers and Rationals

We've taken the natural numbers as given. Supposedly we know what they are: 0, 1, 2, 3, ... We know how to add and multiply natural numbers.

It doesn't seem a major leap to suppose we also know how to subtract natural numbers, so we take the integers as unproblematic: . . . –3, –2, –1, 0, 1, 2, 3, ... Nor does it seem unreasonable to assume we can divide natural numbers. So we take the rational numbers as given, though we cannot so easily list those.

Mathematicians long ago made it clear that these assumptions are justified by showing how we can reduce the arithmetic of the integers and the rationals to the arithmetic of the natural numbers. Given the natural numbers, we can construct the integers and the rationals.

I'll present those constructions here. They are archetypes of what mathematicians mean by saying one mathematical structure or theory can be reduced to another. First I'll show how the rational numbers can be derived from the integers. Then I'll show how to construct the integers from the natural numbers. Finally, we'll see how to define the natural numbers within the theories of integers and of rationals.

A. The Rational Numbers

1. A construction

The rationals are ratios of integers. Given the integers, $\mathbb{Z}$, we can construct the rationals by using pairs of integers, (i, j) such that $j \neq 0$, thinking of the pair (i, j) as the rational $^{i}/_{j}$. Assuming we already have the rationals:

$$(i,j) \approx (k,l) \text{ iff } \frac{i}{j} = \frac{k}{l} \text{ iff } il = kj$$

The relation $\approx$, which you can show is an equivalence relation on pairs of integers, can thus be defined using just the multiplication of the integers.

For addition we have:

$$(i,j) + (k,l) = (r,s) \text{ iff } \frac{i}{j} + \frac{k}{l} = \frac{r}{s} \text{ iff } \frac{il+kj}{jl} = \frac{r}{s}$$

$$\text{iff } s(il+kj) = r(jl) \text{ iff } (il+kj, jl) \approx (r,s)$$

So we can set $(i,j) + (k,l) \equiv_{\text{Def}} (il+kj, jl)$.

Similarly, we have:

$$(i,j) \cdot (k,l) = (r,s) \text{ iff } \frac{i}{j} \times \frac{k}{l} = \frac{r}{s} \text{ iff } \frac{ik}{jl} = \frac{r}{s} \text{ iff } (ik)s = (il)r$$

So we can set: $(i,j) \cdot (k,l) \equiv_{\text{Def}} (ik, jl)$.

Similarly, we can set: $(i,j) < (k,l) \equiv_{\text{Def}} il < kj$.

Using the notation from Chapter VI.D (pp. 117–118) for equivalence relations and classes, we now have the structure $< \mathbb{Z}^2 - \{(i,0): i \in \mathbb{Z}\}; <, +, \cdot >/_{\approx}$. We write $[(i,j)]$ for the equivalence class of (i,j). I'll let you show the following, using the map $\varphi(^i/_j) = [(i,j)]$.

Theorem 1 $<\mathbb{Q}; <, +, \times>$ is isomorphic to $<\mathbb{Z}^2 - \{(i,0): i \in \mathbb{Z}\}; <, +, \cdot>/_{\approx}$.

To prove Theorem 1, we have to assume we already have the rational numbers. We've only constructed the rationals in the sense that we have constructed an isomorphic structure. There are other constructions that would work equally well, in that they use no more complicated tools to establish Theorem 1. For example, we could identify the rational $^i/_j$ with the triple $(i, j, 0)$ such that $j \neq 0$, letting the final 0 "idle" in the construction.

To the extent that the construction of the rationals given above is considered the right one, the obvious, canonical construction of the rationals from the integers, we are justified in saying that we have constructed the rationals from the integers. In that case, Theorem 1 would be vacuous.

2. A translation

Let's now turn to the first-order theories of the integers and the rationals.

Z-Arithmetic, the *theory of the integers* in $L(= ; <, \oplus, *)$
$\quad$ $\text{Th}(\mathbf{Z})$, where $\mathbf{Z} = <\mathbb{Z}; \text{less than}, +, \cdot>$

Q-Arithmetic, the *theory of the rationals* in $L(= ; <, \oplus, *)$
$\quad$ $\text{Th}(\mathbf{Q})$, where $\mathbf{Q} = <\mathbb{Q}; \text{less than}, +, \cdot>$

We want to translate **Q-Arithmetic** in L(= ; <, ⊕, ∗) to **Z-Arithmetic** in L(= ; <, ⊕, ∗). But we can't have a function that takes as values pairs of elements of the universe. We have to convert the function symbols ⊕ and ∗ into defined predicates. We can eliminate those symbols and replace them with predicates $R_{\oplus}(x, y, z)$, which plays the role of '$x \oplus y = z$', and $R_{*}(x, y, z)$, which plays the role of '$x * y = z$', via a translation as in Chapter VIII.E. That is, we have:

φ:L(= ; <, ⊕, ∗) → L(= ; <, $R_{\oplus}$, R_{*})

For any collection of wffs Γ and any A, $\Gamma \vDash A$ iff $\Gamma^{\varphi} \vDash A^{\varphi}$.

Recall that for this translation we need that both of the following are in **Q-Arithmetic** in L(= ; <, $R_{\oplus}$, R_{*}):

(1) $\forall x\, \forall y\, \exists ! z\, R_{\oplus}(x, y, z)$ and $\forall x\, \forall y\, \exists ! z\, R_{*}(x, y, z)$

Now we need predicates in L(= ; <, ⊕, ∗) to which we can translate =, $R_{\oplus}$, R_{*}, and <. We define:

$P_{=}(x, y, z, w) \equiv_{\text{Def}} x * w = y * z$

$P_{<}(x, y, z, w) \equiv_{\text{Def}} x * w < y * z$

$P_{\oplus}(x, y, z, w, r, s) \equiv_{\text{Def}} ((x * w) \oplus (y * z)) * s = (y * w) * r$

$P_{*}(x, y, z, w, r, s) \equiv_{\text{Def}} (x * z) * s = (y * w) * r$

Then we can define a translation ψ from L(= ; <, $R_{\oplus}$, R_{*}) to L(= ; <, ⊕, ∗) inductively:

$\psi(x_i = x_j) = P_{=}(x_{2i}, x_{2i+1}, x_{2j}, x_{2j+1})$

$\psi(x_i < x_j) = P_{<}(x_{2i}, x_{2i+1}, x_{2j}, x_{2j+1})$

$\psi(R_{\oplus}(x_i, x_j, x_k)) = P_{\oplus}(x_{2i}, x_{2i+1}, x_{2j}, x_{2j+1}, x_{2k}, x_{2k+1})$

$\psi(R_{*}(x_i, x_j, x_k)) = P_{*}(x_{2i}, x_{2i+1}, x_{2j}, x_{2j+1}, x_{2k}, x_{2k+1})$

¬, →, ∧, ∨ are translated homophonically.

$\psi(\forall x_i\, A) = \forall x_{2i}\, \forall x_{2i+1}\, (x_{2i+1} \neq 0 \rightarrow \psi(A))$

$\psi(\exists x_i\, A) = \exists x_{2i}\, \exists x_{2i+1}\, (x_{2i+1} \neq 0 \wedge \psi(A))$

Let γ: L(=, <, ⊕, ∗) → L(= ; <, ⊕, ∗) be $\psi \circ \varphi$ and $\Gamma^{\gamma} = \{A^{\gamma}: A \in \Gamma\}$.

Given any assignment of references σ on **Z** such that for all $i \geq 0$, $\sigma(x_{2i+1}) \neq 0$, we can define an assignment of references σ^{γ} on **Q** by:

$$\sigma^{\gamma}(x_i) = \frac{\sigma(x_{2i})}{\sigma(x_{2i+1})}$$

Lemma 2

a. For any assignment of references τ on **Q**, there is some σ on **Z** such that $\sigma^{\gamma} = \tau$.

b. For any assignment of references σ on **Z** such that $\sigma(x_{2i+1}) \neq 0$ for all $i \geq 0$, $\sigma \models A^{\gamma}$ iff $\sigma^{\gamma} \models A$.

Proof (a) Given an assignment of references τ on **Q**, for every i put $\tau(x_i)$ in lowest terms, $\mathsf{a}_i / \mathsf{b}_i$, where a_i is positive, and b_i is negative iff $\tau(x_i)$ is negative. Set $\sigma(x_{2i}) = \mathsf{a}_i$ and $\sigma(x_{2i+1}) = \mathsf{b}_i$. Then $\sigma^{\gamma} = \tau$.

(b) The proof is by induction on the length of A. For atomic wffs this is a straightforward (though long) application of the definitions, using that

$$\begin{aligned} \mathbf{Z} \models\ & \forall x\, \forall y\, \forall z\, \forall w\, \exists r\, \exists s\, \mathrm{P}_{\oplus}(x, y, z, w, r, s) \ \wedge \\ & \forall x\, \forall y\, \forall z\, \forall w\, \forall r\, \forall s\, \forall t\, \forall u \\ & (\mathrm{P}_{\oplus}(x, y, z, w, r, s) \wedge \mathrm{P}_{\oplus}(x, y, z, w, t, u) \rightarrow \mathrm{P}_{=}(r, s, t, u)) \end{aligned}$$

and similarly for multiplication. For the universal quantifier we have:

$\sigma \models (\forall x_i\, A)^{\gamma}$ iff $\sigma \models \forall x_{2i}\, \forall x_{2i+1}\, (x_{2i+1} \neq 0 \rightarrow A^{\gamma})$

iff for every τ that differs from σ at most in what it assigns x_{2i} and x_{2i+1}, $\tau \models (x_{2i+1} \neq 0 \rightarrow A^{\gamma})$

iff for every δ (on **Q**) that differs from σ^{γ} at most in what it assigns x_i, $\delta^{\gamma} \models (x_{2i+1} \neq 0 \rightarrow A^{\gamma})$ — by part (a)

iff for every δ (on **Q**) that differs from σ^{γ} at most in what it assigns x_i, $\delta \models A$ — by induction

iff $\sigma^{\gamma} \models \forall x_i\, A$

The case for the existential quantifier is similar, and I'll leave it to you. ■

Theorem 3 For any wff A and any collection of wffs Γ in $L(= ; <, \oplus, *)$,

a. $\mathbf{Q} \models A$ iff $\mathbf{Z} \models A^{\gamma}$.

b. $A \in$ **Q-Arithmetic** iff $A^{\gamma} \in$ **Z-Arithmetic**.

c. $\Gamma \models_{\textbf{Q-Arithmetic}} A$ iff $\Gamma^{\gamma} \models_{\textbf{Z-Arithmetic}} A^{\gamma}$.

Proof Parts (a) and (b) are straightforward corollaries to Lemma 2. For part (c)

$\Gamma \models_{\textbf{Q-Arithmetic}} A$

iff for some $B_1, \ldots, B_n$ in Γ, $\{B_1, \ldots, B_n\} \models_{\textbf{Q-Arithmetic}} A$ — by compactness

iff for some $B_1, \ldots, B_n$ in Γ, $\models_{\textbf{Q-Arithmetic}} B_1 \rightarrow (\cdots \rightarrow (B_n \rightarrow A) \cdots)$ — by the semantic deduction theorem

iff for some $B_1, \ldots, B_n$ in Γ, $\models_{\textbf{Z-Arithmetic}} (B_1 \rightarrow (\cdots \rightarrow (B_n \rightarrow A) \cdots))^{\gamma}$ — by part (b)

iff for some $B_1^{\gamma}, \ldots, B_n^{\gamma}$ in Γ^{γ}, $\models_{\textbf{Z-Arithmetic}} (B_1^{\gamma} \rightarrow (\cdots \rightarrow (B_n^{\gamma} \rightarrow A^{\gamma}) \cdots))$ — by homophony for $\rightarrow$

iff for some $B_1^\gamma, \ldots, B_n^\gamma$ in Γ^γ, $\{B_1^\gamma, \ldots, B_n^\gamma\} \vDash_{\textbf{Z-Arithmetic}} A^\gamma$

iff $\Gamma^\gamma \vDash_{\textbf{Z-Arithmetic}} A^\gamma$ ■

To translate the arithmetic of the rationals to the arithmetic of the integers, we first took pairs of elements of the universe and chose a subset of those. Then we took equivalence classes on those pairs. That was straightforward. But then we could not simply translate function symbols into function symbols, because in predicate logic we have no notation for functions that take pairs of elements as values. We were forced to convert functions into predicates, and then translate those predicates. This enormously complicates the translation.

We could add to the vocabulary of predicate logic symbols for functions that take as values pairs of the universe, or triples of the universe, or . . . . Then we would need in the formal language vocabulary to pick out pairs, triples, etc., and atomic wffs such as $f(x) = (y, z)$ or $f((x, y), (z, w)) = (r, s, t)$. This would not correspond to the motivation of functions as name-making devices, but perhaps that is just the nature of mathematical abstractions. For the construction of the rationals from the integers we could take $f_*((x, y), (z, w)) = (x * z, y * w)$. So far as I know, no one has thought it appropriate to extend the vocabulary and foundations of predicate logic in this way. Rather, the general method of translating using equivalence relations is used, which I'll now present so we can use it again.

Exercises for Section A

1. a. Show that $(i, j) \approx (k, l)$ iff $il = kj$ is an equivalence relation on $\mathbb{Z}^2$.
   b. List all elements in $[(1, 3)]$.
2. Write out the definition of $P_\oplus(2, 3, 7, 9, 49, 27)$. Is it true in $\mathbf{Z}$?
3. a. Show that the numbers 0 and 1 are definable in **Z-Arithmetic**.
   b. Show that the numbers 0 and 1 are definable in **Q-Arithmetic**.
4. Why do we need that the wffs at (1) are in **Q-Arithmetic** in $L(=; <, R_\oplus, R_*)$?
5. Prove parts (a) and (b) of Theorem 3.

B. Translations via Equivalence Relations

In the last section we showed that in **Z-Arithmetic** there are formulas that define via equivalence classes on $\mathbb{Z}^2$ a model isomorphic to $\mathbf{Q}$. In this section we'll see how to generalize that construction in two ways: to allow for the objects of the new model to be n-tuples for any $n \geq 1$, rather than just pairs, and to allow for theories of more than just one model.

The theory of a class of models Given a theory Σ and a class of models $\boldsymbol{S}$, $\Sigma = \text{Th}(\boldsymbol{S})$ means $\Sigma = \{A: \text{for every } M \text{ in } \boldsymbol{S}, M \vDash A\}$.

By the downward Löwenheim-Skolem theorem, any theory with an infinite model is the theory of its countable models.

To translate $\Sigma = \text{Th}(\boldsymbol{S})$ in the language $L(\Sigma)$ into a theory $\Delta = \text{Th}(\boldsymbol{D})$ in $L(\Delta)$, we need to decide what it means for formulas in $L(\Sigma)$ to define all the models in $\boldsymbol{S}$ via equivalence classes on models in $\boldsymbol{D}$. To make this simpler, I'll assume that the only categorematic symbols in $L(\Sigma)$ are the predicate symbols $Q_0, \ldots, Q_i, \ldots$ where Q_i is n_i-ary. I'll deal with name symbols after that, and for a theory with function symbols, such as **Q-Arithmetic**, we can first translate those into predicate symbols as in Chapter VIII.E.

Equivalence-relation translations There is an *n-ary equivalence-relation translation* of $\Sigma = \text{Th}(\boldsymbol{S})$ to $\Delta = \text{Th}(\boldsymbol{D})$ means that:

A. In $L(\Delta)$ there are wffs:

$U(x_1, \ldots, x_n)$ — This picks out the subset of n-tuples for the universe.

$P_{=}(x_1, \ldots, x_{2n})$ — This establishes the equivalence relation that defines the universe from that subset of n-tuples.

A_i with exactly $n \cdot n_i$ many free variables. — These act as the predicates Q_i.

B. The translation is the mapping $*$ from $L(\Sigma)$ to $L(\Delta)$ defined by:

$(x_i = x_j)^* = P_{=}(\vec{x}_i, \vec{x}_j)$ where $\vec{x}_i$ is the sequence of variables $x_{ni}, x_{ni+1}, \ldots, x_{ni+(n-1)}$

$Q_i(x_{j_1}, \ldots, x_{j_{n_i}})^* = A_i(\vec{x}_{j_1}, \ldots, \vec{x}_{j_{n_i}})$

$\neg, \rightarrow, \wedge, \vee$ are translated homophonically

$(\forall x_i\, A)^* = \forall \vec{x}_i\, (U(\vec{x}_i) \rightarrow A^*)$ $\quad (\exists x_i\, A)^* = \exists \vec{x}_i\, (U(\vec{x}_i) \wedge A^*)$

where $\forall \vec{x}_i$ is $\forall x_{ni} \forall x_{ni+1} \ldots \forall x_{ni+(n-1)}$

$\exists \vec{x}_i$ is $\exists x_{ni} \exists x_{ni+1} \ldots \exists x_{ni+(n-1)}$

C. Given any model M in $\boldsymbol{D}$, with universe U,

The set of n-tuples in M satisfying $U(x_1, \ldots, x_n)$ is a non-empty set $\mathsf{V} \subseteq \mathsf{U}^n$.

A_i is the subset of $n \cdot n_i$-tuples in U that satisfy A_i in M.

$P_{=}(y_1, \ldots, y_{2n})$ is interpreted as an equivalence relation $\approx$ on U that is a congruence relation with respect to $\mathsf{A}_1, \ldots, \mathsf{A}_i, \ldots$.

$\langle \mathsf{V}; \mathsf{A}_1, \ldots, \mathsf{A}_i, \ldots \rangle /_{\approx}$ is in $\boldsymbol{S}$.

D. For any N in $\boldsymbol{S}$, there is an M in $\boldsymbol{D}$ such that $\mathsf{N} \simeq \langle \mathsf{V}; \mathsf{A}_1, \ldots, \mathsf{A}_i, \ldots \rangle /_{\approx}$.

$$L(\Sigma) \xrightarrow{\;*\;} L(\Delta)$$
$$\boldsymbol{S} \xleftarrow{\;*\;} \boldsymbol{D}$$

Parts (C) and (D) together say that the mapping of M to $\mathsf{M}^* = \langle \mathsf{V}; \mathsf{A}_1, \ldots, \mathsf{A}_i, \ldots \rangle /_{\approx}$ is onto from $\boldsymbol{D}$ to $\boldsymbol{S}$.

If U is interpreted as the universal relation in all models of $\boldsymbol{D}$, we can simplify the translation of the quantifiers to: $(\forall x_i\, \mathrm{A})^* = \forall \vec{x}_i\,(\mathrm{A}^*)$ and $(\exists x_i\, \mathrm{A})^* = \exists \vec{x}_i\,(\mathrm{A}^*)$.

Theorem 4 If there is an n-ary equivalence-relation translation $*$ of $\Sigma = \mathrm{Th}(\boldsymbol{S})$ into $\Delta = \mathrm{Th}(\boldsymbol{D})$, then for any wff A, and collection of wffs Γ in $\mathrm{L}(\Sigma)$, and $\mathfrak{M}$ in $\boldsymbol{D}$:

a. $\mathfrak{M}^* \vDash \mathrm{A}$ iff $\mathfrak{M} \vDash \mathrm{A}^*$

b. $\mathrm{A} \in \Sigma$ iff $\mathrm{A}^* \in \Delta$

c. $\Gamma \vDash_\Sigma \mathrm{A}$ iff $\Gamma^* \vDash_\Delta \mathrm{A}^*$

Proof (a) Given any assignment of references σ on $\mathfrak{M}$, there is an assignment of references σ^* on $\langle \mathfrak{V}\, ; \mathfrak{A}_1, \ldots, \mathfrak{A}_i, \ldots \rangle /_{\approx}$ such that for all i,

$$\sigma^*(x_i) = [(\sigma(x_{ni}), \sigma(x_{ni+1}), \ldots, \sigma(x_{ni+(n-1)}))]$$

Moreover, we can show that given any τ on $\mathfrak{M}^*$, there is a σ on $\mathfrak{M}$ such that $\sigma^* = \tau$. Namely, for each $i \geq 0$, choose from the equivalence class $\tau(x_i)$ one representative, $(\mathfrak{a}_1, \ldots, \mathfrak{a}_n)$, and set $\sigma(x_{ni}) = \mathfrak{a}_1, \ldots, \sigma(x_{ni+(n-1)}) = \mathfrak{a}_n$. If we can show that for every σ and A, $\sigma \vDash \mathrm{A}^*$ iff $\sigma^* \vDash \mathrm{A}$, then we will have:

$$\begin{aligned} \mathfrak{M} \vDash \mathrm{A}^* \ &\text{iff for every } \sigma \text{ on } \mathfrak{M},\ \sigma \vDash \mathrm{A}^* \\ &\text{iff for every } \sigma^* \text{ on } \mathfrak{M}^*,\ \sigma^* \vDash \mathrm{A} \\ &\text{iff } \mathfrak{M}^* \vDash \mathrm{A} \end{aligned}$$

To prove $\sigma \vDash \mathrm{A}^*$ iff $\sigma^* \vDash \mathrm{A}$ we proceed by induction on the length of a wff.

$$\begin{aligned} \sigma \vDash (x_i = x_j)^* \ &\text{iff } \sigma \vDash \mathrm{P}_=(\vec{x}_i, \vec{x}_j) \\ &\text{iff } (\sigma(x_{ni}), \sigma(x_{ni+1}), \ldots, \sigma(x_{ni+(n-1)})) \\ &\qquad \approx (\sigma(x_{nj}), \sigma(x_{nj+1}), \ldots, \sigma(x_{nj+(n-1)})) \\ &\text{iff } [(\sigma(x_{ni}), \sigma(x_{ni+1}), \ldots, \sigma(x_{ni+(n-1)}))] \\ &\qquad = [(\sigma(x_{nj}), \sigma(x_{nj+1}), \ldots, \sigma(x_{nj+(n-1)}))] \\ &\text{iff } \sigma^*(x_i) = \sigma^*(x_j) \text{ iff } \sigma^* \vDash x_i = x_j \end{aligned}$$

When A is $\mathrm{A}_i(y_1, \ldots, y_{n \cdot n_i})$, we have (a) because $\approx$ is a congruence relation on $\mathfrak{U}$ with respect to $\mathfrak{A}_i$. I'll leave to you when A is of the form $\neg\mathrm{B}$, $\mathrm{B} \to \mathrm{C}$, $\mathrm{B} \wedge \mathrm{C}$, or $\mathrm{B} \vee \mathrm{C}$. For the universal quantifier:

$$\begin{aligned} \sigma \vDash (\forall x_i\, \mathrm{B})^* \ &\text{iff } \sigma \vDash \forall \vec{x}_i\,(\mathrm{U}(\vec{x}_i) \to \mathrm{B}^*) \\ &\text{iff for every } \tau \text{ that differs from } \sigma \text{ at most in what it assigns } \vec{x}_i, \\ &\qquad \tau \vDash \mathrm{U}(\vec{x}_i) \to \mathrm{B}^* \\ &\text{iff for every } \tau \text{ that differs from } \sigma \text{ at most in what it assigns } \vec{x}_i, \\ &\qquad \text{if } \tau(\vec{x}_i) \in \mathfrak{U}, \text{ then } \tau \vDash \mathrm{B}^* \\ &\text{iff for every } \tau^* \text{ that differs from } \sigma^* \text{ at most in what it assigns } x_i, \\ &\qquad \tau^* \vDash \mathrm{B} \\ &\text{iff } \sigma^* \vDash \forall x_i\, \mathrm{B} \end{aligned}$$

The case when A is of the form $\exists x_i B$ is done similarly.

(b) This follows because the mapping $*: \boldsymbol{D} \to \boldsymbol{S}$ is onto.

(c) This follows as in the proof of Theorem 3.c. ■

We can amend the definition of equivalence-relation translations to include *languages with name symbols*:

(A) (add) For each name symbol c_i in $L(\Sigma)$ there is a sequence $\vec{c}_i = (c_{i_1}, \ldots, c_{i_n})$ of names in $L(\Delta)$.

(B) $Q_i(u_{j_1}, \ldots, u_{j_{n_i}})^* = A_i(\vec{u}_{j_1}, \ldots, \vec{u}_{j_{n_i}})$ where u_i is any term.

C. The Integers

From the natural numbers, $\mathbb{N}$, we can construct the integers by using markers to distinguish positive and negative, for example:

...	−3	−2	−1	0	1	2	3	...
...	(0, 3)	(0, 2)	(0, 1)	(0, 0)	(1, 0)	(2, 0)	(3, 0)	...

It's just a matter of giving the right rules for evaluating addition, multiplication, and the ordering. But it's a lot of work to take care of all the different cases into which we have to split the evaluations. Instead of identifying a subset of the pairs of natural numbers with the integers, it's easier to take all pairs of natural numbers and use an equivalence relation on those. Think of the pair (m, n) as the integer $m - n$.

$$\begin{aligned}(m, n) \approx (p, q) \text{ iff } & m - n = p - q \\ \text{iff } & m + q = p + n\end{aligned}$$

This relation, which uses just addition on the natural numbers in its definition, is an equivalence relation on pairs of natural numbers, as you can show. All we need to do is define functions for addition and multiplication, and a relation for the ordering, that respect this relation.

$$\begin{aligned}(m, n) + (p, q) = (r, s) \text{ iff } & (m - n) + (p - q) = r - s \\ \text{iff } & m + p + s = r + n + q \\ \text{iff } & (m + p, n + q) \approx (r, s)\end{aligned}$$

So we can set $(m, n) + (p, q) \equiv_{\text{Def}} (m + p, n + q)$.

For multiplication we have:

$$\begin{aligned}(m, n) \cdot (p, q) = (r, s) \text{ iff } & (m - n) \times (p - q) = r - s \\ \text{iff } & mp + nq - (mq + np) = r - s \\ \text{iff } & mp + nq + s = mq + np + r\end{aligned}$$

So we can set $(m, n) \cdot (p, q) \equiv_{\text{Def}} (mp + nq,\ mq + np)$.

And we can set: $(m, n) < (p, q) \equiv_{\text{Def}} m + q < p + n$.

The mapping $\varphi(\mathbf{z}) = \begin{cases} [(\mathbf{z}, 0)] & \text{if } 0 \leq \mathbf{z} \\ [(0, \mathbf{z})] & \text{if } \mathbf{z} < 0 \end{cases}$ establishes the following.

Theorem 5 $\langle \mathbb{Z}; <, +, \times \rangle$ is isomorphic to $\langle \mathbb{N}^2; \lessdot, \dotplus, \dot{\cdot} \rangle /_{\approx}$.

Now we can translate the first-order arithmetic of the integers in L(= ; <, ⊕, ∗) into the first-order arithmetic of the natural numbers in L(= ; ⊕, ∗) using the method of the last section. First we translate the functions in the language L(= ; <, ⊕, ∗) into predicates in L(= ; <, $S_\oplus$, S_*) via a mapping φ as in Chapter VIII.E. Then we apply Theorem 4, using Theorem 5 and:

$U(x_1, x_2) \equiv_{\text{Def}} x_1 = x_1$

$W_=(x, y, z, w) \equiv_{\text{Def}} x \oplus w = y \oplus z$

$W_<(x, y, z, w) \equiv_{\text{Def}} x \oplus w < y \oplus z$

$W_+(x, y, z, w, r, s) \equiv_{\text{Def}} y \oplus w \oplus r = x \oplus z \oplus s$

$W_*(x, y, z, w, r, s) \equiv_{\text{Def}}$
$\quad (x * z) \oplus (y * w) \oplus s = (x * w) \oplus (y * z) \oplus r$

In L(= ; ⊕, ∗), **Arithmetic** is Th(**N**), where $\mathbf{N} = \langle \mathbb{N}; +, \cdot \rangle$ (recall that 0 and 1 are definable). Both 0 and 1 are definable in **Z** by the same formulas as in **N**. I'll let you to show the following as a corollary to Theorem 4.

Theorem 6 For any wff A in L(= ; <, ⊕, ∗),

a. **Z**⊨A iff **N**⊨A*.

b. A ∈ **Z-Arithmetic** iff A* ∈ **Arithmetic**.

c. $\Gamma \vDash_{\textbf{Z-Arithmetic}} A$ iff $\Gamma^* \vDash_{\textbf{Arithmetic}} A^*$.

Exercises for Sections B and C

1. Give a definition of addition and multiplication for constructing the integers using the set $\{(n, m): n = 0 \text{ or } m = 0\}$. What equivalence relation is needed?
2. a. In the translation of Section C, what is U interpreted as in **N**?
   b. Using the notation of Section B, what is U for the translation of Section A?
3. In the notation of Section B, prove that for any *n*-ary equivalence-relation translation of Σ into Δ, $\Delta \vDash \exists x_1 \ldots \exists x_n\, U(x_1, \ldots, x_n)$.
4. Prove parts (b) and (c) of Theorem 4 and Theorem 5.
5. Give a translation of **Q-Arithmetic** into **Arithmetic**.
6. Argue either for or against the following:
   a. The rational numbers actually are what is given by the construction in Section A.
   b. The integers actually are what is given by the construction in Section C.

7. Let **C** = Th($\langle \mathbb{C}; +, \cdot \rangle$), where $\mathbb{C}$ is the set of complex numbers.
   Let **R** = Th($\langle \mathbb{R}; +, \cdot \rangle$), where $\mathbb{R}$ is the set of real numbers.
   Exhibit an equivalence-relation translation of **C** into **R**.

8. a. Show that there is no equivalence-relation translation of the arithmetic of the real numbers into **Q-Arithmetic** with respect to the models $\langle \mathbb{R}$; less than, $+, \cdot \rangle$ and $\langle \mathbb{Q}$; less than, $+, \cdot \rangle$. (Hint: Count.)
   b. What does this mean for constructing the real numbers from the rationals?

D. Relativizing Quantifiers and the Undecidability of Z-Arithmetic and Q-Arithmetic

The integers can be constructed from the natural numbers and the theory of the integers can be translated into the theory of the natural numbers. But on the usual understanding, the natural numbers are just part of the integers. To translate the first-order theory of the natural numbers into the first-order theory of the integers, however, we need to give a definition of the set of natural numbers in the theory of the integers. Then we can relativize quantifers in the same way that we relativized quantifiers to predicates in the examples in Chapter VII.A.

A theorem of Lagrange says that an integer is a natural number iff it is the sum of four squares of integers (see Chapter 7 of *Herstein, 1964*). So writing y, z, r, s for x_1, x_2, x_3, x_4, set:

$$\mathrm{N}(x) \equiv_{\mathrm{Def}} \exists z\, \exists y\, \exists r\, \exists s\, ((y * y) \oplus (z * z) \oplus (r * r) \oplus (s * s) = x)$$

Then N(x) defines $\{n\colon n \in \mathbb{N}\}$ in Th($\langle \mathbb{Z}; +, \cdot, 0, 1 \rangle$).

To translate **Arithmetic** into **Z-Arithmetic** we can use Theorem 4, taking $n = 1$, where U is N and the equivalence relation is the identity. But that would require eliminating function symbols, which isn't necessary here. We can set out a simpler kind of translation, and by doing it in a general way we'll be able to use it later. Recall (p. 186) that if M is a model with universe A, and $\varnothing \neq \mathsf{U} \subseteq \mathsf{A}$ such that U is closed under every function in M, then $\mathsf{M}/_{\mathsf{U}}$ is the restriction of M to U.

Translating by relativizing quantifiers Let Σ = Th(***S***) and Δ = Th(***D***), where L(Σ) = L(Δ). There is a *translation by relativizing quantifers to* a wff U(x_1) from Σ into Δ with respect to these classes of models means that:

A. For every model M in ***D***, the subset of elements of the universe that satisfy U is a set $\mathsf{U} \neq \varnothing$, and U is closed under the functions of M.

B. For every model M in ***D***, $\mathsf{M}/_{\mathsf{U}}$ is in ***S***.

C. For every model N in ***S***, there is some M in ***D*** such that $\mathsf{M}/_{\mathsf{U}} \simeq \mathsf{N}$.

D. The translation is the mapping * from L(Σ) to itself via:

$A^* \doteq A$ for every atomic A

$\neg, \rightarrow, \wedge, \vee$ are translated homophonically

$(\forall x_i\, A)^* = \forall x_i\, (U(x_i) \rightarrow A^*)$

$(\exists x_i\, A)^* = \exists x_i\, (U(x_i) \wedge A^*)$

Part (B) says that the mapping M to $\mathsf{M}^* = \mathsf{M}/_{\mathsf{U}}$ is onto from $\boldsymbol{D}$ to $\boldsymbol{S}$.

Theorem 7 If there is a translation by relativizing quantifers of $\Sigma = \mathrm{Th}(\boldsymbol{S})$ into $\Delta = \mathrm{Th}(\boldsymbol{D})$, then for any wff A, any collection of wffs Γ in L(Σ), and any M in $\boldsymbol{D}$,

a. $\mathsf{M}/_{\mathsf{U}} \vDash A$ iff $\mathsf{M} \vDash A^*$

b. $A \in \Sigma$ iff $A^* \in \Delta$

c. $\Gamma \vDash_{\Sigma} A$ iff $\Gamma^* \vDash_{\Delta} A^*$

Corollary 8 There is a translation by relativizing quantifiers to N from **Arithmetic** to **Z-Arithmetic.**

Corollary 8 is important for the following consequences.

Corollary 9 In the language L(= ; ⊕, ∗, o, ł),

a. **Z-Arithmetic** is undecidable.

b. **Z-Arithmetic** is not axiomatizable.

c. If Γ is a theory and $\Gamma \subseteq$ **Z-Arithmetic**, then Γ is undecidable.

Proof (a) We have a translation * of **Arithmetic** to **Z-Arithmetic** such that for every A, $A \in$ **Arithmetic** iff $A^* \in$ **Z-Arithmetic**. So if **Z-Arithmetic** were decidable, **Arithmetic** would be, too. But **Arithmetic** is undecidable.

(b) Every complete axiomatizable theory is decidable (Theorem XV.8). But **Z-Arithmetic** is complete and undecidable. So it is not axiomatizable.

(c) Recall the undecidable theory $\mathbf{Q} \subset$ **Arithmetic** of Chapter XV that is axiomatized by a single axiom Q. Suppose Γ is a theory and $\Gamma \subseteq$ **Z-Arithmetic**. Consider $\mathrm{Th}(\Gamma \cup \mathbf{Q}^*)$. Since $\mathbf{Q}^* \subseteq$ **Z-Arithmetic**, $\mathrm{Th}(\Gamma \cup \mathbf{Q}^*)$ is consistent. Set:

$$\Delta = \{B\colon B^* \in \mathrm{Th}(\Gamma \cup \mathbf{Q}^*)\}$$

Then $\mathbf{Q} \subseteq \Delta$. Since $\Delta^* \subseteq$ **Z-Arithmetic**, $\Delta \subseteq$ **Arithmetic**. Hence by Corollary XV.16, Δ is undecidable. So $\mathrm{Th}(\Gamma \cup \mathbf{Q}^*)$ is undecidable. Suppose that Γ were decidable. Then $\Gamma \cup \mathbf{Q}^* \vDash A$ iff $\Gamma \vDash Q^* \rightarrow A$, and so $\mathrm{Th}(\Gamma \cup \mathbf{Q}^*)$ would be decidable, which is a contradiction. So Γ is undecidable. ∎

The natural numbers are also contained in the rationals. *Julia Robinson, 1949* (and more succinctly in *1965*) exhibited a very complicated predicate P(*x*) that defines the set of natural numbers in **Q-Arithmetic**. So proceeding as for the integers, we have the following.

Corollary 10 In the language L(= ; ⊕, ∗, o, ɫ),

a. There is a translation by relativizing quantifiers to P of **Arithmetic** to **Q-Arithmetic**.

b. **Q-Arithmetic** is undecidable.

c. **Q-Arithmetic** is not axiomatizable.

d. If Γ is a theory and Γ ⊆ **Q-Arithmetic**, then Γ is undecidable.

Many of our abstract mathematical theories are based on abstracting from the structure of the integers or the rationals. If you're familiar with the theory of integral domains and the theory of rings (see, for example, *Herstein, 1964*), it is straightforward to formalize those in first-order predicate logic, and each is contained in **Z-Arithmetic**. Similarly, the theory of fields (Chapter XVII.E.1 below) and the theory of fields with characteristic zero (Exercise 1 of Chapter XVII.E.1) are each contained in **Q-Arithmetic**. So by Corollaries 9 and 10 we have the following.

Corollary 11 In the language L(= ; ⊕, ∗, o, ɫ),

a. The theory of integral domains is undecidable.

b. The theory of rings is undecidable.

c. The theory of fields is undecidable.

b. The theory of fields of characteristic zero is undecidable.

Here, though, the language L(= ; ⊕, ∗, o, ɫ) must be understood as an abbreviation of a formal language, since we are formalizing informal theories, not specific examples.

Exercises for Section D

1. Show that the predicate 'is less than' on the integers is definable in **Z-Arithmetic**. (Hint: Define the positive integers.)
2. Give a translation of **Arithmetic** in L(= ; <, +, ·, 0, 1) to **Z-Arithmetic**.
3. Why can't we conclude from Corollary 9 that the theory of groups is undecidable?

XVII The Real Numbers

A. What Are the Real Numbers?

In ancient times mathematics was about whole numbers and their ratios.

But about the time of Pythagoras someone discovered that lengths of certain line segments could not be expressed as ratios of whole numbers. For example, consider the diagonal of a square whose side has unit length. The length d of the diagonal cannot be a ratio of whole numbers relative to the unit length. That's because by Pythagoras' theorem, $d^2 = 1^2 + 1^2 = 2$. So if $d = m/n$, where m and n have no common factor, then $2 = m^2/n^2$. So $2n^2 = m^2$. Since 2 is a factor of the left-hand side, it is also a factor of the right-hand side. So 2 divides m, and we have $2n^2 = (2r)^2$. But then $n^2 = 2r^2$, and hence 2 is a factor of n, too, which is a contradiction.

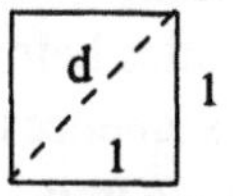

Hence, we have a point on the line that we cannot name with any ratio of whole numbers. Whole numbers and their ratios are inadequate for measurement in geometry.

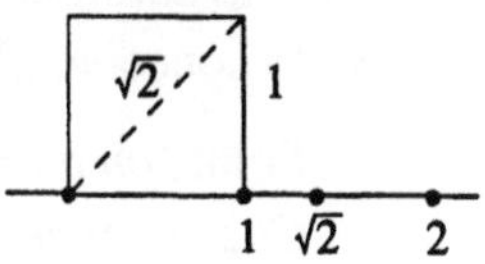

Through two millennia real numbers were understood in terms of geometric constructions and approximations: line segments whose lengths are not ratios of whole numbers can be approximated as closely as we wish with line segments whose lengths are ratios of whole numbers. More and more properties of the real numbers —or points on the line—were "discovered", that is, were shown to follow from some basic assumptions.

But that understanding of the real numbers when coupled with infinitesimal analyses in the calculus of Newton and Leibniz eventually became inadequate: Confusions and mistakes abounded, and many said that the notion of infinitesimals was incoherent. In response, in the 19th century Richard Dedekind and Georg Cantor each proposed a construction of the real numbers from the rationals.

Cantor proposed that the reals be taken to be convergent sequences of rational numbers, with two convergent sequences identified if their difference converges to 0. A rational number q can be identified with the sequence q, q, q, ... With his construction the arithmetical properties of the real numbers are fairly easy to establish. However, it's hard to define an ordering and show, for example, that there is indeed a number corresponding to $\sqrt{2}$.

Dedekind, in contrast, defined the real numbers to fill in the points on the rational line. A *Dedekind cut* is a non-empty subset of rationals, A, such that (i) A is bounded above, that is, there is some r such that for all $p \in A$, $p < r$, and (ii) A is closed downwards under '$\leq$', that is if $p \in A$, then for any rational q such that $q \leq p$, $q \in A$, and (iii) there is no greatest element in A, that is, if $p \in A$, then there is an $r \in A$ such that $p < r$. A rational number q can be identified with the set of rationals strictly less than q. And $\{x: x \leq 0\} \cup \{x: x^2 < 2\}$ can be identified with $\sqrt{2}$. The set of real numbers is taken to be the set of all Dedekind cuts. It's not hard to establish an ordering of the real numbers that has the properties we expect, but with this construction it is hard to establish the arithmetical properties of the reals.

Since the rationals were considered unproblematic, the theory of the real numbers was thought to be put on a solid foundation. But these constructions, done in an informal theory of sets, were seen to require a new assumption not used in mathematics before: Infinite collections, such as $\{x: x \leq 0\} \cup \{x: x^2 < 2\}$, or infinite sequences had to be considered completed, not in the process of construction but a single unit. Formalizing or clarifying the assumptions on which those constructions were based was an important motive in the development of set theory. Nowadays the constructions are still done by mathematicians in an informal theory of sets. The two constructions are equivalent, in that an isomorphism of them relative to addition, multiplication, and the ordering can be given. So we have four ways to conceive of the real numbers:

Points on a line.
Characterized by their properties.
Constructed: *As sequences of rationals*; *as Dedekind cuts of rationals.*

We want to use classical mathematical logic to clarify the relations among these conceptions. We cannot answer the question of what the real numbers are, but we can show that the answers that have been proposed are in some sense equivalent.

In this chapter I'll develop an axiomatic characterization of the real numbers. We could take the resulting axiomatization as all there is to the real numbers: Any set that satisfies the axiomatization has a right to be called 'the real numbers'. But we need that there is at least one model of the axiomatization, so I give a construction of the reals via Dedekind cuts in an appendix to this chapter. In what follows, when I say that the real line has some property *by construction* I am referring to that. For the construction of the reals by sequences, see, for example, *Mendelson, 1973.*

I will first present an axiomatic theory of addition, then an axiomatization of the ordering of the reals, then an axiomatization for addition with an ordering. Finally, I'll present an axiomatic theory of addition, multiplication, and an ordering that characterizes Th(<$\mathbb{R}$; <, +, ·, 0, 1>). Studying this progression of theories will allow us to see many of the ideas and methods of the previous chapters applied and refined, and will also be used in our characterization of the reals via Euclidean geometry in the next two chapters.

In what follows, I'll use x, y, z, w for x_1, x_2, x_3, x_4. So, for example, $\forall x\, \exists y\, (x < y)$ is a wff, not a scheme of wffs. I will use n, m, i, j, k to range over natural numbers.

Exercises for Section A

1. a. Relate the representation of real numbers as decimals, which we learn in school, to the constructions described above.
   b. Show that $\sqrt{2}$ can be given in decimal notation.
2. Using proofs similar to that for $\sqrt{2}$, show the following:
   a. $\sqrt{3}$ is not rational.
   b. $\sqrt{p}$ is not rational for any prime p.
   c. $\sqrt[3]{2}$ is not rational.
   d. $\sqrt[n]{p}$ is not rational for any prime p and any natural number $n > 1$.
   e. $\sqrt{6}$ is not rational.
   f. $\sqrt{2} + \sqrt{3}$ is not rational.
3. Show that there are two numbers r, s that are not rational, yet $r \cdot s$ is rational.
4. Determine whether the following sets of rationals are Dedekind cuts:

   a. $\{x: x = 0\}$
   b. $\{x: x \leq 2/3\}$
   c. $\{x: x < 2/3\}$
   d. $\{x^2: x \leq 2\}$
   e. $\{x^2: x > 2\}$
   f. $\{x: x^2 - 1 < 0\}$
   g. $\{x: x^3 - 1 < 0\}$
   h. $\{x: -4x^3 + 2x^2 + 7 < 0\}$

B. Divisible Groups

We start with a theory of addition in the theory of groups. Writing '⊕' in place of '∘', we have the following theory.

AbelianG, the theory of *abelian groups* in $\mathrm{L}(=;\oplus,\mathrm{o})$

The axioms of $\mathbf{G}^{\#}$ (p. 204) plus:

Abelian $\forall x\, \forall y\, (x \oplus y = y \oplus x)$

In **Abelian G** we define the inverse function, ~, by:

$$\sim t = u \equiv_{\mathrm{Def}} t \oplus u = \mathrm{o}$$

I'll write $t \sim u$ for $t \oplus (\sim u)$.[1] For all terms t and natural numbers n, we define by induction a metalogical abbreviation:

$$0t \equiv_{\mathrm{Def}} \mathrm{o}$$

$$(n+1)t \equiv_{\mathrm{Def}} nt \oplus t$$

I'll let you prove the following using induction on n.

Lemma 1 For all natural numbers n and m, each of the following is a theorem of **AbelianG**.

a. $\forall x\, [n(mx) = (nm)x]$

b. $\forall x\, [(m+n)x = mx \oplus nx]$

c. $\forall x\, \forall y\, [n(x \oplus y) = nx \oplus ny]$

d. $\forall x\, [\sim(nx) = n(\sim x)]$

We need to have enough elements in our groups to divide by any natural number. So we consider the following theories.

Theories of divisible groups

GD*n*	**GD$_{\mathsf{U}}$** for any $\mathsf{U} \subseteq \mathbb{N}$	**GD**, *divisible groups*
AbelianG	**AbelianG**	**AbelianG**
$\exists x\, \exists y\, (x \neq y)$	$\exists x\, \exists y\, (x \neq y)$	$\exists x\, \exists y\, (x \neq y)$
D_n	$\{D_n : n \in \mathsf{U}\}$	$\{D_n : n \geq 1\}$

$D_n \equiv_{\mathrm{Def}} \forall x\, \exists y\, (x = ny)$ for $n \geq 1$ *the division axioms*

A model of **GD*n*** is called an *n-divisible* group. A model of **GD** is called a *divisible* group.[2]

---

[1] *Tarski, 1953* shows that $\mathbf{G}^{\#}$ is undecidable; *Szmieliew, 1948* shows that **AbelianG** is decidable.

[2] Some authors call divisible groups *complete* groups.

Lemma 2

a. **AbelianG** $\vdash D_m \wedge D_n \rightarrow D_{m \cdot n}$ for all natural numbers n and m.

b. If $\mathsf{M} \models \mathbf{GD}n$ and $\mathsf{M} \models \mathbf{GD}m$, then $\mathsf{M} \models \mathbf{GD}n \cdot m$ for all n and m.

Proof Part (b) follows by (a). For (a), we have in **AbelianG**:

$\forall x\, \exists y\, (x = ny) \wedge \forall y\, \exists z\, (y = mz)$	hypothesis
$\forall x\, \exists y\, (x = ny \wedge \exists z\, (y = mz))$	as $\vdash (\exists y\, A \wedge \forall y\, B) \rightarrow \exists y\, (A \wedge B)$
$\forall x\, \exists y\, \exists z\, (x = ny \wedge y = mz)$	Tarski-Kuratowski algorithm
$\forall x\, \exists z\, (x = n(mz))$	substitution of equals
$\forall x\, \exists z\, (x = (n \cdot m) z)$	Lemma 1. ■

Lemma 3 For $\mathsf{U} \subseteq Prime = \{p: p$ is a prime natural number$\}$,

a. $\langle \mathbb{Q}; +, 0 \rangle \models \mathbf{GD}_{\mathsf{U}}$ and $\langle \mathbb{R}; +, 0 \rangle \models \mathbf{GD}_{\mathsf{U}}$.

b. $\mathbf{GD}_{\mathsf{U}}$ is consistent.

c. There is a model M_{U} of $\mathbf{GD}_{\mathsf{U}}$ such that for every prime p, M_{U} is p-divisible iff $p \in \mathsf{U}$.

d. $\mathbf{GD} = \mathbf{GD}_{Prime}$

e. Every model of $\mathbf{GD}_{Prime}$ is infinite.

f. $\mathbf{GD}_{\mathsf{U}}$ is not complete for any $\mathsf{U} \subsetneq Prime$.

Proof The first part of (a) I'll leave to you, and the second part is by the construction of the reals. Since $\mathbf{GD}_{\mathsf{U}}$ has a model, (b) follows.

(c) Let M_{U} be $\langle \mathbb{Q}; +, 0 \rangle$ restricted to:

$$\left\{ \frac{m}{r_0^{n_0} \cdot \ldots \cdot r_k^{n_k}} : m \text{ is an integer, } n_0, \ldots, n_k \text{ are natural numbers, } r_0, \ldots, r_k \in \mathsf{U} \right\}$$

I'll let you show that M_{U} is a model of $\mathbf{GD}_{\mathsf{U}}$ by showing that it is closed under addition. I'll also let you show that M_{U} is p-divisible iff $p \in \mathsf{U}$.

(d) This follows by the Fundamental Theorem of Arithmetic, which says that every natural number $n \geq 2$ can be uniquely written as $n = p_1^{a_1} \cdot \ldots \cdot p_k^{a_k}$, where $p_1 < \ldots < p_k$ are primes, and for all i, $0 < a_i$.

Parts (e) and (f) follow from the construction for (c). ■

In any model of **GD** every element is divisible by every non-zero natural number. So we can make the following definition for any $n \neq 0$ and terms t, u:

$$t/n = u \equiv_{\text{Def}} t = nu$$

So, for example, $3y/8 = 5x/6$ is equivalent to $3y = 40x/6$, which in turn is equivalent to $18y = 40x$. I'll let you prove the following in **GD**.

Lemma 4 For all natural numbers m, n, i, j such that $n \neq 0$ and $j \neq 0$,

a. $\vdash \forall x \forall y\ ({}^{mx}/_{n} = {}^{iy}/_{j}) \leftrightarrow (jm)x = (ni)y$

b. $\vdash \forall x \forall y \forall z\ ({}^{mx}/_{n} \oplus y = z) \leftrightarrow (x = {}^{nz}/_{m} \sim {}^{ny}/_{m})$

c. $\vdash \forall x \forall y\ i\ ({}^{mx}/_{n})/_{j} = {}^{imx}/_{jn}$

Theorem 5 **GD** is not finitely axiomatizable.

Proof Suppose **GD** were finitely axiomatizable. Then it is axiomatizable by a finite subset Σ of the axioms of **GD** (Lemma XV.2), which, as you can show, contradicts Lemma 3.c. ■

Both <$\mathbb{Q}$; +, 0> and <$\mathbb{R}$; +, 0> are models of **GD**. But there other models.

Theorem 6 **GD** is not $\aleph_0$-categorical, nor $\mathfrak{c}$-categorical.

Proof (a) Let $\mathbb{Q}_2$ be the structure with universe {(a, b): a, b rational numbers}, 'o' interpreted as (0, 0), and addition defined via addition on the rationals as: (a, b) + (c, d) = (a + c, b + d). $\mathbb{Q}_2$ is countable (see Exercises 7 and 5 of Chapter IX.B). I'll let you prove that $\mathbb{Q}_2 \vDash$**GD**. Suppose φ is a homorphism from <$\mathbb{Q}$; +, 0> to $\mathbb{Q}_2$. Then for some a and b, $\varphi(1) = (a, b)$. So given any rational ${}^{m}/_{n}$, $\varphi({}^{m}/_{n}) = {}^{m}/_{n}(a, b) = ({}^{m}/_{n}a, {}^{m}/_{n}b)$. So for no $q \in \mathbb{Q}$ do we have $\varphi(q) = (2a, 3b)$. So φ is not an isomorphism.

In Exercise 7 I ask you to show that the same construction reading $\mathbb{R}$ for $\mathbb{Q}$ gives a model of cardinality $\mathfrak{c}$ that is not isomorphic to <$\mathbb{R}$; +, 0>. ■

Nonetheless, we can prove that **GD** is complete and decidable by the method of elimination of quantifiers (Chapter XIII, pp. 221–223). So every model of **GD** is elementarily equivalent to <$\mathbb{Q}$; +, 0>, which is elementarily equivalent to <$\mathbb{R}$; +, 0>.

The decidability and completeness of the theory of divisible groups

Theorem **GD** is complete and decidable.

Proof The proof is by elimination of quantifiers.

Let C be a wff of L(= ; ⊕, o) of the form $\exists z\ (C_1 \wedge \cdots \wedge C_m)$, where each C_i is either an atomic wff or the negation of an atomic wff. I'll show how to eliminate the quantifier by induction on m.

We can assume that z appears in each C_i. By repeated use of Lemma 4 and the substitution of equals, we can assume that every atomic formula is of the form $(z = t)$ where z does not appear in t, or is of the form $(z = z)$. For $m = 1$, we have the following cases, which I will let you justify:

(i) $\exists z\ (z = z)$ Replace this by T.

(ii) $\exists z\ (z \neq z)$ Replace this by F.

(iii) $\exists z\ (z = t)$ Replace this by T.

(iv) $\exists z\ (z \neq t)$ Replace this by T.

Hence, we can eliminate the quantifier when $m = 1$.

Suppose now that we can eliminate the quantifier when there are m conjuncts, and C is of the form $\exists z\,(C_1 \wedge \cdots \wedge C_m \wedge C_{m+1})$. If one of the conjuncts is $(z = z)$, we can eliminate it, since the resulting wff is equivalent to C, and we are done by induction. If one of the conjuncts is $(z = t)$, we can suppose C has the form $((z = t) \wedge C_1 \wedge \cdots \wedge C_m)$. Then C is equivalent to $(C_1 \wedge \cdots \wedge C_m)(t/z)$, and we are done by induction. So suppose that each of the conjuncts is an inequality. If one of them is $(z \neq z)$, replace the entire formula by F. Otherwise, C is $(z \neq t_1) \wedge \cdots) \wedge (z \neq t_m)$, where z does not appear in any of the t_i. Since every model of **GD** is infinite, we can replace C by T.

Now you can show (compare pp. 221–223) that relative to **GD** every wff A is semantically equivalent to either T or F. In the first case, **GD**$\vdash$A; in the second, **GD**$\vdash\neg$A. So **GD** is complete and decidable. ■

Exercises for Section B

1. a. Prove Lemma 1.
   b. Show that $\{G^\# \wedge \exists x\,\exists y\,(x \neq y)\} \nvdash \forall x\,\forall y\; n(x \oplus y) = nx \oplus ny$.
2. Define $\mathbf{GD}_0$ as $\mathbf{GD}n$ except with $n = 0$. Show syntactically that $\mathbf{GD}_0$ is inconsistent.
3. Complete the proof of Lemma 3.c
4. Complete the proof of Lemma 3.f.
5. Prove Lemma 4.
6. Complete the proof of Theorem 5.
7. a. In the proof of Theorem 6, show that $\mathsf{Q}_2 \models$**GD**.
   b. For each $n \geq 2$ define a countable model Q_n with universe the collection of n-tuples of rationals such that if $n \neq m$, Q_n is not isomorphic to Q_m.
   c. In the same manner as in the proof of Theorem 6 define a model R_2 of cardinality $\mathfrak{c}$ that is not isomorphic to $\langle \mathsf{R}; +, 0\rangle$.
   d. For each $n \geq 2$ define a model R_n of cardinality $\mathfrak{c}$ with universe the collection of n-tuples of real numbers such that if $n \neq m$, R_n is not isomorphic to R_m.
8. Show that there is a model M of **GD** such that Q is not isomorphic to any submodel of M. (Hint: Consider the roots of unity in C.)

C. Continuous Orderings

The ordering of the reals is a dense linear ordering. We already have a theory of dense linear orderings, **DLO**, which is $\aleph_0$-categorical, complete, and decidable.

But **DLO** is not $\mathfrak{c}$-categorical. Indeed, there are uncountably many non-isomorphic models of **DLO** of cardinality $\mathfrak{c}$ (Exercises 9 and 10 of Chapter XIII.C). There is nothing we can add to the first-order theory **DLO** to eliminate those unintended models, since **DLO** is already complete. To ensure we have a theory of the ordering of just the reals, we'll make a second-order version of **DLO**.

We want to eliminate the possibility that least upper bounds do not exist, thus filling in the points on the line as Dedekind did. If we have two non-empty subsets of the ordering, X and Y, such that every element in X is less than every element in Y, there must be a point that separates them: the least upper bound of X.

Least upper bounds If <U ; <> is a linearly ordered set and B ⊆ U, then u is an *upper bound* for B iff for every b ∈ B, b < u. And u is a *least upper bound* (*l.u.b.*) for B iff u is an upper bound for B, and if a is an upper bound for B, then u ≤ a

---

If a subset of a linearly ordered set has a least upper bound then, as you can show, there is only one.

Dedekind's axiom (*the continuity axiom*)

$$\forall \ldots [\,\forall X\, \forall Y\, (\exists x\, \exists y\, (X(x) \wedge Y(y)) \wedge \forall x\, \forall y\, ((X(x) \wedge Y(y)) \rightarrow x \leq y)$$
$$\rightarrow \exists z\, \forall x\, \forall y\, ((X(x) \wedge Y(y)) \rightarrow x \leq z \leq y))\,]$$

A (second-order) model of **DLO** plus Dedekind's axiom is called *order-complete* or a *continuous ordering*.

Lemma 7 In any order-complete model <U; <>, every non-empty subset of the universe that is bounded above has a least upper bound.

Proof Suppose A ≠ ∅ and A is bounded above. Set B = {b ∈ U: all a ∈ A, a ≤ b}. By hypothesis, B ≠ ∅. By Dedekind's axiom, there is a c such that for all b ∈ B, all a ∈ A, a ≤ c ≤ b. There can be only one such c, and c is the least upper bound of A. ■

By construction, the real line is order-complete. But there are uncountably many non-isomorphic order-complete models, as I ask you to show in Exercise 4. We need another property of the real line. Every countable model of **DLO** is isomorphic to <Q ; <>, and the rationals are dense in the real line. In Example 16 of Chapter XIV we defined a wff 'Count(X)' that is satisfied by an assignment in a second-order model iff X is assigned a countable subset of the universe. Using that, we can formalize the requirement that there is a dense countable subset.

Countable density

$$\exists X\, (\text{Count}(X) \wedge \forall x\, \forall y\, (x < y \rightarrow \exists z\, [X(z) \wedge (x < z < y)]))$$

DLO plus *countable density* is also not enough to give us a categorical theory, since <Q ; <> and <R ; <> are both models. But combining Dedekind's axiom with countable density we do get a categorical theory.

RLO_2, the theory of the *linear order of the real line* in $L_2(=;<)$

DLO

countable density

Dedekind's axiom

Theorem 8 All models of **RLO_2** are isomorphic to $<\mathbb{R};<>$.

Proof $<\mathbb{R};<>$ is a model of **RLO_2** by construction.

Suppose that $M=<U;<>$ is a model of **RLO_2**. By countable density, there is some countable $B \subseteq U$ that is dense in U. I'll leave to you to show that B is a model of **DLO**. By Theorem XIII.5, there is an isomorphism, φ, from $<B;<>$ to $<\mathbb{Q};<>$.

Given any $c \in U$, define $D_c = \{b \in B: b \leq c\}$. Since B is dense, and the ordering has no endpoints, $D_c \neq \emptyset$. It follows that c is the least upper bound of D_c: There is a least upper bound by Lemma 7, say d, and $d \leq c$. If it were that $d < c$, then since B is dense, there is some $b \in B$ such that $d < b < c$, which is a contradiction on d being the least upper bound. So $c = d$. Now view $<\mathbb{Q};<>$ as a submodel of $<\mathbb{R};<>$. Then, as you can show, by taking $\varphi(c) = \text{l.u.b. } D_c$ we can extend φ to all of U to get an isomorphism from M to $<\mathbb{R};<>$. ■

Exercises for Section C

1. a. Give an example of a subset of the integers that does not have a least upper bound.
   b. Give an example of a bounded subset of the rationals that does not have a least upper bound.
   c. If $<U;<>$ is a linearly ordered set, and $B \subseteq U$, show that there is at most one least upper bound for B.

2. Find the least upper bound in $\mathbb{R}$ of the following sets, or show that none exists:
   a. All positive rationals.
   b. All negative rationals.
   c. $\{1/n: n \in \mathbb{N}\}$.
   d. $\{n/n{+}1: n \in \mathbb{N}\}$

3. a. If $<U;<>$ is a linearly ordered set and $B \subseteq U$, define the greatest lower bound for B.
   b. Find the greatest lower bound, if any, for the sets in Exercise 2.
   c. Show that in any order complete model, every nonempty subset of the universe that is bounded below has a greatest lower bound.

4. a. Let B be a set of cardinality greater than c (see Theorem IX.1), with a linear ordering L (this requires a version of the axiom of choice). Let $\mathbb{R}^+$ = the set of positive real numbers. Define a model M to be $\mathbb{R}$ followed by B-many copies of $\mathbb{R}^+$: For every $b \in B$, set $\mathbb{R}^+_b = \{(r, b): r \in \mathbb{R} \text{ and } r > 0\}$, and $U_M = \bigcup_{b \in B} \mathbb{R}^+_b$; for $r, s \in \mathbb{R}$, $c, d \in \mathbb{R}^+$, $b, b' \in B$, define: (i) $r < s$ iff $r < s$, (ii) $r < (c, b)$, and (iii) $(c, b) < (d, b')$ iff $b < b'$, or $b = b'$ and $c < d$.
      Show that M is order-complete, *countable density* fails in M, and M is not isomorphic to $<\mathbb{R};<>$.
   b. Show that there are uncountably many non-isomorphic order-complete models.

5. a. Formulate a first-order scheme of axioms to replace Dedekind's axiom. (Hint: Compare the first- and second-order formalizations of induction in Chapter XV).
   b. What more can you prove in the first-order theory of **DLO** plus this scheme than you can prove in **DLO**?
6. Complete the proof of Theorem 8 that there is just one way to extend φ to all of U.
7. a. Let $\mathbf{T} = \text{Th}(\mathbf{O1–O5}, \neg\mathbf{O6}, \neg\mathbf{O7})$ be the theory of dense linear orderings with endpoints. Let $\mathbf{RT}_2 \equiv_{\text{Def}} \mathbf{T} \wedge$ *countable density* $\wedge$ *Dedekind's axiom*.
   Let [0, 1] be the unit interval of the real line.
   Show that every model of $\mathbf{RT}_2$ is isomorphic to $\langle [0, 1] ; < \rangle$.
   b. Show that every pair of line segments of the real line are isomorphic.
8. Show that $\mathbf{RLO}_2$ is undecidable. (Hint: Corollary XV.29.)

D. Ordered Divisible Groups

We now put together our theories of addition and orderings.

ODG, the theory of *ordered divisible groups* in $L(=; <, \oplus, o)$

DLO

GD

OA $\forall x \forall y \forall z\, (y < z \rightarrow x \oplus y < x \oplus z)$ *the order-addition axiom*

A model of **ODG** is called an *ordered divisible group*.

The integers $\langle \mathbb{Z} ; <, + 0\rangle$ form an ordered group, but they are not divisible. Both the rationals $\langle \mathbb{Q} ; <, +, 0\rangle$ and the reals $\langle \mathbb{R} ; <, +, 0\rangle$ are ordered divisible groups, so our theory is consistent. A model $M = \langle U; <, +, o\rangle$ is an ordered divisible group iff $\langle U; <\rangle$ is a dense linear ordering without endpoints, $\langle U; +, o\rangle$ is a divisible group, and M satisfies the order-addition axiom. I'll let you prove the next lemma.

Lemma 9 For any natural numbers $n, m > 0$, the following are theorems of **ODG**.

a. $\forall x \forall y\, (x < y \rightarrow o < y \sim x)$

b. $\forall x\, (o < x \rightarrow {\sim} x < o)$

c. $\forall x\, (x < o \rightarrow o < {\sim} x)$

d. $\forall x \forall y\, (o < x \rightarrow y \sim x < y)$

e. $\forall x \forall y\, (x < y \rightarrow nx < nx)$

f. $\forall x \forall y\, (x < y \rightarrow {}^{x}/_{m} < {}^{y}/_{m})$

g. $\forall x\, (o < x \rightarrow o < {}^{nx}/_{m})$

I'll let you now prove the following theorem by combining the decision procedures for **DLO** and **GD** using the method of elimination of quantifiers.

Theorem 10 **ODG** is complete and decidable.

Theorem 11

a. **ODG** = Th(<ℝ; <, +, 0>) = Th(<ℚ; <, +, 0>)

b. **ODG** is not finitely axiomatizable.

c. **ODG** is neither $\aleph_0$-categorical nor c-categorical.

d. Multiplication is not definable in Th(<ℝ; <, +, 0>).

Proof (a) This follows by Theorem 10, since both are models of **ODG**.

(b) The models given in the proof of Lemma 3.c are all densely linearly ordered, and hence they serve to show that **ODG** is not finitely axiomatizable just as they did for **GD** in Theorem 5.

(c) Consider the model given in the proof of Theorem 6. Order it via (a, b) < (b, d) iff a < b or (a = b and c < d), where < is the ordering of the rationals. I'll leave to you to show this is a divisible ordered group. But as in the proof of Theorem 6, it is not isomorphic to <ℚ; <, +, 0>. The similar construction using ℝ in place of ℚ establishes that **ODG** is not c-categorical.

(d) We'll use Padoa's method (Theorem XIII.4, p. 213). The mapping φ: ℝ → ℝ via $\varphi(a) = 2a$ is a homomorphism with respect to <, +, and 0. But φ is not a homomorphism with respect to multiplication: $\varphi(a \cdot b) = 2(a \cdot b) \neq 2a \cdot 2b = \varphi(a) \cdot \varphi(b)$ when $a = b = 1$. So multiplication is not definable. ■

We can give an alternate characterization of **ODG** that replaces the division axioms with a first-order version of Dedekind's axiom..

The first-order scheme of Dedekind's axiom (*Dedekind's axiom*$_1$)

$$\forall \ldots ([\exists x \exists y (A(x) \wedge B(y)) \wedge \forall x \forall y (A(x) \wedge B(y) \rightarrow x \leq y)] \rightarrow$$
$$\exists z \forall x \forall y (A(x) \wedge B(y) \rightarrow x \leq z \leq y))$$

where A is any wff in which x is free, B is any wff in which y is free, and neither y nor z appear in A, and neither x nor z appear in B

OGDed$_1$ in L(=; <, ⊕, o)

DLO

Abelian G

D_2

OA (the order-addition axiom)

Dedekind's axiom$_1$ *scheme*

Theorem 12 **$OGDed_1$ = ODG**

Proof Since **ODG** is complete, **ODG** $\supseteq$ **$OGDed_1$**. To show **$OGDed_1$** $\supseteq$ **ODG**, we need that **$OGDed_1$** $\vdash \exists x \exists y\, (x \neq y)$ and that **$OGDed_1$** $\vdash D_n$ for each prime n. The former is true because every model of **DLO** is infinite.

So let M $\models$ **$OGDed_1$**; we need to show that given any a in M, there is some b such that nb = a. In *Dedekind's axiom*$_1$ take $A(x)$ to be $(nx \leq w)$ and $B(y)$ to be $(w \leq ny)$. If a > o, then na > a and n(−a) < a, and the reverse if a < o, and hence the interpretations of A and B when w takes a as reference are non-empty. So by *Dedekind's axiom*$_1$, there is some b such that for all c, d in M, (i) if nc ≤ a, then c ≤ b, and (ii) if a ≤ nd, then b ≤ d. Suppose nb < a. Then by D_2 (abusing notation), (a − nb)/$2n$ is in M. Now b < b + (a − nb)/$2n$. Yet, as you can show, n(a − nb)/$2n$ < (a − nb), and so n (b + (a − nb)/$2n$) < a, which contradicts (i). So a ≤ nb. Similarly, using (ii), we have that nb ≤ a. Hence a = nb. ■

For a categorical theory of ordered divisible groups we'll go to second-order logic, as we did for dense linear orderings. But here we don't need countable density.

$RODG_2$, the theory of *real ordered divisible groups* in $L_2(=; <, \oplus, o)$

DLO

AbelianG

D_2

OA

Dedekind's axiom

Theorem 13

a. **ODG** $\subseteq$ **$RODG_2$**

b. Every model of **$RODG_2$** is isomorphic to <$\mathbb{R}$; <, +, 0>.

c. **$RODG_2$** is undecidable.

Proof (a) This follows by Theorem 12, since all instances of *Dedekind's axiom*$_1$ are semantic consequences of Dedekind's axiom (see pp. 250–252).

(b) Let M = <U; = , <, +, 0> be a model of **$RODG_2$**. Let 1 be any element of U such that 0 < 1. Here 0 and 1 will play the roles of 0 and 1 in $\mathbb{Q}$. Let:

Q = {n1/m : n, m are integers and $m \neq 0$}

Q exists because M is a model of **GD**. Also 0 ∈ Q, since 01/m = 0. I'll let you show that < Q; <, +, 0> ≃ <$\mathbb{Q}$; <, +, 0>. If we can prove that Q is dense in U, we'll have that <U; = , <, 0> ≃ <$\mathbb{R}$; <, 0> as in the proof of Theorem 8.

We'll do this in two steps. First, I'll show that M has the *Archimedean property*: For any a, b ∈ U, if o < a, then for some natural number n, b < na.

Given a, let $\mathsf{A} = \{n\mathsf{a} : n \in \mathbb{N}\}$. Suppose that M does not have the Archimedean property. Then b is an upper bound for A. So by Lemma 7, there is a least upper bound c for A. Since $\mathsf{o} < \mathsf{a}$, by Lemma 9.d, $(\mathsf{c} - \mathsf{a}) < \mathsf{c}$, and $\mathsf{c} - \mathsf{a}$ is not an upper bound for A, where $-$ interprets $\sim$. So for some natural number m, $(\mathsf{c} - \mathsf{a}) < m\,\mathsf{a}$. But then $\mathsf{c} < (m + 1)\mathsf{a}$, which contradicts the assumption that c is a least upper bound for A. Hence, M has the Archimedean property.

Now we can show that Q is dense in U. Suppose $\mathsf{a}, \mathsf{b} \in \mathsf{U}$ and $\mathsf{a} < \mathsf{b}$. We need to find some $\mathsf{q} \in \mathsf{Q}$ such that $\mathsf{a} < \mathsf{q} < \mathsf{b}$.

By the Archimedean property, there is some natural number $n > 0$ such that:

(2) $\quad \mathsf{1} < n(\mathsf{b} - \mathsf{a})$

By the Archimedean property of M, there is a natural number m_1 such that $n\mathsf{a} < m_1\mathsf{1}$. Let m be the least such. Then:

(3) $\quad (m - 1)\mathsf{1} \leq n\mathsf{a} < m\mathsf{1}$

Combining (2) and (3), as you can show, we get:

(4) $\quad n\mathsf{a} < m\mathsf{1} \leq \mathsf{1} + n\mathsf{a} < n\mathsf{b}$

Since $n > 0$, by Lemma 9.f we have $\mathsf{a} < {}^{m\mathsf{1}}/_{n} < \mathsf{b}$.

Hence, there is an isomorphism $\varphi: \langle \mathsf{U}; =, <, \mathsf{0}\rangle \to \langle \mathbb{R}; <, 0\rangle$, and $\varphi({}^{n\mathsf{1}}/_{m}) = {}^{n}/_{m}$. Finally, $\varphi(\mathsf{a} + \mathsf{b}) = \varphi(\mathsf{a}) + \varphi(\mathsf{b})$ follows because that holds for Q, since any real number is the least upper bound of the rationals less than it.

(c) This follows by Corollary XV.29. ∎

Exercises for Section D

1. Prove Lemma 9

2. a. Show that there are countably many non-isomorphic countable models of **ODG**.
   b. Show that there are uncountably many non-isomorphic models of **ODG** of cardinality c.

3. In the proof of Theorem 13, complete the proof that $\varphi(\mathsf{a} + \mathsf{b}) = \varphi(\mathsf{a}) + \varphi(\mathsf{b})$.

4. *The rational points on a circle*
   Show that the following is a model of **ODG**:
   Take the unit circle with center the origin of the Cartesian plane. For the universe take $\{\mathsf{a}$: the distance in radians α of a from $(1, 0)$ is a rational number$\}$. Set $\mathsf{a} < \mathsf{b}$ iff $\alpha < \beta$. Set $\mathsf{a} + \mathsf{b} = \mathsf{c}$ iff $\alpha + \beta = \gamma$. (Show that no point is represented by two different distances.)

5. Use the method of elimination of quantifiers, combining the proofs for **DLO** and **GD**, to prove that **ODG** is complete and decidable.

E. Real Closed Fields

1. Fields

We now combine the group axioms for multiplication with those for addition.

F, the *theory of fields* in $L(=; \oplus, *, \mathrm{o}, \mathfrak{l})$

AbelianG

$\forall x\,(x * \mathfrak{l} = x)$

$\forall x\,\forall y\,\forall z\,(x * (y * z) = (x * y) * z)$

$\forall x\,\forall y\,(x * y = y * x)$

$\forall x\,(x \neq \mathrm{o} \rightarrow \exists y\,(x * y = \mathfrak{l}))$

$\forall x\,\forall y\,\forall z\,(x * (y \oplus z) = (x * y) \oplus (x * z))$

$\mathfrak{l} \neq \mathrm{o}$

A model of **F** is called a *field*. By Corollary XVI.11 **F** is undecidable.

Both $<\mathbb{Q}; +, \cdot, 0, 1>$ and, by construction, $<\mathbb{R}; +, \cdot, 0, 1>$ are fields. But there are also finite fields, as described in Exercise 1.

We can adopt the same definitions for $\oplus$ as before, since if $\mathsf{M} = <\mathsf{U}; +, \bullet, \mathsf{o}, 1>$ is a field, then $<\mathsf{U}; +, \mathsf{o}>$ is an abelian group. Also, if $\mathsf{M} = <\mathsf{U}; +, \bullet, \mathsf{o}, 1>$ is a field, then $<\mathsf{U} - \{\mathsf{o}\}; \bullet, 1>$ is an abelian group, so we can adopt similar definitions for $*$. We define for natural numbers n and terms t, u:

$t^0 = \mathfrak{l}$ $\qquad\qquad$ $\underline{1} = \mathfrak{l}$

$t^{n+1} = t * t^n$ $\qquad\qquad$ $\underline{n+1} = \mathfrak{l} \oplus \underline{n}$

$t^{-1} = u \equiv_{\text{Def}} t * u = \mathfrak{l} \wedge t \neq \mathrm{o}$

${}^{t}/_{u} \equiv_{\text{Def}} t * u^{-1}$

For example, $\underline{3} = \mathfrak{l} \oplus (\mathfrak{l} \oplus \mathfrak{l})$.

We can now establish some basic facts about fields. Parts (a)–(d) are the multiplicative versions of Lemma 1; the other parts I leave as Exercise 2.

Lemma 14 For all natural numbers n and m, the following are theorems of **F**:

a. $\forall x\,\forall y\,((x^m)^n = x^{(m \cdot n)})$

b. $\forall x\,(x^{(m \cdot n)} = x^m * x^n)$

c. $\forall x\,\forall y\,((x * y)^n = x^n * y^n)$

d. $\forall x\,(x^n)^{-1} = (x^{-1})^n$

e. $\forall x\,(\underline{n} * x = nx)$

f. $\forall x\,\forall y\,(x \neq \mathrm{o} \rightarrow \exists z\,(x * z = y))$

g. $\forall x\,\forall y\,(x * (\sim y) = \sim(x * y))$

h. $\forall x\,\forall y\,((\sim x) * (\sim y) = x * y)$

i. $\forall x\,\exists y\,(x = n\,y)$

j. $\forall x\,(\mathrm{o} * x = \mathrm{o})$

By part (i), if $\mathsf{M} = \langle \mathsf{U}; +, \bullet, \mathsf{o}, \mathsf{1} \rangle$ is a field, then $\langle \mathsf{U}; +, \mathsf{o} \rangle$ is a divisible group.

We know the theory of fields is not complete, as $\exists x\, (x^2 = 2)$ is true of the reals but not the rationals. We need to extend our theory.

Exercises for Section E.1

1. *Finite and infinite fields*
   a. Define $\mathsf{N}_p = \langle \{0, 1, \ldots, p\}; +, \bullet, 0, 1 \rangle$, the *natural numbers modulo p*, where
      $n + m =$ the remainder of $(n + m)$ when divided by p
      $n \bullet m =$ the remainder of $(n \cdot m)$ when divided by p
      Show that if p is prime, then N_p is a field.
   b. Show that for every finite field there is a minimum number $n > 0$, called the *characteristic* of the field, such that $\forall x\, (n x = 0)$ is true in that field.
   c. Show that the characteristic of a finite field is prime.
      (Hint: Consider multiples of 1 and suppose the characteristic were not prime.)
   d. Show that $\forall x\, (p x = \mathrm{o})$ is true in a field where p is a prime iff the field is isomorphic to N_p.
   e. Show that for every prime p, $\mathbf{F} \wedge \forall x\, (p x = \mathrm{o})$ is categorical and is complete.
   f. A field has *characteristic* 0 iff it is a model of :
      $\mathbf{IF} \equiv_{\mathrm{Def}} \mathbf{F} \cup \{\neg \forall x\, (p x = \mathrm{o})\colon p \text{ is prime}\}$
      Show that a field has characteristic 0 iff it is infinite.
   g. Show that **IF** is not finitely axiomatizable.
   h. Show that **IF** is not complete.
   i. Show that **IF** is not decidable. (Hint: Corollary XVI.11.)
2. Prove parts (e)–(j) of Lemma 14.
3. Show that in **IF** the following are theorems.
   a. $\forall x\, \forall y\, (x * y = \mathrm{o} \leftrightarrow x = \mathrm{o} \vee y = \mathrm{o})$
   b. $\forall x\, \forall y\, \forall z\, ((x \neq \mathrm{o} \wedge x * y = x * z) \rightarrow y = z)$
   (Hint: Multiply by x^{-1}.)

2. Ordered fields

Now we'll combine our theories of addition, multiplication, and ordering.

OF, the *theory of ordered fields* in $\mathrm{L}(=; <, \oplus, *, \mathrm{o}, \mathrm{I})$

F

LO

OA

$\forall x\, \forall y\, ((\mathrm{o} < x \wedge \mathrm{o} < y) \rightarrow \mathrm{o} < x * y)$ *the order-multiplication axiom*

Both $\langle \mathbb{Q}; <, +, \cdot, 0, 1 \rangle$ and, by construction, $\langle \mathbb{R}; <, +, \cdot, 0, 1 \rangle$ are models of **OF**. I'll let you establish the next lemma.

Lemma 15 The following are theorems of **OF**:

a. $\forall x\, (\mathrm{o} < x \rightarrow {\sim} x < \mathrm{o})$

b. $\forall x\, (x < \mathrm{o} \rightarrow \mathrm{o} < {\sim} x)$

c. $\forall x\, \forall y\, (x < y \rightarrow \mathrm{o} < y \sim x)$

d. $\forall x\, \forall y\, (x < y \rightarrow {\sim} y < {\sim} x)$

e. $\forall x\, \forall y\, (\mathrm{o} < y \rightarrow x < x \oplus y)$

f. $\forall x\, \forall y\, (\mathrm{o} < y \rightarrow x \sim y < x)$

g. $\forall x\, \forall y\, \forall z\, ((\mathrm{o} < x \wedge y < z) \rightarrow x * y < x * z)$

h. $\forall x\, \forall y\, \forall z\, ((x < \mathrm{o} \wedge y < z) \rightarrow x * z < x * y)$

i. $\forall x\, (\mathrm{o} \neq x \rightarrow \mathrm{o} < x^2)$

j. $\mathrm{o} < \text{ł}$

k. ${\sim}\text{ł} < \mathrm{o}$

l. $\mathrm{o} < \underline{n}$ for $n > 0$

m. $\forall x\, (x < x \oplus \text{ł})$

n. $\forall x\, (x \sim \text{ł} < x)$

o. $\forall x\, (\mathrm{o} < x \rightarrow \mathrm{o} < \text{ł}/x)$

p. $\forall x\, \forall y\, ((\mathrm{o} < x \wedge x < y) \rightarrow ((\mathrm{o} < \text{ł}/y) \wedge (\text{ł}/y < \text{ł}/x)))$

q. $\forall x\, \forall y\, (x < y \rightarrow ((x < x \oplus (y \sim x)/2) \wedge (x \oplus (y \sim x)/2 < y)))$

r. $\forall x\, \forall y\, ((x = \mathrm{o} \vee y = \mathrm{o}) \leftrightarrow x * y = \mathrm{o})$

s. $\forall x\, (\mathrm{o} < x < \text{ł} \rightarrow x * x < x)$

Theorem 16

a. **DLO** $\subseteq$ **OF**

b. **OF** has no finite models.

Proof By parts (m), (n), and (q) of Lemma 15. ■

To investigate models of **OF**, we'll need some definitions and some algebra. A *polynomial*, p(x), is any term of the form:

$$t_0 \oplus (t_1 * x) \oplus \cdots \oplus (t_n * x^n)$$

(In Exercise 2 I ask you to give an inductive definition.) If t_i is a closed term for each i, we call p(x) a *rational polynomial*. The term t_i is called the *coefficient* of x^i. The number n is called the *degree* of the polynomial.

In $\langle \mathbb{R}; <, +, \cdot, 0, 1 \rangle$ the coefficients of a rational polynomial are interpreted as rationals. A rational polynomial is *non-zero* if some t_i is not interpeted as 0 in that model, and a number r is a *root* of a rational polynomial p(x) means that for every assignment of references σ, if $\sigma(x) = r$, then $\sigma \models \mathrm{p}(x) = \mathrm{o}$. A real number r is *algebraic over the rationals* iff it is the root of some non-zero rational polynomial. If r is algebraic over the rationals, the *degree* of r is the smallest number n such that r is a root of a polynomial of degree n over the rationals where the interpretation of $t_n \neq 0$. For example, 3 has degree 0 over the rationals; $\sqrt{2}$ is algebraic over the rationals with degree 2, because it is a root of x^2; and $\frac{-1+\sqrt{5}}{2}$ is algebraic of degree 2 over the rationals because it is a root of $x^2 + x - 1$; and $\sqrt[5]{47}$ is algebraic over the rationals with degree 5. Generally, if p and q are prime, then $\sqrt[p]{q}$ has degree p over the rationals (see, for example, *Herstein, 1964*). A real number is *transcendental* if it is not algebraic over the rationals.

More generally, if B is a subfield of A and B $\neq$ A, we say that A is an *extension* of B. An element a of A is *algebraic over* B iff there is a polynomial p(x) and an assignment of references σ such that for each coefficient t_i, $\sigma(t_i)$ is in B, and $\sigma(x)$ = a, and $\sigma \models$ p(x) = o. Finally, A is an *algebraic extension* of B iff it is an extension and every element in A is algebraic over B.

Theorem 17

a. There are uncountably many countable models of **OF**, each a submodel of $<\mathbb{R}\,; <, +, \cdot, 0, 1>$, no two of which are isomorphic.

b. **OF** is not complete.

Proof There are countably many rational polynomials. Each has at most finitely many roots (see, for example, *van der Waerden, 1953*). So there are countably many numbers algebraic over the rationals. Given any collection of numbers algebraic over the rationals, B, the closure of B $\cup$ $\mathbb{Q}$ under the operations of + and $\cdot$ (see p. 162) is a subfield of $<\mathbb{R}\,; <, +, \cdot, 0, 1>$ which inherits the ordering of $\mathbb{R}$, and hence which satisfies **DLO**, since $\mathbb{Q}$ is dense in $\mathbb{R}$. Since the closure can be defined inductively, the closure of B $\cup$ $\mathbb{Q}$ is countable (Exercises 6 and 7 of Chapter IX.B).

Now for each prime p, let r_p be a root of $x^p - \underline{47}$. Then the closure of each subset of B = $\{r_p\colon p \text{ is a prime}\}$ yields a distinct subfield (see *Herstein, 1964*), and no two are isomorphic because they are not elementarily equivalent (consider the formulas $\exists x\,(x^p - \underline{47} = \text{o})$). This also shows that **OF** is not complete. ■

Theorem 18 **OF** is undecidable.

Proof The work in Chapter XVI.D applies equally to L(=; <, ⊕, ∗, o, ł). In particular, the proof here follows as for Corollary XVI.11. ■

Exercises for Section E.2

1. Prove Lemma 15.
(Hints: (a)–(c) Add $\sim x$. (d) Use (a). (g) Use that o $< x * (z \sim y)$ and add $x * y$. (i) Split into two cases where $x <$ o or o $< x$ and use (g) and (h). (o) Use that o $<$ ł $= y *$ ł$/y$. (p) Multiply both sides by ł$/y$ and by ł$/x$. (q) Use (k) and (n).)

2. Give an inductive definition of 'polynomial of degree n'.
3. a. Define inductively the closure of $\mathbb{Q} \cup \{\sqrt{2}\}$ under addition and multiplication.
   b. If B is countable and $\mathsf{B} \subseteq \mathbb{R}$, define inductively the closure under addition and multiplication of $\mathsf{B} \cup \mathbb{Q}$ and show that it is countable.
4. a. Show that $\mathsf{F} = \{p + q\sqrt{2} : p, q \in \mathbb{Q}\}$ is a subfield of $\mathbb{R}$.
   b. Show that F = the closure of $\mathbb{Q} \cup \{\sqrt{2}\}$ under addition and multiplication.
   c. Let $\mathsf{G} = \{p + q\sqrt{3} : p, q \in \mathbb{Q}\}$. Show that F is not elementarily equivalent to G. (Hint: Suppose for some $p, q \in \mathbb{Q}$, $(p + q\sqrt{2})^2 = 3$. Get a contradiction by splitting into cases: $p = 0$, $q = 0$, or $p, q \neq 0$. In the last case show that $\sqrt{2} = (3 - p^2 - 2q^2)/_{2pq}$.)

3. Real closed fields

Our goal has been to axiomatize Th($\langle \mathbb{R} ; <, +, \cdot, 0, 1 \rangle$) in L($= ; <, \oplus, *, \text{o}, \text{ł}$). There are several ways to do that. *Artin and Schreier, 1927*, characterized *real closed fields* in two steps. First, a field is *formally real* means ~ 1 is not the sum of squares. Then a real closed field is one that is formally real and has no algebraic extension that is formally real. Condition (ii) cannot be formalized in our first-order language. But Artin and Schreier also showed there that those same fields can be characterized by the following conditions:

The field is formally real.
Every element is a square or the negative of a square.
Every polynomial of odd degree has a root.

RCF, the *theory of real closed fields* in L($= ; <, \oplus, *, \text{o}, \text{ł}$)

OF

$\forall x \, \exists y \, (x = y^2 \lor x = \sim y^2)$

$\forall x_0 \, \forall x_1 \ldots \forall x_{2n} \, \exists y \, (x_0 \oplus (x_1 * y) \oplus \cdots \oplus (x_{2n} * y^{2n}) \oplus y^{2n+1} = \text{o})$

(a scheme of axioms, one for each natural number n, writing y for x_{2n+1})

In Exercise 1 I ask you to prove the following.

Lemma 19

a. For every natural number $n > 0$,
RCF$\vdash \forall x_0 \, \forall x_1 \ldots \forall x_n \, \neg(x_0{}^2 \oplus x_1{}^2 \oplus \cdots \oplus x_n{}^2 = \sim \text{ł})$

b. In **RCF**:
'o' is definable by: $x \oplus x = x$.
'ł' is definable by: $x * x = x \land \neg(x \oplus x = x)$.
'$x < y$' is definable by: $\exists z \, (z \neq \text{o} \land z^2 = y \sim x)$.

By construction, $\langle \mathbb{R}; <, +, \cdot, 0, 1\rangle$ is a real closed field.

Theorem 20 **RCF** is not finitely axiomatizable.

Proof If **RCF** were finitely axiomatizable, then by Lemma XV.2 there would be a finite subset of the axioms of **RCF** that axiomatizes it. Choose such a subset, and let n be the largest number that is the degree of an instance of the axiom scheme that is in that subset. Let $\mathbb{A}$ = the closure under addition and multiplication of $\{r : r \in \mathbb{R}$ and r is algebraic over $\mathbb{Q}$ of degree $\leq n\}$. This is a field and each element in it has degree $\leq n$ over the rationals (see, for example, Chapter 5.1 of *Herstein, 1964*). Hence, by the supposed finite axiomatization, it is a model of **RCF**. Let $p > n$ be a prime; then, as you can show, every root of $\exists x\ (x^p = 47)$ has degree p over the rationals. So $\exists x\ (x^p = 47)$ is false in this model, yet it is a theorem of **RCF**. Hence, no finite axiomatization exists. ■

Kreisel and Krivine, 1967, in a proof that requires too much algebra to include here, apply the method of elimination of quantifiers to show the following.[3]

Theorem 21 **RCF** is complete and decidable.

Corollary 22

a. $\mathbf{RCF} = \mathrm{Th}(\langle \mathbb{R}; <, +, \cdot, 0, 1\rangle)$.

b. Every model of **RCF** is elementarily equivalent to $\langle \mathbb{R}; <, +, \cdot, 0, 1\rangle$.

c. $\mathrm{Th}(\langle \mathbb{R}; +, \cdot\rangle)$ in $L(=; \oplus, *)$ is complete and decidable, and **RCF** is an extension by definitions of that theory.

d. None of $\mathbb{N}$, $\mathbb{Z}$, or $\mathbb{Q}$ is definable in $\mathrm{Th}(\langle \mathbb{R}; <, +, \cdot, 0, 1\rangle)$.

Proof (a) and (b) follow by Theorem 21, since $\langle \mathbb{R}; <, +, \cdot, 0, 1\rangle$ is a model of **RCF**. Part (c) follows by Lemma 19.b and Theorem 21. For part (d), if any of the sets were definable, then by the same method as in the proof of Corollary XVI.9 we would have that $\mathrm{Th}(\langle \mathbb{R}; <, +, \cdot, 0, 1\rangle)$ is undecidable, which it is not. ■

Not all models of **RCF** are isomorphic to $\langle \mathbb{R}; <, +, \cdot, 0, 1\rangle$, for by the Löwenheim-Skolem theorem there are countable models. In particular, the closure of $\mathbb{Q} \cup \{r : r$ is algebraic over the rationals$\}$ is a subfield of $\langle \mathbb{R}; <, +, \cdot, 0, 1\rangle$ and is a model of **RCF** (see, for example, Chapter 5.1 of *Herstein, 1964*). By Exercise 8 of Chapter IX.B it is countable. But **RCF** is not $\aleph_0$-categorical (Exercise 12), nor is it $\mathfrak{c}$-categorical (Exercises 13).

Tarski, 1940 gives a very different axiomatization of $\mathrm{Th}(\mathbb{R}; <, +, \cdot, 0, 1\rangle)$. He requires that a model be an ordered field and that greatest lower bounds exist for bounded sets definable by predicates in the language.

[3] The theorem is originally due to Tarski, as discussed below. A very different proof that **RCF** is complete can be found in *Abraham Robinson, 1956*, from which decidability also follows.

TRCF, *Tarski's theory of real closed fields* in L(= ; <, ⊕, ∗, o, ł)

OF plus the scheme:

$$\forall \ldots \ [(\exists x\, A(x) \wedge \exists w\, \forall x\, (A(x) \rightarrow w < x)) \rightarrow \exists z\, \forall y\, (\forall x\, (A(x) \rightarrow y < x) \rightarrow y \leq z)]$$

where A is any wff in which x is free and in which neither y nor z appear

---

Tarski proves by elimination of quantifers that this theory is complete and decidable.[4] Since <ℝ ; <, +, ·, 0, 1> is a model of **TRCF**, Th(<ℝ ; <, +, ·, 0, 1>) = **TRCF**, and hence **TRCF** = **RCF**. Tarski also shows that his axiom scheme can be replaced by a simpler scheme that is the restriction to polynomials of the Weierstrass Nullstellensatz (Lemma H of the appendix below):

$$\forall \ldots \forall x\, \forall y\, ((o < p(x) \wedge p(y) < o \wedge y < x) \rightarrow \exists z\, (p(z) = o \wedge y < z < x))$$

where p(x) is a polynomial in which neither y nor z appear.

Alternatively, we can use the first-order scheme of Dedekind's axiom to axiomatize Th(<ℝ ; <, +, ·, 0, 1>).

DRCF, the *Dedekind theory of real closed fields* in L(= ; <, ⊕, ∗, o, ł)

OF

Dedekind's axiom$_1$ (scheme)

I'll let you show that each instance of Tarski's scheme is a theorem of **DRCF**, and hence that **DRCF** = **TRCF** = **RCF**.

We now have a complete theory of the real numbers. That's as good as we can get in first-order logic. In this theory we can formalize a lot of the algebra of the real numbers. We can talk about polynomials, equations, and inequalities of any degree. We can state and solve questions about the divisibility of polynomials and their roots. But we cannot define the natural numbers in this theory, and we have no way to use a single wff to talk about all polynomials, so we can't even state that there are transcendental numbers. Nor can we talk about all rational polynomials of a particular degree, since the set of rationals is not definable. Nor can we discuss functions on the real line except for those that are definable in the theory, and there are only countably many of those. Hence, we cannot state or prove any of the theorems about continuous functions that are a staple of the study of the real numbers. The second-order version of the Dedekind axiomatization overcomes these limitations.

---

4 In *Tarski, 1951* he gives a proof that Th(<ℝ; <, +, ·, 0, 1> is decidable by the method of elimination of quantifiers, but he gives no axiomatization there.

$\mathbf{RCF_2}$, the *second-order theory of real closed fields* in $L_2(=; <, \oplus, *, o, 1)$

OF

Dedekind's axiom

That is, $\mathbf{RCF_2}$ is the set of semantic consequences of this finite list of axioms.

Theorem 23

a. All models of $\mathbf{RCF_2}$ are isomorphic to $<\mathbb{R}; <, +, \cdot, 0, 1>$.

b. $\mathbf{RCF_2}$ is undecidable.

Proof (a) Given any model of $\mathbf{RCF_2}$, $<\mathbb{Q}; <, +, \cdot, 0, 1>$ is isomorphic to the set of interpretations of $\{\underline{n}/\underline{m} : n, m$ are natural numbers with $m \neq 0\}$. So the proof of Theorem 13.b can be used here to show that $<\mathbb{R}; <, +, \cdot, 0, 1>$ is isomorphic to the model.

(b) By Corollary XV.29. ■

The real numbers can also be characterized by requiring every nonempty set bounded above to have a least upper bound.

The least upper bound principle $\forall X\,([\exists x\, X(x) \wedge \exists y\, \forall x\,(X(x) \rightarrow x \leq y)] \rightarrow$
$\exists z\,(\forall x\,(X(x) \rightarrow x \leq z) \wedge \forall y\,(\forall x\,(X(x) \rightarrow x \leq y) \rightarrow z \leq y)))$

In Theorem M of the appendix to this chapter I show:

$\mathbf{RCF_2} = \mathrm{Th}_2(\mathbf{OF} \wedge$ *the least upper bound principle*$)$

We have accomplished our goal of characterizing the real numbers in terms of their properties. We can now formalize a great deal more of the traditional calculus of the real numbers, such as 'Every continuous function on a closed interval has a greatest value on that interval.' But integration and differentiation are functions of functions on the real numbers, so to state claims about integrals and derivatives, such as 'Every differentiable function is continuous' we need a third-order language. And there is much in the theory of functions on the real line that requires still higher-order formalizations.

Exercises for Section E.3

1. Prove Lemma 19. (Hint: For part (a) prove it for $n = 1$ and proceed by induction.)
2. Show that $\mathbf{OF} \cup \{\forall x\,(o < x \rightarrow \exists y\,(y^2 = x))\} \vdash \forall x\, \exists y\,(x = y^2 \vee x = \sim y^2)$.
3. Give an axiomatization of $\mathrm{Th}(<\mathbb{R}; +, \cdot>)$ in $L(=; \oplus, *)$.
4. Show that in the real numbers every root of $(x^p = 47)$ where p is prime has degree p.

5. Show that in **RCF** we can prove that a second-degree polynomial has a root iff its discriminant is not negative. That is,
   $\mathbf{RCF} \vdash \forall x\, \forall y\, \forall z\, \forall w\, ((y * x^2) \oplus (z * x) \oplus w = \mathrm{o} \leftrightarrow \mathrm{o} \leq z^2 \sim 4(y * w))$
   (Hint: Show it's true for the real numbers.)
6. Show that every real closed field has a subfield isomorphic to the rationals.
7. Show directly that in every model of **TRCF** if $\mathcal{A}$ is the non-empty extension of a unary predicate in the language and $\mathcal{A}$ is bounded below, then $\mathcal{A}$ has a greatest lower bound.
8. Show that **TRCF** = **DCRF**.
9. Show that $\mathbf{RCF_2}$ is undecidable without using Corollary XV.29 by defining the set of natural numbers in Th(<$\mathbb{R}$; <, +, ·, 0, 1>) and translating **Arithmetic** into $\mathbf{RCF_2}$.
10. In the language $L_{2F}(=; <, \oplus, *, \mathrm{o}, \mathrm{I})$ formalize and/or discuss the following claims (see p. 329 below for the definition of 'continuous function'):
    a. There are unbounded continuous functions on (0, 1).
    b. Every continuous function on a closed interval has a greatest value on that interval.
    c. Every continuous function can be approximated by polynomials.
    (Hint: See the proof of Theorem 13 for how to quantify over natural numbers.)
11. Show that the least upper bound principle is a semantic consequence of $\mathbf{RCF_2}$.
12. **RCF** *is not* $\aleph_0$*-categorical*
    The set of numbers algebraic over $\mathcal{B}$ is a field. The real number e is not algebraic over the rationals (see, for example, *Herstein, 1964*, Chapters 5.1 and 5.2).
    Let $\mathcal{A}$ be the field of numbers algebraic over $\mathbb{Q}$. Let $\mathcal{B}$ be the closure of $\mathcal{A} \cup \{e\}$ under addition and multiplication.
    a. Show that $\mathcal{B}$ is countable.
    b. Let $\mathcal{A}(e)$ be the algebraic closure of $\mathcal{B} = \{\mathrm{r}: \mathrm{r}$ is algebraic over $\mathcal{B}\}$. Show that $\mathcal{A}(e)$ is countable and is a real closed field.
    c. Show that $\mathcal{A}(e)$ is not isomorphic to $\mathcal{A}$.
    (Hint: Suppose φ were an isomorphism. The image $\varphi(\mathrm{q}) = \mathrm{q}$, for $\mathrm{q} \in \mathbb{Q}$. Consider a such that $\varphi(\mathrm{a}) = e$.)
13. Show that **RCF** is not $\mathfrak{c}$-categorical.
    (Hint: As in Section 63 of *van der Waerden, 1953*, Volume 1, define the quotient field of the polynomial ring over the reals, formally adjoin roots to every odd-degree polynomial in that field, and show that the resulting field is not isomorphic to $\mathbb{R}$, yet has cardinality $\mathfrak{c}$).

The theory of fields in the language of name quantification

In Chapter XV.F we saw that we could characterize the arithmetic of the natural numbers without the assumptions of second-order logic using quantifications over names. Similarly we can give a finite characterization of the arithmetic of the rationals in the language of name quantification.

First we have to formulate the theory of ordered fields in the language using both '$^{-1}$' and '$\sim$' as functions, since we need to use those to form names.

OF-n, the theory of *ordered fields* in $L_{qn}(=; <, \oplus, *, \sim, ^{-1}, \text{o}, \text{ł})$

OF

$\forall x\, (x \oplus (\sim x) = \text{o})$

$\forall x\, (x \neq \text{o} \rightarrow x * x^{-1} = \text{ł}) \wedge (\text{o}^{-1} = \text{o})$

We need a default value for o^{-1} because the interpretation of '$^{-1}$' must be defined for every object in the universe.

In every model of **OF** the objects that are named form a field isomorphic to $<\mathbb{Q}; <, +, \cdot, -, {}^1/, 0, 1>$.

Recall that *AN* is the wff $\forall x_1\, \exists a_0\, (x_1 = a_0)$.

Rational N-Arithmetic in $L_{qn}(=; <, \oplus, *, \sim, ^{-1}, \text{o}, \text{ł})$

OF-n

AN

Theorem

a. Each model of **Rational N-Arithmetic** is isomorphic to $<\mathbb{Q}; <, +, \cdot, -, {}^1/, 0, 1>$.

b. **Rational N-Arithmetic** = Th($<\mathbb{Q}; <, +, \cdot, -, {}^1/, 0, 1>$).

We cannot categorically axiomatize the reals in the logic of name quantification because every consistent theory has a countable model. But we can give a theory of the real closed subfields of the real numbers.

Recall from Chapter XVI.D that there is a predicate $\text{P}(x)$ with just x free that defines the set of natural numbers in the first-order theory of the rationals. Let $P(x)$ be that predicate with each variable x_i other than x replaced by the name variable a_i. Then $P(x)$ defines the set of natural numbers in **Rational N-Arithmetic** in the sense that in any model of **Rational N-Arithmetic** the set of objects satisfying $P(x)$ are isomorphic to the natural numbers, and indeed are the interpretations of $\underline{n}$ for $n \geq 0$. So within **Rational N-Arithmetic** we can quantify over the rationals by quantifying over names, and quantify over the natural numbers using quantification relative to $P(x)$. So we can say that every object in the universe can be approximated by rationals. I'll write q for the name variable a_1.

Real N-Arithmetic, the *theory of real closed subfields of the reals* in $L_{qn}(=; <, \oplus, *, \sim, ^{-1}, \text{o}, \text{ł})$

RCF (with **OF/n** replacing **OF**)

AN

$\forall x\, \forall y\, (\text{P}(y) \rightarrow \exists q\, (\text{o} < (q \sim x) \wedge (q \sim x) < y^{-1}))$ *approximation*

Theorem Every model of **Real N-Arithmetic** is isomorphic to a real closed field F where $\mathbb{Q} \subseteq \mathsf{F} \subseteq \mathbb{R}$.

Appendix: Real Numbers as Dedekind Cuts

In this appendix I'll present a construction of the real numbers from the rationals using Dedekind cuts, roughly following *Rudin, 1964*.

I'll assume that the rational numbers, $\mathbb{Q}$, are given, and I will use p, q, r, s, t, u as variables ranging over rationals. Note that $\mathbb{Q}$ has the *Archimedean property*:

For any p, q, if $0 < p$, there is some natural number n such that $q < np$.

Dedekind cut A Dedekind cut is any set $\alpha \subseteq \mathbb{Q}$ satisfying:

D1. $\alpha \neq \varnothing$ and $\alpha \neq \mathbb{Q}$.

D2. For every $p \in \alpha$, if $q < p$, then $q \in \alpha$.

D3. For every $p \in \alpha$, there is some $r \in \alpha$, such that $p < r$.

I will use $\alpha, \beta, \gamma, \delta$ to range over Dedekind cuts. Let $\mathbb{R}$ be the set of all Dedekind cuts. Define $\mathrm{D}(p) = \{q: q < p\}$ to be the *Dedekind cut of* p. I'll let you show the following.

Lemma A For all rationals p, q, r and every Dedekind cut α:

a. $\mathrm{D}(p)$ is a Dedekind cut.

b. If $p \in \alpha$ and $q \notin \alpha$, then $p < q$.

c. If $r \notin \alpha$ and $r < s$, then $s \notin \alpha$.

d. For any Dedekind cut α, there is some r such that for all $p \in \alpha$, $p < r$.

To define an ordering on $\mathbb{R}$ that has the appropriate properties, set:

$\alpha < \beta \equiv_{\mathrm{Def}} \alpha \subset \beta$ and $\alpha \neq \beta$

I'll also write '$\beta > \alpha$' for '$\alpha < \beta$'. Now we can establish that $\mathbb{R}$ is an ordered set satisfying the least upper bound principle.

Lemma B

a. The relation '<' is a linear order on $\mathbb{R}$.

b. Given any $\mathcal{B} \subseteq \mathbb{R}$, if $\mathcal{B} \neq \varnothing$ and $\mathcal{B}$ is bounded above, then there is a least upper bound for $\mathcal{B}$ in $\mathbb{R}$.

Proof (a) For any α, $\alpha \nless \alpha$. Also if $\alpha < \beta$, then $\beta \nless \alpha$. Transitivity of the ordering follows from the transitivity of $\subset$. To show trichotomy, suppose that $\alpha \nless \beta$ and $\alpha \neq \beta$. Then $\alpha \not\subset \beta$. Hence, there is some $p \in \alpha$ such that $p \notin \beta$. So by Lemma A, if $q \in \beta$, then $q < p$. Hence, $q \in \alpha$. So $\beta \subset \alpha$, and $\beta < \alpha$.

For (b), suppose $\varnothing \subset \mathcal{B} \subset \mathbb{R}$, and β is an upper bound for B. Set:

$\gamma = \bigcup \{\alpha: \alpha \in \mathcal{B}\}$

That is, $p \in \gamma$ iff for some $\alpha \in \mathcal{B}$, $p \in \alpha$. I will show that γ is a Dedekind cut and $\gamma =$ l.u.b. $\mathcal{B}$. I'll let you show $\gamma \neq \varnothing$. For every $\alpha \in \mathcal{B}$, $\alpha \subset \beta$, so $\gamma \subset \beta$. Hence, $\gamma \neq \mathbb{Q}$. So γ satisfies D1.

Now suppose that $p \in \gamma$. Then for some $\delta \in \mathcal{B}$, $p \in \delta$. If $q < p$, then $q \in \delta$, hence $q \in \gamma$. Further, since $p \in \delta$, there is some $r \in \delta$ such that $p < r$. But $r \in \gamma$, too. So γ satisfies D2 and D3 and is a Dedekind cut.

For every $\alpha \in \mathcal{B}$, $\alpha \leq \gamma$. Suppose that $\delta < \gamma$. Then there is some $p \in \gamma$ such that $p \notin \delta$. Since $p \in \gamma$, there is some $\alpha \in \mathcal{B}$ such that $p \in \alpha$. Hence by Lemma A, $\delta < \alpha$, and δ is not an upper bound for $\mathcal{B}$. Thus, $\gamma =$ l.u.b. $\mathcal{B}$. ■

We now define addition on $\mathbb{R}$, using addition of the rationals, '+'.

$\alpha + \beta \equiv_{\text{Def}} \{r+s : r \in \alpha \text{ and } s \in \beta\}$

$0^* \equiv_{\text{Def}} \text{D}(0) = \{p : p < 0\}$

$-\alpha \equiv_{\text{Def}} \{-p : p \text{ is an upper bound for } \alpha,\ p \neq \text{l.u.b. } \alpha\}$

I'll now show that <R; +, 0*> is an ordered abelian group, using the axiomatization of groups by Kleene (p. 199).

Lemma C For every α, β, γ,

a. $\alpha + \beta$ is a Dedekind cut.

b. $\alpha + 0^* = \alpha$

c. $-\alpha$ is a Dedekind cut, and $\alpha + (-\alpha) = 0^*$.

d. $\alpha + (\beta + \gamma) = (\alpha + \beta) + \gamma$

e. $\alpha + \beta = \beta + \alpha$

f. If $\beta < \gamma$, then $\alpha + \beta < \alpha + \gamma$.

Proof (a) Given α, β, $\alpha + \beta \neq \varnothing$. Let $r \notin \alpha$ and $s \notin \beta$. Then for any $p \in \alpha$, $q \in \beta$, $p < r$ and $q < s$, so $p + q < r + s$, so $r + s \notin \alpha + \beta$, and so $\alpha + \beta \neq \mathbb{R}$. So $\alpha + \beta$ satisfies D1.

For D2 and D3, let $r + s \in (\alpha + \beta)$, where $r \in \alpha$, $s \in \beta$. If $q < r + s$, then $q - s < r$, so $q - s \in \alpha$. And $q = (q - s) + s \in (\alpha + \beta)$. So D2 is satisfied. Finally, choose $t \in \alpha$ such that $r < t$. Then $r + s < t + s \in (\alpha + \beta)$. So D3 is satisfied.

(b) If $r \in \alpha$ and $s \in 0^*$, then $r + s < r$, and hence $r + s \in \alpha$, so $\alpha + 0^* \subseteq \alpha$. To show that $\alpha \subseteq \alpha + 0^*$, let $p \in \alpha$, and pick $r \in \alpha$ with $p < r$. Then $(p - r) \in 0^*$, and $p = (r + (p - r)) \in (\alpha + 0^*)$. So $\alpha + 0^* = \alpha$.

(c) To show that $-\alpha$ is a Dedekind cut, there is some $r \notin \alpha$ such that $r \neq \text{l.u.b. } \alpha$. So $-r \in \alpha$, and hence $\alpha \neq \varnothing$. Any $r \in \alpha$ is not in $-\alpha$, so $-\alpha \neq \mathbb{Q}$, and we have D1.

For D2, suppose that $p \in (-\alpha)$, and $q < p$. Then $-q > -p$, and $-q$ is an upper bound for α that is not the least upper bound of α. Hence, $q \in \alpha$.

For D3, if $p \in (-\alpha)$, then $-p$ is an upper bound for α and $p \neq \text{l.u.b. } \alpha$. So there is a $q < -p$ that is an upper bound for α. Let $r = q + {}^{(-p-q)}/_{2}$. Then $q < r < -p$, and r is an upper bound of α and $r \neq \text{l.u.b. } \alpha$. Hence, $p < r \in (-\alpha)$.

To show that $\alpha + (-\alpha) \subseteq 0^*$, suppose $p \in \alpha$ and $q \in (-\alpha)$. Then $p < -q$. So $p + q < 0$, and hence $(p + q) \in 0^*$. So $\alpha + (-\alpha) \subseteq 0^*$.

To show that $0^* \subseteq \alpha + (-\alpha) \subseteq$, let $p < 0$. We need to show that $p \in \alpha + (-\alpha)$. If $p \in \alpha$, then by the Archimedean property of the rationals there is a minimal $n \geq 0$ such that $p + (n+1)(-p)$ is an upper bound for α. So $p + n(-p) \in \alpha$, and $-(p + (n+1)(-p)) \in -\alpha$. So $p \in \alpha + (-\alpha)$. If $p \notin \alpha$, then there is a minimal $n \geq 1$ such that $(n+1)p \in \alpha$. Then $-(n+1)p \in -\alpha$, and $p + -(n+1)p \in \alpha + (-\alpha)$. So $p \in \alpha + (-\alpha)$.

(d) This follows because addition is associative in $\mathbb{Q}$.

(e) This follows because addition is commutative in $\mathbb{Q}$.

(f) By definition, $\alpha + \beta \subseteq \alpha + \gamma$. If $\alpha + \beta = \alpha + \gamma$, then by adding $-\alpha$ to both sides we would have $\beta = \gamma$, which is not true. So $\alpha + \beta < \alpha + \gamma$. ■

To define multiplication on $\mathbb{R}$, we can't use $\alpha \times \beta = \{p \cdot q : p \in \alpha \text{ and } q \in \beta\}$ because the product of two negative numbers is positive. Instead, we'll first define multiplication on the positive real numbers, $\mathbb{R}^+ = \{\alpha : 0^* < \alpha\}$ using the multiplication we have for the rationals.

For $\alpha, \beta \in \mathbb{R}^+$

$\alpha \times \beta \equiv_{\text{Def}} \{p : p \leq 0\} \cup \{p \cdot q : p \in \alpha, q \in \beta, \text{ and } 0 < p \text{ and } 0 < q\}$

$1^* \equiv_{\text{Def}} \text{D}(1) = \{p : p < 1\}$

For $\alpha \neq 0^*$,

$\alpha^{-1} \equiv_{\text{Def}} \{p : p \leq 0\} \cup \{1/p : p \text{ is an upper bound for } \alpha \text{ and } p \neq \text{l.u.b. } \alpha\}$

The proof of the following is sufficiently similar to the proof of Lemma C that I'll leave it to you. You'll need to note that because of the two-part definitions of multiplication and inverse, the proofs break into separate cases.

Lemma D For any $\alpha, \beta, \gamma \in \mathbb{R}^+$,

a. $\alpha \times \beta \in \mathbb{R}^+$

b. $\alpha \times 1^* = \alpha$

c. If $\alpha \neq 0$, then α^{-1} is a Dedekind cut and $\alpha \times \alpha^{-1} = 1^*$.

d. $\alpha \times (\beta \times \gamma) = (\alpha \times \beta) \times \gamma$

e. $\alpha \times \beta = \beta \times \alpha$

f. $\alpha \times (\beta + \gamma) = (\alpha \times \beta) + (\alpha \times \gamma)$

g. $0^* < \alpha \times \beta$

Now we can define multiplication on $\mathbb{R}$.

$$\alpha \times \beta \equiv_{\text{Def}} \begin{cases} \alpha \times \beta & \text{if } 0^* < \alpha \text{ and } 0^* < \beta \\ (-\alpha) \times (-\beta) & \text{if } \alpha < 0^* \text{ and } \beta < 0^* \\ -((-\alpha) \times \beta) & \text{if } \alpha < 0^* \text{ and } \beta > 0^* \\ -(\alpha \times (-\beta)) & \text{if } \alpha > 0^* \text{ and } \beta < 0^* \\ 0^* & \text{if } \alpha = 0^* \text{ or } \beta = 0^* \end{cases}$$

I'll let you show, using the identity $-(-\delta) = \delta$, that Lemma 15.a–e now hold for $\mathbb{R}$. Using that and Lemmas B.a, C, and D.g, we have the following.

Theorem E $\langle \mathbb{R}; +, \times, 0^*, 1^* \rangle$ is an ordered field.

I'll now show that an isomorphic image of the rationals is contained in the real numbers. Let $\mathbb{Q}^* \equiv_{\text{Def}} \{\text{D}(p) : p \in \mathbb{Q}\}$.

Lemma F

a. $\mathbb{Q}$ is isomorphic to $\mathbb{Q}^*$.

b. $\mathbb{Q}^*$ is dense in $\mathbb{R}$. That is, for any $\alpha < \beta$, there is some p such that $\alpha < \text{D}(p) < \beta$.

Proof (a) I'll first show that $\text{D}(p + q) = \text{D}(p) + \text{D}(q)$. If $r \in (\text{D}(p) + \text{D}(q))$, then $r = s + t$, where $s < p$ and $t < q$, so $r < p + q$, and hence $r \in \text{D}(p + q)$. In the other direction, if $r \in \text{D}(p + q)$, then $r < p + q$. Set $u = (p + q - r)/2$ and $s = p - u$, and

$t = q - u$. Then $s \in \mathrm{D}(p)$ and $t \in \mathrm{D}(q)$, and $r = s + t$, so $r \in (\mathrm{D}(p) + \mathrm{D}(q))$.

The proof that $\mathrm{D}(p \cdot q) = \mathrm{D}(p) \times \mathrm{D}(q)$ is similar, and I'll leave it to you.

To establish that $p < q$ iff $\mathrm{D}(p) < \mathrm{D}(q)$, first suppose that $p < q$. Then $p \in \mathrm{D}(q)$. But $p \notin \mathrm{D}(p)$. So $\mathrm{D}(p) < \mathrm{D}(q)$. If $\mathrm{D}(p) < \mathrm{D}(q)$, then there is some r such that $r \in \mathrm{D}(q)$ and $r \notin \mathrm{D}(p)$. Hence, $p \leq r < q$, so $p < q$. Thus, $\mathbb{Q}$ is isomorphic to $\mathbb{Q}^*$.

(b) If $\alpha < \beta$, then there is some p such that $p \in \beta$ and $p \notin \alpha$. Let r be such that $p < r$ and $r \in \beta$. Since $r \in \beta$, and $r \notin \mathrm{D}(r)$, we have $\mathrm{D}(r) < \beta$. Since $p \in \mathrm{D}(r)$ and $p \notin \alpha$, $\alpha < \mathrm{D}(r)$. Hence, $\mathbb{Q}^*$ is dense in $\mathbb{R}$. ■

We need to show that $\mathbb{R}$ is a real closed field. The lemmas needed for that require only that $\mathbb{R}$ is an ordered field satisfying the least upper bound principle (Lemma B.b); they will not rely on the particular definition we gave for $\mathbb{R}$.

To simplify the exposition, since we have an isomorphism of $\mathbb{Q}$ into $\mathbb{R}$, I am going to assume that $\mathbb{Q} \subset \mathbb{R}$. I'll write '+' in place of '+', '$\beta\gamma$' in place of '$\beta \times \gamma$', 0 in place of 0^*, and 1 in place of 1^*. I will also use the abbreviation β^n from p. 316 above. I'll invoke various equalities and inequalities that are either proved in lemmas earlier in the chapter for all ordered fields or follow easily from those. We also need some definitions.

The *absolute value* of α is $|\alpha| = \begin{cases} \alpha & \text{if } \alpha \geq 0 \\ -\alpha & \text{if } \alpha < 0 \end{cases}$

A function f from the real numbers to the real numbers is *continuous at* α means:

Given any $\varepsilon > 0$, there is a $\delta > 0$ such that for all γ such that $|\gamma| < \delta$, $|\mathsf{f}(\alpha + \gamma) - \mathsf{f}(\alpha)| < \varepsilon$.

If f and g are functions, define their sum and product by, for all α:

$(\mathsf{f} + \mathsf{g})(\alpha) = \mathsf{f}(\alpha) + \mathsf{g}(\alpha)$ the *sum* of f and g

$(\mathsf{f} \cdot \mathsf{g})(\alpha) = \mathsf{f}(\alpha) \cdot \mathsf{g}(\alpha)$ the *product* of f and g

I'll let you prove the following.

Lemma G

a. If f and g are continuous at α, then their sum and product are continuous at α.

b. The identity function, $\mathsf{f}(\alpha) = \alpha$ for all α, is continuous at all α.

c. Every constant function $\mathsf{f}(\alpha) = \beta$ for all α is continuous at all α.

d. Every polynomial is continuous at all α.

Lemma H Weierstrass' Nullstellensatz If f is continuous at every δ such that $\alpha \leq \delta \leq \beta$, and $\mathsf{f}(\alpha) < 0$ and $0 < \mathsf{f}(\beta)$, then there is some γ such that $\alpha < \gamma < \beta$ and $\mathsf{f}(\gamma) = 0$.

Proof Let $\mathsf{B} = \{\delta : \alpha \leq \delta \leq \beta \text{ and } \mathsf{f}(\delta) < 0\}$. Then $\mathsf{B} \neq \varnothing$ and B has an upper bound. Hence, by Lemma B.b, there is a least upper bound for B, call it γ, and $\alpha \leq \gamma \leq \beta$. It remains to show that $\mathsf{f}(\gamma) = 0$.

Suppose to the contrary that $\mathsf{f}(\gamma) > 0$. Then there is a δ such that for any λ, if $0 < \lambda < \delta$ then, $|\mathsf{f}(\gamma - \lambda) - \mathsf{f}(\gamma)| < \mathsf{f}(\gamma)$. Hence, since γ is the l.u.b. of B, $\mathsf{f}(\gamma) - \mathsf{f}(\gamma - \lambda) < \mathsf{f}(\gamma)$. And hence $0 < \mathsf{f}(\gamma - \lambda)$. So for all κ such that $\gamma - \delta < \kappa \leq \gamma$ we have $0 < \mathsf{f}(\kappa)$. But this contradicts that γ is the l.u.b. for B.

On the other hand, if $\mathsf{f}(\gamma) < 0$, then there is a δ such that for any λ, if $0 < \lambda < \delta$,

$f(\gamma+\lambda) - f(\gamma) < -f(\gamma)$. So $f(\gamma+\lambda) < 0$. But that contradicts that γ is an upper bound for B. Thus by trichotomy, we have that $f(\gamma) = 0$. And so $\gamma \neq \alpha$ and $\gamma \neq \beta$. ■

Lemma I If $\alpha > 0$, then there is some δ such that $\delta^2 = \alpha$.

Proof Consider the polynomial $p(x) = x^2 - \alpha$. We have $p(0) < 0$, and $p(1+\alpha) > 0$. Hence, for some δ such that $0 < \delta < 1 + \alpha$, $p(\delta) = 0$. That is, $\delta^2 = \alpha$ ■

Lemma J Every odd degree polynomial in $\mathbb{R}$ has a root in $\mathbb{R}$.

Proof First note: α is a root of the nth-degree polynomial

$$q(x) = \beta_0 + \beta_1 x + \cdots + \beta_n x^n$$

iff α is a root of a polynomial of the form

$$p(x) = \alpha_0 + \alpha_1 x + \cdots + \alpha_{n-1} x^{n-1} + x^n.$$

Hence, we need to prove the lemma only for such polynomials. For those I will show that there is some M such that $p(M) > 0$ and $p(-M) < 0$, from which the lemma will follow by the Nullstellensatz.

Let M be a real number such that $M > 1$ and $M > |\alpha_0| + |\alpha_1| + \cdots |\alpha_{n-1}|$. Then:

$$\begin{aligned} M^n &> (|\alpha_0| + |\alpha_1| + \cdots + |\alpha_{n-1}|)^n \\ &\geq |\alpha_0| + |\alpha_1| M + \cdots + |\alpha_{n-1}| M^{n-1} \\ &\geq |\alpha_0 + \alpha_1 M + \cdots + \alpha_{n-1} M^{n-1}| \end{aligned}$$

So $p(M) > 0$. Similarly, $p(-M) < 0$, since $(-M)^n < 0$, as n is odd. ■

We thus have that $\mathbb{R}$ is a real closed field. To show that $\mathbb{R}$ is a model of the second-order theory of real closed fields, we need to show that Dedekind's axiom holds.

Lemma K Given any nonempty subsets A, B of $\mathbb{R}$, such that for all $\alpha \in A$, $\beta \in B$, $\alpha \leq \beta$, there is a γ such that for all $\alpha \in A$, $\beta \in B$, $\alpha \leq \gamma$ and $\gamma \leq \beta$.

Proof Let γ = l.u.b. A. Then for all $\alpha \in A$, $\alpha \leq \gamma$. Since any $\beta \in B$ is an upper bound for A, $\gamma \leq \beta$. ■

By Theorem E and Lemmas I, J, and K, we now have the following.

Theorem L $\langle \mathbb{R}; +, \times, 0^*, 1^* \rangle$ is a model of $\mathbf{RCF_2}$.

Our proof of Theorem L used only that $\mathbb{R}$ is an ordered field satisfying the least upper bound principle. So we also have the following.

Theorem M $\mathbf{RCF_2} = \mathrm{Th}_2(\mathbf{OF} \wedge$ *the least upper bound principle*$)$

We've defined $\mathbb{R}$ and have shown that it has all the properties we invoked in the previous sections. But we could just as easily do the entire construction with '>' in place of '<' to obtain a model of $\mathbf{RCF_2}$. Or we could define the set of sequences of rationals as in Section A and show that it has all the properties we need, as *Mendelson, 1973* does; and there are many modifications of that definition that will serve, too. Or we could take the set of numbers defined in Euclidean plane geometry (Chapter XIX below). We have equally good reason to call any one of these 'the real numbers'. All we've done in this appendix is show that there is a model of $\mathbf{RCF_2}$; we have not defined the real numbers.

XVIII One-Dimensional Geometry

in collaboration with **Lesław Szczerba**

A. What Are We Formalizing?

Formalizing the geometry of flat surfaces has a long tradition. Euclid's *Elements* (ca. 300 B.C.) was already a summary of many generations' work (see *Heath*, 1956). Euclid, and others until quite recently, saw their work not as a formalization but as a discovery of truths and an attempt to reduce those many truths to a few from which all others could be deduced. But what were those truths about?

Points and lines are the basic "things" in geometry. Euclid defined them as:

A *point* is that which has no parts.
A *curve* is a length without width.
A *line* is a curve that lies evenly with respect to all its parts.

Whatever points, curves, and lines are, they certainly aren't something we can see in our daily lives. Often we make diagrams in geometry:
We refer to the triangle xyz, with sides xy, xz, and yz.
But no claim about geometry can be about this figure.
These lines have width. These points have size.

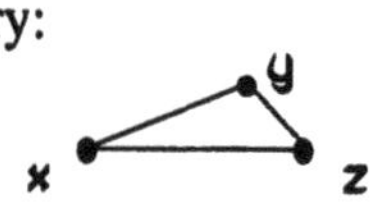

We cannot prove that between any two points like these there is always a third point.

Plato, and many others since, said that such pictures are only suggestions for our imagination, leading us to think of the abstract world of forms and "real" lines, points, and triangles. Such pictures are not part of geometry, but only aids to our perception of abstract objects that are not accessible to our senses.

A different view takes seriously the origin of geometry in plotting and measuring the earth ("geo"-"metry"). Such diagrams, it is said, are the basis of geometry in the sense that we begin with them and then abstract away what we consider inessential: the width of the lines, the size of the black dots that represent points. Plane geometry arises by abstraction from experience, and it is the art of abstracting—closely related to reasoning by analogy—that defines the nature of not only geometry but mathematics and much of the sciences (see *Epstein, 2004*).

Whichever view we take, it seems that we will not have the same confidence about our knowledge of points and lines that we have about the natural numbers. We will need more reason to think that the axiom systems we develop about points and lines are consistent.

It was René Descartes who provided a way to "see" geometry in terms of arithmetic, though not the arithmetic of the counting numbers. The Cartesian plane is now considered the fundamental interpretation of plane geometry, reducing the consistency of the axioms to the consistency of our theories of the real numbers. The real numbers can be defined and understood in terms of the natural numbers and subsets of those, as we have done in the previous chapter. We do not need to accept some new kinds of objects in order to show that plane geometry has a model.

But we will not start with the real numbers. Geometry provides another route to the real numbers, for in axiomatizing Euclidean plane geometry we will find that we can interpret the arithmetic of the real numbers within that theory.

To that end, we need to start with geometric notions. We could take as primitive two sorts of objects, points and lines, using a many-sorted logic as described in Chapter XIV.H.1 But that is inelegant and unnecessary. We can use just one type of variable meant to range over points. From points we will define all other geometric notions, such as lines, triangles, and angles, using just two fundamental notions:

Betweenness formalizing that one point lies between two others.

Congruence formalizing that two line segments have the same length, where a line segment is determined by the points at its ends.

In this chapter we'll axiomatize the geometry of the line. From the betweenness relation we will derive a dense linear ordering on the line. From the congruence relation we will define an addition of points, which will allow us to interpret the theory of divisible groups. Combining the theories of betweenness and congruence, we will have a theory equivalent to the theory of ordered divisible groups. By extending that theory to a second-order one using a betweenness version of Dedekind's axiom, we will get a categorical axiomatization of the geometry of the real line.

But we cannot define multiplication of real numbers in our first-order theory of the geometry of the line. For that we need an axiomatization of two-dimensional geometry, which we'll do in the next chapter. At the end of the next chapter I'll discuss the history of modern formalizations of geometry.

B. The One-Dimensional Theory of Betweenness

1. An axiom system for betweenness, B1

For the formal language of betweenness, we can take $L(=; P_0^3)$. For legibility I'll write $B(x, y, z)$ for $P_0^3(x, y, z)$, meant to formalize *y lies between x and z*.

I'll write $x, y, z, w, r, s, t, p, q$ for $x_1, x_2, x_3, x_4, x_5, x_6, x_7, x_8, x_9$.

B1, the *theory of one-dimensional betweenness* in $L(=; B)$

B1-1 $\exists x \exists y \; x \neq y$

B1-2 $\forall x \forall y \; B(x, y, x) \rightarrow x = y$ *the one-point set is convex*

B1-3 $\forall x \forall y \forall z \; B(x, y, z) \rightarrow B(z, y, x)$ *direction is irrelevant*

B1-4 $\forall x \forall y \forall z \forall w \; B(x, y, z) \wedge B(x, z, w) \rightarrow B(y, z, w)$ *transitivity*

B1-5 $\forall x \forall y \exists z \; y \neq z \wedge B(x, y, z)$ *extension axiom*

B1-6 $\forall x \forall z \; x \neq z \rightarrow \exists y \, (x \neq y \wedge y \neq z \wedge B(x, y, z))$ *density*

B1-7 $\forall x \forall y \forall z \; B(x, y, z) \vee B(y, z, x) \vee B(z, x, y)$ *dimension axiom*

Our goal now is to show how to interpret **B1** within the theory of dense linear orderings, **DLO**, and to interpret **DLO** within **B1**.

Exercises for Sections A and B.1

1. Describe two different views of what points and lines are.
2. How does the choice of view of what points and lines are affect how one might justify that an axiom system is consistent?
3. Write out axiom *B1-3* in the fully formal language.
4. Draw a picture for axiom *B1-4*.
5. Which axioms of **B1** are true of the plane as well as of the line, understanding $B(x, y, x)$ to assert that y lies between x and z on the line through x and z?

2. Some basic theorems of B1

In this chapter we'll need to show that certain formulas are theorems of our theories. Syntactic derivations are long and difficult to follow, so I will give semantic proofs instead. This is how mathematicians typically work (compare the discussion in Chapter XII.A, p. 196). I will show that a wff is true in every model of the theory; then by the completeness of first-order logic the wff is a theorem. To conserve space I'll write these semantic proofs in smaller type, leaving without comment many proofs as exercises. A statement of a wff in a lemma is meant to indicate that the wff is a theorem of the theory.

A model of **B1** will be $\langle \mathsf{U}; \mathsf{B} \rangle$. When working within a model I'll designate points of the universe as x, y, z, w, etc.

Lemma 1

a. $\forall x\, \forall y\ \ B(x, x, y)$

b. $\forall x\, \forall y\ \ B(x, y, y)$

c. $\forall x\, \forall y\, \forall z\ (x = y \wedge y = z \rightarrow B(x, y, z))$

d. $\forall x\, \forall y\, \forall z\ \ B(x, y, z) \wedge B(x, z, y) \rightarrow y = z$

e. $\forall x\, \forall y\, \forall z\ \ B(x, y, z) \wedge B(y, x, z) \rightarrow x = y$

f. $\forall x\, \forall y\, \forall z\, \forall w\ \ B(x, y, w) \wedge B(y, z, w) \rightarrow B(x, y, z)$

g. $\forall x\, \forall y\, \forall z\, \forall w\ \ B(x, y, z) \wedge B(x, z, w) \rightarrow B(x, y, w)$

h. $\forall x\, \forall y\, \forall z\, \forall w\ \ B(x, y, w) \wedge B(y, z, w) \rightarrow B(x, z, w)$

Proof (a) By *B1-7*, for every $\mathsf{x}, \mathsf{y} \in \mathsf{U}$, (i) $\mathsf{B}(\mathsf{x}, \mathsf{x}, \mathsf{y})$ or (ii) $\mathsf{B}(\mathsf{x}, \mathsf{y}, \mathsf{x})$ or (iii) $\mathsf{B}(\mathsf{y}, \mathsf{x}, \mathsf{x})$. If (ii), by axiom *B1-2* we have $\mathsf{x} = \mathsf{y}$, and we're done by B1-7. For (iii), use *B1-3*. For (b), by (a) we have $\mathsf{B}(\mathsf{y}, \mathsf{y}, \mathsf{x})$, so by *B1-3*, we have $\mathsf{B}(\mathsf{x}, \mathsf{y}, \mathsf{y})$. Part (c) follows from (a), since $\mathsf{B}(\mathsf{x}, \mathsf{y}, \mathsf{z})$ iff $\mathsf{B}(\mathsf{x}, \mathsf{y}, \mathsf{y})$. ■

Lemma 2 *The line is smooth*

a. $\forall x\, \forall y\, \forall z\, \forall w \quad x \neq y \wedge \mathrm{B}(x, y, z) \wedge \mathrm{B}(x, y, w) \rightarrow (\mathrm{B}(x, z, w) \vee \mathrm{B}(x, w, z))$

b. $\forall x\, \forall y\, \forall z\, \forall w \quad x \neq y \wedge \mathrm{B}(x, y, z) \wedge \mathrm{B}(x, y, w) \rightarrow (\mathrm{B}(y, z, w) \vee \mathrm{B}(y, w, z))$

c. $\forall x\, \forall y\, \forall z\, \forall w \quad \mathrm{B}(x, y, z) \wedge \mathrm{B}(x, w, z) \rightarrow \mathrm{B}(z, y, w) \vee \mathrm{B}(z, w, y)$

d. $\forall x\, \forall y\, \forall z\, \forall w \quad (y \neq z \wedge \mathrm{B}(x, y, z) \wedge \mathrm{B}(y, z, w)) \rightarrow \mathrm{B}(x, y, w)$

e. $\forall x\, \forall y\, \forall z\, \forall w \quad (y \neq z \wedge \mathrm{B}(x, y, z) \wedge \mathrm{B}(y, z, w)) \rightarrow \mathrm{B}(x, z, w)$

Proof (a) By *B1-7*, (i) $\mathrm{B}(x, z, w)$ or (ii) $\mathrm{B}(z, w, x)$ or (iii) $\mathrm{B}(w, x, z)$. If (i), we're done. If (ii), by *B1-3* we have $\mathrm{B}(x, w, z)$ and we are done. If (iii), since $\mathrm{B}(x, y, w)$, by *B1-3* we have $\mathrm{B}(w, y, x)$. From $\mathrm{B}(w, x, z)$ by *B1-4* we have $\mathrm{B}(y, x, z)$. From this and the assumption that $\mathrm{B}(x, y, z)$ by *B1-3* we have $\mathrm{B}(z, x, y)$ and $\mathrm{B}(z, y, x)$. But then by Lemma 1.d, we have $x = y$, a contradiction.

(b) By (a) we have (i) $\mathrm{B}(x, z, w)$ or (ii) $\mathrm{B}(x, w, z)$. If (i), by *B1-4* we have $\mathrm{B}(y, z, w)$. If (ii), by *B1-4* we have $\mathrm{B}(y, w, z)$.

(c) By *B1-5* there is a p such that $\mathrm{B}(x, z, p)$ and $z \neq p$. From $\mathrm{B}(x, y, z)$ and $\mathrm{B}(x, w, z)$, by *B1-4* we have $\mathrm{B}(y, z, p)$ and $\mathrm{B}(w, z, p)$, so by *B1-3*, $\mathrm{B}(p, z, y)$ and $\mathrm{B}(p, z, w)$. Since $z \neq p$, by (b) we have $\mathrm{B}(z, y, w)$ or $\mathrm{B}(z, w, y)$. ∎

Lemma 3 *Any four points are in a segment*

$\forall x\, \forall y\, \forall z\, \forall w\, \exists r\, \exists s \quad (\mathrm{B}(r, x, s) \wedge \mathrm{B}(r, y, s) \wedge \mathrm{B}(r, z, s) \wedge \mathrm{B}(r, w, s))$

Proof From B1-7 we have $\mathrm{B}(x, y, z)$ or $\mathrm{B}(y, z, x)$ or $\mathrm{B}(z, x, y)$. We can assume $\mathrm{B}(x, y, z)$ and the other parts follow by relabeling. Using *B1-7* again, we have (i) $\mathrm{B}(x, z, w)$ or (ii) $\mathrm{B}(z, w, x)$ or (iii) $\mathrm{B}(w, x, z)$. If (i), using Lemma 1.g we may take x for r and w for s. If (ii), we take x for r and z for s. If (iii), we can use Lemma 1.h and take w for r and z for s. ∎

3. Vectors in the same direction

To give a direction on the line, we first use a four-place betweenness operation.

$\mathrm{B}(x, y, z, w) \equiv_{\mathrm{Def}} \mathrm{B}(x, y, z) \wedge \mathrm{B}(x, z, w)$

Lemma 4

a. $\forall x\, \forall y\, \forall z\, \forall w \quad \mathrm{B}(x, y, z, w) \leftrightarrow (\mathrm{B}(x, y, z) \wedge \mathrm{B}(y, z, w))$

b. $\forall x\, \forall y \quad \mathrm{B}(x, x, y, y)$

c. $\forall x\, \forall y\, \forall z\, \forall w \quad \mathrm{B}(x, y, z, w) \leftrightarrow \mathrm{B}(w, z, y, x)$

d. $\forall x\, \forall y\, \forall z\, \forall w\, \forall r\, \forall s \quad \mathrm{B}(x, y, z, w) \wedge \mathrm{B}(y, r, s, z) \rightarrow \mathrm{B}(x, r, s, w)$

e. $\forall x\, \forall y\, \forall z\, \forall r\, \forall s \quad \mathrm{B}(r, x, y, s) \wedge \mathrm{B}(r, y, z, s) \rightarrow \mathrm{B}(r, x, z, s)$

f. $\forall x\, \forall y\, \forall z\, \forall w \quad \mathrm{B}(x, y, w) \wedge \mathrm{B}(x, z, w) \rightarrow \mathrm{B}(x, y, z, w) \vee \mathrm{B}(x, z, y, w)$

Proof (a) By Lemma 1. (b) By Lemma 1. (c) By (a) and *B1-3*.

(d) Assuming B(x, y, z, w) and B(y, r, s, z), by definition we have (i) B(x, y, z), (ii) B(x, z, w), and (iii) B(y, r, s), and (iv) B(y, s, z). We need B(x, r, s) and B(x, s, w). For the former, by (i) and (iv) and Lemma 1.h, B(x, s, w). Using that and (iii), by Lemma 1.h we are done. To establish B(x, s, w), by (i) and (iv) and Lemma 1.h we have (v) B(x, s, z). By (v) and (ii) we have, by Lemma 1.g, B(x, s, w). Part (e) is proved similarly.

(f) Given B(x, y, w) and B(x, z, w), by B1-3, B(w, y, x) and B(w, z, x). So by Lemma 2.c we have (i) B(x, y, z) or (ii) B(x, z, y). If (i), then B(x, y, z, w). If (ii), then B(x, z, y, w). ■

Now we can define directions.

Two vectors are in the same direction

$xy \uparrow zw \equiv_{\text{Def}} \exists r \exists s \; \text{B}(r, x, y, s) \wedge \text{B}(r, z, w, s)$

The idea is that in a model, $xy \uparrow zw$ iff the segment pointing from x to y is in the same direction as the segment pointing from z to w.

Lemma 5

a. $\forall x \forall y \; (xy \uparrow xy)$

b. $\forall x \forall y \; (xy \uparrow yx \leftrightarrow x = y)$

c. $\forall x \forall y \forall z \forall w \; (xy \uparrow zw \rightarrow zw \uparrow xy)$

d. $\forall x \forall y \forall z \forall w \; (xy \uparrow zw \rightarrow yx \uparrow wz)$

e. $\forall x \forall y \forall z \; (xy \uparrow yz \rightarrow xy \uparrow xz)$

f. *Any two vectors have the same or opposite direction*
$\forall x \forall y \forall z \forall w \; (xy \uparrow zw \vee xy \uparrow wz)$

Proof For part (f) Use Lemma 3 and Lemma 4.f. ■

Lemma 6 *Transitivity*

$\forall x \forall y \forall z \forall w \forall r \forall s \; (xy \uparrow rs \wedge zw \uparrow rs \wedge r \neq s \rightarrow xy \uparrow zw)$

Proof Assume $xy \uparrow rs$, and $zw \uparrow rs$, and $r \neq s$. Then there are points x', y', z', and w' such that:

(i) B(x′, x, y, y′) and B(x′, r, s, y′)

(ii) B(z′, z, w, w′) and B(z′, r, s, w′)

By Lemma 5.f we have (iii) $x'y' \uparrow z'w'$ or (iv) $xy \uparrow w'z'$. If (iii), there are points p, p′ with B(p, x′, y′, p′) and B(p, z′, w′, p′). So from (i) and (ii), we get by Lemma 4.d, B(p, x, y, p′) and B(p, z, w, p′), that is $xy \uparrow zw$. If (iv) we get similarly that there are points q, q′ such that B(q, x, y, q′) and B(q, w′, z′, q′). So B(q, r, s, q′) and B(q, s, r, q′). But then $rs \uparrow sr$, which by Lemma 5.b means r = s, a contradiction. So we have (i), and we're done. ■

4. An ordering of points and $\mathbf{B1}_{01}$

In the last section we saw how to say that two vectors point in the same direction. But what direction that is we cannot say, for the line as we have it has no inherent direction. In any model we can give it a standard direction: Pick out two points (by axiom *B1-1*), 0 and 1, and stipulate the direction to be from 0 to 1.

In the formal language we will add two new names, c_0, c_1, which I'll write as 'o' and 'ł'. These parameters will not be definable, but we'll see shortly how we can eliminate them.

$\mathbf{B1}_{01}$, *one-dimensional betweenness with parameters* in L(=; B, o, ł)

B1

o ≠ ł

Now we can define an order relation on the line.

An order relation on the line

$x < y \equiv_{\text{Def}} xy \uparrow$ oł $\wedge\ x \neq y$

A model of **B1** will be <U; B, 0, 1>, in which I'll write the interpretation of the defined predicate < as <, with ≤ and ≤ the usual abbreviations.

Lemma 7

a. $\forall x \neg (x < x)$

b. $\forall x \forall y \ \ x < y \rightarrow \neg (y < x)$

c. $\forall x \forall y \forall z \ \ x < y \wedge y < z \rightarrow x < z$

d. $\forall x \forall y \ \ x < y \vee y < x \vee x = y$

e. o < ł

Proof (b) By Lemma 5.b. (c) If x < y and y < z, then xy↑o1 and yz↑01, so by Lemma 6, xy↑yz. By Lemma 5.e we have xy↑xz. So xz↑01. And x ≠ z by Lemma 5.b. (d) By Lemma 5.f. ∎

Theorem 8 *< is a dense linear ordering without endpoints*

If <U; B, 0, 1>⊨**B1**, the defined predicate < is a dense linear ordering and U is infinite.

Proof By Lemma 7, < is a linear ordering. That it is dense without endpoints comes from axioms *B1-6* and *B1-5*. That U is infinite then follows because every dense linear ordering is infinite. ∎

Theorem 9 *Betweenness in terms of the ordering*
$\forall x\, \forall y\, \forall z\ \ B(x, y, z) \leftrightarrow (x \leq y \leq z) \vee (z \leq y \leq x)$

Proof If $x \leq y$ and $y \leq z$, then $xy \uparrow 01$ and $yz \uparrow 01$. Thus $xy \uparrow yz$. So there are p, q such that $B(p, x, y, q)$ and $B(p, y, z, q)$. So $B(x, y, q)$ and $B(y, z, q)$. So by Lemma 1.f, $B(x, y, z)$.

Suppose $B(x, y, z)$. By Lemma 7.d, (i) $x \leq z$ or (ii) $z \leq x$. If (i), then by Lemma 7.d, we have (iii) $y \leq x \leq z$ or (iv) $x \leq y \leq z$ or (v) $x \leq z \leq y$. If (iv), we are done. If (iii), then by what we just proved, $B(y, x, z)$ and if (v) we have $B(x, z, y)$. So by Lemma 1, we have $x = y$ or $y = z$ and we are done. Case (ii) is proved similarly. ■

5. Translating between B1 and the theory of dense linear orderings

In this section I'll show how to translate between the theory of one-dimensional betweenness and the theory of dense linear orderings without endpoints, **DLO**. We first map **B1** in L(=; B) to **DLO** in L(=; <). I'll write $x \leq y \leq z$ for $x \leq y \wedge y \leq z$.

A mapping $\varphi : L(=; B) \rightarrow L(=; <)$ is given by:

$\varphi(B(x, y, z))$ is $(x \leq y \leq z) \vee (z \leq y \leq x)$

$\neg, \rightarrow, \wedge, \vee, \forall, \exists$, and $=$ are translated homophonically

---

To show that this is an equivalence-relation translation (Chapter XVI.B) we'll look at the induced mapping of structures.

In a model $N = \langle U; < \rangle$ of **DLO**, set:

$B_{\leq}(a, b, c) \equiv_{\text{Def}}\ a \leq b \leq c$ or $c \leq b \leq a$

$B(N) \equiv_{\text{Def}} \langle U; B_{\leq} \rangle$

$$\begin{array}{ccc} L_{\mathbf{B1}} & \xrightarrow{\ \varphi\ } & L_{\mathbf{DLO}} \\ \text{models of } \mathbf{B1} & \xleftarrow{\ B\ } & \text{models of } \mathbf{DLO} \end{array}$$

Lemma 10 If $N \vDash$ **DLO**, then $B(N) \vDash$ **B1**.

Proof Just check that each axiom of **B1** holds in $B(N)$. ■

To show that φ is an equivalence-relation translation we need to show that for every model M of **B1**, there is some model N of **DLO** such that $M = B(N)$. But because of the parameters we need to take two steps to define the ordering.

A mapping ψ: L(=; <) → L(=; B, o, ł) is given by:

$\psi(x < y)$ is $x < y$

¬, →, ∧, ∨, ∀, ∃, and = are translated homophonically

In a model $M = \langle U; B \rangle$ of **B1** and $0 \neq 1 \in U$, set:

$a <_B b \equiv_{Def} a < b$ — That is, $a <_B b$ iff $a \neq b$ and for some p, q, B(p, a, b) and B(p, b, q) and B(p, 0, 1) and B(p, 1, q).

$O(M)_{0,1} \equiv_{Def} \langle U; <_B \rangle$

L_{DLO} —ψ→ $L_{B1_{01}}$

models of **DLO** ←$O(\;)_{01}$— models of $\mathbf{B1_{01}}$

Lemma 11 If $M \models$ **B1** and $0 \neq 1$ are elements of the universe of M, then $O(M)_{0,1} \models$ **DLO**.

Theorem 12 *The representation theorem for* **B1** *and* **DLO**

a. If $N \models$ **DLO**, then there is a model $M \models$ **B1** such that $O(M)_{0,1} = N$.
In particular, if $N = \langle U; < \rangle$, and $0, 1 \in U$ and $0 < 1$, then $O(B(N))_{0,1} = N$.

b. If $M \models$ **B1**, then there is a model $N \models$ **DLO** such that $B(N) = M$.
In particular, if $M = \langle U; B \rangle \models$ **B1** and $0 \neq 1$ are in U, then $B(O(M)_{0,1}) = M$.

Proof (a) We start with < and define ≤. Then B(a, b, c) iff $a \leq b \leq c$ or $c \leq b \leq a$. Then by definition, $a <_B b$ iff $ab \uparrow 01$ and $a \neq b$ iff there are p, q such that B(p, a, b) and B(p, b, q) and B(p, 0, 1) and B(p, 1, q) and $a \neq b$. But that's iff there are p, q such that ($p \leq a \leq b$ and $p \leq a \leq q$ and $p \leq 0 \leq 1$ and $p \leq 1 \leq q$) or ($b \leq a \leq p$ and $q \leq a \leq p$ and $1 \leq 0 \leq p$ and $q \leq 1 \leq p$) and $a \neq b$. Since $0 < 1$, it must be the first, so $a < b$.

(b) We start with B. Then $a \leq b$ iff $ab \uparrow 01$. Then $B_{\leq}(a, b, c)$ iff $a \leq b \leq c$ or $c \leq b \leq a$. So by Theorem 9, $B_{\leq}(a, b, c)$ iff B(a, b, c). ■

In part (a) of the theorem, if we do not require that $0 < 1$, then we have only that $O(B(N))_{0,1}$ is isomorphic to N, for the ordering can be reversed if $1 < 0$, as you can note in the proof. That is:

If $M = \langle U; B \rangle \models$ **B1**, and $0 \neq 1$ and $0' \neq 1'$, then $O(B(N))_{0,1} \simeq O(B(N))_{0',1'}$.

If in addition $01 \uparrow 0'1'$, then $O(B(N))_{0,1} = O(B(N))_{0',1'}$.

But part (b) shows that all choices of $0 \neq 1$ yield the same result if we start with B.

***Corollary 13** Completeness, decidability, and categoricity of* **B1**

a. **B1** is $\aleph_0$-categorical, complete, and decidable.

b. There are uncountably many non-isomorphic models of **B1** of cardinality c.

Proof (a) Given any two countable models M_1, M_2 of **B1**, there are countable models N_1, N_2 of **DLO** such that $B(N_1) = M_1$ and $B(N_2) = M_2$. By Theorem XIII.5, $N_1 \simeq N_2$, and hence $B(N_1) \simeq B(N_2)$. So **B1** is $\aleph_0$-categorical. Since every model of **B1** is infinite (Theorem 8), by Theorem XIII.8 **B1** is complete. Since **B1** complete and axiomatizable, it's decidable (Theorem XV.8). Part (b) follows from Exercise 10 of Chapter XIII.C. ■

Now we eliminate parameters from $\mathbf{B1}_{01}$.

A mapping $\rho : L(=; B, o, ł) \to L(=; B)$ is given by:

$\rho(A)$ is $\forall x \forall y\, (x \neq y \to A(x/o, y/ł))$
where x and y are the least variables not appearing in A

$$L_{DLO} \xrightarrow{\ \psi\ } L_{\mathbf{B1}_{01}} \xrightarrow{\ \rho\ } L_{\mathbf{B1}}$$

$$\text{models of } \mathbf{DLO} \xleftarrow[\ O(\ )_{01}\ ]{} \text{models of } \mathbf{B1}_{01}$$

In Section E below I'll show that this is a translation. That is, $\Gamma \vDash_{\mathbf{B1}_{01}} A$ iff $\rho(\Gamma) \vDash_{\mathbf{B1}} \rho(A)$. So we can invoke Theorem XVI.4 to show that we have translations.

***Theorem 14** The translation theorem for* **B1** *and* **DLO**

a. The mappings φ and ψ are equivalence-relation translations.

b. $\Gamma \vDash_{\mathbf{B1}} A$ iff $\varphi(\Gamma) \vDash_{\mathbf{DLO}} \varphi(A)$.

c. $\Gamma \vDash_{\mathbf{DLO}} A$ iff $(\rho \circ \psi)(\Gamma) \vDash_{\mathbf{B1}} (\rho \circ \psi)(A)$.

6. The second-order theory of betweenness

For a categorical theory of one-dimensional betweenness, we add axioms corresponding to the second-order axioms we added to **DLO** in Chapter XVII.C.

Countable density-B

$\exists X\, (\text{Count}(X) \wedge \forall x \forall y\ x \neq y \to \exists z\, (X(z) \wedge B(x, z, y))$

Dedekind's axiom-B *(continuity axiom for the line)*

$\forall \ldots \forall X \forall Y\, ([\exists x \exists y\, (X(x) \wedge Y(y)) \wedge \exists w \forall x \forall y\, (X(x) \wedge Y(y) \to B(w, x, y))]$
$\quad \to \exists z \forall x \forall y\, (X(x) \wedge Y(y) \to B(x, z, y)))$

Theorem 15

a. If $\mathsf{M} \models_2 (\mathbf{B1} \wedge \textit{countable density-B})$, and $\mathsf{0} \neq \mathsf{1}$ are elements of the universe of M, then $O(\mathsf{M})_{\mathsf{0},\mathsf{1}} \models_2 \textit{countable density}$.

b. If $\mathsf{N} \models (\mathbf{DLO} \wedge \textit{countable density})$, then $B(\mathsf{N}) \models \textit{countable density-B}$.

c. If $\mathsf{M} \models_2 (\mathbf{B1} \wedge \textit{Dedekind's axiom-B})$, and $\mathsf{0} \neq \mathsf{1}$ are elements of the universe of M, then $O(\mathsf{M})_{\mathsf{0},\mathsf{1}} \models_2 \textit{Dedekind's axiom}$.

d. If $\mathsf{N} \models_2 (\mathbf{DLO} \wedge \textit{Dedekind's axiom})$, then $B(\mathsf{N}) \models_2 \textit{Dedekind's axiom-B}$.

B1*, the *second-order theory of one-dimensional betweenness* in $L_2(=; B)$

B1
countable density-B
Dedekind's axiom-B

Recall that $\mathbf{RLO_2}$ is the second-order categorical theory of the ordering of the real line (Chapter XVII.C, p. 311). That is, $\mathbf{RLO_2}$ is to **DLO** as **B1*** is to **B1**. I'll let you prove the following, using Theorem XVII.8 and Corollary XV.29.

Theorem 16

a. $\Gamma \models_{2\,\mathbf{B1*}} A$ iff $\varphi(\Gamma) \models_{2\,\mathbf{RLO_2}} \varphi(A)$

b. $\models_{2\,\mathbf{RLO_2}} A$ iff $\models_{2\,\mathbf{B1*}} (\rho \circ \psi)(A)$

c. **B1*** is categorical, and every model is isomorphic to $B(\mathbb{R}; <)$, where $<$ is the usual ordering of the reals.

d. **B1*** is undecidable.

Exercises for Section B

1. Prove in $\mathbf{B1_{01}}$:
   a. $\forall x\, \forall y\, \forall z\ (B(x, y, z) \wedge y \lesssim z) \rightarrow x \lesssim y$
   b. $\forall x\, \forall y\, \forall z\ (B(x, y, z) \wedge x \lesssim y) \rightarrow y \lesssim z$
2. Write out the translation $\psi : L(=; <) \rightarrow L(=; B, \mathsf{o}, \mathsf{ł})$ in primitive notation.
3. Prove: $\sigma \models_{\mathsf{M}} B(x, y, z)$ iff $\sigma \models_{O(\mathsf{M})_{\mathsf{0},\mathsf{1}}} x \leq y \leq z \vee z \leq y \leq x$.
4. Prove that **B1** is complete and decidable directly from Theorem 14 and not the representation theorem. What advantage does this have for showing decidability?
5. *The axioms of* **B1** *are independent*
   Show that each axiom of **B1** is independent (Chapter II.D.3) by showing that it fails in the designated model while all the other axioms hold in that model:
   B1-1 $\langle \varnothing; \varnothing \rangle$

B1-2	$<\mathsf{U};\mathsf{U}^3>$	$\mathsf{U} = \{0, 1, 2\}$
B1-3	$<\mathsf{R};\mathsf{B}>$	$\mathsf{B}(\mathsf{a}, \mathsf{b}, \mathsf{c})$ iff $(\mathsf{a} \leq \mathsf{b} \leq \mathsf{c}$ or $\mathsf{c} \leq \mathsf{b} \leq \mathsf{a})$ or $(\mathsf{b} \leq \mathsf{a} \leq 0 \leq \mathsf{c}$ or $\mathsf{c} \leq 0 \leq \mathsf{a} \leq \mathsf{b})$
B1-4	$<\mathsf{U};\mathsf{B}>$	$\mathsf{U} = \{0, 1, 2\}$ and $\mathsf{B}(0, 1, 2)$, $\mathsf{B}(0, 2, 1)$, but not $\mathsf{B}(1, 2, 1)$.
B1-5	$B([0, 1]; <)$	$[0, 1]$ is the unit interval of R and $<$ is the usual ordering
B1-6	$B(\mathsf{Z}; <)$	Z is the set of integers and $<$ is its usual ordering
B1-7	$<\mathsf{R}^2;\mathsf{B}>$	$\mathsf{B}(\mathsf{a}, \mathsf{b}, \mathsf{c})$ iff for some t such that $0 \leq \mathsf{t} \leq 1$, $\mathsf{b} = \mathsf{a}(1-\mathsf{t}) + \mathsf{c}\mathsf{t}$ (This is why we call *B1-7* 'the dimension axiom'.)

6. a. Show that in **B1*** Dedekind's axiom is independent.
   (Hint: Consider $B(<\mathsf{Q}; <>)$.)
   b. Show that in **B1*** the axiom of countable density is independent.
   (Hint: Take $B(\mathsf{N})$, where N is the model given in Exercise 4 of Chapter XVII.C.)

C. The One-Dimensional Theory of Congruence

1. An axiom system for congruence, C1

Our formal language will be $L(=; P_0^4)$. For legibility I'll write $xy \equiv zw$ in place of $P_0^4(x, y, w, z)$, which is meant to formalize *the line segment xy is congruent to (has the same length as) the line segment wz*.

C1, a *theory of one-dimensional congruence* in $L(=; \equiv)$

C1-1 $\exists x \exists y\ (x \neq y)$

C1-2 $\forall x \forall y\ (xy \equiv yx)$ — *reflexivity*

C1-3 $\forall x \forall y\ (xy \equiv zz \rightarrow x = y)$

C1-4 $\forall x \forall y \forall z \forall w \forall r \forall s\ (xy \equiv rs \wedge zw \equiv rs \rightarrow xy \equiv zw)$ — *transitivity*

C1-5 $\forall x \forall z \exists y\ (xy \equiv yz)$ — *midpoint*

C1-6 $\forall x \forall y\ (x \neq y \rightarrow \exists z\,(x \neq z \wedge xy \equiv yz))$ — *the extension axiom*

C1-7 $\forall x \forall y \forall z \forall w \forall r\ (x \neq z \wedge y \neq w \wedge rx \equiv rz \wedge ry \equiv rw \rightarrow xy \equiv zw)$ — *the dimension axiom*

A model of **C1** will be $\mathsf{M} = <\mathsf{U}; \approx>$.

Our goal now will be to establish an equivalence between **C1** and the theory of 2-divisible groups. Then we'll add axioms to **C1** to get a theory equivalent to the theory of divisible groups, **GD**.

Lemma 17

a. $\forall x \forall y\ \ xy \equiv xy$

b. $\forall x \forall y \forall z \forall w\ \ xy \equiv zw \rightarrow zw \equiv xy$

c. $\forall x \forall y \forall z \forall w\ \ xy \equiv zw \rightarrow xy \equiv wz$

d. $\forall x \forall y \forall z \forall w\ \ xy \equiv zw \rightarrow yx \equiv zw$

e. $\forall x \forall y \forall z \forall w\ \ xy \equiv zw \rightarrow zw \equiv yx$

f. $\forall x \forall y \forall z \forall w\ \ xy \equiv zw \rightarrow zw \equiv yx$

g. $\forall x \forall y\ \ xx \equiv yy$

h. *There are at most two points equidistant from a given point*
$\forall x \forall y \forall z \forall w\ (xy \equiv xz \wedge xy \equiv xw) \rightarrow (y = z \vee z = w \vee w = y)$

Proof Parts (a)–(f) require only axioms C1-2 and C1-4. (g) There are two cases. If $\mathbf{x} = \mathbf{y}$, then by axiom C1-2 we are done. If $\mathbf{x} \neq \mathbf{y}$, then by axiom C1-5, there is a $\mathbf{z}$ such that $\mathbf{xz} \approx \mathbf{zy}$. By (d) and axiom C1-7 (if $\mathbf{x} \neq \mathbf{y}$ and $\mathbf{x} \neq \mathbf{y}$ and $\mathbf{zx} \approx \mathbf{zy}$ and $\mathbf{zx} \approx \mathbf{zy}$, then $\mathbf{xx} \approx \mathbf{yy}$), we have (g). For (h), assume we are given $\mathbf{xy} \approx \mathbf{xz} \approx \mathbf{xw}$, and further that $\mathbf{y} \neq \mathbf{z}$ and $\mathbf{z} \neq \mathbf{w}$. By (b) and axiom C1-7, $\mathbf{yz} \approx \mathbf{ww}$, so by axiom C1-3, $\mathbf{w} = \mathbf{y}$. ∎

2. Point symmetry

We now define a midpoint relation.

---

y is the midpoint of the segment xz

$\mathrm{M}(x, y, z) \equiv_{\mathrm{Def}} (x = y \wedge y = z) \vee (x \neq z \wedge xy \equiv yz)$

I'll write $\mathfrak{m}$ for the interpretation of M in a model. I'll leave the following to you, noting only that (e) follows by *C1-7*.

Lemma 18

a. $\forall x \forall y \forall z\ \ \mathrm{M}(x, y, z) \leftrightarrow \mathrm{M}(z, y, x)$

b. $\forall x \forall y\ \ \mathrm{M}(x, y, y) \leftrightarrow x = y$

c. $\forall x \forall y\ \ \mathrm{M}(x, y, x) \leftrightarrow x = y$

d. $\forall x \forall y \forall z\ \ \mathrm{M}(x, y, z) \rightarrow xy \equiv yz$

e. $\forall x \forall y \forall z \forall w \forall r\ \ \mathrm{M}(x, r, z) \wedge \mathrm{M}(y, r, w) \rightarrow xy \equiv zw$

Lemma 19 *The midpoint relation is a function of its first two variables*
$\forall x \forall y \exists! z\ \mathrm{M}(x, y, z)$

Proof For existence, use *C1-6*. For uniqueness, if $\mathfrak{m}(\mathbf{x}, \mathbf{y}, \mathbf{z})$ and $\mathfrak{m}(\mathbf{x}, \mathbf{y}, \mathbf{w})$, then by Lemma 18.e, $\mathbf{xx} \approx \mathbf{zw}$, and hence by *C1-3* and Lemma 17.b, $\mathbf{z} = \mathbf{w}$. ∎

By Lemma 19 we can introduce a function symbol for $\mathrm{M}(x, y, -)$.

z is the symmetrical image of x with respect to y

$S_y(x) = z \equiv_{Def} M(x, y, z)$

In a model, any point y can be taken as reference for y, and then the interpretation of S_y in a model is a function of a single variable, which we can write as $\mathsf{S_y}$, *the point symmetry with origin* y. I'll write $\mathsf{S_x S_y}$ for $\mathsf{S_x} \circ \mathsf{S_y}$.

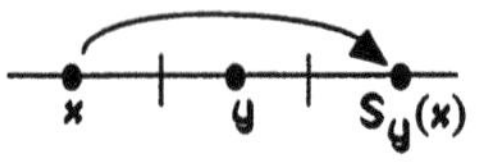

Lemma 20

a. $\forall x\, \forall y\, \forall z\;\; S_y(x) = z \rightarrow xy \equiv yz$

b. $\forall x\, \forall y\, \forall z\;\; xy \equiv yz \rightarrow (x = z \vee S_y(x) = z)$

c. $\forall x\, \forall y\;\; M(x, y, S_y(x))$

I'll now show that point symmetries are automorphisms. So they *preserve* (are homorphisms with respect to) all defined notions, too.

Lemma 21 *Point symmetries are automorphisms*

a. Point symmetries are involutions
$\forall x\, \forall y\;\; S_y(S_y(x)) = x$

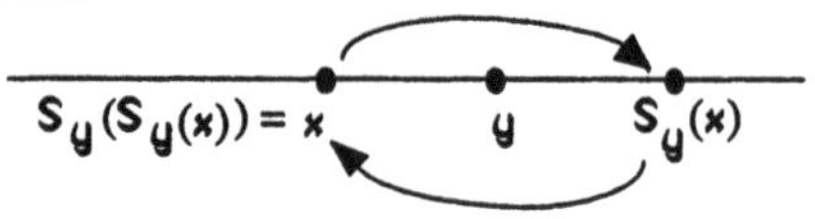

b. Point symmetries are 1-1
$\forall p\, \forall x\, \forall y\;\; S_p(x) = S_p(y) \leftrightarrow x = y$

c. Point symmetries are onto
$\forall p\, \forall y\, \exists x\;\; S_p(x) = y$

d. Point symmetries are isometries
$\forall p\, \forall x\, \forall y\;\; xy \equiv S_p(x)S_p(y)$

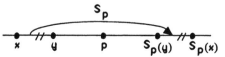

e. Point symmetries preserve congruence
$\forall p\, \forall x\, \forall y\, \forall z\, \forall w\;\; xy \equiv zw \leftrightarrow S_p(x)S_p(y) \equiv S_p(z)S_p(w)$

f. Point symmetries preserve the midpoint relation
$\forall p\, \forall x\, \forall y\, \forall z\;\; M(x, y, z) \leftrightarrow M(S_p(x), S_p(y), S_p(z))$

g. The origin of a symmetry is its only fixed point
$\forall x\, \forall y\;\;\; S_y(x) = x \leftrightarrow x = y$

Proof (a) $\mathsf{S_y(S_y(x)) = x}$ iff $\mathsf{m(S_y(x), y, x)}$, which follows by Lemma 20.c.

(b) Suppose $\mathsf{m(x, p, z)}$ and $\mathsf{m(y, p, z)}$. Then by Lemma 18.a and Lemma 19, $\mathsf{x = y}$. Part (c) is by *C1-5*, and part (d) is by *C1-7*.

(e) Suppose $\mathsf{S_p(x) = a}$, $\mathsf{S_p(y) = b}$, $\mathsf{S_p(z) = c}$, $\mathsf{S_p(w) = d}$. Then $\mathsf{xp \approx pa}$ and $\mathsf{yp \approx pb}$, so by *C1-7*, $\mathsf{xy \approx ab}$. Similarly, $\mathsf{zw \approx cd}$. So if $\mathsf{xy \approx zw}$, then $\mathsf{ab \approx cd}$.

(f) Since each point symmetry is an automorphism, each preserves defined notions.

(g) By Lemma 18.b. ■

Lemma 22 *An equality of functions*

$\forall x\, \forall y\, \forall z\;\; S_{S_x(y)}(z) = S_x(S_y(S_x(z)))$ In a model $S_{S_x(y)}$ is the same function as $S_x S_y S_x$.

Proof Given any w, $m(w, y, S_y(w))$. Hence by Lemma 21.f, for any w, $m(S_x(w), S_x(y), S_xS_y(w))$. Thus, $S_xS_y(w) = S_{S_x(y)}(S_x(w))$. So in particular, if we let $w = S_x(z)$, then $S_xS_y(S_x(z)) = S_{S_x(y)}(S_x(S_x(z)))$. But then by Lemma 21.a, $S_xS_yS_x(z) = S_{S_x(y)}(z)$. ■

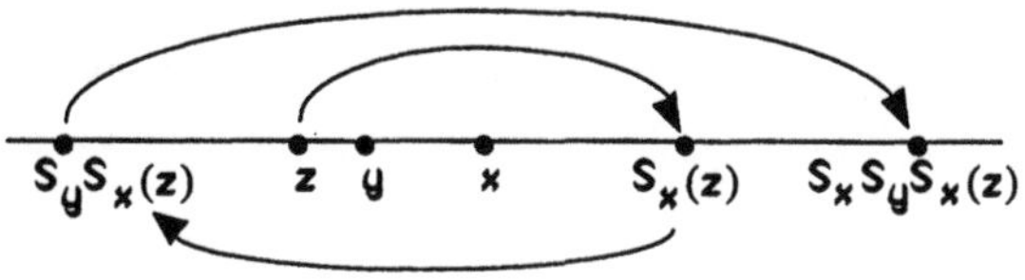

Lemma 23 $\forall x\, \forall y\, \forall z\;\; M(x, y, z) \rightarrow M(S_x(y), y, S_z(y))$

Proof Assume $m(x, y, z)$. By Lemma 20.c, $m(y, z, S_z(y))$. So by Lemma 21.e., $m(S_y(y), S_y(z), S_yS_z(y))$. So by Lemma 21.g, $m(y, S_y(z), S_yS_z(y))$. We also have $m(z, y, x)$, so $S_y(z) = x$. Hence, $m(y, x, S_yS_z(y))$. So by definition, $S_x(y) = S_y(S_z(y))$. So by definition, $m(S_z(y), y, S_x(y))$, so by Lemma 18.a we're done. ■

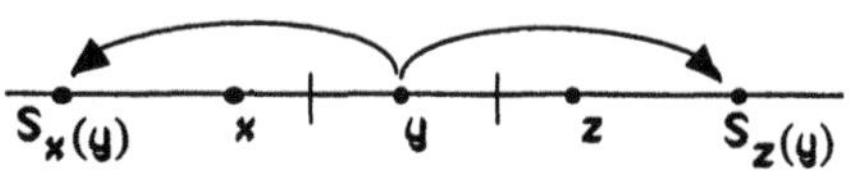

I'll now show that if two point symmetries agree on even one point, they are the same. Hence in a model, $S_x(y)$ is a function of two arguments, x and y.

Lemma 24 $\forall x\, \forall y\; (\exists z\; S_x(z) = S_y(z)) \rightarrow x = y$

Proof Assume that for some z, (i) $S_x(z) = S_y(z)$. We need to show that $x = y$. We have $m(z, x, S_x(z))$, so by definition, $zx \approx xS_x(z)$. So by *C1-2*, Lemma 17.b, and *C1-4*, $zx \approx S_x(z)x$. By Lemma 21.d $zx \approx S_y(z)S_y(x)$, so by (i), we have (ii) $zx \approx S_x(z)S_y(x)$.

By Lemma 21.d and (ii) we have $zx \approx S_{S_y(x)}(z)S_{S_y(x)}(x)$. Hence by Lemma 22, $zx \approx (S_yS_xS_y(z))(S_yS_xS_y(x))$. Yet,

$$\begin{aligned} S_yS_xS_y(z) &= S_yS_xS_x(z) && \text{by (i)} \\ &= S_y(z) && \text{Lemma 21.a} \\ &= S_x(z) && \text{by (i)} \end{aligned}$$

So we have (iii) $zx \approx S_x(z)(S_yS_xS_y(x))$.

We also have (iv) $zx \approx S_x(z)x$ by definition. Combining (ii), (iii), and (iv) by Lemma 17.h, we have (v) $x = S_y(x)$ or (vi) $S_y(x) = S_yS_xS_y(x)$ or (vii) $S_yS_xS_y(x) = x$. In case (v), $x = y$ by Lemma 21.g (y is the only fixed point of S_y). In case (vi), by Lemma 21.b (the function is 1-1), we have (viii) $x = S_xS_y(x)$. Now by Lemma 21.g, $S_x(x) = x$ and $S_xS_xS_y(x) = S_y(x)$. So we have by Lemma 21.a, $x = S_y(x)$. Hence, by Lemma 21.g, $x = y$. In case (vii), by applying S_y to both sides, by Lemma 21.a, $S_xS_y(x) = S_y(x)$. So by Lemma 21.g, $x = S_y(x)$, and hence again by Lemma 21.g, $x = y$. ■

Lemma 25 *Commuting of point symmetries*

a. $\forall x\, \forall y\, \forall z\;\; S_x S_y(z) = S_y S_x(z) \rightarrow x = y$

b. $\forall x\, \forall y\, \forall z\, \forall w\;\; S_x S_y S_z(w) = S_z S_y S_x(w)$

Proof (a) Assume $S_x S_y(z) = S_y S_x(z)$. By Lemma 20.c, $m(S_y(z), x, S_x S_y(z))$, so $m(S_y(z), x, S_y S_x(z))$. As point symmetries are homomorphisms, $m(S_y S_y(z), S_y(x), S_y S_y S_x(z))$. Since point symmetries are involutions, $m(z, S_y(x), S_x(z))$. So by definition, $S_x(z) = S_{S_y(x)}(z)$. Hence, by Lemma 24, $x = S_y(x)$, which by Lemma 21.g gives us $x = y$.

(b) First we'll show

(i) $S_x S_y S_z(x) = S_z S_y S_x(x)$

We have:

$(S_y S_z(x))\, x$	$\approx (S_y S_y S_z(x))\, S_y(x)$	since point symmetries are isometries
	$\approx S_z(x)\, S_y(x)$	since point symmetries are involutions
	$\approx (S_z S_z(x))\,(S_z S_y(x))$	isometry
	$\approx x\,(S_z S_y(x))$	involution

So by Lemma 20.b, (ii) $S_y S_z(x) = S_z S_y(x)$ or (iii) $S_x S_y S_z(x) = S_z S_y(x)$.

For (ii), by (a) $y = z$, hence we have (i) by involutions. For (iii) by Lemma 21.g, $S_x S_z S_y(x) = S_z S_y S_x(x)$. Thus (i) is established.

Now set:

(iv) $x' = S_x S_y S_z(x) =$ (by i) $S_z S_y S_x(x)$

(v) $z' = S_x S_y S_z(z) =$ (by i) $S_z S_y S_x(z)$

Since point symmetries are isometries, we have for any w,

$x'z' \approx xz$ and $x' S_x S_y S_z(w) \approx xw$ and $z' S_x S_y S_z(w) \approx zw$
and $x' S_z S_y S_x(w) \approx xw$ and $z' S_z S_y S_x(w) \approx zw$

So by Lemma 20.b, (vi) $S_x S_y S_z(w) = S_z S_y S_x(w)$ or (vii) $S_{x'} S_x S_y S_z(w) = S_z S_y S_x(w) = S_{z'} S_x S_y S_z(w)$ or (viii) $x = z'$. If (vi), we are done. If (vii) or if (viii) (by Lemma 24) $S_x S_y S_z(x) = S_x S_y S_z(z)$, so $x = z$, from which (b) also follows. ■

3. Addition of points

Lemma 26 *The midpoint relation picks out a midpoint from the endpoints*

$\forall x\, \forall z\, \exists!\, y\;\; \mathrm{M}(x, y, z)$

Proof For existence, use *C1-5*. For uniqueness, suppose $m(x, y, z)$ and $m(x, w, z)$. By definition, $S_y(x) = z = S_w(x)$, so by Lemma 24, $y = w$. ■

The midpoint function

$x \vdots z = y \equiv_{\mathrm{Def}} \mathrm{M}(x, y, z)$

I'll write : for the interpretation of $\vdots$ in a model.

Lemma 27

a. $\forall x\,\forall z\;\; \mathrm{M}(x, x \vdots z, z)$

b. $\forall x\,\forall z\;\; x(x \vdots z) = (x \vdots z)z$

c. $\forall x\,\forall z\;\; x \vdots z = z \vdots x$

d. $\forall x\;\; x \vdots x = x$

e. $\forall x\,\forall z\;\; \mathrm{S}_{x \vdots z}(x) = z$

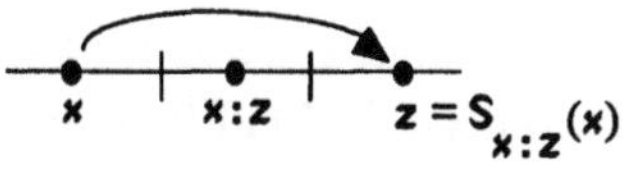

f. $\forall x\,\forall z\;\; x \vdots \mathrm{S}_z(x) = z$

g. $\forall x\,\forall z\,\forall p\;\; \mathrm{S}_p(x \vdots z) = \mathrm{S}_p(x) \vdots \mathrm{S}_p(z)$

Proof Part (g) follows because every point symmetry is an automorphism. ∎

To define an addition of points, we need to pick out one point as an origin. Any point on the line will do. Naming that parameter 'o', we'll take $\mathbf{C1}_0$ to be **C1** formulated in the language L(= ; ≡, o). And we make the following definitions.

$x + y \equiv_{\mathrm{Def}} \mathrm{S}_{x \vdots y}(\mathrm{o})$

$\sim x \equiv_{\mathrm{Def}} \mathrm{S}_{\mathrm{o}}(x)$

$1/_2 x \equiv_{\mathrm{Def}} x \vdots \mathrm{o}$

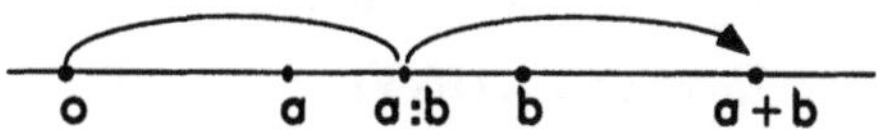

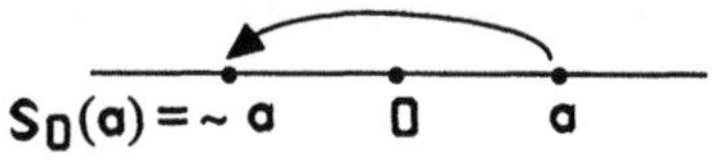

I'll write the interpretation of + in a model as +, and of ~ as ~.

Lemma 28 $\forall x\,\forall y\,\forall z\;\; \mathrm{S}_{x+y}(z) = \mathrm{S}_x \mathrm{S}_{\mathrm{o}} \mathrm{S}_y(z)$

Proof This is an identity of functions. We have:

S_{x+y}	$= \mathsf{S}_{\mathsf{S}_{x:y}(0)}$	definition
	$= \mathsf{S}_{x:y}\mathsf{S}_0\mathsf{S}_{x:y}$	Lemma 22
	$= \mathsf{S}_{x:y}\mathsf{S}_y\mathsf{S}_y\mathsf{S}_0\mathsf{S}_{x:y}$	point symmetries are involutions
	$= \mathsf{S}_{x:y}\mathsf{S}_y\mathsf{S}_{x:y}\mathsf{S}_0\mathsf{S}_y$	Lemma 25.b
	$= \mathsf{S}_{\mathsf{S}_{x:y}(y)}\mathsf{S}_0\mathsf{S}_y$	Lemma 22
	$= \mathsf{S}_x\mathsf{S}_0\mathsf{S}_y$	Lemma 27.e, c

∎

Lemma 29 *The axioms for 2-divisible groups* **GD2** *hold for* +

a. $\forall x\,\forall y\,\forall z\;\; (x+y)+z = x+(y+z)$

b. $\forall x\;\; x + \mathrm{o} = x$

c. $\forall x\;\; x + (\sim x) = \mathrm{o}$

d. $\forall x\,\forall y\;\; x+y = y+x$

e. $\forall x\,\forall y\;\; 1/_2 x = y \leftrightarrow x = y+y$

f. $\forall x\;\; 1/_2 x + 1/_2 x = x$

Proof (a) $\mathsf{S}_{(x+y)+z}$ = (Lemma 28) $\mathsf{S}_x\mathsf{S}_0\mathsf{S}_y\mathsf{S}_0\mathsf{S}_z$ = (by associativity of composition of functions) $\mathsf{S}_{x+(y+z)}$. So by Lemma 24, $(x + y) + z = x + (y + z)$.

(b) S_{x+0} = (Lemma 28) $\mathsf{S}_x\mathsf{S}_0\mathsf{S}_0$ = (point symmetry is an involution) = S_x, and by Lemma 24 we're done.

(c) S_{x+-x} = (Lemma 28) $\mathsf{S}_x\mathsf{S}_0\mathsf{S}_{-x} = \mathsf{S}_x\mathsf{S}_0\mathsf{S}_{\mathsf{S}_0(x)}$ = (Lemma 22) $\mathsf{S}_x\mathsf{S}_0\mathsf{S}_0\mathsf{S}_x\mathsf{S}_0$ = (involution) S_0. So by Lemma 24, $x + -x = 0$.

(d) S_{x+y} = (Lemma 28) $\mathsf{S}_x\mathsf{S}_0\mathsf{S}_y$ = (Lemma 25.b) $\mathsf{S}_y\mathsf{S}_0\mathsf{S}_x = \mathsf{S}_{y+x}$, and by Lemma 24 we are done.

(e) $^1/_2\, x = y$ iff (definition) $x \dot{:} 0 = y$ iff (definition) $\mathsf{m}(x, y, 0)$. But $\mathsf{m}(x, y, 0)$ iff (definition) $x = \mathsf{S}_y(0)$. But $x = \mathsf{S}_y(0)$ iff (Lemma 27.d) $x = \mathsf{S}_{y:y}(0)$. So by definition, $x = y + y$. ■

Lemma 30 If $\mathsf{M} = \langle \mathsf{U}; \approx, 0\rangle$ is a model of $\mathbf{C1}_0$ and + is the interpretation of + in M, then $\mathsf{M} = \langle \mathsf{U}; +\rangle$ is a 2-divisible group.

The next lemma now follows by group theory (Chapter XII).

Lemma 31

a. $\forall x \forall y \;\; ^1/_2(x+y) = {}^1/_2 x + {}^1/_2 y$

b. $\forall x \forall y \;\; ^1/_2(\sim x) = \sim({}^1/_2 x)$

c. $\forall x \forall y \;\; \sim(x+(\sim y)) = y+(\sim x)$

4. Congruence expressed in terms of addition

Lemma 32

a. $\forall x \forall y \forall z \;\; \mathrm{M}(x, y, z) \leftrightarrow x+z = y+y$

b. $\forall x \forall y \;\; x \dot{:} y = {}^1/_2(x+y)$

c. $\forall x \forall y \;\; \mathrm{S}_y(x) = y+y+(\sim x)$

d. $\forall x \forall y \;\; xo \equiv oy \rightarrow (x = y) \vee (x = \sim y)$

Proof (a) We have $\mathsf{m}(x, y, z)$ iff (definition) $x{:}z = y$ iff (Lemma 27.d) $x{:}z = y{:}y$ iff (Lemma 24) $\mathsf{S}_{x:z}(0) = \mathsf{S}_{y:y}(0)$ iff (definition) $x + z = y + y$.

(c) S_{y+y-x} = (Lemma 29) $\mathsf{S}_y\mathsf{S}_0\mathsf{S}_y\mathsf{S}_0\mathsf{S}_{-x}$ = (definition) $\mathsf{S}_y\mathsf{S}_0\mathsf{S}_y\mathsf{S}_0\mathsf{S}_{\mathsf{S}_0(x)}$ = (Lemma 22) $\mathsf{S}_y\mathsf{S}_0\mathsf{S}_y\mathsf{S}_0\mathsf{S}_0\mathsf{S}_x\mathsf{S}_0$ = (involution) $\mathsf{S}_y\mathsf{S}_0\mathsf{S}_y\mathsf{S}_x\mathsf{S}_0$ = (Lemma 25.b) $\mathsf{S}_y\mathsf{S}_0\mathsf{S}_0\mathsf{S}_x\mathsf{S}_y$ = (involution) $\mathsf{S}_y\mathsf{S}_x\mathsf{S}_y$ = (Lemma 22) $\mathsf{S}_{\mathsf{S}_y(x)}$. So by Lemma 24 we have $\mathsf{S}_y(x) = y + y - x$.

(d) $x\,0 = 0\,y$ iff (Lemma 20.b) $(x = 0) \vee (\mathsf{S}_0(x) = y)$. In the first case we are done. The second case is iff (definition) $-x = y$. Now use group theory to get that $x = -y$. ■

Theorem 33

$\forall x \forall y \forall z \forall w \;\; xy \equiv zw \leftrightarrow (x+z = y+w) \vee (x+w = y+z)$

Proof Since point symmetries are isometries, by *C1-4*,

$xy \approx zw$ iff $S_{x:0}(x)S_{x:0}(y) \approx S_{z:0}(z)S_{z:0}(w)$
iff $0S_{1/2x}(y) \approx 0S_{1/2z}(w)$ — Lemma 27.e, definition
iff $S_{1/2x}(y) = S_{1/2z}(w)$ or $S_{1/2x}(y) = -S_{1/2z}(w)$ — Lemma 32.d
iff $1/2x + 1/2x - y = 1/2z + 1/2z - w$ or $1/2x + 1/2x - y = -(1/2z + 1/2z - w)$ — Lemma 32.c
iff $x - y = z - w$ or $x - y = -(z - w)$ — Lemma 29.f

Now use group theory to derive $x + w = y + z$ or $x + z = y + w$. ∎

Theorem 33 shows that the choice of interpretation of 'o' is inessential for deriving '≡' from addition.

5. Translating between C1 and the theory of 2-divisible groups

In this section we'll translate between **C1** and the theory of 2-divisible groups, **GD2** (Chapter XVII.B). We start with a map from **C1** to **GD2**.

A mapping α: L(=; ≡) → L(=; ⊕, o) is given by:

$\alpha(xy \equiv zw)$ is $(x \oplus z = y \oplus w) \vee (x \oplus w = y \oplus z)$

¬, →, ∧, ∨, ∀, ∃, and = are translated homophonically

In a model N =<U; +> of **GD2**, set:

$ab \approx_+ cd \equiv_{Def} (a + c = b + d)$ or $(a + d = b + c)$

$C(N) \equiv_{Def} \langle U; \approx_+ \rangle$

In the other direction, we again take two steps to get from **GD2** to **C1** in order to eliminate the parameter.

A mapping β :L(=; ⊕, o) → L(=; ≡, o) is given by:

$\beta(x \oplus y)$ is $x + y$

¬, →, ∧, ∨, ∀, ∃, and = are translated homophonically

In a model M = <U; ≈> of **C1** and $0 \in U$, for the same universe and identity set:

$a +_0 b = c \equiv_{Def} a + b = c$

That is, $a +_0 b = c$ iff there is some p such that ([(0 = c = p) or (0 ≠ c and 0p ≈ pc)] and [(a = p = b) or (a ≠ b and ap ≈ pb)])

$G(M)_0 \equiv_{Def} \langle U; +_0, 0 \rangle$

Lemma 34

a. If $\mathsf{M} \vDash \mathbf{C1}$ and $\mathsf{0}$ is an element of the universe of M, then $G(\mathsf{M})_{\mathsf{0}} \vDash \mathbf{GD2}$.

b. If $\mathsf{N} \vDash \mathbf{GD2}$, then $C(\mathsf{N}) \vDash \mathbf{C1}$.

c. If $\mathsf{0}' \in \mathsf{U}$, then $G(\mathsf{M})_{\mathsf{0}} \simeq G(\mathsf{M})_{\mathsf{0}'}$ via the mapping $\mathsf{S}_{\mathsf{0}:\mathsf{0}'}$.

Proof (c) The point symmetry $\mathsf{S}_{\mathsf{0}:\mathsf{0}'}$ is an automorphism of $\langle \mathsf{U}; \approx \rangle$ (Lemma 21). By Lemma 27.e, $\mathsf{S}_{\mathsf{0}:\mathsf{0}'}(\mathsf{0}) = \mathsf{0}'$. So we need only show $\mathsf{S}_{\mathsf{0}:\mathsf{0}'}$ is a homorphism of $+_{\mathsf{0}}$ to $+_{\mathsf{0}'}$.

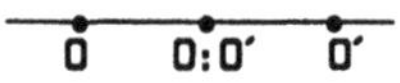

$$
\begin{aligned}
\mathsf{S}_{\mathsf{0}:\mathsf{0}'}(\mathsf{a}) +_{\mathsf{0}'} \mathsf{S}_{\mathsf{0}:\mathsf{0}'}(\mathsf{b}) &= \mathsf{S}_{\mathsf{S}_{\mathsf{0}:\mathsf{0}'}(\mathsf{a}):\mathsf{S}_{\mathsf{0}:\mathsf{0}'}(\mathsf{b})}(\mathsf{0}') && \text{definition} \\
&= \mathsf{S}_{\mathsf{S}_{\mathsf{0}:\mathsf{0}'}(\mathsf{a}:\mathsf{b})}(\mathsf{0}') && \mathsf{S}_{\mathsf{0}:\mathsf{0}'} \text{ is an automorphism} \\
&= \mathsf{S}_{\mathsf{0}:\mathsf{0}'} \mathsf{S}_{\mathsf{a}:\mathsf{b}} \mathsf{S}_{\mathsf{0}:\mathsf{0}'}(\mathsf{0}') && \text{Lemma 22} \\
&= \mathsf{S}_{\mathsf{0}:\mathsf{0}'} \mathsf{S}_{\mathsf{a}:\mathsf{b}}(\mathsf{0}) && \text{Lemma 27.e, c} \\
&= \mathsf{S}_{\mathsf{0}:\mathsf{0}'}(\mathsf{a} + \mathsf{b})
\end{aligned}
$$

■

Theorem 35 *The representation theorem for* **C1** *and* **GD2**

a. If $\mathsf{M} \vDash \mathbf{C1}$, then there is a model $\mathsf{N} \vDash \mathbf{GD2}$ such that $C(\mathsf{N}) = \mathsf{M}$.
In particular, if $\mathsf{M} = \langle \mathsf{U}; \approx \rangle \vDash \mathbf{C1}$, and $\mathsf{0} \in \mathsf{U}$, then $C(G(\mathsf{M})_{\mathsf{0}}) = \mathsf{M}$.

b. If $\mathsf{N} \vDash \mathbf{GD2}$, then there is a model $\mathsf{M} \vDash \mathbf{C1}$ such that $G(\mathsf{M})_{\mathsf{0}} = \mathsf{N}$.
In particular, if $\mathsf{N} = \langle \mathsf{U}; +, \mathsf{0} \rangle$, then $G(C(\mathsf{N}))_{\mathsf{0}} = \mathsf{N}$.

A mapping $\gamma : L(=;\equiv,\mathrm{o}) \to L(=;\equiv)$ is given by:

$\gamma(\mathrm{A}) = \forall x\, \mathrm{A}(x/\mathrm{o})$
where x is the least variable not appearing in A,
and $\gamma(\mathrm{A})$ is A if 'o' does not appear in A.

In Section E below I'll show that eliminating parameters from $\mathbf{C1_0}$ is a translation. That is, $\Gamma \vDash_{\mathbf{C1_0}} \mathrm{A}$ iff $\gamma(\Gamma) \vDash_{\mathbf{C1}} \gamma(\mathrm{A})$. So we can show that our mappings are translations, using Theorem XVI.4.

Theorem 36 *The translation theorem for* **C1** *and* **GD2**

a. The mappings α and β are equivalence-relation translations.

b. $\Gamma \vDash_{\mathbf{C1}} \mathrm{A}$ iff $\alpha(\Gamma) \vDash_{\mathbf{GD2}} \alpha(\mathrm{A})$.

c. $\Gamma \vDash_{\mathbf{GD2}} \mathrm{A}$ iff $\beta(\Gamma) \vDash_{\mathbf{C1}} (\gamma \circ \beta)(\mathrm{A})$.

Corollary 37 **C1** is not complete.

Proof **GD2** is not complete (Lemma XVII.3.f). ■

6. Division axioms for C1 and the theory of divisible groups

Axiom *C1-5* guarantees that models of **C1** correspond to models of 2-divisible groups. We now formulate axioms to extend **C1** to a theory equivalent to the theory of divisible groups.

$A_2 \equiv_{\text{Def}} \exists x_1 (x_0 x_1 \equiv x_1 x_2 \wedge x_0 \neq x_2)$

$A_{n+1} \equiv_{\text{Def}} \exists x_n (A_n \wedge x_{n-1} x_n \equiv x_n x_{n+1} \wedge x_{n-1} \neq x_{n+1})$ for $n \geq 2$

Division axioms $C1\text{-}5_n \equiv_{\text{Def}} \forall x_0\, \forall x_n\, (x_0 \neq x_n \rightarrow A_n)$ for $n \geq 2$

$\mathbf{C1_{Div}}$, the *complete theory of one-dimensional congruence* in $L(=;\equiv)$

C1

$C1\text{-}5_p$ such that p is a prime and $p > 2$

I'll let you show $\models C1\text{-}5_2 \leftrightarrow C1\text{-}5$. I'll also let you show the next lemma, using Lemma XVII.3.d.

Lemma 38 If N is an interpretation for $L(=;\oplus, \circ)$ and $n \geq 2$, then

a. N is an n-divisible group iff $C(\mathsf{N}) \models C1\text{-}5_n$.

b. N is a divisible group iff $C(\mathsf{N}) \models \{C1\text{-}5_p : p \text{ is a prime}\}$.

Since any model of $\mathbf{C1_{Div}}$ is also a model of **C1**, we can use the same translations as before. I'll write $\mathbf{C1o_{Div}}$ for $\mathbf{C1_{Div}}$ in the language $L(=;\equiv, \circ)$.

Theorem 39 *The translation theorem for* $\mathbf{C1_{Div}}$ *and* **GD**

a. The mapping α is an equivalence-relation translation of $\mathbf{C1_{Div}}$ to **GD**.

b. The mapping β is an equivalence-relation translation from **GD** to $\mathbf{C1o_{Div}}$.

c. $\Gamma \models_{\mathbf{C1_{Div}}} A$ iff $\alpha(\Gamma) \models_{\mathbf{GD}} \alpha(A)$

d. $\Gamma \models_{\mathbf{GD}} A$ iff $(\gamma \circ \beta)(\Gamma) \models_{\mathbf{C1_{Div}}} (\gamma \circ \beta)(A)$

The following now comes from the corresponding claims about **GD** in Chapter XVII.B.

Theorem 40 *Completeness and decidability of* $\mathbf{C1_{Div}}$

a. $\mathbf{C1_{Div}}$ is complete and decidable.

b. $\mathbf{C1_{Div}}$ is not finitely axiomatizable.

c. $\mathbf{C1_{Div}}$ is neither $\aleph_0$-categorical nor $\mathfrak{c}$-categorical.

Exercises for Section C

1. Prove in **C1**: $\forall x\, \forall y\, \forall z\, \forall w\;\; xy \approx zw \rightarrow (x \vdots w = y \vdots z \vee x \vdots z = y \vdots w)$
(Hint: Use Lemma 36 in **C1** in L(=; =, o) and eliminate the parameter.)
2. Show that if f is an automorphism on a model, then it is a homomorphism with respect to all defined notions in the model, too.
3. Complete the proof of Corollary 37.
4. Prove $\vDash C1\text{-}5_2 \leftrightarrow C1\text{-}5$.
5. Prove a representation theorem for **C1** and **GD**. (Compare Theorem 35.)
6. *Independence of the axioms of* **C1**
Show that each axiom of **C1** is independent by showing that it fails in the designated model below while all the other axioms hold in that model.

C1-1	<∅; ∅>	
C1-2	<ℝ; ≈>	ab ≈ cd iff a − b = c − d (put a = 0, b = 1)
C1-3	<ℝ; $ℝ^4$>	(≡ is interpreted as the universal relation)
C1-4	<ℝ; ≈>	ab ≈ cd iff |a − b| = |c − d| or a = b = d (intepret *x*, *y*, and *r* as 2, *z* as 3, *w* as 1, and *s* as 4)
$C1\text{-}5_p$	C(M)	M is the model of $\mathbf{GD}_p$ given in the proof of Lemma XVII.3
C1-6	<[0, 1]; ≈>	[0, 1] is the unit interval of ℝ, and ab ≈ cd iff |a − b| = |c − d|
C1-7	<$ℝ^2$; ≈>	ab ≈ cd iff the distance from a to b is the same as the distance from c to d in the usual metric for $ℝ^2$. (This is why we call *C1-7* 'the dimension axiom'.)

D. One-Dimensional Geometry

1. An axiom system for one-dimensional geometry, E1

E1, *one-dimensional geometry* in L(=; B, ≡)

B1

$\mathbf{C1}_{\mathbf{Div}}$

BC1 $\forall x \forall y \forall z \forall r \forall s \forall t\, ((xy \equiv rs \wedge yz \equiv st \wedge zx \equiv tr) \wedge \mathrm{B}(x, y, z)) \rightarrow \mathrm{B}(r, s, t)$
isometries preserve betweenness

2. The monotonicity of addition

Lemma 41

a. $\forall x\, \forall y\, \forall z\;\; (\mathrm{B}(x, y, z) \wedge xy \equiv xz) \rightarrow y = z$

b. $\forall x\, \forall y\, \forall z\;\; \mathrm{M}(x, y, z) \rightarrow \mathrm{B}(x, y, z)$

c. $\forall x\, \forall y\;\; \mathrm{B}(x, y, \mathrm{S}_y(x))$

d. Point symmetries are homomorphism of betweenness
$\forall x\, \forall y\, \forall z\, \forall p\;\; \mathrm{B}(x, y, z) \to \mathrm{B}(\mathrm{S}_p(x), \mathrm{S}_p(y), \mathrm{S}_p(z))$

Proof (a) Assume $\mathsf{B}(\mathsf{x}, \mathsf{y}, \mathsf{z})$ and $\mathsf{xy} \approx \mathsf{xz}$. By axiom *C1-2*, we also have $\mathsf{yz} \approx \mathsf{zy}$. By Lemma 17.f, we also have $\mathsf{zx} \approx \mathsf{yx}$. In sum, we have $\mathsf{B}(\mathsf{x}, \mathsf{y}, \mathsf{z})$ and $\mathsf{xy} \approx \mathsf{xz}$ and $\mathsf{yz} \approx \mathsf{zy}$ and $\mathsf{zx} \approx \mathsf{yx}$, so (axiom *BC1*) $\mathsf{B}(\mathsf{x}, \mathsf{z}, \mathsf{y})$. From $\mathsf{B}(\mathsf{x}, \mathsf{y}, \mathsf{z})$ and $\mathsf{B}(\mathsf{x}, \mathsf{z}, \mathsf{y})$ we have by axiom *B1-4*, $\mathsf{B}(\mathsf{y}, \mathsf{z}, \mathsf{y})$. Hence by axiom *B1-2*, $\mathsf{y} = \mathsf{z}$.

(b) Assume $\mathsf{xy} \approx \mathsf{yz}$ and $\mathsf{x} \neq \mathsf{z}$, and that $\mathsf{B}(\mathsf{x}, \mathsf{y}, \mathsf{z})$ fails. In that case, by axiom *B1-7*, we have either (i) $\mathsf{B}(\mathsf{y}, \mathsf{z}, \mathsf{x})$ or (ii) $\mathsf{B}(\mathsf{z}, \mathsf{x}, \mathsf{y})$. From Lemma 17.b we have $\mathsf{yz} \approx \mathsf{yx}$. From (i), by part (a) $\mathsf{z} = \mathsf{x}$. From (ii) we have $\mathsf{x} = \mathsf{z}$. Both are contradictions, so $\mathsf{B}(\mathsf{x}, \mathsf{y}, \mathsf{z})$. If $\mathsf{x} = \mathsf{y} = \mathsf{z}$, we also have by Lemma 1.c $\mathsf{B}(\mathsf{x}, \mathsf{y}, \mathsf{z})$.

(c) From Lemma 20.c, $\mathsf{m}(\mathsf{x}, \mathsf{y}, \mathsf{S}_{\mathsf{y}}(\mathsf{x}))$. So by (c), $\mathsf{B}(\mathsf{x}, \mathsf{y}, \mathsf{S}_{\mathsf{y}}(\mathsf{x}))$.

(d) By Lemma 21.d, $\mathsf{xy} \approx \mathsf{S}_{\mathsf{p}}(\mathsf{x})\mathsf{S}_{\mathsf{p}}(\mathsf{y})$, and $\mathsf{yz} \approx \mathsf{S}_{\mathsf{p}}(\mathsf{y})\mathsf{S}_{\mathsf{p}}(\mathsf{z})$, and $\mathsf{zx} \approx \mathsf{S}_{\mathsf{p}}(\mathsf{z})\mathsf{S}_{\mathsf{p}}(\mathsf{x})$. From $\mathsf{B}(\mathsf{x}, \mathsf{y}, \mathsf{z})$ we have by axiom *BC1*, $\mathsf{B}(\mathsf{S}_{\mathsf{p}}(\mathsf{x}), \mathsf{S}_{\mathsf{p}}(\mathsf{y}), \mathsf{S}_{\mathsf{p}}(\mathsf{z}))$. ∎

Thus, *point symmetries are automorphisms of models* $\langle \mathsf{U}; \mathsf{B}, \approx \rangle$ of **E1** and so preserve defined notions here, too. But they reverse the ordering. Still, that's all we'll need to show that addition respects the ordering. But first we need to distinguish two points.

$\mathbf{E1_{01}}$ in L(=; B, ≡, o, ɫ)

E1

$\mathrm{o} \neq ɫ$

Lemma 42

a. $\forall x\, \forall y\, \forall z\;\; x < y \to \mathrm{S}_z(y) < \mathrm{S}_z(x)$

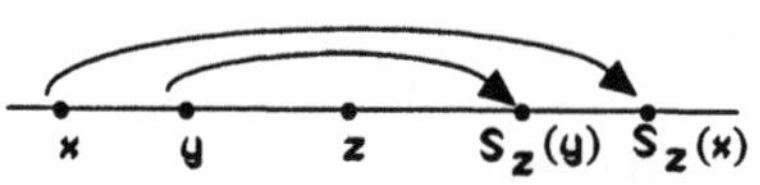

b. $\forall x\, \forall y\;\; x < y \to {\sim}y < {\sim}x$

c. $\forall x\, \forall y\, \forall z\;\; x < y \to x + z < y + z$

Proof (a) We assume $\mathsf{x} < \mathsf{y}$. By axiom *B1-7*, we have (i) $\mathsf{B}(\mathsf{x}, \mathsf{y}, \mathsf{z})$ or (ii) $\mathsf{B}(\mathsf{y}, \mathsf{z}, \mathsf{x})$ or (iii) $\mathsf{B}(\mathsf{z}, \mathsf{x}, \mathsf{y})$. We'll show the conclusion follows for (i); the other cases can be done similarly. From (i) and Theorem 9, $\mathsf{x} < \mathsf{y} \leq \mathsf{z}$. By Lemma 41.c, $\mathsf{B}(\mathsf{x}, \mathsf{z}, \mathsf{S}_{\mathsf{z}}(\mathsf{x}))$. Hence, we have (iv) $\mathsf{z} \leq \mathsf{S}_{\mathsf{z}}(\mathsf{x})$. By (i) via Lemma 41.d, $\mathsf{B}(\mathsf{S}_{\mathsf{z}}(\mathsf{x}), \mathsf{S}_{\mathsf{z}}(\mathsf{y}), \mathsf{S}_{\mathsf{z}}(\mathsf{z}))$, so $\mathsf{B}(\mathsf{S}_{\mathsf{z}}(\mathsf{z}), \mathsf{S}_{\mathsf{z}}(\mathsf{y}), \mathsf{S}_{\mathsf{z}}(\mathsf{x}))$ by axiom *B1-3*. Hence by Lemma 21.g, $\mathsf{B}(\mathsf{z}, \mathsf{S}_{\mathsf{z}}(\mathsf{y}), \mathsf{S}_{\mathsf{z}}(\mathsf{x}))$. Hence by (iv), $\mathsf{S}_{\mathsf{z}}(\mathsf{y}) \leq \mathsf{S}_{\mathsf{z}}(\mathsf{x})$. But since $\mathsf{y} \neq \mathsf{x}$, by Lemma 21.b $\mathsf{S}_{\mathsf{z}}(\mathsf{y}) \neq \mathsf{S}_{\mathsf{z}}(\mathsf{x})$, so $\mathsf{S}_{\mathsf{z}}(\mathsf{y}) < \mathsf{S}_{\mathsf{z}}(\mathsf{x})$.

(b) By (a) and definition.

(c) From $\mathsf{x} < \mathsf{y}$ we have by (b), $-\,\mathsf{y} < -\,\mathsf{x}$, and so by (a), $\mathsf{S}_{\mathsf{0}:\mathsf{z}}(-\,\mathsf{x}) < \mathsf{S}_{\mathsf{0}:\mathsf{z}}(-\,\mathsf{y})$. It now suffices to show that $\mathsf{S}_{\mathsf{0}:\mathsf{z}}(-\,\mathsf{x}) = \mathsf{x} + \mathsf{z}$ and $\mathsf{S}_{\mathsf{0}:\mathsf{z}}(-\,\mathsf{y}) = \mathsf{y} + \mathsf{z}$ to get $\mathsf{x} + \mathsf{z} < \mathsf{y} + \mathsf{z}$. We'll show the first, from which the second follows as the choice of x is arbitrary.

$\mathsf{S}_{\mathsf{0}:\mathsf{z}}(-\,\mathsf{x})$	$= \mathsf{S}_{1/2(\mathsf{0}+\mathsf{z})}(-\,\mathsf{x})$	Lemma 32.b
	$= \mathsf{S}_{1/2\,\mathsf{z}}(-\,\mathsf{x})$	Lemma 29.b,d

We now show $\mathsf{S}_{\mathsf{S}_{1/2\,\mathbf{z}}(-\,\mathsf{x})} = \mathsf{S}_{\mathsf{x}+\mathbf{z}}$, from which $\mathsf{S}_{\mathsf{0}:\mathbf{z}}(-\,\mathsf{x}) = \mathsf{x} + \mathbf{z}$ follows by Lemma 21.b.

$$
\begin{aligned}
\mathsf{S}_{\mathsf{S}_{1/2\,\mathbf{z}}(-\,\mathsf{x})} &= \mathsf{S}_{1/2\,\mathbf{z}}\mathsf{S}_{-\mathsf{x}}\mathsf{S}_{1/2\,\mathbf{z}} && \text{Lemma 22}\\
&= \mathsf{S}_{1/2\,\mathbf{z}}\mathsf{S}_{\mathsf{S}_{\mathsf{0}}(\mathsf{x})}\mathsf{S}_{1/2\,\mathbf{z}} && \text{definition}\\
&= \mathsf{S}_{1/2\,\mathbf{z}}\mathsf{S}_{\mathsf{0}}\mathsf{S}_{\mathsf{x}}\mathsf{S}_{\mathsf{0}}\mathsf{S}_{1/2\,\mathbf{z}} && \text{Lemma 22}\\
&= \mathsf{S}_{1/2\,\mathbf{z}+\mathsf{x}}\mathsf{S}_{\mathsf{0}}\mathsf{S}_{1/2\,\mathbf{z}} && \text{Lemma 28}\\
&= \mathsf{S}_{1/2\,\mathbf{z}+\mathsf{x}+1/2\,\mathbf{z}} && \text{Lemma 28}\\
&= \mathsf{S}_{\mathsf{x}+\mathbf{z}} && \text{Lemma 29}
\end{aligned}
$$
■

3. Translating between E1 and the theory of ordered divisible groups

In this section we'll translate between the theory of one-dimensional geometry **E1** and the theory of ordered divisible groups, **ODG** (Chapter XVII.D). We begin by translating **E1** to **ODG**.

A mapping $\delta: L(=; \mathrm{B}, \equiv) \to L(=; <, \oplus, \mathrm{o})$ is given by:

$\delta(\mathrm{B}(x, y, z))$ is $(x \leq y \wedge y \leq z) \vee (z \leq y \wedge y \leq x)$

$\delta(xy \equiv zw)$ is $(x \oplus z = y \oplus w) \vee (x \oplus w = y \oplus z)$

$\neg, \rightarrow, \wedge, \vee, \forall, \exists$, and $=$ are translated homophonically

Given a model $\mathsf{N} = \langle \mathsf{U}; <, +, \mathsf{o}\rangle$ of **ODG** we define:

$\mathsf{B}_{\leq}(\mathsf{a}, \mathsf{b}, \mathsf{c}) \equiv_{\mathrm{Def}} \mathsf{a} \leq \mathsf{b} \leq \mathsf{c}$ or $\mathsf{c} \leq \mathsf{b} \leq \mathsf{a}$

$\mathsf{ab} \approx_{+} \mathsf{cd} \equiv_{\mathrm{Def}} (\mathsf{a} + \mathsf{c} = \mathsf{b} + \mathsf{d})$ or $(\mathsf{a} + \mathsf{d} = \mathsf{b} + \mathsf{c})$

$E_1(\mathsf{N}) \equiv_{\mathrm{Def}} \langle \mathsf{U}; \mathsf{B}_{\leq}, \approx_{+}\rangle$

Lemma 43 If $\mathsf{N} \models$ **ODG**, then $E_1(\mathsf{N}) \models$ **E1**.

Proof By Lemma 10 and Lemma 34, we have: $E_1(\mathsf{N}) \models$ **B1**, $E_1(\mathsf{N}) \models$ **C1**, and $E_1(\mathsf{N}) \models \forall x\, \forall y\, \forall z\; (x < y \rightarrow x + z < y + z)$. It remains to show that $E_1(\mathsf{N}) \models$ *BC1*.

Assume $\mathsf{xy} \approx \mathsf{rs}$ and $\mathsf{yz} \approx \mathsf{st}$ and $\mathsf{zx} \approx \mathsf{tr}$ and $\mathsf{B}(\mathsf{x}, \mathsf{y}, \mathsf{z})$. We need $\mathsf{B}(\mathsf{r}, \mathsf{s}, \mathsf{t})$.

From $\mathsf{B}(\mathsf{x}, \mathsf{y}, \mathsf{z})$ we have by Theorem 9, $\mathsf{x} \leq \mathsf{y} \leq \mathsf{z}$ or $\mathsf{z} \leq \mathsf{y} \leq \mathsf{x}$. If $\mathsf{x} = \mathsf{y}$ or $\mathsf{y} = \mathsf{z}$ or $\mathsf{x} = \mathsf{z}$ we are done, since then $\mathsf{r} = \mathsf{s}$ or $\mathsf{s} = \mathsf{t}$ or $\mathsf{r} = \mathsf{t}$. We'll assume $\mathsf{x} < \mathsf{y} < \mathsf{z}$, and the other case can be done similarly. Theorem 33 gives each of the following pairs:

(i) $\mathsf{x} + \mathsf{r} = \mathsf{y} + \mathsf{s}$ or (ii) $\mathsf{x} + \mathsf{s} = \mathsf{y} + \mathsf{r}$

(iii) $\mathsf{y} + \mathsf{s} = \mathsf{z} + \mathsf{t}$ or (iv) $\mathsf{y} + \mathsf{t} = \mathsf{z} + \mathsf{s}$

(v) $\mathsf{z} + \mathsf{t} = \mathsf{x} + \mathsf{r}$ or (vi) $\mathsf{z} + \mathsf{r} = \mathsf{x} + \mathsf{t}$

We check each possibility. If (i), then $\mathsf{r} - \mathsf{s} = \mathsf{y} - \mathsf{x} > \mathsf{0}$, so $\mathsf{r} > \mathsf{s}$. If (iii), then $\mathsf{s} > \mathsf{t}$, and hence $r > s > t$. So if both (i) and (iii), then by Theorem 9, $B(r, s, t)$.

If (ii) and (iv), we are done similarly. If (i) and (v), then also (iii) and we are done. If (ii) and (vi), then we have (iv), and we are done.

If (iv) and (vi), then $t - z = r - x = s - y$, so we have (ii), and we are done as before. The other cases are done similarly. ■

In the other direction, we again take two steps.

A mapping $\mu : L(=; <, \oplus, o) \rightarrow L(=; B, \equiv, o, 1)$ is given by:

$\mu(x < y)$ is $x < y$

$\mu(x \oplus z)$ is $x + z$

$\neg, \rightarrow, \wedge, \vee, \forall, \exists$, and $=$ are translated homophonically

In a model $M = \langle U; B, \approx \rangle$ of **E1** and $0 \neq 1 \in U$, define:

$a <_B b \equiv_{Def} a \neq b$ and for some p, q,
$B(p, a, b)$ and $B(p, b, q)$ and $B(p, 0, 1)$ and $B(p, 1, q)$

$a +_0 b = c \equiv_{Def}$ there is some p such that $[(0 = p = c)$ or $(0 \neq c$ and $0p \approx pc)]$ and $[(a = p = b)$ or $(a \neq b$ and $ap \approx pb)]$

$OG(M)_{0,1} \equiv_{Def} \langle U; <_B, +_0, 0 \rangle$

Lemma 44 If $M \vDash$ **E1** and $0 \neq 1$ are elements of the universe of M, then

a. $OG(M)_{0,1} \vDash$ **ODG**.

b. If $0' \neq 1' \in U$, then $OG(M)_{0,1} \simeq OG(M)_{0',1'}$ via:

$S_{o:o'}$ in case $01 \uparrow 1'0'$.

$S_{o'}S_{o:o'}$ in case $01 \uparrow 0'1'$.

Proof (a) This follows from Lemma 11, Lemma 34, and Lemma 42.

(b) By Lemma 34.c, we have that $S_{o:o'} : G(M)_{0,1} \rightarrow G(M)_{0',1'}$ is an isomorphism. But by Lemma 42.a, it reverses the ordering on $G(M)_{0,1}$. This is exactly what we want if $01 \uparrow 1'0'$ (see the remarks under Theorem 12).

Suppose, though, that $01 \uparrow 0'1'$. Then if we take $S_{o:o'}$, we need to reverse the ordering back to what we started with. By Lemma 42.a, we can do this with $S_{o'}$. ■

Theorem 45 *The representation theorem for* **E1** *and* **ODG**

a. If $M \vDash$ **E1**, then there is a model $N \vDash$ **ODG** such that $E_1(N) = M$.
In particular, if $M = \langle U; B, \approx \rangle \vDash$ **E1** and $0 \neq 1$ are in U, then $E_1(OG(M)_{0,1}) = M$.

b. If $N \vDash$ **E1**, then there is a model $M \vDash$ **B1** such that $OG(M)_{0,1} = N$.
In particular, if $N = \langle U; < \rangle$, and $0, 1 \in U$ and $0 < 1$, then $OG(E_1(N))_{0,1} = N$.

Note that from (a) we have that for any model $M = \langle U; B, \approx \rangle$ of **E1** and $0 \neq 1$ and $0' \neq 1'$, $E_1(OG(M)_{0,1}) = E_1(OG(M)_{0',1'})$.

Now using Theorems XVII.10 and 11, we can prove the following.

Theorem 46

a. **E1** is complete and decidable.

b. **E1** $=$ Th($E_1(\langle \mathbb{R}; <, +, 0 \rangle)$) $=$ Th($E_1(\langle \mathbb{Q}; <, +, 0 \rangle)$).

c. **ODG** = Th($OG(\langle \mathbb{R}; B, \approx \rangle)_{0,1}$)
where B, $\approx$ are the usual betweenness and congruence relations on the real line.

d. **E1** is not finitely axiomatizable.

e. **E1** is neither $\aleph_0$-categorical nor $\mathfrak{c}$-categorical.

Now we eliminate parameters from $\mathbf{E1_{01}}$.

A mapping λ: L(=; B, ≡, o, ł) → L(=; B, ≡) is given by:

λ(A) is $\forall x \forall y (x \neq y \rightarrow A(x/o, y/ł))$

where x and y are the least variables not appearing in A
and λ(A) = A if neither o nor ł appear in A.

In Section E below I'll show that $\Gamma \vDash_{\mathbf{E1_{01}}} A$ iff $\gamma(A) \vDash_{\mathbf{E1}} \gamma(A)$.

Theorem 47 *The translation theorem for* **E1** *and* **ODG**

a. The mappings δ and μ are equivalence-relation translations.

b. $\Gamma \vDash_{\mathbf{E1}} A$ iff $\delta(\Gamma) \vDash_{\mathbf{ODG}} \delta(A)$.

c. $\Gamma \vDash_{\mathbf{ODG}} A$ iff $(\lambda \circ \mu)(\Gamma) \vDash_{\mathbf{E1}} (\lambda \circ \mu)(A)$.

From Theorem XVII.12, in the axiomatization of **ODG** we can replace the division axioms by the first-order scheme of *Dedekind's axiom*$_1$.

ODG = Th({**DLO**, **AbelianG**, *order-addition axiom*, D_2, *Dedekind's axiom*$_1$ })

Correspondingly, in L(=; B) we have the first-order scheme of Dedekind's axiom-B, from which we can obtain an alternate axiomatization of **E1**.

Dedekind's axiom$_1$-B

$$\forall \ldots ([\exists x \exists y (C(x) \wedge D(y)) \wedge \exists w \forall x \forall y (C(x) \wedge D(y) \rightarrow B(w, x, y))]$$
$$\rightarrow \exists z \forall x \forall y (C(x) \wedge D(y)) \rightarrow B(x, z, y))$$

where C is any wff in which x is free, D is any wff in which y is free, and neither y nor z appear in C, and neither x nor z appear in D

E1-D, the *Dedekind axiomatization of one-dimensional geometry* in L(=; B, ≡)

B1

C1

BC1

Dedekind's axiom$_1$-B

Theorem 48

a. M⊨**E1-D** and 0 ≠ 1 are elements of the universe of M iff $E_1(\mathsf{M})_{0,1}$⊨**ODG**.

b. **E1-D** = **E1**.

We've shown that we can define addition and an ordering in $\mathbf{E1_{01}}$ in the sense that we can define predicates that translate to ⊕ and < in **ODG**. But what about multiplication? Recall from Chapter XVII.E.3 that **RCF** is the theory of real closed fields in L(=; ⊕, ∗, o, ɨ), and **RCF** = Th(<ℝ ; <, +, ·, 0, 1>).

Theorem 49 *Multiplication is not definable in* $\mathbf{E1_{01}}$

There is no predicate P definable in $\mathbf{E1_{01}}$ such that we can extend the mapping δ to an equivalence-relation translation from $\mathbf{E1_{01}}$ to **RCF**.

Proof If there were such a translation, then we could use it to define multiplication in **ODG**, which we know is not possible by Theorem XVII.11.d. ■

4. Second-order one-dimensional geometry

We can get a categorical axiomatization of the geometry of the real line equivalent to the theory of real ordered divisible groups, $\mathbf{RODG_2}$ (Chapter XVII.D) by adding Dedekind's axiom-B to our first-order axiomatization. By Theorem 48, we need not include the division axioms.

E1*, the second-order theory of *one-dimensional geometry* in L_2(=; B, ≡)

B1

C1

BC1

Dedekind's axiom-B

Theorem 50

a. M$\models_2$**E1*** iff M ≃ E_1(<ℝ; <, +, 0>).

b. N$\models_2$$\mathbf{RODG_2}$ iff N ≃ OG(<ℝ; B, ≈>)$_{0,1}$
where B, ≈ are the usual betweenness and congruence relations on the real line

c. **E1*** is categorical and undecidable.

Proof (a) Recall that $\mathbf{RODG_2}$ is Th(**ODG** $\wedge$ *Dedekind's axiom*). By Theorem XVII.13, every model of $\mathbf{RODG_2}$ is isomorphic to $\langle\mathbb{R};<,+,0\rangle$, the real numbers with their usual ordering and addition. Suppose that $\mathsf{M}\models_2\mathbf{E1^*}$. Every instance of the first-order scheme of Dedekind's axiom-B is a consequence of Dedekind's axiom-B, hence $\mathsf{M}\models\mathbf{E1}$. Hence, there is some $\mathsf{N}\models\mathbf{ODG}$ such that $E_1(\mathsf{N})\simeq\mathsf{M}$. By Theorem 15, $\mathsf{N}\models$*Dedekind's axiom*. So $\mathsf{N}\models_2\mathbf{RODG_2}$. Hence, $\langle\mathbb{R};<,+,0\rangle\simeq\mathsf{N}$. Hence, $E_1(\langle\mathbb{R};<,+,0\rangle)\simeq E_1(\mathsf{N})\simeq\mathsf{M}$.

(c) The undecidability of **E1*** follows from Corollary XV.29. ■

Exercises for Section D

1. Let A(x), B(y) be formulas with one free variable each. Let M be a model of **E1**. Let $\mathsf{A}=\{\mathsf{x}:\mathsf{x}$ satisfies $\mathsf{A}\}$ and $\mathsf{B}=\{\mathsf{y}:\mathsf{y}$ satisfies $\mathsf{B}\}$.
   a. Show that if A and B satisfy the hypothesis of *Dedekind's axiom*$_1$*-B*, then $\mathsf{A}=\mathsf{B}$ iff for some a, $\mathsf{A}=\{\mathsf{a}\}=\mathsf{B}$.
   In that case what point z satisfies the conclusion of *Dedekind's axiom*$_1$*-B*?
   b. Show that if A and B satisfy the hypothesis of *Dedekind's axiom*$_1$*-B*, then there is some q such that $\mathsf{q}\notin\mathsf{A}\cup\mathsf{B}$ and for all $\mathsf{a}\in\mathsf{A}$ and $\mathsf{b}\in\mathsf{B}$, $\mathsf{B}(\mathsf{q},\mathsf{a},\mathsf{b})$.

2. *Alternate translation of* **E1** *into* **ODG**
   In L(=; <, ⊕, o) set: $|x|=y \equiv_{\text{Def}} (x\geq \text{o}\wedge y=x)\vee(x<\text{o}\wedge y=\sim x)$.
   Define a mapping μ: L(=; B, ≡) → L(=; <, ⊕, o) via:

   $\theta(\text{B}(x,y,z))$ is $|x\sim y|+|y\sim z|=|x\sim z|$

   $\theta(xy\equiv zw)$ is $|x\sim y|=|z\sim w|$

   ¬, →, ∧, ∨, ∀, ∃, and = are translated homophonically

   Prove: **E1**$\models$A iff **ODG**$\models\theta$(A).

3. a. Complete the proof of Theorem 46 using the representation theorem.
   b. Complete the proof of Theorem 46 using Theorem 47.

4. *Dependence of some of the axioms of* **E1**
   Let $\mathbf{E1_{ind}}$ in L(=; B, ≡) be the axiom system **E1** with *C1-1, C1-3, B1-5, B1-6.* and the division axioms deleted. Prove the following:
   a. $\mathbf{E1_{ind}}\models$*C1-1*.
   b. $\mathbf{E1_{ind}}\models$*C1-3*.
   (Hint: Assume $\mathsf{xb}\approx\mathsf{zz}$. Use *C1-2* and Lemma 17.g, d and *C1-4* to get $\mathsf{xb}\approx\mathsf{xb}$ and $\mathsf{xb}\approx\mathsf{bx}$ and $\mathsf{bx}\approx\mathsf{xx}$. Then use Lemma 1.a and *BC-1* to get $\mathsf{B}(\mathsf{x},\mathsf{b},\mathsf{x})$, from which by *B1-2*, $\mathsf{x}=\mathsf{b}$.)
   c. $\mathbf{E1_{ind}}\models$*B1-5*.
   (Hint: Prove (i) If $\mathsf{B}(\mathsf{x},\mathsf{y},\mathsf{z})$ and $\mathsf{xy}\approx\mathsf{xz}$, then $\mathsf{y}=\mathsf{z}$ (use *C1-2*, Lemma 17, and *BC-1* to get $\mathsf{B}(\mathsf{x},\mathsf{z},\mathsf{y})$, and then *B1-4*). For (c), we need to show that given any x, y, there is some z such that $\mathsf{B}(\mathsf{x},\mathsf{y},\mathsf{z})$ and $\mathsf{z}\neq\mathsf{y}$. Suppose $\mathsf{x}\neq\mathsf{y}$. By *C1-6*, some z, $\mathsf{xy}\approx\mathsf{yz}$. If not $\mathsf{B}(\mathsf{x},\mathsf{y},\mathsf{z})$, then by *B1-7*, either $\mathsf{B}(\mathsf{y},\mathsf{z},\mathsf{x})$ or $\mathsf{B}(\mathsf{z},\mathsf{x},\mathsf{y})$. For each alternative derive a contradiction using (i).)

d. $\mathbf{E1}_{\mathbf{ind}} \vDash$ *B1-6.*
(Hint: We need to show that given any x, z, if $x \neq z$, then there is some y such that $x \neq y \neq z$ and $B(x, y, z)$. By *C1-5*, some y, $xy \approx yz$. Use Lemma 17 and *C1-3* to show that $x \neq y \neq z$. If not $B(x, y, z)$, then by *B1-7*, $B(y, z, x)$ or $B(z, x, y)$. Derive a contradiction from each alternative.)

5. *The axiom system* $\mathbf{E1}_{\mathbf{ind}}$ *is independent*
Show that each axiom of $\mathbf{E1}_{\mathbf{ind}}$ is independent by showing that it fails in the designated model while all the other axioms hold in that model:

B1-1	$\langle \varnothing ; \varnothing, \varnothing \rangle$	
C1-2	$\langle \mathbb{R} ; B, \approx \rangle$	where $B(a, b, c)$ iff ($a \leq b \leq c$ or $c \leq b \leq a$) $ab \approx cd$ iff $a - b = c - d$
C1-4	$\langle \{0, 1\} ; B, \approx \rangle$	where $B(a, b, c)$ iff $a = b$ or $b = c$ $ab \approx cd$ iff $a - b = c - d$ or $a - b = d - c$ or $c = d$
C1-5	$\langle \mathbb{Z} ; B, \approx \rangle$	where $B(a, b, c)$ iff ($a \leq b \leq c$ or $c \leq b \leq a$) $ab \approx cd$ iff $\lvert a - b \rvert = \lvert c - d \rvert$
C1-6	$\langle [0, 1] ; B, \approx \rangle$	where $B(a, b, c)$ is the usual betweenness relation on the line $ab \approx cd$ iff $\lvert a - b \rvert = \lvert c - d \rvert$
C1-7	$\langle U ; B, \approx \rangle$	U is the union of the non-negative *x*-axis and the non-positive *y*-axis in $\mathbb{R}^2$. Set $a < b$ iff in the usual metric on $\mathbb{R}^2$, a is closer to the origin than b. $B(a, b, c)$ iff ($a \leq b \leq c$ or $c \leq b \leq a$) $ab \approx cd$ iff in the usual metric for $\mathbb{R}^2$, the distance from a to b is the same as the distance from c to d.
B1-2	$\langle U ; U^3, U^4 \rangle$	where $U = \{0, 1, 2\}$
B1-3	$\langle (0, 1) ; B, \approx \rangle$	where $B(a, b, c)$ iff $b = c$ $ab \approx cd$ iff $\lvert a - b \rvert = \lvert c - d \rvert$
B1-4	$\langle \mathbb{Z}_3 ; B, \approx \rangle$	Here $\mathbb{Z}_3$ is the integers modulo 3, that is $\{0, 1, 2\}$ where $a + b = c$ iff the remainder of $a + b$ divided by 3 is c. This is a 2-divisible but not 3-divisible group. Define $ab \approx cd$ iff ($a + c = b + d$) or ($a + d = b + c$). Define B as the set of ordered triples: $\{(a, a, b), (a, b, b) : a, b \in \mathbb{Z}\} \cup \{(0, 1, 2), (1, 2, 0), (2, 0, 1), (2, 1, 0), (1, 0, 2), (0, 2, 1)\}$. That is, $B(a, b, c)$ iff ($a \neq b$ and $b \neq c$ and $a \neq c$) or ($a = b$ or $b = c$).
B1-7	$\langle U ; \varnothing, U^4 \rangle$	where $U = \{0, 1, 2\}$
BC1	$\langle \mathbb{R} ; B, \mathbb{R}^4 \rangle$	where B is the usual betweenness relation on the real line

It is open whether adding *Dedekind's axiom*$_1$*-B* yields an independent set of axioms, because it's not known if *B-4* is then independent.

6. *The independence of the primitive notions of* **E1***

The primitives ≡ and B are independent in the sense that neither can be defined in **E1***, and hence not in **E1**. To prove this, show the following.

a. If there are two models <U ; B, ≈> and <U ; B, ≈′> of **E1*** that differ only in their congruence relations, then ≡ is not definable in **E1***.
b. The function $a \mapsto a^3$ is a 1-1 correspondence of the real line that preserves the betweenness relation but does not preserve equidistance.
c. ≡ is not definable in **E1***.
d. If there are two models <U ; B, ≈> and <U ; B′, ≈> of **E1*** that differ only in their betweenness relations, then B is not definable in **E1***.
e. For the real line, $ab \approx cd$ iff $(a + d = b + c)$ or $(a + c = b + d)$.
f. We say that a 1-1 onto mapping f of the real line to itself is a *linear transformation* iff for every a, b, $f(a + b) = f(a) + f(b)$. Every linear transformation of the real line preserves congruence.
g. If there is a linear transformation of the real line that does not preserve betweenness, then B is not definable in **E1***.
h. Every linear transformation of the real line that is of the form $f(a) = \lambda a$ for some real number $\lambda \neq 0$ preserves betweenness.
i. From algebra, $\mathbb{R}$ can be viewed as a vector space over $\mathbb{Q}$. Any linearly independent set of elements of $\mathbb{R}$ can be extended to a basis of $\mathbb{R}/\mathbb{Q}$. So there is a basis of $\mathbb{R}/\mathbb{Q}$ of the form $B = \{1, \sqrt{2}\} \cup C$, where the cardinality of C is c and for any $a \in \mathbb{R}$, there are finitely many elements $\{b_1, \ldots, b_n\} \subset B$ and $\{q_1, \ldots, q_n\} \subset \mathbb{Q}$, such that $a = q_1 b_1 + \ldots + q_n b_n$. Define $f: B \to B$ by $f(1) = -1$, and $f(b) = b$ if $b \neq 1$. Then there is a unique way that f can be extended to a mapping g of $\mathbb{R}$ to itself. Show that g is linear transformation.
(Hint: It's an involution.)
j. Show that g does not preserve the betweenness relation on $\mathbb{R}$.
k. B is not definable in **E1***.

E. Named Parameters

Suppose we're reasoning about some things. We want to pick out one or a pair of those things, or some of those things that satisfy some condition, and give them names. The names are only a convenience for our reasoning, and any collection of things satisfying the condition will serve equally well as references for the names: Any model is homogeneous with respect to collections of objects satisfying the conditions. That is, given any model and any two collections of objects that could serve as references for the names, there is an automorphism of the model taking the one collection to the other. We call the particular things that we name *parameters*, though we also use that word for the names we add to our theory.

We can always add or eliminate parameters in our theories, as I'll now show.

Named parameters Let Σ be a theory in a language L, and B a formula in L such that:

a. $\Sigma \models \exists z_1 \ldots \exists z_n \, B(z_1, \ldots, z_n)$
where $z_1, \ldots, z_n$ is the list of all variables free in B.

b. For any model M of Σ, if $\{\mathsf{a}_1, \ldots, \mathsf{a}_n\}$ and $\{\mathsf{b}_1, \ldots, \mathsf{b}_n\}$ satisfy B, then there is an automorphism f of M such that for each $i \le n$, $\mathsf{f}(\mathsf{a}_i) = \mathsf{b}_i$.

Let $a_1, \ldots, a_n$ be any name symbols that do not appear in L.

The *extension by parameters of* Σ *by condition* B is:

$$\Sigma(a_1, \ldots, a_n) = \text{Th}(\Sigma \cup \{B(a_1/z_1, \ldots, a_n/z_n)\})$$

The mapping $\pi : L(a_1, \ldots, a_n) \to L$ *eliminating parameters from* $\Sigma(a_1, \ldots, a_n)$ is:

$$\pi(A) = \forall y_1 \ldots \forall y_n \, (B(y_1/a_1, \ldots, y_n/a_n) \to A(y_1/a_1, \ldots, y_n/a_n))$$

where $y_1, \ldots, y_n$ are the first n variables not appearing in A or B;
$\pi(A) = A$ if none of $a_1, \ldots, a_n$ appear in A

We also allow the case when nothing is assumed about the parameters, that is, there is no formula B and $\Sigma(a_1, \ldots, a_n)$ is just Σ in $L(a_1, \ldots, a_n)$. In that case, $\pi(A) = A(y_1/a_1, \ldots, y_n/a_n)$, where $y_1 \ldots y_n$ are the first n variables not appearing in A.

Theorem 51 Adding and eliminating parameters

a. If $\Sigma(a_1, \ldots, a_n)$ is an extension by parameters of Σ, then $\Sigma(a_1, \ldots, a_n)$ is a conservative extension of Σ. That is, for Γ, A in $L(\Sigma)$,
$\Gamma \models_{\Sigma} A$ iff $\Gamma \models_{\Sigma(a_1, \ldots, a_n)} A$.

b. If π is a map eliminating parameters, then
$\Gamma \models_{\Sigma(a_1, \ldots, a_n)} A$ iff $\pi(\Gamma) \models_{\Sigma} \pi(A)$.

Proof (a) I'll leave this to you.

(b) Let $\Sigma+ = \Sigma(a_1, \ldots, a_n)$ be the extension by parameters of Σ by condition B. Suppose first that $\pi(\Gamma) \not\models_{\Sigma} \pi(A)$. Then for some model N of Σ, $\mathsf{N} \models \pi(\Gamma)$ and $\mathsf{N} \not\models \pi(A)$. So there is some assignment of references σ on N such that:

$$\sigma \not\models \forall y_1 \ldots \forall y_n \, B(y_1/a_1, \ldots, y_n/a_n) \to A(y_1/a_1, \ldots, y_n/a_n)$$

Hence, $\sigma \models \exists y_1 \ldots \exists y_n \, (B(y_1/a_1, \ldots, y_n/a_n) \wedge \neg A(y_1/a_1, \ldots, y_n/a_n))$. So there is some τ such that

$$\tau \models B(y_1/a_1, \ldots, y_n/a_n) \text{ and } \tau \not\models A(y_1/a_1, \ldots, y_n/a_n).$$

Let N^* be N with a_i interpreted as $\tau(y_i)$ for $1 \le i \le n$. I'll leave to you to show that $\mathsf{N}^* \models \Gamma$ and N^* is a model of $\Sigma+$. But $\mathsf{N}^* \not\models A$. Hence, $\Gamma \not\models_{\Sigma+} A$.

Now suppose that $\pi(\Gamma) \vDash_{\Sigma} \pi(A)$. Since $\vDash_{\Sigma}$+B, by universal instantiation we have that $\pi(A) \vDash_{\Sigma}$+A. So to prove that $\Gamma \vDash_{\Sigma}$+A it is enough to show that $\Gamma \vDash_{\Sigma}+\pi(\Gamma)$. To do that we will show for any C in L, $C \vDash_{\Sigma}+\pi(C)$.

Suppose not. Then there is some model M of $\Sigma(a_1, \ldots, a_n)$, some formula C in L, and some assignment of references σ on M such that $M \vDash C(a_1, \ldots, a_n)$, $\sigma \vDash B(y_1/a_1, \ldots, y_n/a_n)$, and $\sigma \nvDash C(y_1/a_1, \ldots, y_n/a_n)$. Suppose that each a_i is interpreted as a_i in M and $\sigma(y_i) = \mathsf{b}_i$. Then $\{\mathsf{a}_1, \ldots, \mathsf{a}_n\}$ and $\{\mathsf{b}_1, \ldots, \mathsf{b}_n\}$ both satisfy B, so there is an automorphism f of M such that for each i, $\mathsf{f}(\mathsf{b}_i) = \mathsf{a}_i$. Let τ satisfy for each x, $\tau(x) = \mathsf{f}(\sigma(x))$. Since $\sigma \nvDash C(y_1/a_1, \ldots, y_n/a_n)$ and isomorphisms preserve truth (Theorem XIII.3), $\tau \nvDash C(y_1/a_1, \ldots, y_n/a_n)$. But $\tau \vDash C(y_1/a_1, \ldots, y_n/a_n)$ iff $M \vDash C(a_1, \ldots, a_n)$, so $M \nvDash C(a_1, \ldots, a_n)$, which is a contradiction, and we are done. ■

To apply this theorem to the mappings in this chapter we need only show that the appropriate automorphisms exist.

For the mapping eliminating parameters from $\mathbf{B1}_{01}$, since every model of that theory is infinite, by the Löwenheim-Skolem theorem and the proof above it is enough for condition (b) to apply to countable models. So given any countable model M = <U; B> of **B1** and $\mathsf{0} \neq \mathsf{1} \in \mathsf{U}$ and $\mathsf{a} \neq \mathsf{b} \in \mathsf{U}$, modify the proof that all countable models of **DLO** are isomorphic (Theorem XIII.5) to show that there is an isomorphism of M to itself taking 0 to a and 1 to b.

For the mappings eliminating parameters from $\mathbf{C1}_0$ the automorphisms are established in Lemma 34, and for $\mathbf{E1}_{01}$ the automorphisms are given in Lemma 44.

XIX Two-Dimensional Euclidean Geometry

in collaboration with **Lesław Szczerba**

A. The Axiom System E2

In this chapter we'll continue our geometric analysis of the real numbers by formalizing the geometry of flat surfaces. Our goal is to give a theory that is equivalent to the theory of real numbers presented in Chapter XVII.

Our axiomatization of two-dimensional geometry will use the same primitives as for one-dimension: points and the relations of betweenness and congruence. Lines and other geometric figures and relations, which others often take as primitive, will be definable. Roughly, since two points determine a line, we can define a line as all those points lying in the betweenness relation with respect to two given points. Then we can quantify over lines as "pseudo-variables" by quantifying over pairs of points.

So, as in Chapter XVIII, our formal language will be $L(=; P_0^3, P_0^4)$, which again we can write as $L(=; B, \equiv)$ with the same notational conventions as before.

E2, *basic two-dimensional Euclidean geometry* in L(=; B, ≡)

E1 $\exists x \exists y \; x \neq y$

E2 $\forall x \forall y \exists r \exists s \; \mathrm{B}(x, y, r) \wedge yx \equiv ys \wedge yr \equiv ys \wedge sx \equiv sr$

E3 $\forall x \forall z \exists y \; \mathrm{B}(x, y, z) \wedge yx \equiv yz$

E4 $\forall x \forall y \; \mathrm{B}(x, x, y)$

E5 $\forall x \forall y \; \mathrm{B}(x, y, x) \rightarrow x = y$

E6 $\forall x \forall y \forall z \forall r \forall s \; \mathrm{B}(x, y, z) \wedge \mathrm{B}(r, z, s) \rightarrow \exists p\, (\mathrm{B}(r, y, p) \wedge \mathrm{B}(x, p, s))$

E7 $\forall x \forall y \forall z \forall w \; \mathrm{B}(x, y, z) \wedge \mathrm{B}(x, z, w) \rightarrow \mathrm{B}(y, z, w)$

E8 $\forall x \forall y \forall z \forall w \forall r \forall s \; (xy \equiv rs \wedge zw \equiv rs) \rightarrow xy \equiv zw$

E9 $\forall x \forall y \forall z \forall r \forall s \forall t$
$$((xy \equiv rs \wedge yz \equiv st \wedge zx \equiv tr) \wedge \mathrm{B}(x, y, z)) \rightarrow \mathrm{B}(r, s, t)$$

E10 $\forall x \forall y \forall z \forall r \forall s \forall t$
$$(\mathrm{B}(x, y, z) \wedge \mathrm{B}(r, s, t) \wedge yx \equiv yz \wedge sx \equiv sz \wedge yr \equiv yt \wedge sr \equiv st) \rightarrow xr \equiv zt$$

E11 $\forall x \forall y \forall z \exists r \exists s \; \mathrm{B}(x, y, r) \wedge \mathrm{B}(y, x, s) \wedge zr \equiv zs$

E12 $\forall x \forall y \forall z \forall r \forall s \; (x \neq y \wedge \mathrm{B}(x, y, z) \wedge xr \equiv xs \wedge yr \equiv ys) \rightarrow zr \equiv zs$

E13 $\forall x \forall y \forall z \forall r \forall s \; (r \neq s \wedge xr \equiv xs \wedge yr \equiv ys \wedge zr \equiv zs) \rightarrow$
$$\mathrm{B}(x, y, z) \vee \mathrm{B}(y, z, x) \vee \mathrm{B}(z, x, y)$$

E14 $\forall x \forall y \forall z \; \neg\mathrm{B}(x, y, z) \wedge \neg\mathrm{B}(y, z, x) \wedge \neg\mathrm{B}(z, x, y) \rightarrow \exists r\, (rx \equiv ry \wedge ry \equiv rz)))$

A model M of **E2** will be $\langle \mathsf{U}; =, \mathsf{B}, \approx \rangle$.

Motivation of the axiom system

A short informal motivation of the axioms may help in understanding the proofs. I'll assume that we have a model in which B is interpreted as B and ≡ is interpreted as ≈.

E1 $\exists x \exists y\, (x \neq y)$

This is the only axiom that begins with an existential quantifier. It ensures that our models are not trivial.

E2 $\forall x\, \forall y\, \exists r\, \exists s\, (\mathrm{B}(x, y, r) \wedge yx \equiv ys \wedge yr \equiv ys \wedge sx \equiv sr)$

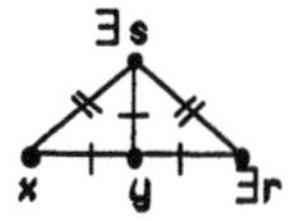

Given **x** and **y**, there are **r** and **s** such that we can draw an isoceles right triangle and duplicate it. This is a dimension axiom:
It's true in two or more dimensions, but false of the line.
(Congruent line segments are pictured with the same number of slashes through them.)

E3 $\forall x\, \forall z\, \exists y\ \mathrm{B}(x, y, z) \wedge yx \equiv yz$
This is a midpoint axiom.

E4 $\forall x\, \forall y\ \mathrm{B}(x, x, y)$
This is Lemma 1.a of our axiomatization of betweenness in one-dimension.

E5 $\forall x\, \forall y\ \mathrm{B}(x, y, x) \rightarrow x = y$
This is *B1-2* from our axiomatization of betweenness in one-dimension.

E6 $\forall x\, \forall y\, \forall z\, \forall r\, \forall s\ \mathrm{B}(x, y, z) \wedge \mathrm{B}(r, z, s) \rightarrow \exists p\, (\mathrm{B}(r, y, p) \wedge \mathrm{B}(x, p, s))$

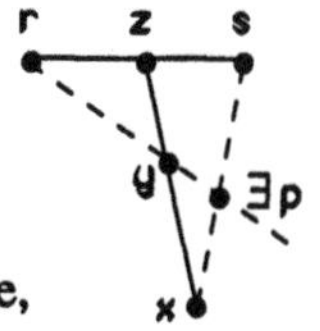

This is equivalent to: *Given any (non-degenerate) triangle, there is no line passing through all three sides.*

This is a weak version of Pasch's axiom: *Given a triangle, if a line passes through one side, it passes through another side.* Pasch's axiom contains dimension, since in three dimensions you could miss a side. This one, being weaker, does not contain dimension, since the line **rp** passes through the extension of **zs**, all the lines must lie on the same plane, even in three dimensions.

E7 $\forall x\, \forall y\, \forall z\, \forall w\ \ \mathrm{B}(x, y, z) \wedge \mathrm{B}(x, z, w) \rightarrow \mathrm{B}(y, z, w)$
This is axiom *B1-4*.

E8 $\forall x\, \forall y\, \forall z\, \forall w\, \forall r\, \forall s\ \ (xy \equiv rs \wedge zw \equiv rs) \rightarrow xy \equiv zw$
This is axiom *C1-4*.

E9 $\forall x\, \forall y\, \forall z\, \forall r\, \forall s\, \forall t\ \ [\,(xy \equiv rs \wedge yz \equiv st \wedge zx \equiv tr) \wedge \mathrm{B}(x, y, z)] \rightarrow \mathrm{B}(r, s, t)$
This is axiom *BC1*.

E10 $\forall x\, \forall y\, \forall z\, \forall r\, \forall s\, \forall t\ \ (\mathrm{B}(x, y, z) \wedge \mathrm{B}(r, s, t) \wedge$
$yx \equiv yz \wedge sx \equiv sz \wedge yr \equiv yt \wedge sr \equiv st) \rightarrow xr \equiv zt$

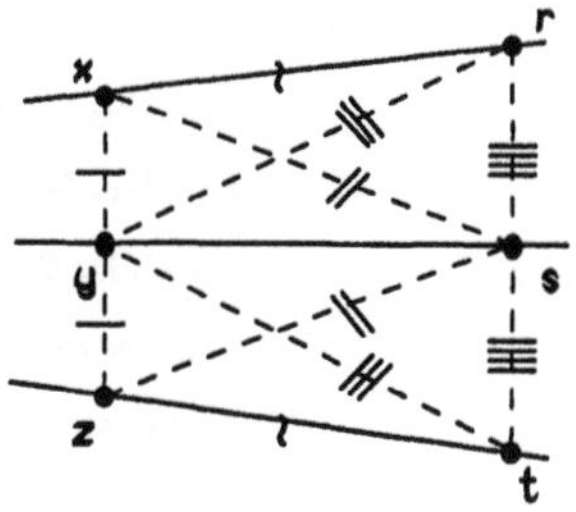

This says that line symmetry is an isometry.
In one dimension it yields that point symmetry is an isometry.

E11 $\forall x\, \forall y\, \forall z\, \exists r\, \exists s\ \ \mathrm{B}(x, y, r) \wedge \mathrm{B}(y, x, s) \wedge zr \equiv zs$
Given a line segment, we can draw a circle that encloses it.
This replaces an axiom of Euclid: *Given any point and any length, we can draw a circle with that point as center and with that length as radius.*

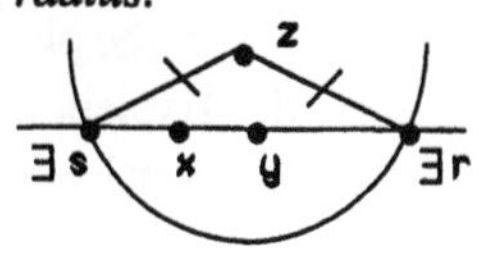

E12 $\forall x \forall y \forall z \forall r \forall s\, (x \neq y \wedge B(x, y, z) \wedge xr \equiv xs \wedge yr \equiv ys) \rightarrow zr \equiv zs$
If x and y all lie on the bisector of the line through r and s, then z must lie on it, too

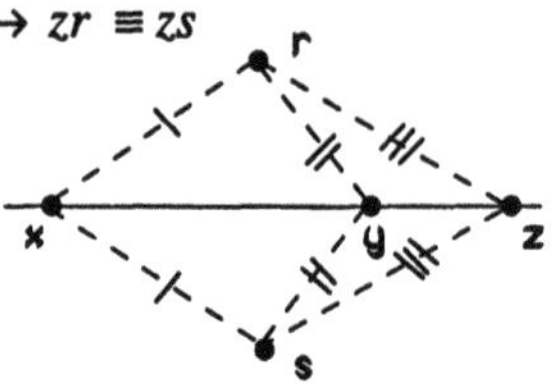

E13 $\forall x \forall y \forall z \forall r \forall s\, (r \neq s \wedge xr \equiv xs \wedge yr \equiv ys \wedge zr \equiv zs) \rightarrow$
$B(x, y, z) \vee B(y, z, x) \vee B(z, x, y)$
If x, y, z, all lie on the bisector of the line through r and s, then x, y, z are collinear. This is a dimension axiom: It's not true in three dimensions because the bisector is then a plane.

E14 $\forall x \forall y \forall z\, (\neg B(x, y, z) \wedge \neg B(y, z, x) \wedge \neg B(z, x, y)) \rightarrow \exists r\, (rx \equiv ry \wedge ry \equiv rz))$
Given a triangle, there is a circle through its vertices.
This is equivalent to Euclid's original parallel postulate, which uses more complex notions: *If a straight line falling on two straight lines makes the interior angles on the same side less than two right angles, the two straight lines, if produced indefinitely, meet on that side on which the angles are less than the two right angles.* (*Boyer, 1968*, p. 116).
In the 17th century Euclid's axiom was replaced by: *Given a line and a point not on the line, there is exactly one line through the point parallel to the line.* We'll derive that in our system.

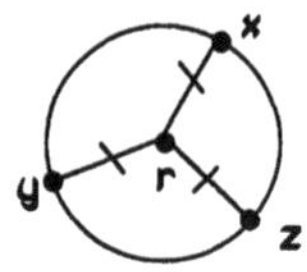

Exercises for Section A

1. Write out axiom E3 in the fully formal language.
2. a. Draw a picture for *E9*. Explain in your own words why it's true of the plane.
   b. Is it true of the line?
   c. Is it true of three dimensional space?
   d. How did you justify your answers to (b) and (c)?
3. a. Draw a picture for *E13*. Explain in your own words why it's true of the plane.
   b. Show that it's true of the line, but not true of three-dimensional space.
   c. How did you justify your answer to (b)?
4. What axiom is used in place of Euclid's parallel postulate?
5. a. Define the Cartesian plane over the reals to be $\mathbf{R}^2 = \langle R^2; B, \approx \rangle$ where if $a = (a_1, a_2)$, $b = (b_1, b_2)$, and $c = (c_1, c_2)$, and $d = (d_1, d_2)$, then:
   $B(a, b, c)$ iff there is some $t \in R$ with $0 \leq t \leq 1$ and $(1 - t)a + tc = b$
   $ab \approx cd$ iff $(b_1 - a_1)^2 + (b_2 - a_2)^2 = (d_1 - c_1)^2 + (d_2 - c_2)^2$
   Verify that the Cartesian plane is a model of **E2**.
   b. Justify the definition of congruence in the Cartesian plane over the reals in terms of the Pythagoras' theorem.
   c. Justify the definition of betweenness in the Cartesian plane in terms of the geometry of the plane.
   d. Explain why you do or do not find (a–c) a more convincing demonstration of the consistency of **E2** than the informal motivation of the axioms.

B. Deriving Geometric Notions

1. Basic properties of the primitive notions

Lemma 1

a. $\forall x\, \forall y\, \forall z\;\; \mathrm{B}(x, y, z) \rightarrow \mathrm{B}(z, y, x)$

b. $\forall x\, \forall y\;\; xy \equiv xy$

c. $\forall x\, \forall y\, \forall z\, \forall w\;\; xy \equiv zw \rightarrow xy \equiv zw$

d. $\forall x\, \forall y\;\; xy \equiv yx$

e. $\forall x\, \forall y\, \forall z\, \forall w\, \forall r\, \forall s\;\; (xy \equiv rs \wedge zw \equiv rs) \rightarrow zw \equiv xy$

f. $\forall x\, \forall y\;\; xx \equiv yy$

g. $\forall x\, \forall y\, \forall z\;\; xy \equiv zz \rightarrow x = y$

h. $\forall x\, \forall y\, \forall z\, \forall w\;\; xy \equiv zw \rightarrow xy \equiv wz$

i. $\forall x\, \forall y\, \forall z\, \forall w\;\; \mathrm{B}(x, y, w) \wedge \mathrm{B}(y, z, w) \rightarrow \mathrm{B}(x, y, z)$

j. $\forall x\, \forall y\, \exists z\;\; \mathrm{B}(x, y, z) \wedge xy \equiv yz$

k. $\forall x\, \forall y\, \forall z\;\; (x \neq y \wedge yx \equiv yz) \rightarrow y \neq z$

Proof (a) Assume $\mathsf{B}(\mathsf{x}, \mathsf{y}, \mathsf{z})$. By *E4* we have $\mathsf{B}(\mathsf{z}, \mathsf{z}, \mathsf{x})$. By *E6* there is then some w such that $\mathsf{B}(\mathsf{z}, \mathsf{y}, \mathsf{w})$ and $\mathsf{B}(\mathsf{x}, \mathsf{w}, \mathsf{x})$. By *E5*, this means that $\mathsf{x} = \mathsf{w}$. So we have $\mathsf{B}(\mathsf{z}, \mathsf{y}, \mathsf{x})$.

(b) By *E2*, there is a z such that $\mathsf{xy} \approx \mathsf{xz}$. So by *E8*, $\mathsf{xy} \approx \mathsf{xy}$. (c) is similar.

(d) By *E3*, there is a z such that $\mathsf{B}(\mathsf{x}, \mathsf{z}, \mathsf{y})$ and $\mathsf{zx} \approx \mathsf{zy}$. So by (a) and (c), $\mathsf{B}(\mathsf{y}, \mathsf{z}, \mathsf{x})$ and $\mathsf{zy} \approx \mathsf{zx}$. So by *E10*, $\mathsf{xy} \approx \mathsf{yx}$.

(e) This follows by *E8* and (c).

(f) By *E3*, there is a z such that $\mathsf{B}(\mathsf{x}, \mathsf{z}, \mathsf{y})$ and $\mathsf{zx} \approx \mathsf{zy}$. So by *E10*, $\mathsf{xx} \approx \mathsf{yy}$.

(g) Suppose $\mathsf{xy} \approx \mathsf{zz}$. By (c), $\mathsf{zz} \approx \mathsf{xy}$ and by (d), $\mathsf{zz} \approx \mathsf{yx}$. By (f), $\mathsf{zz} \approx \mathsf{xx}$. By *E4*, $\mathsf{B}(\mathsf{z}, \mathsf{z}, \mathsf{z})$. So by *E9*, $\mathsf{B}(\mathsf{x}, \mathsf{y}, \mathsf{x})$. So by *E5*, $\mathsf{x} = \mathsf{y}$.

(h) Suppose $\mathsf{zy} \approx \mathsf{zw}$. By Lemma (d), $\mathsf{zw} \approx \mathsf{wz}$. So by *E8* we are done.

(i) Suppose $\mathsf{B}(\mathsf{x}, \mathsf{y}, \mathsf{w})$ and $\mathsf{B}(\mathsf{y}, \mathsf{z}, \mathsf{w})$. Then by (a), $\mathsf{B}(\mathsf{w}, \mathsf{z}, \mathsf{y})$ and $\mathsf{B}(\mathsf{w}, \mathsf{y}, \mathsf{x})$. Hence, by *E7*, $\mathsf{B}(\mathsf{z}, \mathsf{y}, \mathsf{x})$. So by (a), $\mathsf{B}(\mathsf{x}, \mathsf{y}, \mathsf{z})$.

(j) Given x and y, by *E2*, there are z, t such that $\mathsf{B}(\mathsf{x}, \mathsf{y}, \mathsf{z})$ and $\mathsf{yx} \approx \mathsf{yt}$ and $\mathsf{yz} \approx \mathsf{yt}$. So by (e), $\mathsf{yx} \approx \mathsf{yz}$, and so by (d), $\mathsf{xy} \approx \mathsf{yz}$. ■

2. Lines

The betweenness relation allows us to define what it means for three points to lie in a line, namely, $\mathrm{B}(x, y, z) \vee \mathrm{B}(y, z, x) \vee \mathrm{B}(z, x, y)$. So given any two points in a model, x and y, we can call all those points that are in this relation to them a line. But our choice of axioms makes it easier to prove the properties of lines by first considering bisectors of line segments and then showing that every bisector is a line and every line is a bisector.

The bisector of the segment xy

$C_{xy}(z) \equiv_{\text{Def}} xz \equiv yz \wedge x \neq y$

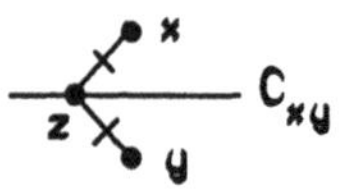

Given $\mathbf{x}$, $\mathbf{y}$ in a model as references for x and y, the interpretation of C_{xy} is a set, $\mathsf{C}_{\mathbf{xy}} = \{\mathbf{z} : \mathbf{xz} \approx \mathbf{yz}\}$.

Pseudo-variables for quantifying over bisectors We can use the letters C, D, E, . . . as pseudo-variables. If A(C) is a wff in which C appears as a unary predicate,

$\forall C\, A(C)$ stands for $\forall x \forall y\ (x \neq y \rightarrow A(C_{xy}))$
where x and y are the least variables not appearing in A

$\exists C\, A(C)$ stands for $\exists x \exists y\ (x \neq y \wedge A(C_{xy}))$
where x and y are the least variables not appearing in A

$C = D$ stands for $\forall z\ (C(z) \leftrightarrow D(z))$
where z is the least variable not appearing in C or D

Quantifying over bisectors this way amounts to quantifying over their extensions, that is, over certain subsets of the universe, and I'll write $z \in C$ for $C(z)$. I'll use $\mathsf{C}, \mathsf{D}, \mathsf{E}, \ldots$ for the interpretations of these pseudo-variables.

Lemma 2

a. $\forall x \forall y\ \ C_{xy} = C_{yx}$

b. $\forall x \forall y\ \ \neg C_{xy}(y)$

c. *Every pair of points lies on a bisector*
$\forall x \forall y\ \exists C\ (x \in C \wedge y \in C)$

d. *There are at least two points on every bisector*
$\forall C\, \exists x \exists y\ (x \neq y \wedge x \in C \wedge y \in C)$

Proof (c) Without the pseudo-variable part (c) is: $\forall x \forall y\ (\exists z \exists w\ (z \neq w \wedge C_{zw}(x) \wedge C_{zw}(y)))$. And this means: $\forall x \forall y\ (\exists z \exists w\ (z \neq w \wedge zx \equiv wx \wedge zy \equiv wy))$.

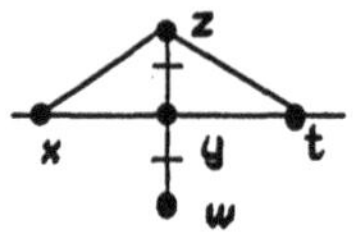

Suppose $\mathbf{x} \neq \mathbf{y}$. Then by *E2*, there are $\mathbf{t}$, $\mathbf{z}$ such that $B(\mathbf{x}, \mathbf{y}, \mathbf{t})$ and $\mathbf{yx} \approx \mathbf{yz} \approx \mathbf{yt}$ and $\mathbf{zx} \approx \mathbf{zt}$. Since $\mathbf{x} \neq \mathbf{y}$ and $\mathbf{yx} \approx \mathbf{yz}$, we have $\mathbf{y} \neq \mathbf{z}$. Now by Lemma 1.j, there is a $\mathbf{w}$ such that $\mathsf{B}(\mathbf{z}, \mathbf{y}, \mathbf{w})$ and $\mathbf{yz} \approx \mathbf{yw}$. We need that $\mathbf{xz} \approx \mathbf{xw}$. By *E10* and Lemma 1, we have $\mathbf{wx} \approx \mathbf{zt}$, so $\mathbf{wx} \approx \mathbf{zx}$, and so $\mathbf{xz} \equiv \mathbf{xw}$. Hence $\mathbf{x}, \mathbf{y} \in \mathsf{C}_{\mathbf{zw}}$.

Now suppose $\mathbf{x} = \mathbf{y}$. By *E1* there are points $\mathbf{z} \neq \mathbf{w}$. If $\mathbf{x} \neq \mathbf{z}$, then by what we've just proved, there is a C such that $\mathbf{x}, \mathbf{z} \in \mathsf{C}$. If $\mathbf{x} \neq \mathbf{w}$, there is a C such that $\mathbf{x}, \mathbf{w} \in \mathsf{C}$.

(d) Let $\mathsf{C} = \mathsf{C}_{\mathbf{zw}}$ where $\mathbf{z} \neq \mathbf{w}$. By *E3* there is an $\mathbf{x}$ such that $\mathsf{B}(\mathbf{z}, \mathbf{x}, \mathbf{w})$ and $\mathbf{zx} \approx \mathbf{zw}$. Hence, $\mathbf{x} \neq \mathbf{z}$. By *E2*, there are $\mathbf{y}$, $\mathbf{q}$ such that $\mathsf{B}(\mathbf{z}, \mathbf{x}, \mathbf{q})$ and $\mathbf{xz} \approx \mathbf{xy} \approx \mathbf{xq}$ and $\mathbf{yz} \approx \mathbf{yq}$. By *E10*, $\mathbf{zz} \approx \mathbf{wq}$, so $\mathbf{w} = \mathbf{q}$. Hence, $\mathbf{yz} \approx \mathbf{yw}$, and both $\mathbf{y}, \mathbf{x} \in \mathsf{C}_{\mathbf{zw}}$. Also, since $\mathbf{xy} \approx \mathbf{xq} \approx \mathbf{xw}$ and $\mathbf{x} \neq \mathbf{w}$, we have $\mathbf{x} \neq \mathbf{y}$. ∎

Lemma 3 ***Bisectors are closed under betweenness***

a. $\forall C\, \forall x\, \forall y\, \forall z\ (x \in C \wedge y \in C \wedge x \neq y \wedge B(x, y, z)) \rightarrow z \in C$

b. $\forall C\, \forall x\, \forall y\, \forall x\ (x \in C \wedge y \in C \wedge B(x, y, z)) \rightarrow z \in C$

Proof (a) Suppose $\mathsf{x} \in \mathsf{C}_{\mathsf{pq}}$. That is, $\mathsf{p} \neq \mathsf{q}$ and $\mathsf{x} \neq \mathsf{y}$ and $\mathsf{xp} \approx \mathsf{xq}$. Suppose also that $\mathsf{py} \approx \mathsf{qy}$ and $\mathsf{B}(\mathsf{x}, \mathsf{y}, \mathsf{z})$. Then by *E12*, $\mathsf{pz} \approx \mathsf{qz}$. That is, $\mathsf{z} \in \mathsf{C}_{\mathsf{pq}}$.

(b) Suppose that $\mathsf{p} \neq \mathsf{q}$ and $\mathsf{x} \in \mathsf{C}_{\mathsf{pq}}$ and $\mathsf{z} \in \mathsf{C}_{\mathsf{pq}}$ and $\mathsf{B}(\mathsf{x}, \mathsf{y}, \mathsf{z})$. If $\mathsf{x} = \mathsf{z}$, then $\mathsf{y} = \mathsf{x}$ by *E5*, so $\mathsf{y} \in \mathsf{C}_{\mathsf{pq}}$. If $\mathsf{x} \neq \mathsf{z}$, then by *E2* there are r, s such that $\mathsf{B}(\mathsf{x}, \mathsf{z}, \mathsf{s})$ and $\mathsf{zx} \approx \mathsf{zs} \approx \mathsf{zr}$. So $\mathsf{z} \neq \mathsf{s}$. By Lemma 1.a, $\mathsf{B}(\mathsf{z}, \mathsf{y}, \mathsf{x})$ and $\mathsf{B}(\mathsf{s}, \mathsf{z}, \mathsf{x})$, so by Lemma 1.i, $\mathsf{B}(\mathsf{s}, \mathsf{z}, \mathsf{y})$. And by (a), $\mathsf{s} \in \mathsf{C}_{\mathsf{pq}}$. But also $\mathsf{z} \in \mathsf{C}_{\mathsf{pq}}$. So by (a), $\mathsf{y} \in \mathsf{C}_{\mathsf{pq}}$. ■

Collinearity

$L(x, y, z) \equiv_{\text{Def}} \exists r\, \exists s\ (r \neq s \wedge C_{rs}(x) \wedge C_{rs}(y) \wedge C_{rs}(z))$

With our conventions this is $L(x, y, z) \equiv_{\text{Def}} \exists C\, (x \in C \wedge y \in C \wedge z \in C)$.

Lemma 4

a. $\forall x\, \forall y\, \forall z\ \ L(x, y, z) \leftrightarrow B(x, y, z) \vee B(y, z, x) \vee B(z, x, y)$

b. $\forall x\, \forall y\, \forall z\, \forall C\, (x \neq y \wedge x \in C \wedge y \in C) \wedge L(x, y, z)) \rightarrow z \in C$

c. *Two points determine a bisector*

$\forall x\, \forall y\, \forall C\, \forall D\, (x \neq y \wedge (x \in C \wedge y \in C) \wedge (x \in D \wedge y \in D)) \rightarrow C = D$

d. $\forall x\, \forall y\ \ L(x, x, y)$

e. $\forall x\, \forall y\, \forall z\ \ L(x, y, z) \rightarrow (L(y, x, z) \wedge L(x, z, y))$

f. $\forall x\, \forall y\, \forall z\, \forall r\, \forall s\ \ (r \neq s \wedge L(r, s, x) \wedge L(r, s, y) \wedge L(r, s, z)) \rightarrow L(x, y, z)$

Proof (a) In a model, let L be the interpretation of the defined predicate L. Suppose $\mathsf{L}(\mathsf{x}, \mathsf{y}, \mathsf{z})$. Then for some p, q with $\mathsf{p} \neq \mathsf{q}$, $\mathsf{C}_{\mathsf{pq}}(\mathsf{x})$ and $\mathsf{C}_{\mathsf{pq}}(\mathsf{y})$ and $\mathsf{C}_{\mathsf{pq}}(\mathsf{z})$. So $\mathsf{px} \approx \mathsf{qx}$, $\mathsf{py} \approx \mathsf{qy}$, and $\mathsf{pz} \approx \mathsf{qz}$. So by *E13*, $\mathsf{B}(\mathsf{x}, \mathsf{y}, \mathsf{z})$ or $\mathsf{B}(\mathsf{y}, \mathsf{z}, \mathsf{x})$ or $\mathsf{B}(\mathsf{z}, \mathsf{x}, \mathsf{y})$.

In the other direction, suppose $\mathsf{B}(\mathsf{x}, \mathsf{y}, \mathsf{z})$ or $\mathsf{B}(\mathsf{y}, \mathsf{z}, \mathsf{x})$ or $\mathsf{B}(\mathsf{z}, \mathsf{x}, \mathsf{y})$. If $\mathsf{x} = \mathsf{y}$, then by Lemma 2.c, some $\mathsf{p} \neq \mathsf{q}$, we have $\mathsf{x}, \mathsf{y}, \mathsf{z} \in \mathsf{C}_{\mathsf{pq}}$. So $\mathsf{L}(\mathsf{x}, \mathsf{y}, \mathsf{z})$. On the other hand, if $\mathsf{x} \neq \mathsf{y}$, then by Lemma 2.c, for some $\mathsf{p} \neq \mathsf{q}$, we have $\mathsf{x}, \mathsf{y} \in \mathsf{C}_{\mathsf{pq}}$, and by Lemma 3 we have $\mathsf{z} \in \mathsf{C}_{\mathsf{pq}}$.

(c) Given $\mathsf{x} \neq \mathsf{y}$, and $\mathsf{p} \neq \mathsf{q}$, and $\mathsf{r} \neq \mathsf{s}$, and $\mathsf{C}_{\mathsf{pq}}(\mathsf{x})$, $\mathsf{C}_{\mathsf{pq}}(\mathsf{y})$, and $\mathsf{C}_{\mathsf{rs}}(\mathsf{x})$, $\mathsf{C}_{\mathsf{rs}}(\mathsf{y})$. Suppose $\mathsf{C}_{\mathsf{pq}}(\mathsf{z})$. Then by definition, $\mathsf{L}(\mathsf{x}, \mathsf{y}, \mathsf{z})$. Hence, by part (b), $\mathsf{C}_{\mathsf{rs}}(\mathsf{z})$. Similarly, if $\mathsf{C}_{\mathsf{rs}}(\mathsf{z})$, then $\mathsf{C}_{\mathsf{pq}}(\mathsf{z})$. ■

Lines

$L_{xy}(z) \equiv_{\text{Def}}\ x \neq y \wedge (B(x, y, z) \vee B(y, z, x) \vee B(z, x, y))$

If $\mathsf{x} \neq \mathsf{y}$ are references for x and y, the interpretation of L_{xy} is a set $\mathsf{L}_{\mathsf{xy}} = \{\mathsf{z}\colon \mathsf{B}(\mathsf{x}, \mathsf{y}, \mathsf{z}) \text{ or } \mathsf{B}(\mathsf{y}, \mathsf{z}, \mathsf{x}) \text{ or } \mathsf{B}(\mathsf{z}, \mathsf{x}, \mathsf{y})\}$, *the line through* x *and* y.

Pseudo-variables for lines We can use K, L, M, N, . . . to range over lines, where:

$\forall L\ A(L)$ stands for $\forall x\, \forall y\, (x \neq y \rightarrow A(L_{xy}))$
where x and y are the least variables not appearing in A

$\exists L\ A(L)$ stands for $\exists x\, \exists y\, (x \neq y \rightarrow A(L_{xy}))$
where x and y are the least variables not appearing in A

$K = L$ stands for $\forall z\, (K(z) \leftrightarrow L(z))$
where z is the least variable not appearing in K or L

I'll write $z \in L$ for $L(z)$ and use $\mathsf{K}, \mathsf{L}, \mathsf{M}, \mathsf{N}, \ldots$ to range over lines in a model, that is, as the extensions L_{xy} of L_{xy} where (x, y) are pairs of points such that $\mathsf{x} \neq \mathsf{y}$.

Lemma 5 *Bisectors and lines are the same*

a. $\forall x\, \forall y\, \forall z\ \ x \neq y \rightarrow (L(x, y, z) \leftrightarrow z \in L_{xy})$

b. $\forall x\, \forall y\ \ L_{xy}(x)$

c. $\forall x\, \forall y\ \ L_{xy} = L_{yx}$

d. $\forall x\, \forall y\, \forall C\ \ (x \neq y \wedge x \in C \wedge y \in C) \rightarrow C = L_{xy}$

e. *Every bisector is a line*
$\forall x\, \forall y\ \ x \neq y \rightarrow \exists z\, \exists w\, \forall p\, (C_{xy}(p) \leftrightarrow L_{zw}(p))$

f. *Every line is a bisector*
$\forall x\, \forall y\ \ x \neq y \rightarrow \exists z\, \exists w\, \forall p\, (L_{xy}(p) \leftrightarrow C_{zw}(p))$

g. $\forall x\, \forall y\, \forall z\ \ (L(x, y, z) \wedge xy \equiv yz \wedge x \neq z) \rightarrow B(x, y, z)$

Proof (a) By Lemma 4.a. (b) From *E4*. (c) By definition and Lemma 1.a.

(d) Suppose $\mathsf{x} \neq \mathsf{y}$ and $\mathsf{x} \in \mathsf{C}$ and $\mathsf{y} \in \mathsf{C}$. If $\mathsf{z} \in \mathsf{C}$, then by definition $\mathsf{L}(\mathsf{x}, \mathsf{y}, \mathsf{z})$, so by (a), $\mathsf{z} \in \mathsf{L}_{\mathsf{xy}}$. If $\mathsf{z} \in \mathsf{L}_{\mathsf{xy}}$, then by (a), $\mathsf{L}(\mathsf{x}, \mathsf{y}, \mathsf{z})$, so by Lemma 4.b, $\mathsf{z} \in \mathsf{C}$.

(e) Given a bisector C, by Lemma 2.d, there are two points on it, x and y. So by (d), $\mathsf{C} = \mathsf{L}_{\mathsf{xy}}$.

(f) Given $\mathsf{x} \neq \mathsf{y}$. By Lemma 2.c, there is some C such that $\mathsf{x}, \mathsf{y} \in \mathsf{C}$. We need to show that $\mathsf{L}_{\mathsf{xy}} = \mathsf{C}$. Suppose $\mathsf{p} \in \mathsf{L}_{\mathsf{xy}}$. Then by (a), $\mathsf{L}(\mathsf{x}, \mathsf{y}, \mathsf{p})$. So by Lemma 4.b, $\mathsf{p} \in \mathsf{C}$. Now suppose $\mathsf{p} \in \mathsf{C}$. If $\mathsf{z} = \mathsf{x}$ we're done. If $\mathsf{p} \neq \mathsf{x}$, since $\mathsf{x}, \mathsf{p} \in \mathsf{C}$, by (e) $\mathsf{L}_{\mathsf{xp}} = \mathsf{C}$, so $\mathsf{p} \in \mathsf{C}$.

(g) Suppose $\mathsf{L}(\mathsf{x}, \mathsf{y}, \mathsf{z})$ and $\mathsf{x} \neq \mathsf{z}$. Then (i) $\mathsf{B}(\mathsf{x}, \mathsf{y}, \mathsf{z})$ or (ii) $\mathsf{B}(\mathsf{y}, \mathsf{z}, \mathsf{x})$ or (iii) $\mathsf{B}(\mathsf{z}, \mathsf{x}, \mathsf{y})$. If (i) we are done. If (ii), then also $\mathsf{B}(\mathsf{x}, \mathsf{z}, \mathsf{y})$. If $\mathsf{xy} \approx \mathsf{yz}$, by *E9*, $\mathsf{B}(\mathsf{z}, \mathsf{x}, \mathsf{y})$. So by Lemma 1.i, $\mathsf{B}(\mathsf{x}, \mathsf{z}, \mathsf{x})$, which by *E5* yields $\mathsf{x} = \mathsf{z}$, a contradiction. If (iii), then by the same reasoning, $\mathsf{y} = \mathsf{z}$, and hence $\mathsf{xz} \approx \mathsf{zz}$, and so by Lemma 1.g, $\mathsf{x} = \mathsf{z}$, a contradiction. So we have (i). ■

3. One-dimensional geometry and point symmetry

We can now show that lines in two-dimensional geometry are models of one-dimensional geometry.

Theorem 6

a. If $<\mathsf{U}; \mathsf{B}, \approx> \models$ **E2**, and $\mathsf{x}, \mathsf{y} \in \mathsf{U}$, and $\mathsf{x} \neq \mathsf{y}$, then $<\mathsf{L}_{\mathsf{xy}}; \mathsf{B}, \approx> \models$ **E1**.

b. If $<\mathsf{U}; \mathsf{B}, \approx> \models_2$(**E2** $\wedge$ *Dedekind's axiom-B*), and $\mathsf{x}, \mathsf{y} \in \mathsf{U}$, and $\mathsf{x} \neq \mathsf{y}$, then $<\mathsf{L}_{\mathsf{xy}}; \mathsf{B}, \approx> \models$ **E1***.

Proof I'll let you show that if $\mathsf{M} = <\mathsf{L}_{\mathsf{xy}}; \mathsf{B}, \approx>$, then each axiom of **E1** holds in M (use Exercise 4 of Chapter XVIII.D to reduce the list of axioms). For part (b), we only need that lines are closed under betweenness, which we have by Lemma 3.b. ∎

Now we can mimic much of what we did for one-dimensional geometry.

The midpoint relation

$\mathrm{M}(x, y, z) \equiv_{\mathrm{Def}} \mathrm{B}(x, y, z) \wedge xy \equiv yz$

I'll write the interpretation of M in a model as m.

Lemma 7

a. $\forall x\, \forall y\, \forall z\ (\mathrm{M}(x, y, z) \wedge x \in \mathrm{L} \wedge y \in \mathrm{L}) \rightarrow z \in \mathrm{L}$

b. $\forall x\, \forall y\, \forall z\ (\mathrm{M}(x, y, z) \wedge x \in \mathrm{L} \wedge z \in \mathrm{L}) \rightarrow y \in \mathrm{L}$

c. $\forall x\, \forall y\, \forall z\ \mathrm{M}(x, y, z) \leftrightarrow \mathrm{M}(z, y, x)$

Lemma 8 Given a model $<\mathsf{U}; \mathsf{B}, \approx>$ of **E2**, if any two of $\mathsf{x}, \mathsf{y}, \mathsf{z}$ are elements of L, then $\mathsf{m}(\mathsf{x}, \mathsf{y}, \mathsf{z})$ iff $\mathsf{m}'(\mathsf{x}, \mathsf{y}, \mathsf{z})$, where m' is the midpoint relation of $<\mathsf{L}_{\mathsf{xy}}; \mathsf{B}, \approx>$.

Lemma 9 $\forall x\, \forall z\, \exists! y\ \mathrm{M}(x, y, z) \wedge \forall x\, \forall y\, \exists! z\ \mathrm{M}(x, y, z)$

Proof From Lemma 8, plus Lemma XVIII.26 and Lemma XVIII.19. ∎

The midpoint function $x \mathbin{\dot{:}} z = y \equiv_{\mathrm{Def}} \mathrm{M}(x, y, z)$

The symmetrical image of x with respect to y $\mathrm{S}_y(x) = z \equiv_{\mathrm{Def}} \mathrm{M}(x, y, z)$

I'll write the interpretation of $\dot{:}$ in a model as $:$, and the interpretation of S_y as S_{y}, the *point symmetry with origin* y.

Lemma 10 $\forall x\, \forall y\ x \in \mathrm{L} \wedge y \in \mathrm{L} \rightarrow x \mathbin{\dot{:}} y \in \mathrm{L} \wedge \mathrm{S}_y(x) \in \mathrm{L}$

Lemma 11 Given a model $<\mathsf{U};\mathsf{B},\approx>$ of **E2**, if $x \neq y$ and $x, y \in \mathsf{L}$, then
$x : y = x :' y$, where $:'$ is the midpoint function of $<\mathsf{L};\mathsf{B},\approx>$
$\mathsf{S}_y(x) = \mathsf{S}'_y(x)$, where S'_y is the point symmetry with respect to y of $<\mathsf{L};\mathsf{B},\approx>$.

Theorem 12 *Point symmetries are automorphisms*

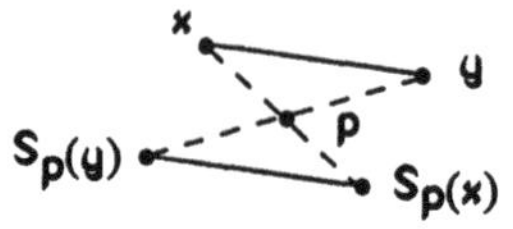

a. *Point symmetries are isometries*
$\forall x\, \forall y\, \forall p\;\; xy \equiv \mathrm{S}_p(x)\mathrm{S}_p(y)$

b. *Point symmetries preserve equivalences*
$\forall x\, \forall y\, \forall z\, \forall w\, \forall p\;\; xy \equiv zw \rightarrow \mathrm{S}_p(x)\mathrm{S}_p(y) \equiv \mathrm{S}_p(z)\mathrm{S}_p(w)$

c. *Point symmetries are onto*
$\forall p\, \forall y\, \exists x\;\; \mathrm{S}_p(x) = y$

d. *Point symmetries are involutions*
$\forall p\, \forall x\;\; \mathrm{S}_p\mathrm{S}_p(x) = x$

e. *Point symmetries are 1-1*
$\forall p\, \forall x\, \exists y\;\; \mathrm{S}_p(x) = \mathrm{S}_p(y) \rightarrow x = y$

f. *Point symmetries preserve betweenness*
$\forall x\, \forall y\, \forall z\, \forall p\;\; \mathrm{B}(x, y, z) \rightarrow \mathrm{B}(\mathrm{S}_p(x), \mathrm{S}_p(y), \mathrm{S}_p(z))$

Proof (a) We cannot use Theorem 6 to prove this because the points may not lie on a line. We use instead axiom *E10*. By definition, $\mathsf{m}(x, \mathsf{p}, \mathsf{S}_\mathsf{p}(x))$ and $\mathsf{m}(y, \mathsf{p}, \mathsf{S}_\mathsf{p}(y))$. Hence, by definition, $\mathsf{B}(x, \mathsf{p}, \mathsf{S}_\mathsf{p}(x))$ and $x\mathsf{p} \approx \mathsf{p}\mathsf{S}_\mathsf{p}(x)$, and $\mathsf{B}(y, \mathsf{p}, \mathsf{S}_\mathsf{p}(y))$ and $y\mathsf{p} \approx \mathsf{p}\mathsf{S}_\mathsf{p}(y)$. So by *E10*, we have $xy \approx \mathsf{S}_\mathsf{p}(x)\mathsf{S}_\mathsf{p}(y)$.

I'll leave (b) to you, and part (c) is by Lemma 9.

(d) By the definition of point symmetries, given any x and p, x, p, and $\mathsf{S}_\mathsf{p}(x)$ are collinear, so x, p, $\mathsf{S}_\mathsf{p}(x)$, and $\mathsf{S}_\mathsf{p}\mathsf{S}_\mathsf{p}(x)$ are collinear, too. So by Lemma 11, point symmetries are involutions.

(e) If we have $\mathsf{S}_\mathsf{p}(x) = \mathsf{S}_\mathsf{p}(y)$, then by (d) applying S_p to both sides, $x = y$.

(f) By (a) and *E9*. ■

Lemma 13

a. $\forall x\, \forall y\, \forall z\, \forall p\;\; \mathrm{L}(x, y, z) \leftrightarrow \mathrm{L}(\mathrm{S}_p(x), \mathrm{S}_p(y), \mathrm{S}_p(z))$

b. $\forall x\, \forall y\, \forall p\;\; \mathrm{S}_p(\mathrm{L}_{xy}) = \mathrm{L}_{\mathrm{S}_p(x)\mathrm{S}_p(y)}$

c. $\forall x\, \forall y\, \forall p\;\; \mathrm{S}_p(\mathrm{C}_{xy}) = \mathrm{C}_{\mathrm{S}_p(x)\mathrm{S}_p(y)}$

d. $\forall x\, \forall y\, \forall z\, \forall p\;\; \mathrm{M}(x, y, z) \leftrightarrow \mathrm{M}(\mathrm{S}_p(x), \mathrm{S}_p(y), \mathrm{S}_p(z))$

e. $\forall x\, \forall y\, \forall p\;\; \mathrm{S}_p(x \vdots y) = \mathrm{S}_p(x) \vdots \mathrm{S}_p(y)$

f. $\forall x\, \forall p\, \forall q\;\; \mathrm{S}_p(\mathrm{S}_q(x)) = \mathrm{S}_{\mathrm{S}_p(q)}(\mathrm{S}_p(x))$

Proof (b) $\mathsf{S}_\mathsf{p}(\mathsf{L}_{xy}) = \{\mathsf{S}_\mathsf{p}(\mathsf{c}) : \mathsf{c} \in \mathsf{L}_{xy}\}$. $\mathsf{L}_{\mathsf{S}_\mathsf{p}(x)\mathsf{S}_\mathsf{p}(y)} = \{\mathsf{d} : \mathsf{L}(\mathsf{S}_\mathsf{p}(x), \mathsf{S}_\mathsf{p}(y), \mathsf{d})\}$. Now use part (a).

(f) We have $m(x, q, S_q(x))$. So $B(x, q, S_q(x))$ and $xq \approx qS_q(x)$. Hence by Theorem 12, $B(S_p(x), S_p(q), S_p(S_q(x)))$ and $S_p(x)S_p(q) \approx S_p(q)S_p(S_q(x))$. Hence, $m(S_p(x), S_p(q), S_p(S_q(x)))$. Hence, $S_{S_p(q)}(S_p(x)) = S_p(S_q(x))$ (draw a picture). ∎

4. Line symmetry

In two-dimensional geometry we can define symmetry with respect to a line. We begin by considering the bisector of a segment.

Lemma 14

a. $\forall x \, \forall y \;\; x \neq y \rightarrow x \dot{:} y \in C_{xy}$

b. *There is a rhombus with a given segment as diagonal*
$\forall x \, \forall y \; (x \neq y \rightarrow \exists z \, \exists w \, (z \neq w \wedge (z \in C_{xy} \wedge w \in C_{xy}) \wedge (x \in C_{zw} \wedge y \in C_{zw})))$

c. $\forall x \, \forall C \;\; x \notin C \rightarrow \exists y \, (C = C_{xy})$

d. $\forall x \, \forall y \, \forall z \;\; x \neq y \rightarrow (C_{xy} = C_{xz} \rightarrow y = z)$

Proof (b) We wish to show the diagram on the left.

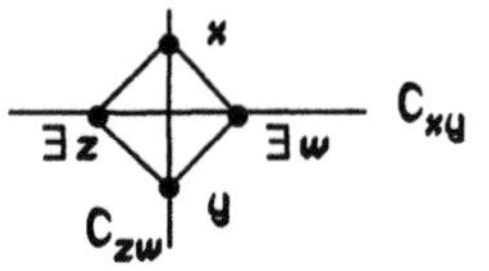

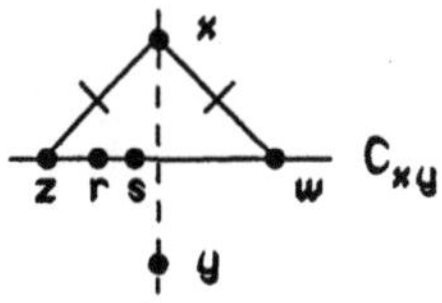

Since $x \neq y$, and since there are two points on every bisector, there are $r, s \in C_{xy}$ and $r \neq s$. So by *E11*, there are z, w such that $B(r, s, w)$ and $B(s, r, z)$ and $xz \approx xw$, as in the diagram on the right. So $z, w \in C_{xy}$ and since $r \neq s$, $z \neq w$. So $xw \approx yw$ and $xz \approx yz$. So $yz \approx yw$, and hence $x, y \in C_{zw}$.

(c) Given C, by Lemma 2.d there are $r, s \in C$ and $r \neq s$. By *E11*, there are z, w such that $B(r, s, w)$ and $B(s, r, z)$ and $xz \approx xw$. Since $r \neq s$, by the one-dimensional case (Lemma XVIII.1 and Theorem 6), $z \neq w$. By Lemma 3, $z, w \in C$. We need to show that $z, w \in C_{xy}$ and we will be done by Lemma 4.c. Consider $y = S_{z:w}(x)$. We have $xz \approx S_{z:w}(x) \, S_{z:w}(z)$ (isometry) $\approx yw$ (definition, Lemma XVIII.27, and Theorem 6). Similarly, $xw \approx yz$. As $xz \approx xw$, we now have $yz \approx xz$ and $yw \approx zw$, so $z, w \in C_{xy}$.

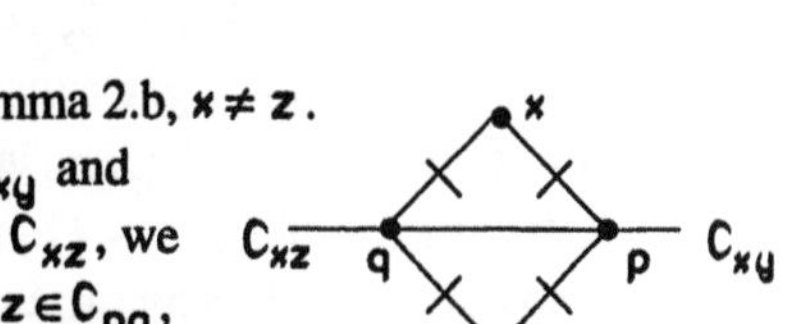

(d) Suppose $x \neq y$ and $C_{xy} = C_{xz}$. Then by Lemma 2.b, $x \neq z$. By part (b), there are p, q such that $p \neq q$ and $p, q \in C_{xy}$ and $x, y \in C_{pq}$. Hence, $xp \approx xq \approx yp \approx yq$. Since $C_{xy} = C_{xz}$, we also have $xp \approx zp$ and $xq \approx zq$. So $zp \approx zq$. Hence, $z \in C_{pq}$, and thus all three of $x, y, z \in C_{pq}$. Since $C_{xy} = C_{xz}$, we have $p{:}q \in C_{xz}$. Hence, $x(p{:}q) \approx y(p{:}q) \approx z(p{:}q)$. So by the one-dimensional case (Lemma XVIII.17.h, Theorem 6), $x = z$ or $x = y$ or $y = z$. Since the first two cases are ruled out, we must have $y = z$. ∎

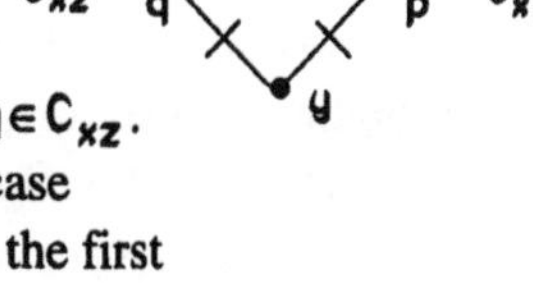

Line symmetry

$S_L(x) = y \equiv_{Def} (x = y \wedge x \in L) \vee (L = C_{xy})$

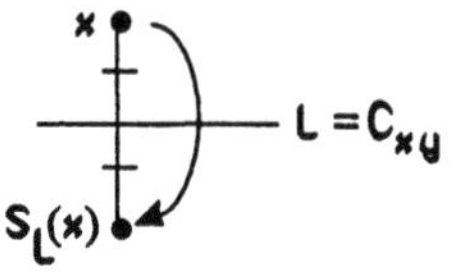

Given line L in a model, $\mathsf{S_L}(\mathsf{x})$ is the *symmetric image of* x *with respect to* L. I'll write $\mathsf{S_L S_K}$ for $\mathsf{S_L} \circ \mathsf{S_K}$.

Theorem 15 *Line symmetries are automorphisms*

a. *Line symmetries are 1-1 onto functions*
$\forall L\, \forall x\, \exists! y\;\; S_L(x) = y \wedge S_L(y) = x$

b. *The fixed points of a line symmetry are the line itself*
$\forall L\, \forall x\;\; (x = S_L(x) \leftrightarrow x \in L)$

c. $\forall L\, \forall x\;\; (x \notin L \rightarrow L = C_{xS_L(x)})$

d. *Line symmetries are involutions*
$\forall L\, \forall x\;\; S_L(S_L(x)) = x$

e. $\forall x\, \forall y\, \forall L\;\; xy \equiv yS_L(x) \leftrightarrow (x \in L \vee y \in L)$

f. $\forall x\, \forall L\;\; (x \vdots S_L(x) \in L)$

g. $\forall x\, \forall y\, \forall L\;\; x \in L \rightarrow xy \equiv xS_L(y)$

h. *Line symmetries are isometries*
$\forall x\, \forall y\, \forall L\;\; xy \equiv S_L(x)S_L(y)$

i. $\forall x\, \forall y\, \forall z\, \forall w\, \forall L\;\; xy \equiv zw \rightarrow S_L(x)S_L(y) \equiv S_L(z)S_L(w)$

j. *Line symmetries are homomorphisms with respect to betweenness*
$\forall x\, \forall y\, \forall z\, \forall L\;\; B(x, y, z) \rightarrow B(S_L(x), S_L(y), S_L(z))$

k. $\forall x\, \forall y\, \forall L\;\; (x \neq y) \rightarrow S_L(C_{xy}) = C_{S_L(x)S_L(y)}$

Proof I will show (a)–(d) all at once. Given L, if $\mathsf{x} \in \mathsf{L}$, then $\mathsf{S_L}(\mathsf{x}) = \mathsf{x}$. If $\mathsf{x} \notin \mathsf{L}$, then since every line is a bisector, by Lemma 14.c there is some y such that $\mathsf{L} = \mathsf{C_{xy}}$. So $\mathsf{x} \neq \mathsf{y}$ and $\mathsf{y} \notin \mathsf{L}$ and $\mathsf{S_L}(\mathsf{x}) = \mathsf{y}$. But as $\mathsf{C_{xy}} = \mathsf{C_{xy}}$ by Lemma 2.a, $\mathsf{S_L}(\mathsf{y}) = \mathsf{x}$.

(e) From left to right, suppose we have a line L and $\mathsf{xy} \approx \mathsf{yS_L}(\mathsf{x})$. If $\mathsf{x} \in \mathsf{L}$, we are done by (b). If not, then $\mathsf{y} \in \mathsf{C_{xS_L(x)}}$, and so by (c), $\mathsf{y} \in \mathsf{L}$. The other direction follows by (b).

(f) If $\mathsf{x} \in \mathsf{L}$, this follows by (b). If $\mathsf{x} \notin \mathsf{L}$, then this follows by (e).

(g) If $\mathsf{y} \in \mathsf{L}$, then by (a), $\mathsf{S_L}(\mathsf{y}) = \mathsf{y}$, and we are done. So suppose $\mathsf{y} \notin \mathsf{L}$. So by definition of bisectors, if $\mathsf{x} \in \mathsf{L}$, then $\mathsf{xy} \approx \mathsf{xS_L}(\mathsf{y})$.

(h) By (f), $\mathsf{x}\!:\!\mathsf{S_L}(\mathsf{x}) \in \mathsf{L}$. So by (g) and Lemma 1.d,

$\mathsf{x}\,(\mathsf{x}\!:\!\mathsf{S_L}(\mathsf{x})) \approx (\mathsf{x}\!:\!\mathsf{S_L}(\mathsf{x}))\,\mathsf{S_L}(\mathsf{x})$

$\mathsf{y}\,(\mathsf{x}\!:\!\mathsf{S_L}(\mathsf{x})) \approx (\mathsf{x}\!:\!\mathsf{S_L}(\mathsf{x}))\,\mathsf{S_L}(\mathsf{y})$

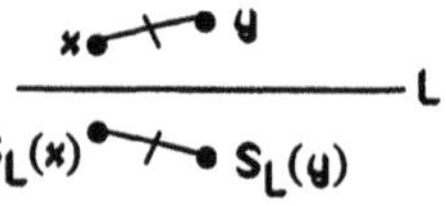

Similarly,

$$x(y{:}S_L(y)) \approx (y{:}S_L(y))S_L(x)$$
$$y(y{:}S_L(y)) \approx (y{:}S_L(y))S_L(y)$$

So by Theorem 6, since each set of triples is on a single line, $B(x, (x{:}S_L(x)), S_L(x))$ and $B(y, (y{:}S_L(y)), S_L(y))$. Hence by *E10*, $xy \approx S_L(x)S_L(y)$.

(i) Use (h).

(j) Use (g) and *E9*.

(k) $z \in S_L(C_{xy})$ iff $z = S_L(w)$ some $w \in C_{xy}$
iff $z = S_L(w)$ and $wx \approx wy$
iff $z = S_L(w)$ and
$S_L(w)S_L(x) \approx S_L(w)S_L(y)$
iff $zS_L(x) \approx zS_L(y)$
iff $z \in C_{S_L(x)S_L(y)}$ ■

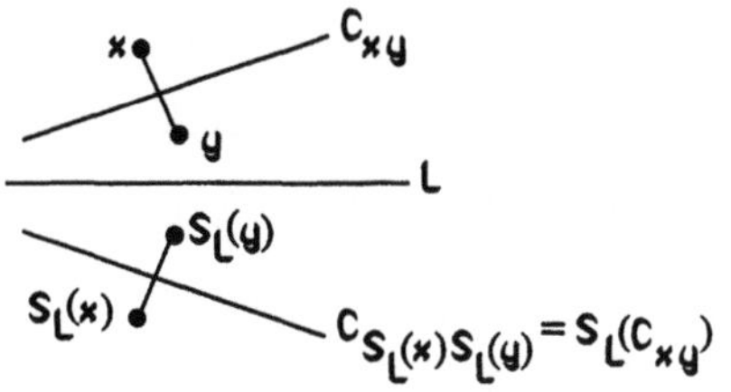

5. Perpendicular lines

We can't define 'two lines are perpendicular' as 'the lines intersect in a right angle', since we haven't defined angles. We use instead the following, which I ask you to show in Exercise 4 is equivalent.

Perpendicular lines, K *is perpendicular to* L

$K \perp L \equiv_{\text{Def}} K \neq L \wedge S_K(L) = L$

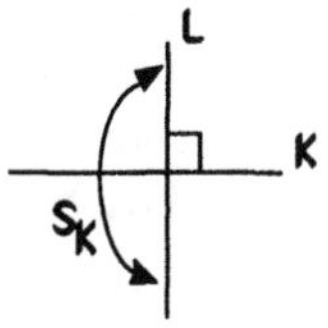

Lemma 16 *Perpendicular lines and bisectors*

a. $\forall L \neg(L \perp L)$

b. $\forall K \forall L \;\; K \perp L \rightarrow L \perp K$

c. $\forall x \forall y \;\; x \neq y \rightarrow C_{xy} \perp L_{xy}$

d. $\forall K \forall L \;\; K \perp L \leftrightarrow \exists x \exists y\,(K = L_{xy} \wedge L = C_{xy})$

Proof (b) Suppose $K \perp L$. We have (i) $S_K(L) = L$. We must show $S_L(K) = K$. By (a), $K \neq L$, so $x \in L$ and $x \notin K$. By Theorem 15.b, (ii) $x \neq S_K(x) \in S_K(L) = L$. So by Theorem 15.c, $K = C_{xS_K(x)}$. So:

$$\begin{aligned} S_L(K) &= S_L(C_{xS_K(x)}) \\ &= C_{S_L(x)S_LS_K(x)} && \text{Lemma 15.k} \\ &= C_{xS_K(x)} && S_L(x) = x \text{ because } x \in L \text{, and} \\ &= K && S_LS_K(x) = x \text{ because } S_K(x) \in L = S_K(L) \end{aligned}$$

(c) Suppose $x \neq y$. Then by Lemma 15.k, $S_{L_{xy}}(C_{xy}) = C_{S_{L_{xy}}(x)S_{L_{xy}}(y)} = C_{xy}$ by Theorem 15.c, as $x, y \in L_{xy}$.

(d) Suppose $K \perp L$. Then $K \neq L$. Let $x \in L$ and $x \notin K$. Let $y = S_K(x) \neq x$, and proceed as in the previous proofs. ■

Theorem 17

a. *There is exactly one point in the intersection of perpendicular lines*

$\forall K\, \forall L\;\; K \perp L \rightarrow \exists! x\, (x \in K \cap L)$

b. *Given a point and a line, there is a perpendicular to that line through that point*

$\forall x\, \forall K\, \exists L\;\; (x \in L \wedge L \perp K)$

c. *Two perpendiculars to the same line have empty intersection*

$\forall K\, \forall L\, \forall M\;\; (K \perp L \wedge L \perp M) \rightarrow (K \cap M = \varnothing \vee K = M)$

d. $\forall K\, \forall L\, \forall M\;\; (K \perp L \wedge L \perp M) \rightarrow \neg(K \perp M)$

Proof (a) By Lemma 16.d, let x, y be such that $K = L_{xy}$ and $L = C_{xy}$. Then $K \cap L = \{ z : zx \approx zy \text{ and } L(x, y, z)\} = \{ x : y \}$ by the one-dimensional case (Theorem 6).

(b) If $x \notin K$, take $L = L_{x S_K(x)}$, and we are done by Lemma 16.d. If $x \in K$, take $y \in K$ such that $y \neq x$ by Lemma 2.d. Then $yx \approx S_x(y)x$. Let $L = C_{y S_x(y)}$. Then $x \in L$. We have $S_x(y) \in K$, so $K = L_{x S_x(y)}$. Hence by Lemma 16.c, $K \perp L$.

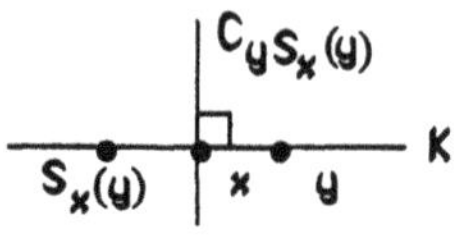

(c) Suppose $K \perp L$ and $L \perp M$ and $K \cap M \neq \varnothing$. We need to show that $K = M$. Let $x \in K \cap M$. If $x \notin L$, then $x \neq S_L(x)$ and, as you can show, $S_L(x) \in K \cap M$. So $K = M$, as two points determine a line.

If $x \in L$, let $y \neq x$ and $y \in L$. Then (i) y, $S_K(y)$, $S_M(y) \in L$ as $S_K(L) = L = S_M(L)$ (using Lemma 16.b). Also $x \in L \cap M \cap K$. So by (a), (ii) $y \notin K$ and $y \notin M$, since two points determine a line. Also, by Theorem 15.c, (iii) $K = C_{y S_K(y)}$ and $M = C_{y S_M(y)}$. Hence, $xy \approx x S_K(y)$ and $xy \approx x S_M(y)$. So via the one-dimensional case ((i) and Theorem 6), $S_K(y) = S_M(y)$ or $S_K(y) = y$ or $S_M(y) = y$. But the latter two can't hold by (ii) and Theorem 15.b. So by (iii) $K = M$.

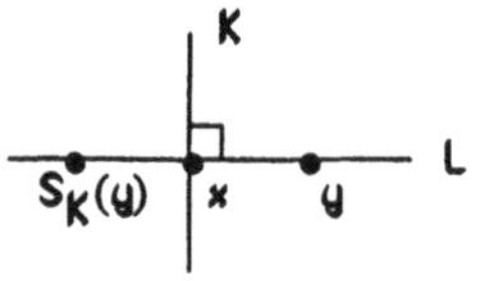

(d) Suppose $K \perp L$ and $L \perp M$. If $K \perp M$, then by (a), $K \cap M \neq \varnothing$. So by (c), $K = M$. Hence, by Lemma 16.a, $\neg(K \perp M)$, a contradiction. So $\neg(K \perp M)$. ∎

Now we are ready to prove a version of Euclid's parallel postulate. This will be our first use of *E14*.

Lemma 18

a. $\forall K\, \forall L\;\; (K \perp L \wedge L \cap M = \varnothing) \rightarrow K \perp M$

b. *A kind of transitivity*

$\forall K\, \forall L\, \forall M\, \forall N\;\; (K \perp L \wedge L \perp M \wedge M \perp N) \rightarrow K \perp N$

Proof (a) Suppose $K \perp L$ and $L \cap M = \varnothing$. By Theorem 17.a, there is a unique $z \in K \cap L$, so $K \neq M$. Let $y \in K$ such that $y \neq z$. Either $y \notin M$ or $y \in K \cap M$. If the latter, then choose w such that $B(z, y, w)$ by Lemma 1.j. Then $w \notin M$, since two points determine a line and $K \neq M$. Thus, we can assume we have some $x \in K$ and $x \notin (L \cup M)$.

I'll first show (i) $L(x, S_L(x), S_M(x))$, and then the lemma follows.

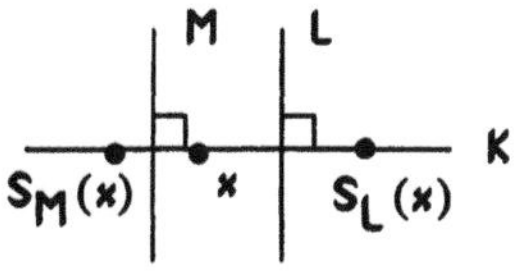

If (i), then $x \neq S_L(x)$, as $x \notin L$. And x, $S_L(x) \in K$ as $K \perp L$. So $K = L_{xS_L(x)}$ and since two points determine a line and (i), x, $S_M(x) \in K$. So by definition, $M = C_{xS_M(x)}$ and by Lemma 16.c, $M \perp K$.

So it remains only to prove (i). Suppose to the contrary that $L(x, S_L(x), S_M(x))$ doesn't hold We will derive a contradiction. Since x, $S_L(x)$, $S_M(x)$ are not collinear, by *E14* we can inscribe a circle around the triangle they form. That is, there is some z such that $S_L(x)z \approx xz \approx S_M(x)z$. By the first congruence and Lemma 15.e, $z \in L$, and by the second congruence, $z \in M$. But that contradicts $L \cap M = \varnothing$, so we are done.

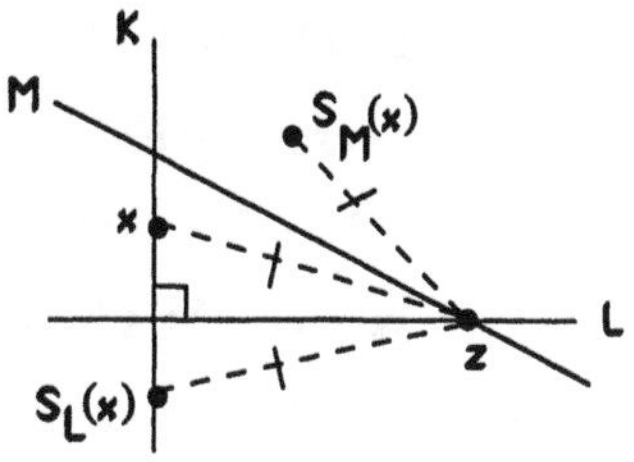

(b) Suppose $K \perp L$ and $L \perp M$ and $M \perp N$. By Theorem 17.c, either $K \cap M = \varnothing$ or $K = M$. In the latter case we are done. In the first case, we are done by (a). ■

Perpendicular line segments

$xy \perp K \quad \equiv_{Def} \quad x \neq y \wedge K \perp L_{xy}$

$K \perp xy \quad \equiv_{Def} \quad xy \perp K$

$xy \perp zw \quad \equiv_{Def} \quad x \neq y \wedge z \neq w \wedge L_{xy} \perp L_{zw}$

Lemma 19 $\forall x \forall y \forall z \forall w$

$$(x \neq y \wedge x \neq z \wedge y \neq z \wedge z \neq w) \wedge (xz \equiv xw \wedge yz \equiv yw) \rightarrow xy \perp zw$$

Proof Suppose $x \neq y$, $x \neq z$, and $y \neq z$, and $z \neq w$. Then x, $y \in C_{zw}$. And by Lemma 16.c, $C_{zw} \perp L_{zw}$. And z, $w \in L_{zw}$. ■

6. Parallel lines

Parallel lines

$K \,//\, M \equiv_{Def} \exists L\ (K \perp L \wedge L \perp M)$

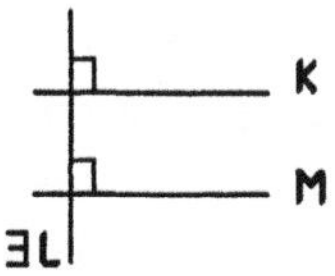

Lemma 20 *Parallelism is an equivalence relation*

a. $\forall K\ K \,//\, K$

b. $\forall K \forall L\ K \,//\, L \rightarrow L \,//\, K$

c. $\forall K \forall L \forall M\ K \,//\, L \wedge L \,//\, M \rightarrow K \,//\, M$

Proof (a) Use Theorems 17.b and 16.b.

(b) By Lemma 16.b.

(c) Suppose $K \mathbin{//} L$ and $L \mathbin{//} M$. So there are L_1 and L_2 such that $K \perp L_1$ and $L_1 \perp L$ and $L \perp L_2$ and $L_2 \perp M$. So $K \perp L_2$ by Lemma 18.b, and so $K \mathbin{//} M$. ■

Theorem 21

a. A standard definition of parallel lines

$$\forall K \forall M \ \ K \mathbin{//} M \leftrightarrow (K = M \lor K \cap M = \varnothing)$$

b. A version of Euclid's parallel postulate

$$\forall x \ \forall K \ \exists! L \ \ (x \in L \land L \mathbin{//} K)$$

c. $\forall K \forall L \forall M \ (K \perp L \land L \mathbin{//} M \to K \perp M)$

Proof (a) Suppose $K \mathbin{//} M$. Then there is some L such that $K \perp L$ and $L \perp M$. So by Theorem 17.c, $K = M$ or $K \cap M = \varnothing$.

Now suppose $K = M$ or $K \cap M = \varnothing$. In the first case we are done by Theorem 20.a. In the second, by Theorem 17.b, there is some L such that $K \perp L$. So by Lemma 18.a, $L \perp M$. So $K \mathbin{//} M$.

(b) Given x and K, by Theorem 17.b, there is some M and $x \in M$ such that $M \perp K$, and hence $K \perp M$ by Lemma 16.b. Again by Theorem 17.b, there is some L and $x \in L$ such that $L \perp K$. So $L \mathbin{//} M$. It remains to show that L is unique. Suppose $L \cap L' \neq \varnothing$ and $L \mathbin{//} M$ and $L' \mathbin{//} M$. Then by Lemma 20.c, $L \mathbin{//} L'$, so by (a), $L = L'$.

(c) Suppose $K \perp L$ and $L \mathbin{//} M$. Then there is some L' such that $L \perp L'$ and $L' \perp M$. So by Lemma 18.b, $K \perp M$. ■

Parallel line segments

$xy \mathbin{//} K \quad \equiv_{\text{Def}} \ x \neq y \land L_{xy} \mathbin{//} K$

$K \mathbin{//} xy \quad \equiv_{\text{Def}} \ xy \mathbin{//} K$

$xy \mathbin{//} zw \quad \equiv_{\text{Def}} \ x \neq y \land z \neq w \land L_{xy} \mathbin{//} L_{zw}$

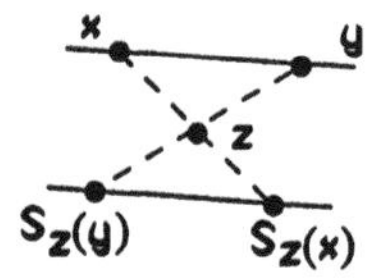

Lemma 22 $\forall x \forall y \forall z \ \ x \neq y \to xy \mathbin{//} S_z(x)\, S_z(y)$

Proof Assume $x \neq y$. Then by Theorem 12 and Lemma 1.g, $S_z(x) \neq S_z(y)$. Suppose it's not the case that $L_{xy} \mathbin{//} L_{S_z(x)S_z(y)}$. By Theorem 21.a, there is some $q \in L_{xy} \cap L_{S_z(x)S_z(y)}$. Since point symmetries are automorphisms, $S_z(q) \in L_{S_z(x)S_z(y)} \cap L_{S_zS_z(x)S_zS_z(y)} = L_{S_z(x)S_z(y)} \cap L_{xy}$. We have two possibilities. If $q \neq S_z(q)$, then since two points determine a line, $L_{xy} = L_{S_z(x)S_z(y)}$, and hence $L_{xy} \mathbin{//} L_{S_z(x)S_z(y)}$. On the other hand, if $q = S_z(q)$, by the one-dimensional case, $z = q$. So $z \in L_{xy}$. By the definition of point symmetries,

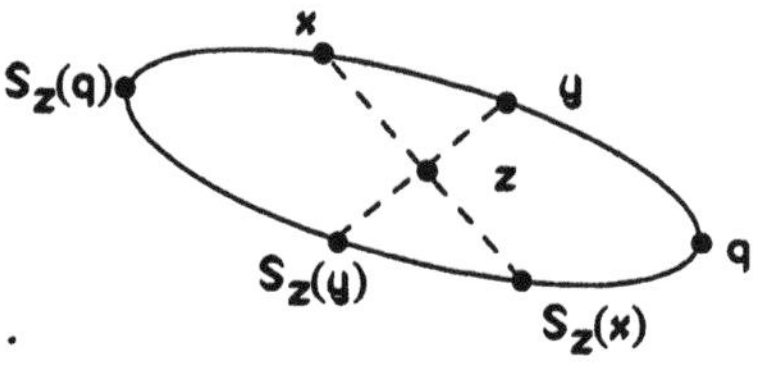

$B(x, z, S_z(x))$, so $z \in L_{xS_z(x)}$. Hence, because two points determine a line, $L_{xy} = L_{xS_z(x)}$. Similarly, $L_{xy} = L_{yS_z(y)}$. So all of x, y, $S_z(x)$, $S_z(y)$ lie on the same line. So $L_{xy} = L_{S_z(x)S_z(y)}$, and again $L_{xy} \,//\, L_{S_z(x)S_z(y)}$. So even if we assume the contrary, we have $L_{xy} \,//\, L_{S_z(x)S_z(y)}$. ■

Lemma 23 *Opposite sides of a parallelogram are congruent*

$\forall x\, \forall y\, \forall z\, \forall w\ (L_{xy} \neq L_{yw} \wedge xy \,//\, zw \wedge xz \,//\, yw) \rightarrow xy \equiv zw$

Proof Suppose we are given x, y, z, w with $L_{xy} \neq L_{yw}$ and $xy \,//\, zw$ and $xz \,//\, yw$. Then not ($L_{xy} \,//\, L_{yw}$), because if they were parallel then they would be equal as they share one point (Theorem 21.a). Since L_{xy} is neither equal to nor parallel to L_{yw}, $L_{xy} \cap L_{yw} = \{y\}$.

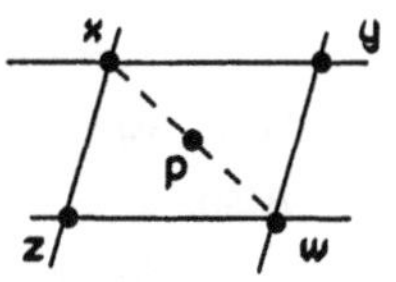

Now let $p = x : w$. We have $S_p(x) = w$ and $S_p(w) = x$. I'll show that $S_p(z) = y$. We have by Lemma 22, $S_p(z)S_p(w) \,//\, zw$, so substituting, $S_p(z)x \,//\, zw$. So by Lemma 20.c, $S_p(z)x \,//\, xy$. So by Theorem 21.a, $S_p(z) \in L_{xy}$. We also have $S_p(z)S_p(x) \,//\, zx$, so $S_p(z)w \,//\, zx$. So $S_p(z)w \,//\, wy$. So $S_p(z) \in L_{yw}$. So $S_p(z) \in L_{xy} \cap L_{yw}$, and hence $S_p(z) = y$. Thus, $S_p(y) = z$. We also have $xy \approx S_p(x)S_p(y)$, while $S_p(x)S_p(y) = wz$, so we are done. ■

Lemma 24

a. $\forall x\, \forall y\, \forall z\, \forall w\ (xy \,//\, zw \wedge xy \equiv zw) \rightarrow x\dot{:}z = y\dot{:}w \vee x\dot{:}w = y\dot{:}z$

b. $\forall x\, \forall y\, \forall z\, \forall w\ (x\dot{:}z = y\dot{:}w) \rightarrow xy \,//\, zw \wedge xy \equiv zw$

Proof (a) We want to show:

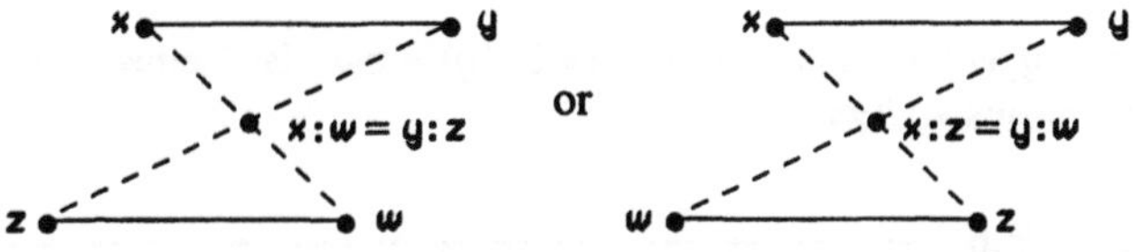

If $L_{xy} = L_{zw}$, we can use the one-dimensional case (Exercise 1 of Chapter XVIII.C). So assume $L_{xy} \neq L_{yw}$ and $xy \,//\, zw$ and $xy \approx zw$. By our notation, we have that $x \neq y$ and $z \neq w$. Set:

$p = x:z$ so $S_p(x) = z$

$q = y:z$ so $S_q(y) = z$

$r = S_p(y)$

$s = S_q(x)$

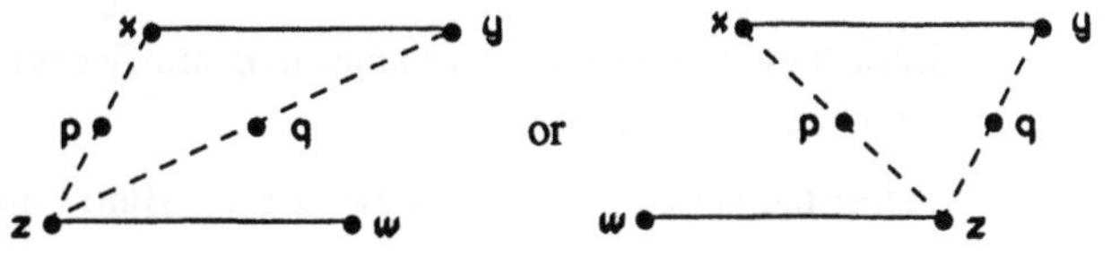

If $r = w$, then $S_p(y) = w$ and $S_p(x) = z$, so $x:z = y:w$. If $s = w$, then $S_q(x) = w$ and $S_q(y) = z$, so $y:z = x:w$. Hence, it remains to prove that $r = w$ or $s = w$. By Lemma 23,

$xy \,//\, S_p(x)S_p(y)$ so $xy \,//\, zr$

$xy \,//\, S_q(x)S_q(y)$ so $xy \,//\, sz$

(by hypothesis) $xy \,//\, zw$

So by Lemma 20.c, $zr \,//\, zw \,//\, sw$. So by Theorem 21.a, (i) all of z, w, r, s lie on the same line. We also have:

$xy \approx S_p(x)S_p(y)$ so $xy \approx zr$
$xy \approx S_q(x)S_q(y)$ so $xy \approx sz$
(by hypothesis) $\quad xy \approx wz$

That is, $zr \approx zs \approx zw$. So by (i) and the one-dimensional case, $r = s$, or $s = w$, or $r = w$.

Thus, we need to show that $r \neq s$ and we are done. Suppose that $r = s$. We will derive a contradiction. If $r = s$, then:

(ii) $S_p(y) = S_q(x)$ and $S_p(x) = S_q(y)$

I'll show that $x = y$, which is the contradiction. We have

$xy \approx S_p(x)S_p(y)$	isometry
$\approx S_q(y)S_p(y)$	by (ii)
$\approx S_qS_q(y)S_qS_p(y)$	isometry
(iii) $xy \approx yS_qS_p(y)$	involution

Also,

$xy // S_p(x)S_p(y)$	by Lemma 22
$S_p(x)S_p(y) // S_q(y)S_p(y)$	by (ii)
$S_q(y)S_p(y) // yS_qS_p(y)$	applying S_q, involution, Lemma 22
$xy // yS_qS_p(y)$	Lemma 20.c

So x, y, $S_qS_p(y)$ are collinear. Hence:

$y = x{:}S_qS_p(y)$	(iii) and the one-dimensional case
$= x{:}S_qS_q(x)$	(ii)
$= x{:}x$	involution
$= x$	

(b) Suppose $x{:}z = y{:}w = p$. Then $xy \approx S_p(x)S_p(y) \approx zw$. By Lemma 22, $xy // S_p(x)S_p(y)$, which means $xy // zw$. ∎

Exercises for Sections B.1–B.6

1. Define a notion of angle in **E2**. (Hint: 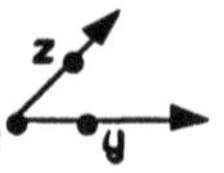

)

2. Show how to use pseudo-variables to quantify over angles in **E2**. Include what it means for two angles to be equal.

3. Define the notion of a right angle in **E2**. (Hint: Equal angles via a symmetry.)

4. Formalize and prove in **E2**: Two lines are perpendicular iff they intersect in a right angle.

5. Formalize and prove in **E2**: Two lines are parallel iff there is another line they both intersect and their opposite interior angles are the same.

6. Formalize and prove in **E2**: Given any two congruent parallel segments, there is a parallelogram with them as sides.

7. Formalize and prove in **E2** the original formulation of Euclid's parallel postulate mentioned in the motivation for axiom *E14*.

7. Parallel projection

Lemma 25 $\forall K\, \forall L\, \forall x\ \neg(K \mathbin{/\!/} L) \rightarrow \exists! y\, (y \in L \wedge xy \mathbin{/\!/} K)$

Proof Suppose not ($K \mathbin{/\!/} L$). By Theorem 21.b, there is a unique M such that $x \in M$ and $M \mathbin{/\!/} K$. So by transitivity, not ($M \mathbin{/\!/} L$). Hence, $M \neq L$ and $M \cap L \neq \varnothing$ by Theorem 21.A. So since two points determine a line, there is a single point, $y \in M \cap L$, and $M = L_{xy}$. ■

The parallel projection of x on L with respect to K

$$P_{K/L}(x) = y \equiv_{Def} \neg(K \mathbin{/\!/} L) \wedge$$
$$(y \in L \wedge xy \mathbin{/\!/} K) \vee (x \in L \wedge x = y)$$

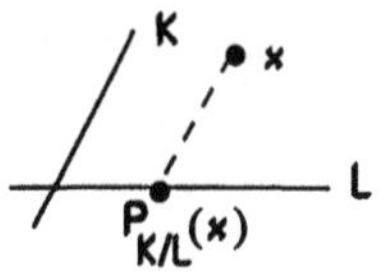

Lemma 26

a. $\forall K\, \forall L\, \forall x\ \ P_{K/L}(x) = x \leftrightarrow (x \in L \wedge \neg(K \mathbin{/\!/} L))$

b. $\forall K\, \forall L\, \forall x\, \forall y\ \ x \neq y \rightarrow\ \ (P_{K/L}(x) = P_{K/L}(y) \leftrightarrow xy \mathbin{/\!/} K)$

Proof (b) Suppose $x \neq y$ and $xy \mathbin{/\!/} K$ and not ($K \mathbin{/\!/} L$). Then $L_{xy} \mathbin{/\!/} K$ and, as in the proof of Lemma 25, there is a z such that $L_{xy} \cap L = \{z\}$. But then $L_{xz} \mathbin{/\!/} K$ and $L_{yz} \mathbin{/\!/} K$, so $z = P_{K/L}(x) = P_{K/L}(y)$.

Now suppose that $x \neq y$, and $P_{K/L}(x) = P_{K/L}(y)$. By transitivity, $xP_{K/L}(x) \mathbin{/\!/} yP_{K/L}(y)$. So $L_{xP_{K/L}(x)} = L_{yP_{K/L}(y)} = L_{xy}$. Hence, $L_{xy} \mathbin{/\!/} K$. ■

Lemma 27 *Parallel projection preserves midpoints*

$\forall K\, \forall L\, \forall x\, \forall y\ \ P_{K/L}(x \!:\! y) = P_{K/L}(x) \!:\! P_{K/L}(y)$

Proof The case when $x = y$ is immediate. So suppose $x \neq y$. If $xy \mathbin{/\!/} K$, then $x(x\!:\!y) \mathbin{/\!/} K$, and we are done by Lemma 26.b. So suppose that not ($xy \mathbin{/\!/} K$).

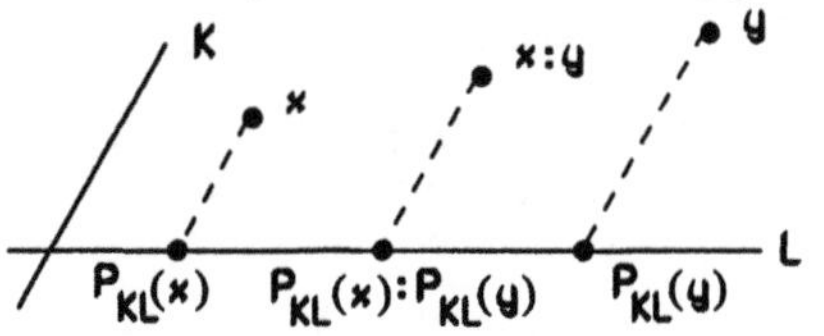

Let $r = P_{K/L}(x)$, $s = P_{K/L}(y)$, $q = x\!:\!y$, and $z = P_{K/L}(x\!:\!y)$. Then, as in the diagram below, $K_1 \mathbin{/\!/} K_2 \mathbin{/\!/} K_3 \mathbin{/\!/} K$ and $x \in K_1$ and $z \in K_2$ and $y \in K_3$.

By Theorem 21.b, let $L_1 \mathbin{/\!/} L$ and $q \in L_1$. As in the proof of Lemma 25, let $\{p\} = K_1 \cap L_1$. By Theorem 21.b, let $L_1 \mathbin{/\!/} L_2 \mathbin{/\!/} L$ and $x \in L_2$. By Lemma 23, $pq \approx rz$. Also, $S_q(p) \in L_1$. By Lemma 22, since $S_q(x) = y$, we have $S_q(K_1) \mathbin{/\!/} K_3$ and so $S_q(K_1) = K_3$. Hence, $S_q(p) \in K_3$. Now as before, by Lemma 23, $qS_q(p) \approx zs$. Hence, as $pq \approx qS_q(p)$, we have $rz \approx zs$. ■

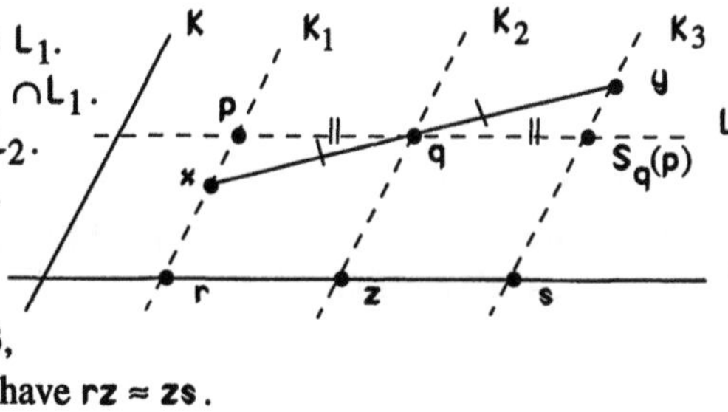

Lemma 28 *Parallel projection preserves congruence of parallel segments*

$$\forall K\, \forall L\, \forall x\, \forall y\, \forall z\, \forall w$$
$$\neg(K \mathbin{//} L) \wedge xy \mathbin{//} zw \wedge xy \equiv zw \rightarrow P_{K/L}(x)P_{K/L}(y) \equiv P_{K/L}(z)P_{K/L}(w)$$

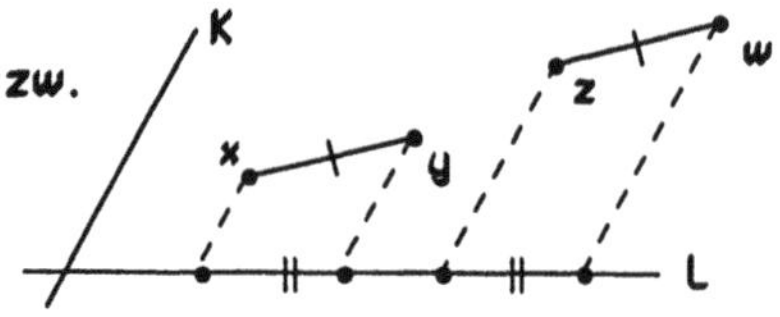

Proof Suppose $x \neq y$ and $z \neq w$ and $xy \mathbin{//} zw$ and $xy \approx zw$. Then by Lemma 24.a, $x:z = y:w$ or $x:w = y:z$. In case $x:z = y:w$ we have by Lemma 27, $P_{K/L}(x):P_{K/L}(z) = P_{K/L}(y):P_{K/L}(w)$. So by Lemma 24.b, $P_{K/L}(x)P_{K/L}(y) \approx P_{K/L}(z)P_{K/L}(w)$. The case when $x:w \approx y:z$ follows analogously. ∎

Lemma 29 *Parallel projection preserves betweenness*

$$\forall K\, \forall L\, \forall x\, \forall y\, \forall z\ \neg(K \mathbin{//} L) \wedge B(x, y, z) \rightarrow B(P_{K/L}(x), P_{K/L}(y), P_{K/L}(z))$$

Proof Assume that $B(x, y, z)$ and not $(K \mathbin{//} L)$. If any two of $x, y, z \in L$, then all three are, and we are done by Lemma 26.a. If not, then we first deal with the case: (‡) $x \in L$.

So we have $P_{K/L}(x) = x$, and $P_{K/L}(y), P_{K/L}(z) \in L$. So either $B(x, P_{K/L}(y), P_{K/L}(z))$ and we are done, or we have:

(i) $B(P_{K/L}(y), x, P_{K/L}(z))$ or (ii) $B(P_{K/L}(y), P_{K/L}(z), x)$.

Suppose (i).

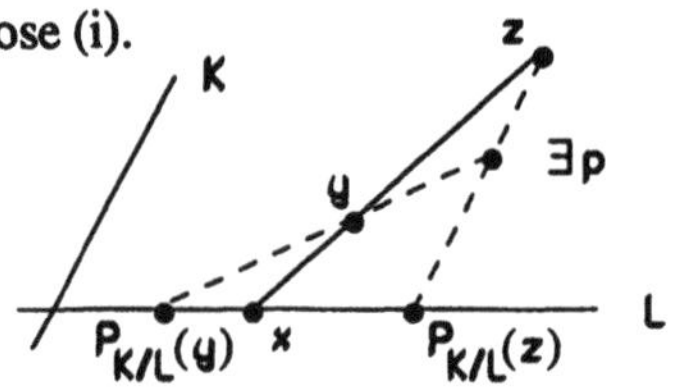

Then we have by *E6* (the Pasch axiom) that there is some p such that $B(z, p, P_{K/L}(z))$ and $B(P_{K/L}(y), y, p)$. Hence, $L_{zP_{K/L}(z)} \cap L_{yP_{K/L}(y)} \neq \varnothing$. These lines are parallel because they are both parallel to K. So by Theorem 21.a, we have $L_{zP_{K/L}(z)} = L_{yP_{K/L}(y)}$. Hence, by Lemma 26.b, $P_{K/L}(y) = P_{K/L}(z)$. Hence, $L_{zP_{K/L}(z)} = L_{yP_{K/L}(y)} = L_{yP_{K/L}(x)}$, all of which are parallel to K, and hence $P_{K/L}(x) = P_{K/L}(y) = P_{K/L}(z)$. So by *E4*, $B(P_{K/L}(x), P_{K/L}(y), P_{K/L}(z))$.

So suppose (ii).

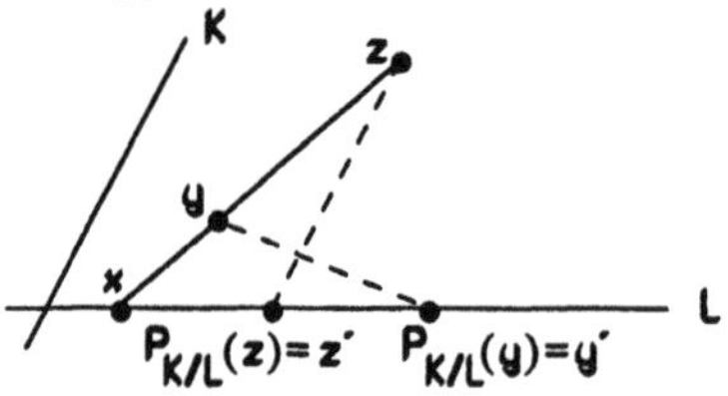

By the one-dimensional case, there is some q such that $B(y, y', q)$. Hence by axiom *E6*, there is some r such that $B(z, r, q)$ and $B(r, y', z')$. So by *E6* again, there is some s such that $B(z, s, z')$ and $B(q, y', s)$.

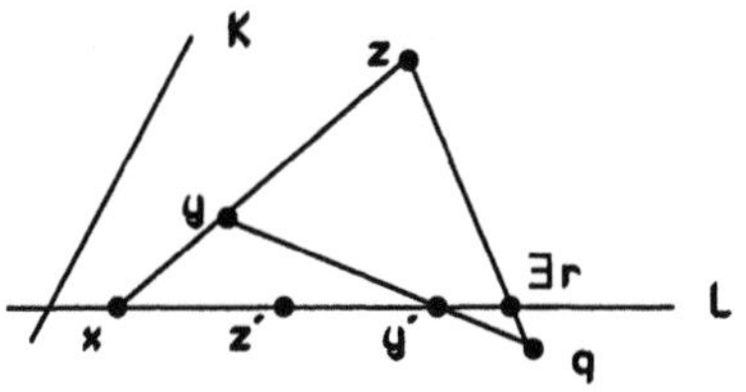

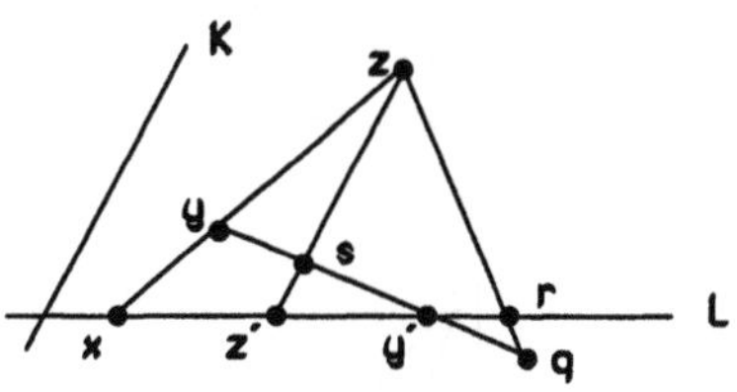

So $s \in L_{zz'} \cap L_{yy'}$. Hence, as these lines are parallel, $L_{zz'} = L_{yy'}$. This implies, as before, $P_{K/L}(x) = P_{K/L}(y) = P_{K/L}(z)$, and we've shown the lemma when $x \in L$.

If $x \notin L$, set $M = L_{xP_{K/L}(z)}$. Then not ($M \mathbin{/\!/} L$), as $M \cap L \neq \varnothing$. If $M \mathbin{/\!/} K$, then $P_{K/L}(x) = P_{K/L}(z)$, so by Lemma 26, $P_{K/L}(x) = P_{K/L}(y) = P_{K/L}(z)$, and we are done. So assume that not ($M \mathbin{/\!/} L$) and not ($M \mathbin{/\!/} K$) as in the diagram.

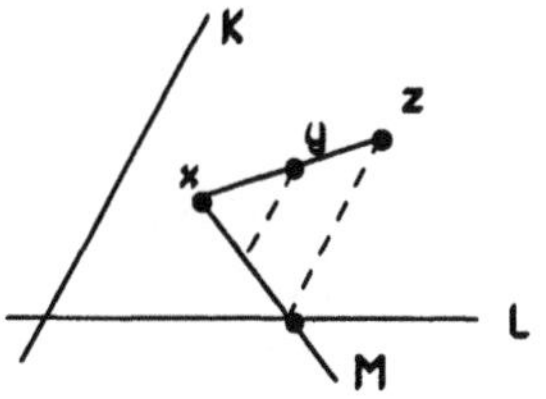

Then by what we have just proved, namely (‡) with appropriate relabeling, $B(P_{K/M}(x), P_{K/M}(y), P_{K/M}(z))$.
As $P_{K/M}(x) = x$ and $P_{K/M}(z) = P_{K/L}(z)$, we have again by (‡), $B(P_{K/L}(x), P_{K/L}(P_{K/M}(y)), P_{K/L}(P_{K/M}(z)))$.
As $L_{yP_{K/M}(y)} \mathbin{/\!/} K$ and $L_{zP_{K/M}(z)} \mathbin{/\!/} K$, we have by Lemma 26, $P_{K/L}(P_{K/M}(y)) = P_{K/L}(y)$ and $P_{K/L}(P_{K/M}(z)) = P_{K/L}(z)$.
Hence, $B(P_{K/L}(x), P_{K/L}(y), P_{K/L}(z))$. ∎

8. The Pappus-Pascal Theorem

Lemma 30

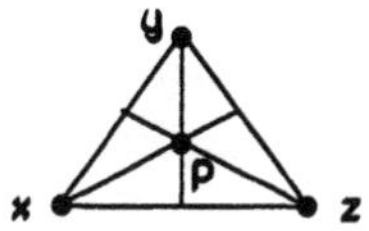

a. The altitudes of a triangle intersect in one point (the orthocentrum of a triangle exists)

$\forall x \,\forall y \,\forall z \;\neg L(x, y, z) \rightarrow \exists!\, p\,(px \perp yz \wedge py \perp zx \wedge pz \perp xy)$

b. $\forall x \,\forall y \,\forall z \;\neg L(x, y, z) \rightarrow \forall p\,((px \perp yz \wedge py \perp zx) \rightarrow pz \perp xy)$

Proof (a) To find the orthocentrum, by *E14* we can circumscribe a circle about the triangle $x'y'z'$ in the diagram below. We will show that the center of that circle is the point p that we want. Using Lemma 22,

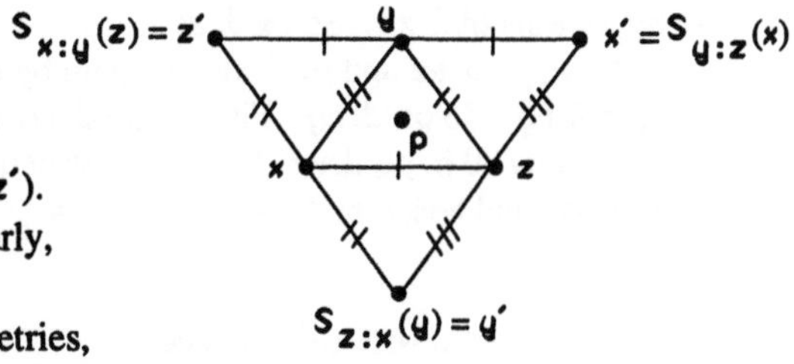

$xz \mathbin{/\!/} S_{x:y}(x)S_{x:y}(z)$

$xz \mathbin{/\!/} S_{y:z}(x)S_{y:z}(z)$

That is, $xz \mathbin{/\!/} yz' \mathbin{/\!/} x'y$. Hence, $L(x', y, z')$. So $x'z' \mathbin{/\!/} x'y \mathbin{/\!/} xz$. So $xz \mathbin{/\!/} x'z'$. Similarly, $xy \mathbin{/\!/} x'y'$ and $yz \mathbin{/\!/} y'z'$.

Also, as point symmetries are isometries, $xz \approx S_{y:z}(x)S_{y:z}(z)$. That is, $xz \approx x'y$. And similarly $xz \approx yz'$. Thus, $x'y \approx yz'$. In the same way, $y'z \approx zx'$ and $z'x \approx xy'$.

By axiom *E14*, there is some p such that $px' \approx py' \approx pz'$. As $pz' \approx py'$ and $xz' \approx xy'$, we have $p, x \in C_{y'z'}$. So $L_{px} = C_{y'z'}$, and also by Lemma 16.c, $C_{y'z'} \perp L_{y'z'}$. So $L_{px} \perp L_{y'z'}$. Since $yz \mathbin{/\!/} y'z'$, by Theorem 21.c, $px \perp yz$. Similarly, $py \perp zx$ and $pz \perp xy$. I'll leave to you to show that p is unique. Part (b) follows from this same proof. ∎

We can now show that if we have two pairs of parallel lines as in the diagram, then the third pair of lines is also a parallel.

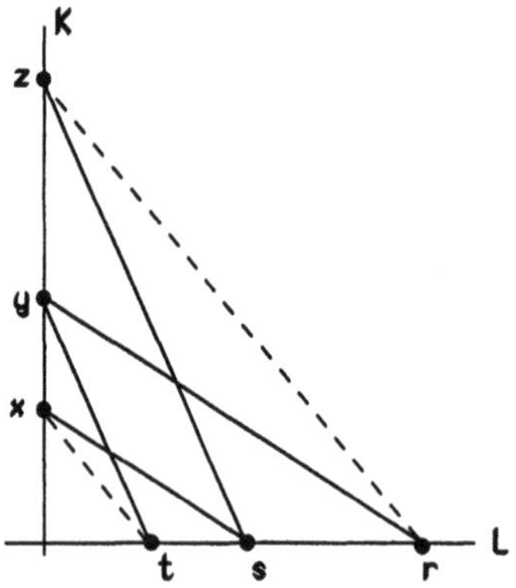

This is a weak version of the Pappus-Pascal theorem because we assume that $K \perp L$. It makes the proof simpler and it's all we'll need. I'll abbreviate $K(x) \wedge K(y) \wedge K(z)$ as $x, y, z \in K$.

Lemma 31 *The Pappus-Pascal theorem*

$\forall K\, \forall L\, \forall x\, \forall y\, \forall z\, \forall r\, \forall s\, \forall t\ [(x, y, z \in K) \wedge (x, y, z \notin L) \wedge$
$(r, s, t \in L) \wedge (r, s, t \notin K) \wedge K \perp L \wedge (xs \,//\, yr \wedge yt \,//\, zs)] \rightarrow xt \,//\, zr$

Proof I'll leave to you the case when $x = y$ or $y = z$. So assume the hypothesis holds in the picture. By Theorem 17.b, let M be such that $y \in M$ and $M \perp L_{xt}$.

Now $M \neq L$, as $y \in M$. Also, we cannot have $M \,//\, L$, for if it were, then $L_{xt} \perp L$, and then $L_{xt} \,//\, K$, so $L_{xt} = K$. But $t \notin K$. Thus, there is a unique p such that $p \in M \cap L$.

Informally, M and L are both altitudes of the triangle y, x, t. So we can use Lemma 30 to show that p is on the third altitude which passes through x, that is, $px \perp yt$. (Note that in this case the orthocentrum is not interior to the triangle.) Spelling this out: $py \perp xt$ and $pt \perp xy$, since $L_{pt} = L$. So by Lemma 30.b, $px \perp yt$. By hypothesis $yt \,//\, zs$. So $px \perp zs$.

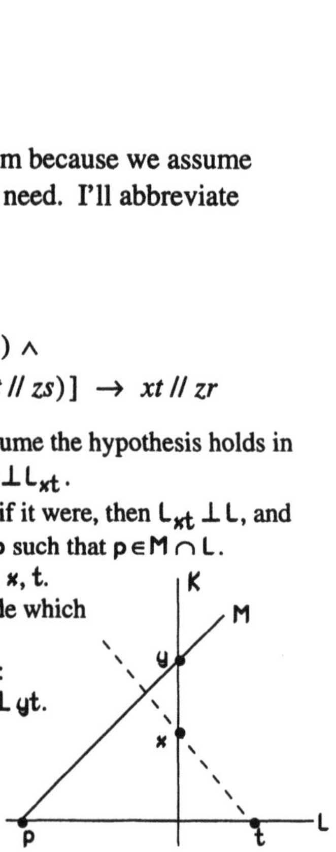

Now, $px \perp sz$ and $ps \perp xz$, so again by Lemma 30.b, $pz \perp xs \,//\, ry$. So $pz \perp ry$. Finally, $pz \perp ry$ and $pr \perp zy$, so by Lemma 30.b, $py \perp rz$. But by construction, $py \perp xt$. So $rz \,//\, xt$, and hence $xt \,//\, rz$. ∎

9. Multiplication of points

Given a model M of **E2**, we can pick any two points, 0, 1, with $0 \neq 1$, to give us a line L_{01}. By Theorem 6, B and ≈ restricted to L_{01} is a model of one-dimensional geometry, **E1**. For addition and an ordering on that line, we can then proceed as in the one-dimensional case and expand the language with two new names, c_0, c_1, which we'll again write as 'o', 'ɫ'.

Addition of points and an ordering in $L(=; B, \equiv, o, \text{ł})$

$x + y = z \equiv_{\text{Def}} (x \in L_{o\text{ł}} \wedge y \in L_{o\text{ł}}) \wedge S_{x:y}(o) = z$

$x < y \equiv_{\text{Def}} (x \in L_{o\text{ł}} \wedge y \in L_{o\text{ł}}) \wedge xy \uparrow o\,\text{ł}$

So $<\mathsf{L_{01}}; <, +, \mathsf{0}, \mathsf{1}>$ is an ordered group. To introduce multiplication for an ordered field, we need to choose another point, $\underline{\mathsf{1}}$.

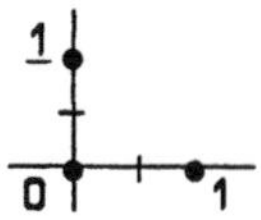

Lemma 32 In $L(=; B, \equiv, o, \text{ł})$,
$\{\mathbf{E2}, o \neq \text{ł}\} \models \exists z\,(o\text{ł} \perp oz \wedge o\text{ł} \equiv oz)$

Proof In a model, by *E2* there are a and z such that $\mathsf{B(a, 0, 1)}$ and $\mathsf{01} \approx \mathsf{0z}$, $\mathsf{0z} \approx \mathsf{oa}$, and $\mathsf{1z} \approx \mathsf{za}$. Hence, z and $\mathsf{0}$ lie on the bisector of $\mathsf{1}$ and a. That is, $\mathsf{L_{0z}} = \mathsf{C_{1a}}$. So by Lemma 16.c, $\mathsf{L_{0z}} \perp \mathsf{L_{1a}}$. But $\mathsf{L_{1a}} = \mathsf{L_{01}}$. ■

$\mathbf{E2}_{\mathbf{01\underline{1}}}$ in $L(=; B, \equiv, o, \text{ł}, \underline{\text{ł}})$

E2

$o \neq \text{ł} \wedge o\text{ł} \perp o\underline{\text{ł}} \wedge o\text{ł} \equiv o\underline{\text{ł}}$

In a model, the interpretation $\mathsf{0}$ of 'o' is called the *origin*.

Axis reflection

$\underline{x} = z \equiv_{\text{Def}} S_{L_{o(\text{ł}:\underline{\text{ł}})}}(x) = z$

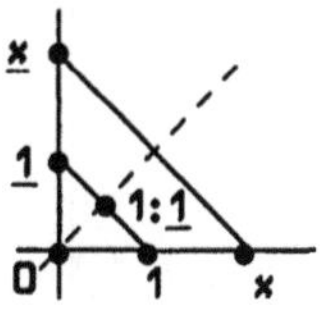

Lemma 33

a. $\underline{\text{ł}} = S_{L_{o(\text{ł}:\underline{\text{ł}})}}(\text{ł})$

b. $\underline{o} = o$

c. $o \neq \underline{\text{ł}} \wedge \text{ł} \neq \underline{\text{ł}}$

d. $\forall x\ \underline{\underline{x}} = x$

e. $\forall x\ (x \in L_{o\text{ł}} \leftrightarrow \underline{x} \in L_{o\underline{\text{ł}}})$

f. $\forall x\, \forall y\ \underline{x} = \underline{y} \rightarrow x = y$

g. $\forall y\ (y \in L_{o\underline{\text{ł}}} \rightarrow \exists! x\,(x \in L_{o\text{ł}} \wedge \underline{x} = y))$

h. $\forall y\ (y \in L_{o\text{ł}} \rightarrow \exists! x\,(x \in L_{o\underline{\text{ł}}} \wedge \underline{x} = y))$

Proof Parts (a) and (b) follow by the definitions, and (c) by Lemma 32.

(d) This is because line symmetries are involutions.

(e) We have that $\mathsf{x} \in \mathsf{L_{01}}$ iff $\mathsf{B(0, 1, x)}$ or $\mathsf{B(1, x, 0)}$ or $\mathsf{B(x, 0, 1)}$, so (d) follows as line symmetries preserve betweenness.

(f) This follows as line symmetries are 1-1.

(g) Given $y \in L_{0\underline{1}}$, let $x = S_{L_{0(1\,:\,\underline{1})}}(y)$. As in the proof of (e), $x \in L_{01}$, and since line symmetries are involutions, $\underline{x} = y$. Uniqueness is by (f). Part (h) follows similarly. ■

Thus, the interpretation of axis reflection in a model gives a 1-1, onto function between L_{01} and $L_{0\underline{1}}$, and our two notations for $\underline{1}$ coincide.

Lemma 34

a. $\forall x\, \forall y\ (x \in L_{o1} \wedge y \in L_{o1}) \rightarrow ((\underline{y}x \mathbin{//} o1 \rightarrow y = o) \wedge (\underline{y}x \mathbin{//} o\underline{1} \rightarrow x = o))$

b. $\forall x\, \forall y\, \forall z\ (x, y, z \in L_{o1} \wedge y \neq o) \rightarrow \exists! w\, (w \in L_{o1} \wedge \underline{y}x \mathbin{//} \underline{z}w))$

c. $\forall x\, \forall y\, \forall z\, \forall w\, \forall r\ (x, y, z, w, r \in L_{o1} \wedge \underline{y}x \mathbin{//} \underline{z}w \wedge \underline{y}x \mathbin{//} \underline{z}r) \rightarrow (\underline{y} = o \vee w = r)$

Proof (a) If $\underline{y}x \mathbin{//} 01$ and $x \in L_{01}$, then $L_{\underline{y}x} = L_{01}$. So $\underline{y} \in L_{01}$. But as $y \in L_{01}$, by Lemma 33.e, $\underline{y} \in L_{0\underline{1}}$. As $L_{01} \cap L_{0\underline{1}} = \{0\}$, $\underline{y} = 0$. The second conjunct is proved similarly.

(b) Suppose we have x, $\underline{y}$, $\underline{z}$ as in the diagram. Then $w = P_{L_{\underline{y}x}/L_{01}}(\underline{z})$.

(c) This follows by (a) and (b). ■

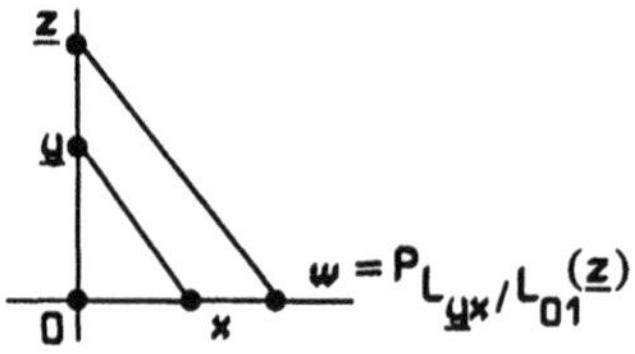

The idea behind our definition of multiplication of points is that in the diagram the two triangles are similar. So we should have $x/\underline{1} = z/\underline{y}$, and $x y = z$.

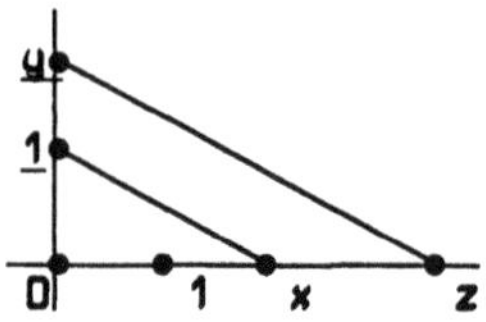

Multiplication of points

$x \mathbin{\times} y = z \ \equiv\!\text{Def}\ x, y, z \in L_{0\underline{1}} \wedge (\underline{1}\, x \mathbin{//} \underline{y}z \vee (y = o \wedge z = o) \vee (x = o \wedge z = o))$

Lemma 35

a. $\forall x\, \forall y\ \ x \times y = o \leftrightarrow x = o \vee y = o$

b. $\forall x\ \ x \times 1 = 1 \times x = x$

c. $\forall y\, \forall z\ \ y \neq o \rightarrow \exists! x\, (x \times y = z)$

d. *Thales' theorem*

$\forall x\, \forall y\, \forall z\, \forall w\ (x \neq o \wedge y \neq o \wedge z \neq o \wedge w \neq 0) \rightarrow (x \times y = z \times w \leftrightarrow \underline{w}x \mathbin{//} \underline{y}z)$

e. $\forall x\, \forall y\, \forall z\ (x \times y) \times z = (x \times z) \times y$

f. $\forall x\, \forall y\ \ x \times y = y \times x$

g. $\forall x\, \forall y\, \forall z\ (x \times y) \times z = x \times (y \times z)$

Proof If $x = 0$ or $y = 0$, then $x \times y = 0$ by definition as the first disjunct does not apply, which you can verify using Lemma 34.a.

If $x \times y = 0$ and $x \neq 0$ and $y \neq 0$, then $\underline{1x} \,//\, \underline{y0}$. So $L_{\underline{1x}} \,//\, L_{\underline{y0}} \,//\, L_{\underline{10}}$, which is impossible. Hence, $x = 0$ or $y = 0$.

(b) We only need to consider when $x \neq 0$. We have $x \times 1 = z$ iff $\underline{1x} \,//\, \underline{1z}$, which implies $x = z$. We have $1 \times x = z$ iff $\underline{11} \,//\, \underline{xz}$. But $\underline{11} \,//\, \underline{xx}$, as $L_{0(1:\underline{1})} = C_{1\underline{1}} = C_{x\underline{x}}$. Hence, $x\underline{x} \perp L_{0(1:\underline{1})}$, and $1\underline{1} \perp L_{0(1:\underline{1})}$. So $x\underline{x} \,//\, 1\underline{1}$. Hence, $1 \times x = x$.

(c) This follows from Lemma 34.b.

(d) Consider the diagram to the right.

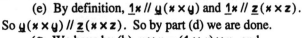

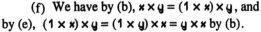

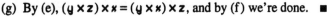

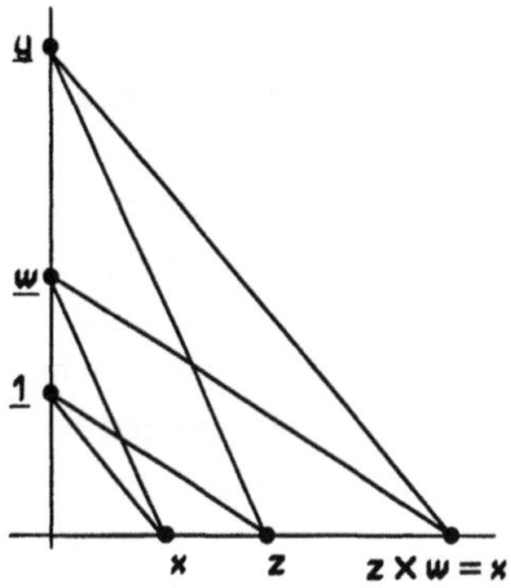

The Pappus-Pascal theorem tells us that if any two of these pairs of lines are parallel, the third pair is parallel, too. We have $\underline{1z} \,//\, \underline{w}(z \times w)$ by definition. If $x \times y = z \times w$, then as $\underline{1x} \,//\, \underline{y}(x \times y)$, we have $\underline{1x} \,//\, \underline{y}(z \times w)$. Hence by the Pappus-Pascal theorem, $\underline{wx} \,//\, \underline{yz}$. In the other direction, if $\underline{wx} \,//\, \underline{yz}$, then $\underline{1x} \,//\, \underline{y}(z \times w)$, and hence by definition, $x \times y = z \times w$.

(e) By definition, $\underline{1x} \,//\, \underline{y}(x \times y)$ and $\underline{1x} \,//\, \underline{z}(x \times z)$. So $\underline{y}(x \times y) \,//\, \underline{z}(x \times z)$. So by part (d) we are done.

(f) We have by (b), $x \times y = (1 \times x) \times y$, and by (e), $(1 \times x) \times y = (1 \times y) \times x = y \times x$ by (b).

(g) By (e), $(y \times z) \times x = (y \times x) \times z$, and by (f) we're done. ■

Thus we have that in any model of $\mathbf{E2_{01\underline{1}}}$, $\langle L_{01} - \{0\}\,;\, \times \rangle$ is a commuative group. We now link addition and multiplication via the distributivity law, using that parallel projection preserves midpoints.

Lemma 36 *The distributivity of multiplication over addition*

a. $\forall x\, \forall y\, \forall z\;\; x \,\check{\times}\, (y \,\vdots\, z) = (x \,\check{\times}\, y) \,\vdots\, (x \,\check{\times}\, z)$

b. $\forall x\, \forall y\, \forall z\;\; S_x(y) \,\check{\times}\, z = S_{x \,\check{\times}\, z}(y \,\check{\times}\, z)$

c. $\forall x\, \forall y\, \forall z\;\; x \,\check{\times}\, (y + z) = (x \,\check{\times}\, y) + (x \,\check{\times}\, z)$

Proof (a) Let $K = L_{\underline{1x}}$. By Lemma 27, $P_{K/L_{01}}(\underline{y : z}) = P_{K/L_{01}}(\underline{y}) : P_{K/L_{01}}(\underline{z})$. So as in the proof of Lemma 34.b, $x \times (y : z) = (x \times y) : (x \times z)$.

(b) Suppose $S_x(y) = p$. So by definition, $x = y : p$, hence by (a), we have (i) $x \times z = (y \times z) : (p \times z)$. Now $S_{x \times z}(y \times z) = q$ iff $q : (y \times z) = (x \times z)$, and there is only one such q. So by (i), $p \times z = q$, and we are done.

(c) Assume $z \neq 0$ and $x + y \neq 0$. I'll show that $(x + y) \times z = (x \times z) + (y \times z)$, from which the result will follow by Lemma 35.f, leaving to you when $z = 0$ or $x + y = 0$.

$(x + y) \times z =$	$S_{x\,:\,y}(0) \times z$	definition
$=$	$S_{(x\,:\,y) \times z}(0)$	by (b)
$=$	$S_{(x \times z)\,:\,(y \times z)}(0)$	by (a), and Lemma 35.f, g
$=$	$(x \times z) + (y \times z)$	by definition

■

Thus in any model of $\mathbf{E2}_{01\underline{1}}$, $\langle \mathsf{L}_{01}; <, +, \times, \mathsf{0}, \mathsf{1} \rangle$ is a field, and by the one-dimensional case $+$ is monotonic with respect to $<$. Now we'll show that multiplication of points is monotonic with respect to $<$, using the Pasch axiom, *E6*.

Lemma 37 $\forall x \forall y \; (\mathrm{o} \leq x) \wedge (\mathrm{o} \leq y) \rightarrow \mathrm{o} \leq x \times y$

Proof Suppose that $\mathsf{0} \leq x$ and $\mathsf{0} \leq y$. Suppose to the contrary that $x \times y < \mathsf{0}$. Then by the one-dimensional case, $\mathsf{B}(x, \mathsf{0}, x \times y)$. Since $\mathsf{0} \leq y$, by definition $\mathsf{B}(\mathsf{0}, y, \mathsf{1})$ or $\mathsf{B}(\mathsf{0}, \mathsf{1}, y)$. As line symmetries preserve betweenness, we have $\mathsf{B}(\underline{\mathsf{0}}, \underline{y}, \underline{\mathsf{1}})$ or $\mathsf{B}(\underline{\mathsf{0}}, \underline{\mathsf{1}}, \underline{y})$. Since $\underline{\mathsf{0}} = \mathsf{0}$, we have (i) $\mathsf{B}(\mathsf{0}, \underline{y}, \underline{\mathsf{1}})$ or (ii) $\mathsf{B}(\mathsf{0}, \underline{\mathsf{1}}, \underline{y})$.

If (ii), the Pasch axiom says $x\underline{\mathsf{1}}$ intersects $\underline{y}\,(x \times y)$. That is, by *E6*, there is a point p such that $\mathsf{B}(x \times y, \mathsf{p}, \underline{y})$ and $\mathsf{B}(x, \underline{\mathsf{1}}, \mathsf{p})$. But $\underline{\mathsf{1}}x \,//\, \underline{y}\,(x \times y)$, which is a contradiction unless $x = \mathsf{0}$. In that case, $x \times y = \mathsf{0}$, which is also contrary to our assumption.

If (i), we proceed similarly to a contradiction, and so $\mathsf{0} \leq x \times y$. ■

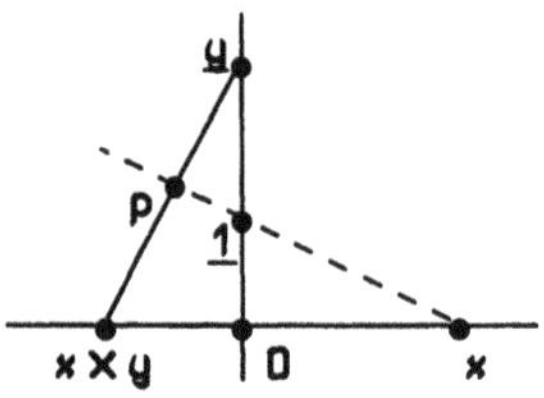

Recall that **OF** is the theory of ordered fields (Chapter XVII.E.2). From Theorem 6, Lemma XVIII.48, and Lemmas 35, 36, and 37, we have the following.

Theorem 38 If $\mathsf{M} \models \mathbf{E2}_{01\underline{1}}$, then $\langle \mathsf{L}_{01}; <, +, \times, \mathsf{0}, \mathsf{1} \rangle \models \mathbf{OF}$.

We'll now proceed as for the representation theorems and translations for one-dimensional geometry and the theory of ordered divisible groups (Chapter XVIII.D). We'll show that every ordered field arises via some model of **E2** as in Theorem 38. And for any ordered field we can define a model of **E2** based on that field, the Cartesian plane over that field (compare Exercise 5 of Section A). First, though, we need to express the geometric primitives in terms of the algebraic operations.

C. Betweenness and Congruence Expressed Algebraically

We want to derive the betweenness and congruence relations in terms of the algebra of the field given by the elements of the interpretation of $\mathrm{L}_{0\underline{1}}$. To accomplish that, we'll create models of $\mathbf{E2}_{01\underline{1}}$ as ordered pairs of those elements.

$\mathrm{N} = \mathrm{L}_{01}$ the *number-axis*

$\underline{\mathrm{N}} = \mathrm{L}_{0\underline{1}}$

We want to take the first co-ordinate of a point x to be the perpendicular projection of it onto N. Since $\underline{\mathsf{N}} \perp \mathsf{N}$, we can take for that the parallel projection with respect to $\underline{\mathsf{N}}$. We'd like to take the second co-ordinate to be $\mathsf{P}_{\mathsf{N}/\underline{\mathsf{N}}}(x)$. But we need an element of the number line N, so we instead take the reflection of that on N.

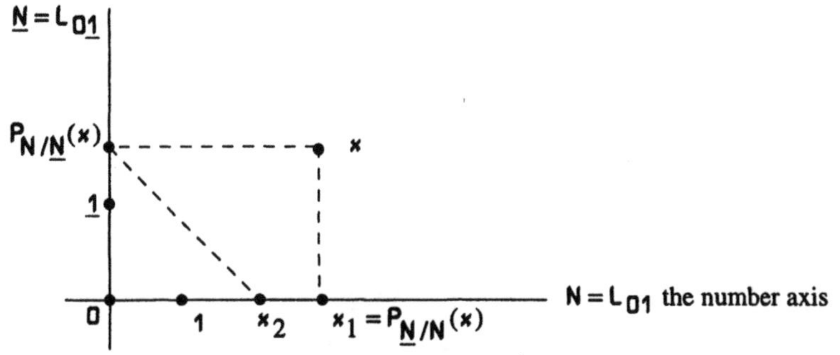

The Cartesian co-ordinates in **E2$_{011}$**

$(x)_1 \equiv_{Def} \mathrm{P}_{\underline{N}/N}(x) \qquad (x)_2 \equiv_{Def} \mathrm{P}_{N/\underline{N}}(x)$

In a model I'll write x_1 for $(x)_1$ and x_2 for $(x)_2$.

Lemma 39

a. $\forall x \,\forall y \;\; x = y \leftrightarrow (x)_1 = (y)_1 \wedge (x)_2 = (y)_2$

b. $\forall x \,\forall y \;\; x \in \mathrm{N} \wedge y \in \mathrm{N} \rightarrow \exists! z \; ((z)_1 = x \wedge (z)_2 = y)$

Proof (a) Suppose that $x^c \neq y^c$. Then either $x_1 \neq y_1$ or $x_2 \neq y_2$. I'll do the first case, and the other is done analogously. If either $x \in \mathrm{N}$ or $y \in \mathrm{N}$ I'll let you do it. If $x \notin \mathrm{N}$ and $y \notin \mathrm{N}$, then $x \neq x_1$ and $y \neq y_1$. So we have $xx_1 \perp \mathrm{N} \perp yy_1$, so $xx_1 \,/\!/\, yy_1$. Since $xx_1 \cap \mathrm{N} = \{x_1\}$ and $yy_1 \cap \mathrm{N} = \{y_1\}$, we must have $xx_1 \neq yy_1$. So $x \notin yy_1$, and hence $x \neq y$.

(b) Given $x, y \in \mathrm{N}$, we have $y \in \underline{\mathrm{N}}$. By Theorem 17.b, let L, M be lines such that $x \in \mathrm{L}$ and $\mathrm{L} \perp \mathrm{N}$, and $y \in \mathrm{M}$ and $\mathrm{M} \perp \underline{\mathrm{N}}$. By Lemma 18.b, $\mathrm{L} \perp \mathrm{M}$. So by Theorem 17.b, there is a unique point $z \in \mathrm{L} \cap \mathrm{M}$. So as you can show, $z_1 = x$ and $z_2 = x$. By (a), z is unique. ■

We can now define algebraic operations on points in terms of their co-ordinates.

$x^c = y^c \quad \equiv_{Def} \; (x)_1 = (y)_1 \wedge (x)_2 = (y)_2$ *co-ordinate equality*

$x^c + y^c = z^c \quad \equiv_{Def} \; ((x)_1 + (y)_1 = (z)_1) \wedge ((x)_2 + (y)_2 = (z)_2)$

$x^c - y^c = z^c \quad \equiv_{Def} \; ((x)_1 - (y)_1 = (z)_1) \wedge ((x)_2 - (y)_2 = (z)_2)$
where $-$ is the defined inverse of addition on points from the one-dimensional case

$x^c \times y^c = z \quad \equiv_{Def} \; ((x)_1 \times (y)_1) + ((x)_2 \times (y)_2) = z$

$(x^c)^2 \equiv_{Def} x^c \times x^c$

$1/2 \equiv_{Def} 0 \vdots 1$

Lemma 40

a. $\forall x \;\; x^c + o^c = x^c = o^c + x^c$

b. $\forall x \, \forall y \;\; (x \vdots y)^c = 1/2\,(x^c + y^c)$

c. $\forall x \, \forall y \;\; (S_x(y))^c = 2(x)^c - (y)^c$

d. $\forall x \;\; (\underline{x})_1 = (x)_2 \wedge (\underline{x})_2 = (x)_1$

e. $\forall x \;\; (S_N(x))_1 = (x)_1 \wedge (S_N(x))_2 = -(x)_2$

f. $\forall x \;\; (S_{\underline{N}}(x))_1 = -(x)_1 \wedge (S_{\underline{N}}(x))_2 = (x)_2$

g. $\forall x \;\; S_N S_{\underline{N}}(x) = S_o(x)$

Proof (b) We have $(x:y)_1 = P_{N/\underline{N}}(x:y) =$ (Lemma 27) $P_{N/\underline{N}}(x) : P_{N/\underline{N}}(y) = x_1 : y_1 = 1/2(x_1 + y_1)$ by the one-dimensional case. The equality for the second co-ordinates follows again by Lemma 27, since line symmetries are automorphisms.

(c) By definition, $S_x(y) : y = x$, and so $(S_x(y) : y)^c = x^c$. So by (a), $1/2(S_x(y))^c + (y)^c = x^c$. By the one-dimensional case the result follows by group theory.

(d) To abuse notation, this part says that $(\underline{x})^c = (x_2, x_1)$. I'll write K for $L_{0(1:\underline{1})}$. Take $K_1 \,//\, K_2 \,//\, \underline{N}$ with $x \in K_2$ and $\underline{x} \in K_1$. Take $L_1 \,//\, L_2 \,//\, N$ with $x \in L_1$ and $\underline{x} \in L_2$. Since line symmetries are automorphisms, as $K_2 \perp N$, we have $S_K(K_2) \perp \underline{N}$, so $S_K(K_2) \,//\, L_2$. Since $x \in K_2$, we have $\underline{x} \in S_K(K_2)$. So, $S_K(K_2) = L_2$. And $S_K(x_1) \in S_K(K_2) = L_2$, and $S_K(x_1) \in \underline{N}$. So $\{S_K(x_1)\} = L_2 \cap \underline{N}$, and $S_K S_K(x_1) = (\underline{x})_2$, and since S_K is an involution, $(\underline{x})_2 = x_1$.

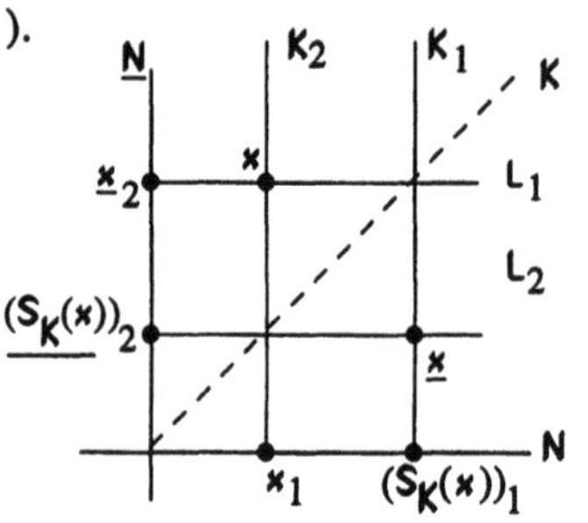

The other co-ordinate is done similarly.

(e) By the definition of line symmetry and Lemma 16.c, $L_{x S_N(x)} \perp N$. Also, $L_{x x_1} \,//\, \underline{N} \perp N$. Hence, $L_{x S_N(x)} \,//\, L_{x x_1}$. Hence, as x is in both, $L_{x S_N(x)} = L_{x x_1}$. So $L_{x S_N(x)} \cap N = L_{x x_1} \cap N = \{x_1\}$, and $x_1 = (S_N(x))_1$.

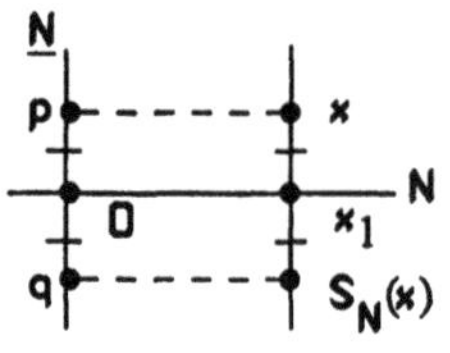

For the second co-ordinate, let $K = L_{0(1:\underline{1})}$, $p = P_{N/\underline{N}}(x) = S_K(x_2)$, and $q = P_{N/\underline{N}}(S_N(x)) = S_K((S_N(x))_2)$. Because $S_N(x_1) = x_1$, we have $x x_1 \approx x_1 S_N(x)$, and also $x x_1 \,//\, x_1 S_N(x)$. So via Lemma 27, since $P_{N/\underline{N}}(x_1) = 0$ AND $P_{N/\underline{N}}(x_1) = 0$, we have $0 = p : q$, and hence $B(q, 0, p)$. As S_K is an automorphism $0 = (S_N(x))_2 : x_2$ and $B((S_N(x))_2), 0, x_2)$. That is $S_N(x)_2 = -x_2$.

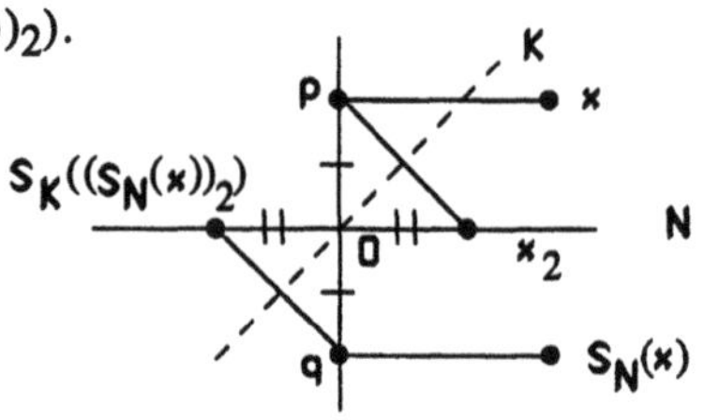

(f) This is proved in the same way as (e).

(g) Use parts (c), (d), (e), and (f). ∎

Lemma 41

a. $\forall x \, \forall y \;\; (x = o \vee y = o \vee xo \perp oy) \leftrightarrow x^c \times y^c = o$

b. $\forall x \, \forall y \;\; (x \neq y \wedge z \neq w \wedge xy \perp zw) \leftrightarrow (x^c - y^c) \times (z^c - w^c) = o$

Proof (a) I'll leave to you the case in both directions when $x = 0$ or $y = 0$

So given $x \neq 0$ and $y \neq 0$. I'll first show that (a) follows from three claims:

(i) $x0 \perp 0S_N(\underline{x})$

(ii) $y0 \,//\, \underline{y_2}(-y_1)$

(iii) $S_{\underline{N}}(x)0 \,//\, (-\underline{x}_1)(-x_2)$

If (i), then:

$x0 \perp 0y$	iff $0S_N(\underline{x}) \,//\, 0y$	by (i), definition, and Theorem 21
	iff $(-x_2)(-\underline{x}_1) \,//\, (-y_1)\underline{y_2}$	by (ii) and (iii)
	iff $y_2 \times (-x_2) = (-x_1) \times (-y_1)$	by Thales' Theorem (Lemma 35.d)
	iff $(x_1 \times y_1) + (x_2 \times y_2) = 0$	group theory
	iff $x^c \times y^c = 0$	

And that is (a).

To prove (i), we have by Lemma 10 that $S_0(x) \in L_{x0}$, so $L_{x0} = L_{xS_0(x)0}$. We also have by Lemma 16.c that $C_{xS_0(x)0} \perp L_{xS_0(x)0}$. So we need only show that $L_{0S_N(\underline{x})0} = C_{xS_0(x)0}$. That is, we need that both $0 \in C_{xS_0(x)0}$ and $S_N(\underline{x}) \in C_{xS_0(x)0}$. We have the first because by isometry $x0 \approx 0S_0(x)$. For the second,

$xS_N(\underline{x})$	$\approx \underline{xS_N(\underline{x})}$	as line symmetry is an isometry
	$\approx \underline{x}S_{\underline{N}}(x)$	as line symmetry is an automorphism
	$\approx S_N(\underline{x})S_N S_{\underline{N}}(x)$	line symmetries preserve congruence
	$\approx S_N(\underline{x})S_0(x)$	by Lemma 40.g

So $S_N(\underline{x}) \in C_{xS_0(x)0}$ and we have (i).

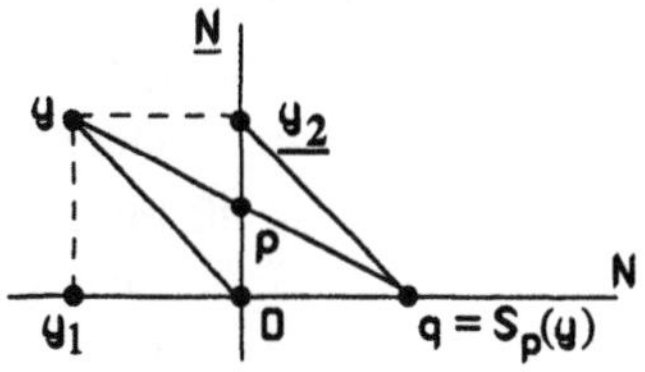

For (ii), consider the diagram. Here $p = \underline{y_2}:0$ and $q = S_p(y)$. As $p = y:q$, by Lemma 24.b, $y0 \,//\, \underline{y_2}q$. So it only remains to show that $q = -y_1$. First, as $yy_2 \,//\, y_10$, and $yy_2 \,//\, S_p(y)S_p(y_2) = q0$, we have $y_10 \,//\, q0$. But $L_{y_10} = N$, so $q \in N$. Also, by isometry, $y\,\underline{y_2} \approx S_p(y)S_p(\underline{y_2}) = q0$. Noting that $P_{\underline{N}/N}(y) = y_1$ and $P_{\underline{N}/N}(\underline{y_2}) = 0$, $P_{\underline{N}/N}(q) = q$, and $P_{\underline{N}/N}(0) = 0$, we have by Lemma 28, $y_10 \approx q0$. So $y_1 : q = 0$, and hence $q = -y_1$. So we have (ii). Using Lemma 41.d,e and the same reasoning, we have (iii). And (a) is proved.

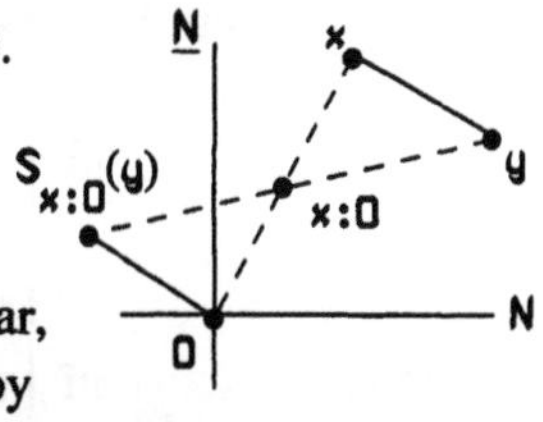

(b) Given $x \neq y$ and $z \neq w$, and $xy \perp zw$, we want to use (a). So we move xy to a parallel segment ending at 0. We have $S_{x:0}(x) = 0$, and by Lemma 22, $xy \,//\, S_{x:0}(x)S_{x:0}(y)$. So $xy \,//\, 0S_{x:0}(y)$, and hence, $xy \,//\, S_{x:0}(y)0$. Similarly, $zw \,//\, 0S_{z:0}(w)$. So as parallels to perpendiculars are perpendicular, $S_{x:0}(y)0 \perp 0S_{z:0}(w)$. So by (a), $S_{x:0}(y)^c \times S_{z:0}(w)^c = 0$. So by Lemma 40.c, $(2(x:0)^c - y^c) \times (2(z:0)^c - w^c) = 0$. Hence by Lemma 40.b, we have $(2(1/2(x^c + 0^c)) - y^c) \times (2(1/2(z^c + 0^c)) - w^c) = 0$. So by Lemma 40.a and group theory, $(x^c - y^c) \times (z^c - w^c) = 0$. For the other direction, reverse these steps. ■

Theorem 42 $\forall x\, \forall y\, \forall z\, \forall w \quad xy \equiv zw \leftrightarrow (x^c - y^c)^2 = (z^c - w^c)^2$

Proof Given x, y, z, w, I'll leave to you the case when $x = y$ or $z = w$. So assume $x \neq y$ and $z \neq w$. First we'll consider the case when $y = 0$ and $w = 0$. I'll let you show that $x0 \approx z0$ iff $L_{0(x:z)} = C_{xz}$. So (leaving the algebra to you):

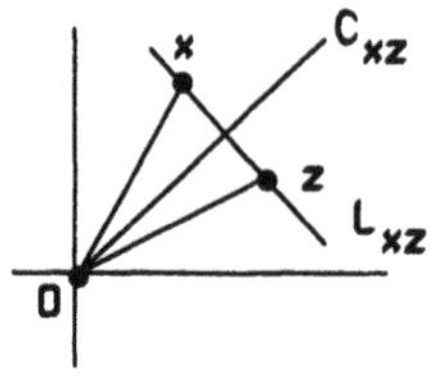

$x0 \approx z0$ iff $0(x:z) \perp xz$
iff $(0^c - (x:z)^c) \times (x^c - z^c) = 0$ Lemma 41.b
iff $-\tfrac{1}{2}(x^c + z^c) \times (x^c - z^c) = 0$ Lemma 40.b
iff $(x^c)^2 - (z^c)^2 = 0$ Lemma 36.c
iff $(x^c)^2 = (z^c)^2$

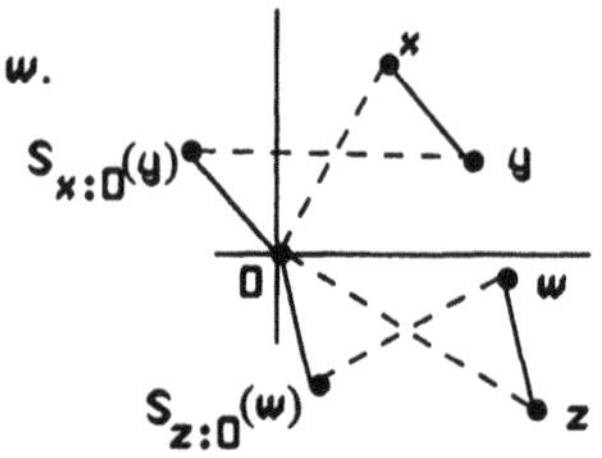

Now we consider any x, y, z, w, such that $x \neq y$ and $z \neq w$. We have $xy \approx S_{x:0}(x)S_{x:0}(y) = 0S_{x:0}(y) \approx S_{x:0}(y)0$. Similarly, $zw \approx 0S_{z:0}(w)$. So (leaving the algebra to you):

$xy \approx zw$ iff $S_{x:0}(y)0 \approx 0S_{z:0}(w)$
iff $(S_{x:0}(y)^c)^2 = (S_{z:0}(w)^c)^2$
iff $(x^c - y^c)^2 = (z^c - w^c)^2$ ■

We need one more definition.

$t \times x^c = z^c \equiv_{Def} (t \times (x)_1 = (z)_1) \wedge (t \times x_2 = (z)_2)$

Theorem 43 $\forall x\, \forall y\, \forall z \quad B(x, y, z) \leftrightarrow \exists t\,(0 \leq t \leq 1 \wedge ((1 - t) \times x^c) + (t \times z^c) = y^c)$

Proof I'll set out only the main points and leave to you the long, though straightforward computations, using the notation of determinants of matrices as shorthand. First, we have:

$xy \,//\, zw$ iff there are p, q such that $xy \perp pq \perp zw$
iff there are p, q such that $p \neq q$ and
$(x^c - y^c) \times (p^c - q^c) = (z^c - w^c) \times (p^c - q^c) = 0^c$

(‡) iff $\begin{vmatrix} x_1 - y_1 & x_2 - y_2 \\ z_1 - w_1 & z_2 - w_2 \end{vmatrix} = 0$ with the calculations left to you

Now we have :

$L(x, y, z)$ iff $xy \,//\, yz$ by Theorem 21.a and Lemma 4.f

iff $\begin{vmatrix} x_1 - y_1 & x_2 - y_2 \\ y_1 - z_1 & y_2 - z_2 \end{vmatrix} = 0$ by above

(‡‡) iff $\begin{vmatrix} 1 & x_1 & x_2 \\ 1 & y_1 & y_2 \\ 1 & z_1 & z_2 \end{vmatrix} = 0$ with the calculations left to you

Now assume that $B(x, y, z)$. Then by definition, $L(x, y, z)$. So, as we've shown, (‡‡).

So via a long calculation, there is some t such that $((1 - t) \times x^c) + (t \times z^c) = y^c$. To show that $0 \leq t \leq 1$, first we have $((1 - t) \times x_1) + (t \times z_1) = y_1$. But we also have by Lemma 29, $B(x_1, y_1, z_1)$. Hence by the one-dimensional case, $x_1 \leq y_1 \leq z_1$ or $z_1 \leq y_1 \leq x_1$. In either case we get via the one-dimensional case, $t = (x_1 - y_1)/(x_1 - z_1)$, which, since L_{01} is an ordered field, gives us $0 \leq t \leq 1$.

Now assume there is some t such that $0 \leq t \leq 1$ and $((1 - t) \times x^c) + (t \times z^c) = y^c$. Then via a long calculation, we have (‡‡). So $L(x, y, z)$. So (i) $B(x, y, z)$ or (ii) $B(y, z, x)$ or (iii) $B(z, x, y)$. If (i), we are done. If (ii), then by Lemma 29, $B(y_1, z_1, x_1)$. So $1 \leq t$, but then $t = 1$, and hence $y = z$, and so $B(x, y, z)$. If (iii), then $B(z_1, x_1, y_1)$, and so $t \leq 0$, which gives that $t = 0$, and so $x = y$, resulting again in $B(x, y, z)$. ■

D. Ordered Fields and Cartesian Planes

We can now translate between the theory of two-dimensional geometry and the theory of ordered fields, **OF** (Chapter XVII.E.2).

We'll translate from **OF** to $\mathbf{E2_{01\underline{1}}}$ in two stages. We first translate the functions of the **OF** into predicates, which we can do as set out in Chapter VIII.E.

A mapping $\gamma: L(=; <, \oplus, *, o, \mathfrak{l}) \to L(=; <, P_\oplus, P_*, o, \mathfrak{l})$ translates functions into predicates. The formulas $\alpha_\oplus, \alpha_*$ are the wffs that ensure that $P_\oplus, P_*$ are predicates that are interpreted as functions (see p. 149).

We have that for any Γ, A, $\Gamma \vDash_{\mathbf{OF}} A$ iff $\gamma(\Gamma) \cup \{\alpha_\oplus, \alpha_*\} \vDash \gamma(A)$.

The mapping $\varphi: L(=; <, P_\oplus, P_*, o, \mathfrak{l}) \to L(=; B, \equiv, o, \mathfrak{l}, \underline{\mathfrak{l}})$ is given in the definition of equivalence-relation translations (Chapter XVI.B, p. 296) by:

$U(x) \equiv_{\mathrm{Def}} x \in L_{o\mathfrak{l}}$

$P_=(x, y) \equiv_{\mathrm{Def}} x = y$

$A_<(x, y) \equiv_{\mathrm{Def}} x < y$

$A_{P_\oplus}(x, y, z) \equiv_{\mathrm{Def}} x + y = z$

$A_{P_*}(x, y, z) \equiv_{\mathrm{Def}} x \times y = z$

$o \equiv_{\mathrm{Def}} o$

$\mathfrak{l} \equiv_{\mathrm{Def}} \mathfrak{l}$

Here $<$, $+$, and $\times$ are the defined notions in $L(=; B, \equiv, o, \mathfrak{l}, \underline{\mathfrak{l}})$.

Lemma 44 $\mathbf{E2}_{01\underline{1}} \vDash \varphi(\alpha_\oplus) \wedge \varphi(\alpha_*)$.

Proof The first conjunct comes from the definition of + and the one-dimensional case (Theorem 6 and Lemma 11). The second conjunct comes from the definition of x and Lemma 34.b. ■

Now we consider mappings of models of $\mathbf{E2}_{01\underline{1}}$ to models of **OF**.

Given any model $\mathsf{E} = \langle \mathsf{V}; \mathsf{B}, \approx \rangle \vDash \mathbf{E2}$ and $\mathsf{0}, \mathsf{1}, \underline{\mathsf{1}} \in \mathsf{V}$, with $\mathsf{0} \neq \mathsf{1}$, $\mathsf{01} \approx \mathsf{0}\underline{\mathsf{1}}$, and $\mathsf{01} \perp \mathsf{0}\underline{\mathsf{1}}$, expand E to the model $\mathsf{E}_{01\underline{1}} = \langle \mathsf{V}; \mathsf{B}, \approx, \mathsf{0}, \mathsf{1}, \underline{\mathsf{1}} \rangle \vDash \mathbf{E2}_{01\underline{1}}$. Define:

$$F(\mathsf{E}_{01\underline{1}}) \equiv_{\text{Def}} \langle \mathsf{L}_{01}; <_{\mathsf{E}}, +_{\mathsf{E}}, \bullet_{\mathsf{E}}, \mathsf{0}, \mathsf{1} \rangle$$

$$\mathrm{L}(=; <, \oplus, *, \mathrm{o}, \mathrm{i}) \xrightarrow{\ \gamma\ } \mathrm{L}(=; <, \mathrm{P}_\oplus, \mathrm{P}_*, \mathrm{o}, \mathrm{i}) \xrightarrow{\ \varphi\ } \mathrm{L}(=; \mathrm{B}, \equiv, \mathrm{o}, \mathrm{i}, \underline{\mathrm{i}})$$

models of **OF** $\xleftarrow{\ F\ }$ models of $\mathbf{E2}_{01\underline{1}}$

When we start with a model $\mathsf{E} \vDash \mathbf{E2}_{01\underline{1}}$, I'll write simply $F(\mathsf{E})$. In this new notation, Theorem 38 is: If $\mathsf{E} \vDash \mathbf{E2}_{01\underline{1}}$, then $F(\mathsf{E}) \vDash \mathbf{OF}$.

To go from **OF** to $\mathbf{E2}_{01\underline{1}}$ we first establish a mapping of models.

Cartesian planes Given $\mathsf{F} = \langle \mathsf{U}; <, +, \bullet, \mathsf{0}, \mathsf{1} \rangle \vDash \mathbf{OF}$, the *Cartesian plane over* F is:

$$C(\mathsf{F}) = \langle \mathsf{U} \times \mathsf{U}; \mathsf{B}_{\mathsf{F}}, \approx_{\mathsf{F}}, (\mathsf{0}, \mathsf{0}), (\mathsf{1}, \mathsf{0}), (\mathsf{0}, \mathsf{1}) \rangle$$

To specify B_{F} and $\approx_{\mathsf{F}}$ we set:

$$(x, y) + (z, w) \equiv_{\text{Def}} (x + z, y + w)$$

$$(x, y) - (z, w) \equiv_{\text{Def}} (x - z, y - w) \quad \text{where } - \text{ is the defined subtraction of } \mathsf{F}$$

$$(x, y) \cdot (z, w) \equiv_{\text{Def}} (x \bullet z) + (y \bullet w)$$

$$(x, y)^2 \equiv_{\text{Def}} (x, y) \cdot (x, y)$$

$$t(x, y) \equiv_{\text{Def}} (t \bullet x, t \bullet y)$$

Then,

$$\mathsf{B}_{\mathsf{F}}((x, y), (p, q), (z, w))$$
$$\equiv_{\text{Def}} \text{there is some } t,\ \mathsf{0} \leq t \leq \mathsf{1} \text{ and } (p, q) = (\mathsf{1} - t)(x, y) + t(z, w)$$

$$(x, y)(z, w) \approx_{\mathsf{F}} (p, q)(r, s) \equiv_{\text{Def}} ((x, y) - (z, w))^2 = ((p, q) - (r, s))^2.$$

Lemma 45 If $\mathsf{F} \vDash \mathbf{OF}$, then $C(\mathsf{F}) \vDash \mathbf{E2}_{01\underline{1}}$.

A mapping $\psi\colon L(=; B, \equiv, \mathsf{o}, \mathit{ł}, \underline{\mathit{ł}}) \to L(=; <, \oplus, *, \mathsf{o}, \mathit{ł})$ is given as in the definition of equivalence-relation translations (p. 296), where:

$\mathrm{U}(x, y) \equiv_{\mathrm{Def}} x = x$ (all pairs are acceptable)

$\mathrm{P}_{=}(x, y, z, w) \equiv_{\mathrm{Def}} x = z \wedge y = w$

$$\mathrm{P_B}(x, y, z, w, p, q) \equiv_{\mathrm{Def}} \exists t\,(\mathsf{o} \le t \le \mathit{ł} \wedge ((\mathit{ł} - t) * x) \oplus (t * p) = z \wedge ((\mathit{ł} - t) * y) \oplus (t * q) = w)$$

$$\mathrm{P}_{\equiv}(x, y, z, w, p, q, r, s) \equiv_{\mathrm{Def}} (x - z)^2 \oplus (y - w)^2 = (p - r)^2 \oplus (q - s)^2$$

$\mathsf{o} \equiv_{\mathrm{Def}} (\mathsf{o}, \mathsf{o})$

$\mathit{ł} \equiv_{\mathrm{Def}} (\mathit{ł}, \mathsf{o})$

$\underline{\mathit{ł}} \equiv_{\mathrm{Def}} (\mathsf{o}, \mathit{ł})$

$$L(=; B, \equiv, \mathsf{o}, \mathit{ł}, \underline{\mathit{ł}}) \xrightarrow{\ \psi\ } L(=; <, \oplus, *, \mathsf{o}, \mathit{ł})$$

$$\text{models of } \mathbf{E2}_{\mathbf{01\underline{1}}} \xleftarrow{\ C\ } \text{models of } \mathbf{OF}$$

Lemma 46 *The representation theorem for* $\mathbf{E2}_{\mathbf{01\underline{1}}}$ *and* **OF**

a. If $\mathsf{E} \models \mathbf{E2}_{\mathbf{01\underline{1}}}$, then there is a model $\mathsf{N} \models \mathbf{OF}$ such that $C(\mathsf{N}) \simeq \mathsf{E}$, namely, $C(F(\mathsf{E})) \simeq \mathsf{E}$.

b. If $\mathsf{F} \models \mathbf{OF}$, then there is a model $\mathsf{M} \models \mathbf{E2}_{\mathbf{01\underline{1}}}$ such that $F(\mathsf{M}) \simeq \mathsf{E}$, namely, $F(C(\mathsf{F})) \simeq \mathsf{F}$.

Proof (a) Suppose $\mathsf{E} = \langle \mathsf{V}; \mathsf{B}, \approx, \mathsf{0}, \mathsf{1}, \underline{\mathsf{1}} \rangle$. By Theorem 38, $F(\mathsf{E}) \models \mathbf{OF}$, and so by Lemma 45, $C(F(\mathsf{E})) \models \mathbf{E2}_{\mathbf{01\underline{1}}}$. Let $\alpha\colon \mathsf{E} \to C(F(\mathsf{E}))$ be defined by: $\alpha(\mathsf{p}) = (\mathsf{p}_1, \mathsf{p}_2)$. By Theorem 39 the mapping is 1-1 and onto. By Theorem 43:

$\mathsf{B}(\mathsf{x}, \mathsf{y}, \mathsf{z})$ iff there is some t with $((\mathsf{1} - \mathsf{t}) \times \mathsf{x}^{\mathsf{c}}) + (\mathsf{t} \times \mathsf{z}^{\mathsf{c}}) = \mathsf{y}^{\mathsf{c}}$ and $\mathsf{o} \le \mathsf{t} \le \mathsf{1}$
iff $\mathsf{B_F}((\mathsf{x}_1, \mathsf{x}_2), (\mathsf{y}_1, \mathsf{y}_2), (\mathsf{z}_1, \mathsf{z}_2))$

Similarly by Theorem 42, $\mathsf{xy} \approx \mathsf{zw}$ iff $(\mathsf{x}_1, \mathsf{x}_2)(\mathsf{y}_1, \mathsf{y}_2) \approx_{\mathsf{F}} (\mathsf{z}_1, \mathsf{z}_2)(\mathsf{w}_1, \mathsf{w}_2)$.

(b) We have that $F(C(\mathsf{F})) \models \mathbf{OF}$. Let $\beta\colon \mathsf{F} \to F(C(\mathsf{F}))$ be given by $\beta(\mathsf{x}) = (\mathsf{x}, \mathsf{0})$. I'll let you show that β is a homomorphism of addition and the ordering using the one-dimensional case. For multiplication, using Thales' theorem (Theorem 35.d):

$(\mathsf{x}, \mathsf{0}) \times (\mathsf{y}, \mathsf{0}) = (\mathsf{z}, \mathsf{0})$ iff $(\mathsf{0}, \mathsf{1})(\mathsf{x}, \mathsf{0}) \parallel (\mathsf{0}, \mathsf{y})(\mathsf{z}, \mathsf{0})$
iff $-(\mathsf{x} \bullet \mathsf{y}) + \mathsf{z} = \mathsf{0}$ by (‡) in the proof of Theorem 43
iff $\mathsf{x} \bullet \mathsf{y} = \mathsf{z}$ ■

Now we can establish that our mappings are translations via Lemma 44.

Lemma 47 The mappings $\varphi \circ \gamma$ and ψ are translations:

$\Gamma \vDash_{\mathbf{OF}} A$ iff $\varphi \circ \gamma(\Gamma) \vDash_{\mathbf{E2}_{01\underline{1}}} \varphi \circ \gamma(A)$.

$\Gamma \vDash_{\mathbf{E2}_{01\underline{1}}} A$ iff $\psi(\Gamma) \vDash_{\mathbf{OF}} \psi(A)$.

All that is left now is to eliminate o, ł, $\underline{ł}$ from $\mathbf{E2}_{01\underline{1}}$ using Theorem XVIII.51.

Theorem 48

a. Let $\mathsf{E} = \langle \mathsf{V}; \mathsf{B}, \approx \rangle \vDash \mathbf{E2}$ and
$\mathsf{0}, \mathsf{1}, \underline{\mathsf{1}} \in \mathsf{V}$, with $\mathsf{0} \neq \mathsf{1}$, and $\mathsf{01} \approx \mathsf{0\underline{1}}$, and $\mathsf{01} \perp \mathsf{0\underline{1}}$
$\mathsf{a}, \mathsf{b}, \mathsf{c} \in \mathsf{V}$, with $\mathsf{a} \neq \mathsf{b}$, and $\mathsf{ab} \approx \mathsf{ac}$, and $\mathsf{ab} \perp \mathsf{ac}$
Then there is an automorphism λ of E such that $\lambda(\mathsf{a}) = \mathsf{0}$, $\lambda(\mathsf{b}) = \mathsf{1}$, and $\lambda(\mathsf{c}) = \underline{\mathsf{1}}$.

b. The names o, ł, $\underline{ł}$ are named parameters for $\mathbf{E2}_{01\underline{1}}$.

Proof I'll sketch how to construct the isomorphism λ in stages, leaving to you to fill in the details. (1) We take a point symmetry to move a to $\mathsf{0}$. Then (2) we take a line symmetry to move b to the line L_{01}. Then (if necessary), (3) we take a line symmetry (or possibly two) with respect to L_{01}.

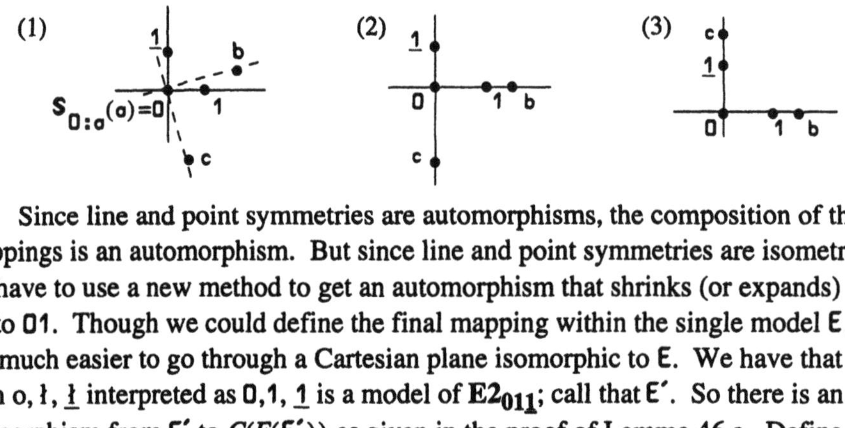

Since line and point symmetries are automorphisms, the composition of these mappings is an automorphism. But since line and point symmetries are isometries, we have to use a new method to get an automorphism that shrinks (or expands) $\mathsf{0b}$ to $\mathsf{01}$. Though we could define the final mapping within the single model E, it's much easier to go through a Cartesian plane isomorphic to E. We have that E with o, ł, $\underline{ł}$ interpreted as $\mathsf{0}, \mathsf{1}, \underline{\mathsf{1}}$ is a model of $\mathbf{E2}_{01\underline{1}}$; call that E'. So there is an isomorphism from E' to $C(F(\mathsf{E}'))$ as given in the proof of Lemma 46.a. Define ρ: $C(F(\mathsf{E}')) \rightarrow C(F(\mathsf{E}'))$ by $\rho((x, y)) = {}^{1}/_{\mathsf{b}}(x, y)$. Here ${}^{1}/_{\mathsf{b}}$ is the multiplicative inverse of b under the defined multiplication, which is possible since $\mathsf{a} \neq \mathsf{b}$ and so $\mathsf{0} \neq \mathsf{b}$ in the mapping of the model (3). I'll leave to you to show that ρ is an automorphism of $C(F(\mathsf{E}'))$. To complete the description of the automorphism of E, use the isomorphism of $C(F(\mathsf{E}'))$ to E given in the proof of Theorem 46.a. ∎

Now we can give translations between **E2** and **OF**.

Let α be the translation eliminating parameters from $\mathbf{E2}_{01\underline{1}}$.

Let ε be the identity mapping from L(=; B, ≡) to L(=; B, ≡, o, ł, $\underline{ł}$).

$$L(=; <, \oplus, *, \text{o}, \text{ł}) \xrightarrow{\gamma} L(=; <, P_{\oplus}, P_{*}, \text{o}, \text{ł}) \xrightarrow{\varphi} L(=; B, \equiv, \text{o}, \text{ł}, \underline{\text{ł}}) \xrightarrow{\alpha} L(=; B, \equiv)$$

$$L(=; B, \equiv) \xrightarrow{\varepsilon} L(=; B, \equiv, \text{o}, \text{ł}, \underline{\text{ł}}) \xrightarrow{\psi} L(=; <, \oplus, *, \text{o}, \text{ł})$$

I'll leave to you the following; part (c) is by Theorems XVII.17 and 18.

Theorem 49

a. The mapping $\alpha \circ \varphi \circ \gamma$ is a translation of **OF** to **E2**:
$\Gamma \vDash_{\mathbf{OF}} A$ iff $\alpha \circ \varphi \circ \gamma(\Gamma) \vDash_{\mathbf{E2}} \alpha \circ \varphi \circ \gamma(A)$.

b. The mapping $\psi \circ \varepsilon$ is a translation of **E2** to **OF**:
$\Gamma \vDash_{\mathbf{E2}} A$ iff $\psi \circ \varepsilon(\Gamma) \vDash_{\mathbf{OF}} \psi \circ \varepsilon(A)$.

c. **E2** is undecidable and is not complete.

E. The Real Numbers

To get a complete geometric theory equivalent to the theory of real closed fields, **RCF** (Chapter XVII.E.3) we need to extend **E2** with the scheme of axioms, *Dedekind's axiom*$_1$*-B* (p. 356).

E2+, *elementary two-dimensional Euclidean geometry* in $L(=; B, \equiv)$

E2

Dedekind's axiom$_1$*-B*

Lemma 50

a. If $\mathsf{F} \vDash \mathbf{RCF}$, then $C(\mathsf{F}) \vDash \mathbf{E2+}$.

b. If $\mathsf{E} = \langle \mathsf{V}; \mathsf{B}, \approx \rangle \vDash \mathbf{E2+}$ and $\mathsf{0}, \mathsf{1}, \underline{\mathsf{1}} \in \mathsf{V}$, with $\mathsf{0} \neq \mathsf{1}$, and $\mathsf{01} \approx \mathsf{0}\underline{\mathsf{1}}$, and $\mathsf{01} \perp \mathsf{0}\underline{\mathsf{1}}$, then $F(\mathsf{E}_{\mathsf{01}\underline{\mathsf{1}}}) \vDash \mathbf{RCF}$.

Proof (a) Suppose $\mathsf{F} \vDash \mathbf{RCF}$. We need that $C(\mathsf{F})$ validates each instance of *Dedekind's axiom*$_1$*-B*.

Let $C(x)$, $D(y)$ be wffs with one free variable each (I'll let you extend the proof to wffs with more than one free variable). Suppose there is some $\mathsf{p} = (\mathsf{p}_1, \mathsf{p}_2)$ in $C(\mathsf{F})$ such that for all $\mathsf{x} = (\mathsf{x}_1, \mathsf{x}_2)$ satisfying C and $\mathsf{y} = (\mathsf{y}_1, \mathsf{y}_2)$ satisfying D, (*) $\mathsf{B}(\mathsf{p}, \mathsf{x}, \mathsf{y})$. Let

$$\mathsf{X} = \{\mathsf{x} : \mathsf{x} \text{ satisfies } C(x)\} \quad \mathsf{Y} = \{\mathsf{y} : \mathsf{y} \text{ satisfies } D(y)\}$$

We may assume that $\mathsf{p} \notin \mathsf{X} \cup \mathsf{Y}$ (Exercise 1 of Chapter XVIII.D). Pick some $\mathsf{d} \in \mathsf{X} \cup \mathsf{Y}$ such that $\mathsf{d} \neq \mathsf{p}$ and $\mathsf{d} \neq \mathsf{0}$. (I'll leave to you the case when $\mathsf{X} = \mathsf{Y} = \{\mathsf{0}\}$.) For all $\mathsf{x} \in \mathsf{X}$ and all $\mathsf{y} \in \mathsf{Y}$, by (*), $\mathsf{x}, \mathsf{y} \in \mathsf{L}_{\mathsf{pd}}$. We may assume that $\mathsf{p} \neq \mathsf{0}$, since if $\mathsf{p} = \mathsf{0}$, by the one-dimensional case (using axiom *B1-5*) we may use instead a point

$z \neq p$ on L_{pd} such that $B(z, 0, d)$.

Now we'll show that for any $q \in X \cup Y$, there is a unique $s_q \in F$ such that:

$$q_1 = p_1 + (s_q \cdot d_1) \text{ and } q_2 = p_2 + (s_q \cdot d_2)$$

First, suppose that $B(p, q, d)$. Then there is some $w \in F$ such that:

$$(1 - w) \cdot p_1 + (w \cdot d_1) = q_1 \text{ and } (1 - w) \cdot p_2 + (w \cdot d_2) = q_2$$

Then $s_q = w - \dfrac{w \cdot p}{d}$, which is possible since $d \neq 0$. If $B(p, q, d)$, I'll let you show that since $w \neq 0$ (as $p \neq q$), we have $s_q = \dfrac{-p}{w \cdot d} + \dfrac{1}{w}$.

By (*), for all $x \in X$ and $y \in Y$, there is a $t \in F$ such that $0 \leq t \leq 1$ and $(1 - t) \cdot p_1 + (t \cdot y_1) = x_1$. So given any $x \in X$ and $y \in Y$, we have:

$$(1 - t) \cdot p_1 + (1 - t) \cdot (s_y \cdot p_1) = s_x \cdot p_1$$

Hence, $t = s_x / s_y$. Since $0 \leq t \leq 1$, we have $s_x \leq s_y$.

Now let X' and Y' be subsets of the universe of F defined by:

$X'(s)$ iff $s = s_x$ and $x \in X$ $\qquad$ $Y'(s)$ iff $s = s_y$ and $y \in Y$

I'll let you show that there are wffs $C'(x)$ and $D'(y)$ in the language of **OF** such that X' is the interpretation of C' and Y' is the interpretation of D'. We have shown that for every w, z in the universe of F, if $X'(w)$ and $Y'(z)$, then $w \leq z$. Hence, since **RCF** validates every wff of *Dedekind's axiom*$_1$, there is some r in F such that for all w, z with $X'(w)$ and $Y'(z)$, we have $w \leq r \leq z$. I'll let you show now that if $s = (r \cdot p_1, r \cdot p_2)$, then for all $x \in X$ and $y \in Y$, we have $B(x, s, y)$.

(b) Suppose that $E \models$ **E2+**. We need that $F(E_{01\underline{1}})$ validates every wff of *Dedekind's axiom*$_1$. Let $C(x)$, $D(y)$ be wffs with with one free variable such that for all x that satisfy C, and all y that satisfy D, $x \leq y$. Note that in this case x, $y \in L_{01}$. Let C', D' in the language of **E2+** be the translations of C and D as in the previous section. Then by the one-dimensional case, there is some p such that for all x that satisfy C' and all y that satisfy D', $B(p, x, y)$. Hence, by *Dedekind's axiom*$_1$*-B*, there is some q such that for all x that satisfy C' and all y that satisfy D', $B(x, q, y)$. So by the one-dimensional case $x \leq q \leq y$. ∎

The representation theorem for elementary two-dimensional Euclidean geometry and real closed fields now follows from Lemma 46; the theorem on translations follows from Theorem 49 using the same mappings.

Theorem 51 *The representation theorem for* **E2+** *and* **RCF**

a. If $E = \langle V; B, \approx \rangle \models$ **E2+** and $0, 1, \underline{1} \in V$, with $0 \neq 1$, and $01 \approx 0\underline{1}$, and $01 \perp 0\underline{1}$, then there is a model $N \models$ **RCF** such that $C(N) \simeq E$, namely, $C(F(E_{01\underline{1}})) \simeq E$.

b. If $F \models$ **RCF**, then there is a model $M \models$ **E2+** such that $F(M) \simeq E$, namely, $F(C(F)) \simeq F$.

Theorem 52 *The translation theorem for* **E2+** *and* **RCF**

a. The mapping $\alpha \circ \varphi \circ \gamma$ is a translation of **RCF** to **E2+**. That is,
$\Gamma \vDash_{\mathbf{RCF}} A$ iff $\alpha \circ \varphi \circ \gamma(\Gamma) \vDash_{\mathbf{E2+}} \alpha \circ \varphi \circ \gamma(A)$

b. The mapping $\psi \circ \varepsilon$ is a translation of **E2+** to **RCF**. That is,
$\Gamma \vDash_{\mathbf{E2+}} A$ iff $\psi \circ \varepsilon(\Gamma) \vDash_{\mathbf{RCF}} \psi \circ \varepsilon(A)$

And we now have the following via Theorems XVII.20–22 for **RCF**.

Corollary 53

a. **E2+** is complete.

b. **E2+** is decidable.

c. **E2+** is not finitely axiomatizable.

d. **E2+** = Th(<ℝ ; <, +, ·, 0, 1>).

But **E2+** is not categorical, for it has a countable model: the Cartesian plane over the closure of ℚ ∪ {r : r is algebraic over the rationals} (see the proof of Theorem XVII.17). For a categorical theory of two-dimensional Euclidean geometry, we go to second-order logic, as for the theory of the real numbers. Recall that $\mathbf{RCF_2}$ is the second-order theory of **OF** plus the single wff *Dedekind's axiom*, and (Theorem XVII.23) all models of $\mathbf{RCF_2}$ are isomorphic to <ℝ ; <, +, ·, 0, 1>.

E2*, the *second-order theory of two-dimensional Euclidean geometry*
in L_2(=; B, ≡)
E2
Dedekind's axiom-B

I'll let you show that **E2+** ⊆ **E2***. The next theorem now follows as in Lemma 50, Theorem 51, and Theorem 52, with the last part by Corollary XV.29.

Theorem 54

a. E ⊨ **E2*** iff E ≃ *C*(<ℝ ; <, +, ·, 0, 1>), the Cartesian plane over the reals.

b. M ⊨ $\mathbf{RCF_2}$ iff M ≃ *F*(*C*(<ℝ ; <, +, ·, 0, 1>)).

e. **E2*** is undecidable.

We have shown, via the representation theorems and translations, that the abstractions of our intuition of the geometry of flat surfaces and our intution of approximating with rational numbers are equivalent, in that from either we can derive the other. But what real numbers are, what points in the plane are, that we have not settled.

Exercises for Sections C–E

1. Prove Lemma 45.

2. Fill in the details in the proof of Theorem 48.a with an explicit formulation of the map.

3. a. Prove: There is a mapping $\alpha: L_2(=; B, \equiv) \to L_2(=; <, \oplus, *, \text{o}, \text{ł})$ such that $\Gamma \models_{\mathbf{E2*}} A$ iff $\alpha(\Gamma) \models_{\mathbf{RCF_2}} \alpha(A)$.

   b. Prove: There is a mapping $\beta: L_2(=; <, \oplus, *, \text{o}, \text{ł}) \to L_2(=; B, \equiv)$ such that $\Gamma \models_{\mathbf{RCF_2}} A$ iff $\beta(\Gamma) \models_{\mathbf{E2*}} \beta(A)$.

4. *Congruence is not definable*

   a. Show that congruence is not definable in $\mathbf{E2^+}$.
   (Hint: Consider the mapping in the Cartesian plane over the reals given by $\delta(x, y) = (x, 2y)$, and use Padoa's method, Theorem XIII.4.)

   b. Show that congruence is not definable in **E2**.

   c. Extend the notion of definability to second-order theories and use the proof of (a) to show that congruence is not definable in $\mathbf{E2^*}$.

5. *Betweenness is definable*
   Define $xy \leq zu \equiv_{\text{Def}} \forall v\,(zv \equiv uv \to \exists w\,(xw \equiv yw \wedge yw \equiv uv))$.
   Show that $\mathbf{E2^+} \vdash B(x, y, z) \leftrightarrow \forall u\,(ux \leq xy \wedge uz \leq zy \to u = y)$.
   (Hint: Draw a picture. Show that in the Cartesian plane over the reals, $xy \leq zu$ iff the length of xy is less than the length of zu. Show the wff is true in that model, and modify that proof to apply to the Cartesian plane over any real closed field.)

6. a. Formalize and prove Pythagoras' theorem in $\mathbf{E2_{011}}$.

   b. Consider points as in the picture in a model $\mathcal{E}$ of $\mathbf{E2_{011}}$.
   Why doesn't (a) show that square roots exist in $\langle L_{01}; <_E, +_E, \cdot_E, 0, 1 \rangle$?
   (Hint: See the first page of Chapter XVII.)

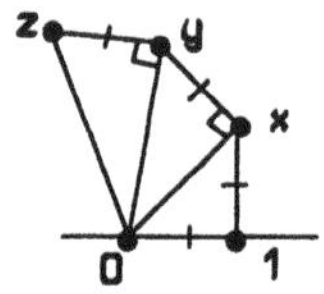

7. Using Pythagoras' theorem, we can show that there are uncountably many non-elementarily equivalent extensions of **E2**. Recall from the proof of Theorem XVII.17 that for each collection $\mathcal{B}$ of primes, we can let $\mathbb{Q}(\mathcal{B})$ = the closure of $\mathbb{Q} \cup \{\sqrt{p} : p \in \mathcal{B}\}$, and then $\mathbb{Q}(\mathcal{B}) \models \mathbf{OF}$. Moreover, if $\mathcal{B} \neq \mathcal{B}'$, then $\text{Th}(\mathbb{Q}(\mathcal{B}))$ is not elementarily equivalent to $\text{Th}(\mathbb{Q}(\mathcal{B}'))$. There is a countable infinity of primes, so there are uncountably many $\mathcal{B}$. In $L(=; B, \equiv, \text{o}, \text{ł}, \underline{\text{ł}})$, using the defined notions, set:

   $A_2(x_0, x_1, x_2) \equiv_{\text{Def}} x_0 \neq x_1 \wedge x_0x_1 \perp x_1x_2 \wedge x_0x_1 \equiv x_1x_2$

   $A_{n+1}(x_0, x_1, \ldots, x_{n+1}) \equiv_{\text{Def}} A_n(x_0, x_1, \ldots, x_n) \wedge x_0x_n \perp x_nx_{n+1} \wedge x_0x_n \equiv x_nx_{n+1}$

   $D_n(x_0, x_1, \ldots, x_n) \equiv_{\text{Def}} \exists x_0\, \exists x_1 \ldots \exists x_n\, (A_n(x_0, x_1, \ldots, x_n) \wedge$
   $\exists y\,(B(x_0, x_1, y) \wedge x_0x_n \equiv x_0y))$
   where y is the least variable not appearing A_n

   a. In the Cartesian plane over $\mathbb{R}$, interpret x_0 as (0, 0) and x_1 as (1, 0) in these wffs.
      i. Draw pictures of points that satisfy A_2, A_3, and A_4. (Hint: See Exercise 6.b.)
      ii. What is the length of the interpretation of x_0x_n?

   b. Show that for every collection $\mathcal{B}$ of primes, $C(\mathbb{Q}(\mathcal{B})) \models D_n$ iff $n \in \mathcal{B}$.

   c. Show that if $\mathcal{B} \neq \mathcal{B}'$, $\text{Th}(C(\mathbb{Q}(\mathcal{B})))$ is not elementarily equivalent to $\text{Th}(C(\mathbb{Q}(\mathcal{B}')))$.

Historical Remarks

Though older systems are reported, the oldest extant axiomatic system of geometry is by Euclid of Alexandria in his *Elements* from about 300 B.C. (see *Heath, 1956* or *Boyer, 1968*). That system was so well developed that it overshadowed all previous systems and established a standard for excellence in reasoning in mathematics for more than two thousand years.

In the 5th Century, though, Proclus claimed that he had found a proof that the fifth axiom of Euclid is superfluous (see the discussion of Axiom *E14*, p. 366 above). Though Proclus' claim was later shown to be false, it started a series of efforts to prove what became known as 'Euclid's parallel postulate' from Euclid's other axioms. The first to understand that the axiom might be independent were Carl Friedrich Gauss (about 1816), Nicolai Ivanovitch Lobachevsky (about 1826), and Janos Bolyai (1829) (see *Boyer, 1968*).

The ideas of these mathematicians, as well as actual independence proofs by Eugenio Beltrami and Felix Klein, called attention to the problem of the axiomatic exposition of geometry. During this period a much higher standard of precision began to be expected of a system of mathematics. The first attempt to meet those standards in geometry was by Moritz Pasch in his *Vorlesungen über neuere Geometrie*, published in 1882. In that work he discovered the first real flaw in Euclid's system, for he had to enrich the system with what is now called *Pasch's axiom* (see the discussion of Axiom *E6*, p. 365 above). Pasch's work was soon superseded by the full development of axiomatic geometry in *Hilbert, 1899*.

The first formal axiomatization of Euclidean geometry in the tradition of modern logic was presented by Alfred Tarski during a course given at the University of Warsaw during the academic year 1926–1927. There he showed that his first-order system was complete and decidable. At the beginning of the Second World War his work was about to be published in France (as *Tarski, 1940*) when all copies were destroyed. A galley proof was discovered much later and a limited number of copies were printed for Tarski's 70th birthday.

Tarski's original approach was to base the exposition on the theory of proportions and the properties of similar triangles. Tarski later returned to that system and with contributions by Eva Kallin, Scott Taylor, H. N. Gupta, Wanda Szmielew, and Wolfram Schwabhäuser made considerable simplifications, which appeared in *Schwabhäuser, Tarski, and Szmielew, 1983*. The different development in the chapters here also came out of that research and was first presented (in Polish) by Szczerba and Marek Kordos in *Kordos and Szczerba, 1976*. The story of this work, with much commentary on the relation of the current axiomatization to others, can be found in *Szczerba, 1986*, *Moszynska, 1986*, *Tarski, 1959*, and *Tarski and Givant, 1999*.

The formal development of Euclidean geometry on the lines of Hilbert's approach was done somewhat earlier in *Borsuk and Szmielew, 1960*, which also contains a history of geometry from Euclid's time.

The axiomatization of the one-dimensional betweenness relation was later developed by Szczerba and Grzegorz Lewandowski in *Lewandowski and Szczerba, 1980*, and the full system of one-dimensional geometry in *Kordos and Szczerba, 1976*.

XX Translations within Classical Predicate Logic

We've seen many examples of translations in the text. Here we'll make some general definitions and investigate some of the properties of translations. Then we'll return to the translations we've already seen in the text as examples.[1]

A. What Is a Translation?

Since we're generally interested in translating named theories, such as **Q**, I'll use boldface capitals to stand for theories here.

Translations within classical predicate logic Let **T** and **R** be theories in first-order classical predicate logic and * a constructive mapping (Chapter XV.B.4) from the language of **T** to the language of **R**. Write $\Gamma^* = \{A^*: A \in \Gamma\}$.

* is *weak validity-preserving* iff for every A, if $\vDash_{\mathbf{T}} A$ then $\vDash_{\mathbf{R}} A^*$.

* is *validity-preserving* iff for every A, $\vDash_{\mathbf{T}} A$ iff $\vDash_{\mathbf{R}} A^*$.

* is a *translation of* **T** *into* **R** iff for every Γ and A, $\Gamma \vDash_{\mathbf{T}} A$ iff $\Gamma^* \vDash_{\mathbf{R}} A^*$.

---

[1] The definitions and analyses here are part of a larger study of translations: Translations between propositional logics are analyzed in *Propositional Logics*. In *Predicate Logic*, formalizations of reasoning from ordinary language to the language of predicate logic are treated as translations, and the reduction of general second-order logic to first-order logic is analyzed as a translation.

We require the maps to be constructive because if we allow nonconstructive maps, then we can map any theory **T** into any theory **R** with a 1-1 mapping that takes the wffs of **T** to the wffs of **R**, and the wffs not in **T** to the wffs that are not in **R**.

Theorem 1

a. The composition of weak validity-preserving maps is weak validity-preserving.

b. The composition of validity-preserving mappings is validity-preserving.

c. The composition of translations is a translation.

d. There is a weak validity-preserving mapping that is not validity-preserving.

e. There is a validity-preserving mapping that is not a translation.

Proof: I'll leave (a)–(c) to you.

(d) I'll actually prove a stronger result. Given any decidable theories **T** and **R**, define a mapping φ from the language of **T** to the language of **R** via:

The nth wff in **T** is mapped to the $2n^{\text{th}}$ wff of **R**.

The nth wff not in **T** is mapped to the $2n+1^{\text{st}}$ wff of **R**.

Since **T** and **R** are decidable, the mapping is constructive. For any A, if $\vDash_{\mathbf{T}} A$ then $\vDash_{\mathbf{R}} \varphi(A)$, so φ is weak validity-preserving. But it is not validity-preserving.

(e) Recall from Chapter XI.A that $E_{\geq n}$ and $E_{!n}$ are wffs of L(=), and the first is true in a model iff there are at least n objects in the model, and the second is true iff the model has exactly n objects. Let **T** be Th($E_{!46}$) in L(=). Up to isomorphism it has only one model, with exactly 46 elements. Hence, it is decidable. I'll exhibit a validity-preserving mapping of **T** to the theory of fields, **F**.

Each $E_{\geq n}$ is in the language of **T** and in the language of **F**. For $n \geq 47$, $E_{\geq n}$ is in neither **T** nor **F** (see Exercise 1 of Chapter XVII.E.1). Consider the mapping:

$(E_{\geq 47})^* = E_{\geq 47} \quad (E_{\geq 53})^* = E_{\geq 53}$

The nth formula in **T** is mapped to $\forall x_1\, (x_1 = x_1)$.

The nth formula not in **T** that is neither $E_{\geq 47}$ nor $E_{\geq 53}$ is mapped to $\neg\forall x_1\, (x_1 = x_1)$.

Since **T** is decidable, the mapping is constructive. And $\vDash_{\mathbf{T}} A$ iff $\vDash_{\mathbf{F}} A^*$. Hence, * is a validity-preserving map.

Since $\mathbf{T} \vDash \neg E_{\geq 47}$, we have $E_{\geq 47} \vDash_{\mathbf{T}} E_{\geq 53}$. But $E_{\geq 47} \nvDash_{\mathbf{F}} E_{\geq 53}$, since there are fields of size 47 and fields of size 53 (Exercise 1 of Chapter XVII.E.1). Hence, * is not a translation. ∎

The maps given in the proof above did not preserve any of the structure of the formulas being translated. Generally, though, we are interested in translations that preserve at least some part of the structure of wffs.

Homophonic mappings A mapping * of predicate logic languages is *homophonic with respect to*:

$\neg$ iff $(\neg A)^* = \neg(A^*)$	$\vee$ iff $(A \vee B)^* = A^* \vee B^*$
$\rightarrow$ iff $(A \rightarrow B)^* = A^* \rightarrow B^*$	$\forall$ iff $(\forall x\, A)^* = \forall x\, (A^*)$
$\wedge$ iff $(A \wedge B)^* = A^* \wedge B^*$	$\exists$ iff $(\exists x\, A)^* = \exists x\, (A^*)$

The mapping is *homophonic with respect to the connectives* if it is homophonic with respect to each of $\neg$, $\rightarrow$, $\wedge$, $\vee$. It is *homophonic* if it is homophonic with respect to all of the above. We also say a mapping is homophonic with respect to *identity* if $(x = y)^*$ is $(x = y)$.

Theorem 2 Every validity-preserving map homophonic with respect to $\rightarrow$ is a translation.

Proof: Let * be a validity-preserving mapping of **T** to **R** that is homophonic with respect to $\rightarrow$. Then for any Γ, A we have:

$\Gamma \vDash_{\mathbf{T}} A$ iff for some $A_1, \ldots, A_n$ in Γ, $\{A_1, \ldots, A_n\} \vDash_{\mathbf{T}} A$ — compactness

iff $\vDash_{\mathbf{T}} A_1 \rightarrow (A_2 \cdots (\rightarrow A_n \rightarrow A)$ — the deduction theorem

iff $\vDash_{\mathbf{R}} (A_1 \rightarrow (A_2 \cdots (\rightarrow A_n \rightarrow A))^*$ — * is validity preserving

iff $\vDash_{\mathbf{R}} A_1^* \rightarrow (A_2^* \cdots (\rightarrow A_n^* \rightarrow A^*)$ — * is homophonic for $\rightarrow$

iff $\{A_1^*, \ldots, A_n^*\} \vDash_{\mathbf{R}} A^*$ iff $\Gamma^* \vDash_{\mathbf{R}} A^*$ ■

Mathematicians typically conceive of reductions of one theory to another as transforming models of one to models of the other, rather than as linguistic mappings. In establishing in the text that mappings were translations, we did that too. Recall for a theory **T** and a class of models $\mathcal{T}$, $\mathrm{Th}(\mathcal{T}) = \{A\text{: for every } \mathsf{M} \text{ in } \mathcal{T},\ \mathsf{M} \vDash A\}$.

Model-preserving maps Let $\mathbf{T} = \mathrm{Th}(\mathcal{T})$ and $\mathbf{R} = \mathrm{Th}(\mathcal{R})$ be two theories in classical predicate logic, and * a constructive (recursive) mapping from the language of **T** to the language of **R**. Then * is *model-preserving with respect to $\mathcal{T}$ and $\mathcal{R}$* if there is an onto mapping * from $\mathcal{R}$ to $\mathcal{T}$ such that for every A in the language of **T** and every model M in $\mathcal{R}$, $\mathsf{M}^* \vDash A$ iff $\mathsf{M} \vDash A^*$.

$$L_{\mathbf{T}} \xrightarrow{\ *\ } L_{\mathbf{R}}$$

$$\mathcal{T} \xleftarrow{\ *\ } \mathcal{R}$$

The mapping is *strongly model-preserving* if there is in addition for each model M in $\mathcal{R}$ a mapping, also designated *, from assignments of references on M to assignments of references on M^* such that for every A in the language of **T**, $\sigma \vDash A^*$ iff $\sigma^* \vDash A$.

Theorem 3 Every model-preserving mapping is a translation.

Proof: Let * be a model-preserving mapping of **T** to **R** with respect to the classes of models $\mathcal{T}$ and $\mathcal{R}$. Then,

$$\Gamma \models_{\mathbf{T}} A \text{ iff for every model } \mathsf{M} \text{ in } \mathcal{T}, \text{ if } \mathsf{M} \models \Gamma, \text{ then } \mathsf{M} \models A$$

$$\text{iff for every model } \mathsf{N} \text{ in } \mathcal{R}, \text{ if } \mathsf{N}^* \models \Gamma, \text{ then } \mathsf{N}^* \models A \qquad \text{* is onto}$$

$$\text{iff for every model } \mathsf{N} \text{ in } \mathcal{R}, \text{ if } \mathsf{N} \models \Gamma^*, \text{ then } \mathsf{N} \models A^* \qquad \text{* is model-preserving}$$

$$\text{iff } \Gamma^* \models_{\mathbf{R}} A^*$$ ■

Translations not only clarify what we mean by one theory being reduced to another, they also allow us to investigate properties of theories.

Lemma 4

a. If there is a validity-preserving mapping of **T** to **R**, and **T** is consistent, then **R** is consistent.

b. If there is a weak validity-preserving mapping of **T** to **R** that is homophonic with respect to $\neg$, and **R** is consistent, then **T** is consistent.

c. If there is a validity-preserving mapping of **T** to **R** that is homophonic with respect to $\neg$, then **T** is consistent iff **R** is consistent.

d. If * is a validity-preserving mapping of **T** to **R** that is homophonic with respect to $\neg$, then for every Δ in $L_{\mathbf{T}}$ such that $\mathbf{T} \subseteq \Delta$, Δ is consistent iff Δ^* is consistent.

e. If there is a validity-preserving mapping of **T** to **R**, and **R** is decidable, then **T** is decidable.

f. If there is a validity-preserving mapping of **T** to **R**, and **T** is undecidable, then **R** is undecidable.

Proof: (a) Suppose **R** is not consistent and * is a validity-preserving mapping of **T** to **R**. Then for every A, $\models_{\mathbf{R}} A^*$. Hence, for every A, $\models_{\mathbf{T}} A$. So **T** is inconsistent.

(b) Suppose that **T** is inconsistent. Then for some A, $\models_{\mathbf{T}} A$ and $\models_{\mathbf{T}} \neg A$. Hence, $\models_{\mathbf{R}} A^*$ and $\models_{\mathbf{R}} \neg(A^*)$, so **R** is inconsistent.

Part (c) is by (a) and (b), and part (d) is proved similarly. I'll leave (e) and (f) to you. ■

Now we can generalize our work on undecidable theories.

Hereditarily and essentially undecidable theories A theory **T** is *hereditarily undecidable* if it is undecidable and for every theory $\Gamma \subseteq \mathbf{T}$ in the language of **T**, Γ is undecidable.

A theory **T** is *essentially undecidable* if it is undecidable and for every consistent theory Γ in the language of **T** such that $\mathbf{T} \subseteq \Gamma$, Γ is undecidable.

In Corollary XV.16 we saw that **Arithmetic** is hereditarily undecidable. Theorem XV.11 showed that **Q** is essentially undecidable. Note that every essentially undecidable theory is consistent, since inconsistent theories are decidable.

Lemma 5 If **T** is consistent and has a finitely axiomatizable and essentially undecidable subtheory, then **T** is hereditarily undecidable.

Proof: Suppose $\mathbf{R} \subseteq \mathbf{T}$ and **R** is a finitely axiomatizable, essentially undecidable theory. Let A be the conjunction of a finite axiomatization of **R**. Suppose that $\Gamma \subseteq \mathbf{T}$ is a theory. Consider $\Sigma = \text{Th}(\Gamma \cup \{A\})$. Since $\Gamma \subseteq \mathbf{T}$, Σ is consistent. So Σ is undecidable as $\mathbf{R} \subseteq \Sigma$. But $\Sigma \vDash B$ iff $\Gamma \vDash A \rightarrow B$, so Γ is undecidable, too.

Theorem 6 Suppose * is a translation of the theory **T** to the theory **R** that is homophonic with respect to $\neg$.

a. If **T** is essentially undecidable, then **R** is essentially undecidable.

b. If **T** is hereditarily undecidable, then **R** is hereditarily undecidable.

c. If **T** is finitely axiomatizable and essentially undecidable, then **R** is hereditarily undecidable.

Proof: (a) Suppose Γ is a consistent theory with $\mathbf{R} \subseteq \Gamma$. Consider the collection $\Delta = \{A\colon A \text{ is in } L(\mathbf{T}) \text{ and } A^* \in \Gamma\}$. Then $\mathbf{T} \subseteq \Delta$. By Lemma 4.d, Δ is consistent as Γ is consistent. To show that Δ is a theory, suppose $\Delta \vDash D$. Then $\Delta^* \vDash D^*$. Hence, $\Gamma \vDash D^*$, so $D^* \in \Gamma$, and $D \in \Delta$. Since **T** is essentially undecidable, Δ is undecidable, too. Hence, Γ is undecidable.

(b) Suppose that **T** is hereditarily undecidable. Let $\Gamma \subseteq \mathbf{R}$ be a theory. Consider again $\Delta = \{A\colon A \text{ is in } L_{\mathbf{T}} \text{ and } A^* \in \Gamma\}$. As in part (a), Δ is a theory. Also $\Delta \subseteq \mathbf{T}$. So Δ is undecidable. Hence, Γ is undecidable.

Part (c) follows by (b) and Lemma 5. ■

Exercises for Section A

1. Prove Theorem 1.a–c.
2. Prove that every validity-preserving map homophonic with respect to both $\neg$ and $\wedge$ is a translation.
3. Show that the composition of model-preserving mappings is model-preserving.
4. Give an example of a translation * from **T** to **R** such that **T*** is not a theory. (Hint: Successor in different languages.)
5. Prove that if there is a validity-preserving mapping of **T** to **R**, and **R** is decidable, then **T** is decidable.
6. Give an example of a translation from **T** to **R** where **R** is undecidable and **T** is decidable. (Hint: See Exercise 4.)

7. Give an example of a weak validity-preserving mapping from **T** to **R** where **T** is consistent and **R** is inconsistent.

8. Give an example of a translation * from a theory **T** to a theory **R** such that **T** is axiomatized by a single wff A, yet $\mathbf{T}^* \neq \mathrm{Th}(A^*)$. (Hint: Chapter XII.)

9. List the theories in Chapters XV and XVI that are hereditarily undecidable.

10. List the theories in Chapters XV and XVI that are essentially undecidable.

11. A theory Γ is a *finite extension* of Δ if there are $C_1, C_2, \ldots, C_n$ such that $\Gamma = \mathrm{Th}(\Delta \cup \{C_1, C_2, \ldots, C_n\})$. Prove that if Γ is a finite extension of Δ in the same language as Δ, and Γ is undecidable, then Δ is undecidable.

12. Give an example of an identity mapping that is a translation from a decidable theory to an undecidable one. (Hint: **S**.)

13. a. Prove that if Γ is essentially undecidable, then Γ is consistent and no complete extension of Γ in the language of Γ is axiomatizable.
    b. Prove that every consistent and decidable theory has a complete and decidable extension in the same language.
    (Hint: Let Δ be consistent and decidable. Use the construction of a complete and consistent theory from Lemma II.8, taking $\Gamma_0 = \Delta$. Show that $\mathrm{Th}(\Gamma_n)$ is decidable for every n. To decide if $\Gamma \vDash A$, note that A is A_n for some n.)
    c. Prove that if Γ is consistent and no complete extension of Γ in the language of Γ is axiomatizable, then Γ is essentially undecidable.

14. Suppose there is a translation of the theory **T** to **R** homophonic with respect to the connectives. Prove: If **T** has a finitely axiomatizable subtheory that is essentially undecidable, then so, too, has **R**.

<u>15</u>. Let * be the mapping from the language of predicate logic to itself that takes each wff to an equivalent wff in conjunctive prenex normal form in the proof of the normal form theorem p. 108. Show that * is a strongly model-preserving translation. With respect to which connectives and quantifiers is it homophonic?

<u>16</u>. Analyze the reduction of 2-sorted logic to first-order logic presented in Chapter XIV.H in terms of the definitions above.

<u>17</u>. Characterize the mappings between second-order theories of the reals and one- and two-dimensional geometry in terms of translations between second-order theories.

B. Examples

1. Translating between different languages of predicate logic

We can formulate our semantic assumptions for classical predicate logic in various formal languages in which $\neg$ plus at least one of $\wedge$, $\vee$, $\rightarrow$ and at least one of $\forall$, $\exists$ appear (Chapter X). So long as the same collection of categorematic symbols appear in the languages (predicate symbols, function symbols, and name symbols), the translations are strongly model-preserving. Though they are not homophonic, they are homomorphisms of the languages.

2. Converting functions into predicates

In Chapter VIII.E we saw how to translate from a language that contains function symbols to a language that has only predicates and names. If a theory Σ is in a language with function symbols $f_1, \ldots, f_n$, then:

(1) $\Gamma \models_{\Sigma} A$ iff $\Gamma^* \models_{\Sigma^* \cup \Delta} A^*$

Here $\Delta = \{\alpha_{f_i} : 1 \leq i \leq n\}$, where α_{f_i} is a wff that asserts that the predicate to which f_i is translated satisfies the appropriate existence assumption to replace a function symbol. Thus, the mapping is a translation from Σ to $\text{Th}(\Sigma^* \cup \Delta)$. We showed (1) by showing that the mapping is strongly model-preserving. And it is homophonic. Here is an example from the text:

The theory of ordered fields **OF** in $L(=; <, \oplus, *, o, 1)$ to $L(=; <, P_{\oplus}, P_{*}, o, 1)$. §XIX.D

Exercise 4 of Chapter VIII.E showed how to translate names into predicates similarly.

3. Translating predicates into formulas

a. Extensions by definitions

In Chapter XII.B we saw how, given a theory **T**, we could translate it to a theory **R** with fewer categorematic symbols so long as the remaining categorematic symbols of **T** could be defined in **R**. In Theorem XII.21 we said that such translations are model-preserving, though we actually showed that they are strongly model-preserving (compare the proof of Lemma XII.9).

Here are the translations by definitions we've seen in the text:

The theory of groups **G** in $L(=; \circ, ^{-1}, e)$ to **G*** in $L(=; \circ, ^{-1})$. §XII.B

G in $L(=; \circ, ^{-1}, e)$ to $\mathbf{G}^{\#}$ in $L(=; \circ, e)$. §XII.B

G in $L(=; \circ, ^{-1}, e)$ to $\mathbf{G}^{+}$ in $L(=; \circ)$. §XII.B

Arithmetic in $L(=; S, \oplus, *, o)$ to any of the languages:
$L(=, \oplus, *, o, 1)$ $L(=; <, *)$ $L(=; \oplus, *)$ $L(=; S, *)$ §XV.C.2

The theory of real closed fields **RCF** in $L(=; <, \oplus, *, o, 1)$ to $L(=; \oplus, *)$. §XVII.E.3

The theory of elementary two-dimensional Euclidean geometry **E2**+ in $L(=; B, \equiv)$ to $L(=; \equiv)$. Exercise 5, §XIX.E

b. Predicates translated into formulas

We've also seen translations in which only the atomic predicates were translated, with the mapping then extended homophonically. (The general form of those is set out in Chapter XVI.B using no equivalence relation.) These are all strongly model-preserving. Here are the examples from the text:

A theory of successor **S** in L(=; S, 0) to a theory of discrete linear orderings in L(=; <). Exercise 11, §XV.A

The theory of one-dimensional betweenness **B1** in L(=; B) to the theory of dense linear orderings without endpoints **DLO** in L(=; <). §XVIII.B.5

DLO in L(=; <) to $\mathbf{B1_{01}}$ in L(=; B, 0, 1). §XVIII.B.5

A theory of one-dimensional congruence **C1** in L(=; ≡) to the theory of 2-divisible groups **GD2** in L(=; ⊕, 0). §XVIII.C.5

GD2 in L(=; ⊕, 0) to **C1** in L(=; ≡, 0). §XVIII.C.5

The theory of one-dimensional congruence $\mathbf{C1_{Div}}$ in L(=; ≡) to **GD** in L(=; ⊕, 0). §XVIII.C.6

GD in L(=; ⊕, 0) to $\mathbf{C1_{0Div}}$ in L(=; ≡, 0). §XVIII.C.6

The theory of one-dimensional geometry **E1** in L(=; B, ≡) to the theory of ordered divisible groups **ODG** in L(=; <, ⊕, 0). §XVIII.D.3

ODG in L(=; <, ⊕, 0) to $\mathbf{E1_{01}}$ in L(=; B, ≡, 0, 1). §XVIII.D.3

4. Relativizing quantifiers

We can also translate by relativizing quantifiers: atomic wffs are left unchanged, the connectives are translated homophonically, and the quantifers are translated relative to a predicate.

$$(\forall x\, B)^* = \forall x\, (U(x) \rightarrow B^*)$$
$$(\exists x\, B)^* = \forall x\, (U(x) \wedge B^*)$$

Those are also strongly model-preserving, as you can show. Here are the examples of relativizing quantifiers from the text:

Arithmetic in L(=; ⊕, ∗, 0, 1) to the arithmetic of the integers **Z-Arithmetic** in L(=; ⊕, ∗, 0, 1). §XVI.D

Arithmetic in L(=; ⊕, ∗, 0, 1) to the arithmetic of the rationals **Q-Arithmetic** in L(=; ⊕, ∗, 0, 1). §XVI.D

5. Establishing equivalence-relations

In Chapter XVI.B we saw a more general method of translating. In addition to allowing for predicates and names to be translated into defined formulas and allowing for quantifiers to be relativized, we can take single variables to be translated to *n*-tuples of variables, with the equality predicate translated to an equivalence relation on *n*-tuples, which we saw is model-preserving. Examples in the text are:

The arithmetic of the rationals **Q-Arithmetic** in L(=; ⊕, ∗, 0, 1) to the arithmetic of the integers **Z-Arithmetic** in L(=; ⊕, ∗, 0, 1). §XVI.A.2

Z-Arithmetic in L(=; ⊕, ∗, o, ł) to **Arithmetic** in L(=; ⊕, ∗, o, ł). §XVI.C

The theory of ordered fields **OF** in L(=; <, ⊕, ∗, o, ł) to two-dimensional Euclidean geometry **E2**$_{\mathbf{011}}$ in L(=; B, ≡). §XIX.D

E2$_{\mathbf{011}}$ in L(=; B, ≡) to **OF** in L(=; <, ⊕, ∗, o, ł). §XIX.D

6. Adding and eliminating parameters

In Chapter XVIII.E we defined the notion of a named parameter and showed how to eliminate named parameters via translations. The examples in the text are:

The theory of one-dimensional betweenness **B1**$_{\mathbf{01}}$ in L(=; B, o, ł) to **B1** in L(=; B). §XVIII.B.5

A theory of one-dimensional congruence **C1** in L(=; ≡, o) to **C1** in L(=; ≡). §XVIII.C.5

The complete theory of one-dimensional congruence **C1**$_{\mathbf{0Div}}$ in L(=; ≡, o) to **C1**$_{\mathbf{Div}}$ in L(=; ≡). §XVIII.C.6

The theory of one-dimensional geometry **E1**$_{\mathbf{01}}$ in L(=; B, ≡, o, ł) to **E1** in L(=; B, ≡). §XVIII.D.3

Basic two-dimensional geometry **E2**$_{\mathbf{011}}$ in L(=; B, ≡, o, ł, $\underline{ł}$) to **E2** in L(=; B, ≡). §XIX.D

We also showed in Chapter XVIII.E that when we add named parameters to a theory, the identity map is a translation from the original theory to the one with parameters. That is, adding parameters is a conservative extension of a theory.

7. Composing translations

In our formalizations of geometry we needed to compose different kinds of translations described above. Here are the examples from the text.

Equivalence-relation translations followed by elimination of parameters

The theory **DLO** in L(=; <) to **B1** in L(=; B, o, ł). §XVIII.B.5

The theory **GD2** in L(=; ⊕, o) to **C1** in L(=; ≡). §XVIII.C.5

The theory **GD** in L(=; ⊕, o) to **C1**$_{\mathbf{Div}}$ in L(=; ≡). §XVIII.C.6

The theory **ODG** in L(=; <, ⊕, o) to **E1** in L(=; B, ≡). §XVIII.D.3

Adding parameters followed by an equivalence-relation translation

The theory **E2** in L(=; B, ≡) to **OF** in L(=; <, ⊕, ∗, o, ł). §XIX.D

E2+ in L(=; B, ≡) to **RCF** in L(=; <, ⊕, ∗, o, ł). §XIX.E

Converting functions into predicates followed by an equivalence-relation translation followed by elimination of parameters

OF in $L(=; <, \oplus, *, o, 1)$ to **E2** in $L(=; B, \equiv)$. §XIX.D

RCF in $L(=; <, \oplus, *, o, 1)$ to **E2+** in $L(=; B, \equiv)$. §XIX.E

8. The general form of translations?

Model-preserving seems to be what mathematicians mean when they say they can reduce one theory or "subject" to another, say two-dimensional geometry to the theory of the real numbers. All the translations above are model-preserving. Lesław Szczerba *1977*, *1979*, *1980* suggests that all the semantic methods mathematicians use to reduce one theory to another are one or a combination of those above.[2]

Defining atomic predicates.

Establishing an equivalence relation on the universe.

Relativizing quantifiers (picking out a subuniverse).

Eliminating or adding parameters.

Converting functions into predicates.

The theory of translations of this chapter is meant to link these semantic methods with syntactic procedures. It is more general in that it includes the translations of classical predicate logic to itself in different languages and the conversion of wffs to prenex normal form (Exercise 15 above), though both those are trivial semantically (the identity map on models establishes that they are model preserving). It also extends the notion of interpretability of one theory in another used in *Tarski, Mostowski, and Robinson, 1953* to establish undecidability results. What is lacking, though, is an example of a model-preserving translation that is not strongly model-preserving, or even one that is not model-preserving, as well as criteria for a translation to preserve meaning comparable to the criteria for translating between logics in *Propositional Logics* and from ordinary language to predicate logic in *Predicate Logic.*

---

2 Szczerba doesn't include translating functions into predicates because he considers only models for theories that are formulated in languages without function symbols. Nor does he have a theorem covering eliminating and adding parameters.

XXI Classical Predicate Logic with Non-Referring Names

When we gave semantics for classical predicate logic, we took as a simplification that every name must refer. In this chapter we'll see how we can lift that restriction. This will allow us to model reasoning about partial functions in mathematics.

A. Logic for Nothing

Lifting the restriction that names refer is often considered part of a program called *free logic*: ridding logic of existential assumptions that are built into its semantics. Along with no longer requiring names to refer, it's sometimes suggested, we should also lift the assumption that the universe of a model must contain at least one object. It is not for logic to assume that there is anything.

How would we interpret 'Everything is a dog' with an empty universe? We said before (Example 1, Chapter VII.A) that 'All dogs bark' is to be interpreted as

'If there is anything that's a dog, then it barks,' where 'If . . . then . . .' is interpreted classically. We followed the mathematicians' view that universal quantification should have no existential component: 'all' does not include 'and there exists'. So 'Everything is a dog' should be interpreted as 'If there is anything, then it is a dog.' Thus 'Everything is a dog', and hence '$\forall x$ (x is a dog)', is true of the empty universe.

We don't need any complicated semantics for the logic of the empty universe. We don't need to worry how to make the semantics consistent with our interpretation of variables using assignments of references when there are no assignments. Simply: Every closed wff beginning with a universal quantifier is true; every closed wff beginning with an existential quantifier is false. The resulting formal system bears little resemblance to classical predicate logic. For example, $\forall x\ (x \neq x)$ is true, so $\neg \forall x\ (x \neq x)$ is no longer a tautology.

We continue to require that every universe is non-empty.

B. Non-Referring Names in Classical Predicate Logic?

There is a well-known way due to Bertrand Russell of formalizing descriptive names in classical predicate logic, including descriptive names that do not refer such as 'The cat that Richard L. Epstein likes'. Every apparently atomic wff in which a non-referring descriptive name appears is converted into a false proposition.[1]

W. V. O. Quine suggested that we can use the same method to eliminate all names, replacing, for example, 'Pegasus' with the predicate '— pegasizes'. That, he said, would allow us to formalize reasoning with atomic non-referring names, that is, ones that have no structure: *singular names*. But as I discussed in Chapter VIII.D of *Predicate Logic*, Quine's suggestion miscontrues the role of names in reasoning.

In any case, with Russell's analysis we can never formalize a claim we believe is true that uses a non-referring name. For example, many say the following is true:

(1) Pegasus is a winged horse.

What reasons can be given for that? If the only semantic property of a name is whether it has a reference, and if so, what that reference is, it is hard to see how we can justify (1) as true. The most natural place to formalize reasoning with non-referring names is in a predicate logic in which further semantic values of language beyond truth-values and reference are taken into account, such as subject matter or referential content. In *Propositional Logics* I present a range of propositional logics that take into account such semantic values. In a subsequent volume I will present predicate logics based on those propositional logics in which the formalization of claims such as (1) might seem natural.

---

1 See Chapter VIII.D of *Predicate Logic* for a presentation of Russell's analysis with a comparison to Peter Strawson's analysis of descriptive names. Chapter IX.E of that book describes other approaches due to Hilbert and Bernays as well as others.

Still, we can develop semantics for non-referring names relative to the assumptions of classical predicate logic, viewing closed wffs with non-referring names as propositions, by taking as primitive whether a particular atomic proposition is true, just as we did in the pre-mathematical development of the semantics in Chapter IV. This will serve as a reference for predicate logics based on other semantic values and will allow us to formalize reasoning with partial functions, which is important in mathematics.

C. Semantics for Classical Predicate Logic with Non-Referring Names

I'll first consider languages without function symbols and without the equality predicate. I'll extend the semantics to cover the equality predicate later in this section and treat languages with function symbols in Section F.

1. Assignments of references and atomic predications

Consider the semantics given for classical predicate logic before we made any mathematical abstractions. We start with a universe and a complete set of assignments of references for variables. Variables are meant to stand for objects of the universe, so if σ is an assignment of references, then $\sigma(x)$ is an object in the universe. We can extend our semantics to include assignments for names just as we did before, only no longer requiring that each name have a reference. We need not assume that an assignment of reference assigns any value to 'Pegasus'; indeed, since the only value it could assign in these semantics is a reference, it cannot assign a value. Thus, an assignment of references is a mapping from terms to objects of the universe satisfying the following.

---

(2) ***Assignments of references***

i. For every variable x, $\sigma(x)$ is defined.

ii. For every name c, if $\sigma(c)$ is defined, then for every τ, $\tau(c)$ is defined and $\tau(c)$ is the same object as $\sigma(c)$.

Condition (ii) reflects our requirement that the use of names must not be ambiguous.

In Chapter IV we took as primitive whether $\sigma \models$ 'x is a dog'. Doing so did not depend on any particular view of why or how atomic predications are true. Similarly, when we have non-referring names, we don't care how or why a claim such as (1) is true or false. All that matters is that (1) has a truth-value in the model.

Now consider a binary predication:

(3) Pegasus is bigger than x

If $\sigma(x)$ is Fred Kroon, how are we to determine the truth-value of (3)? Here, too,

(3) is an atomic predication. It does not matter to the logician what truth-value is given to (3), so long as some truth-value is given. Thus, we have:

(4) For every assignment of references σ, whether $\sigma \vDash P(t_1, \ldots, t_n)$ is taken as primitive.

This allows for the widest possible application of our logic, depending on how truth-values are assigned. It does not allow us to model the view that (1) has no truth-value; but that view is more naturally modeled within a many-valued logic than within classical predicate logic.[2]

We need to make a restriction on (4). In classical predicate logic we assume that all predicates in our language are extensional: How we refer to an object in a predication does not matter for the truth-value of that predication. Given further semantic values, we could reason with non-extensional predicates and drop that requirement. But given truth-values and references as the only semantic values, we are stuck with the assumption that predicates are extensional. We formulate a condition on consistency and extensionality of predications by modifying the one we adopted in Chapter IV.E.2 (p. 85) to note that for any closed terms t and u all we can take into account is whether $\sigma(t)$ is the same object as $\tau(u)$ or whether t is u:

(5) For all atomic wffs $Q(t_1, \ldots, t_n)$ and $Q(u_1, \ldots, u_n)$ and assignments σ and τ, if for each $i \leq n$, either $\sigma(t_i)$ is the same object as $\tau(u_i)$ or t_i is a closed term and t_i is u_i, then $\sigma \vDash Q(t_1, \ldots, t_n)$ iff $\tau \vDash Q(u_1, \ldots, u_n)$.

2. The quantifiers

The interpretations I'll present here are what I consider the most reasonable in the context of what we've assumed previously in this text. Using these we'll see in Section E that we can model many other views of how to interpret the quantifiers.

a. The universal quantifier

For the universal quantifier we only have to note that an assignment of references does not take into account names that do not refer.

Evaluation of the universal quantifier

(6) $\sigma \vDash \forall x\, A$ iff both: i. For every assignment of references τ that differs from σ at most in what it assigns as reference to x, $\tau \vDash A(x)$.
ii. For every name (symbol) c, $\sigma \vDash A(c)$.

Clause (ii) adds nothing when c refers; it is non-redundant only for non-referring names. Hence (6) *gives the same evaluation as we previously used for*

[2] Some logicians think that using supervaluations makes a no-truth-value approach compatible with the semantics of classical logic, as described in *Bencivenga, 1986*. See also *Morscher and Simon, 2001* for a survey of approaches to free logic.

models in which all names refer. We are only making explicit a distinction between variables and names that is not needed when all names refer.

b. The existential quantifier

Suppose we take (1) to be true. Do we then conclude the following is true?

(7) $\exists x$ (x is a winged horse)

Opinion divides. Some say (7) is true *because* (1) is true. There is something of which 'x is a winged horse' is true, but not an "existent" thing.[3] Others say (7) is false, because whatever role 'Pegasus' plays in our language, it is not a referring name, so it does not pick out something that exists.

But we always took '$\exists$' to mean 'there exists', and that was considered unequivocal. If we talk about different kinds of existence, then we should consider how atoms exist as different from dogs, numbers as different from tables. We didn't do that in establishing the foundations of predicate logic, though we can model such views within predicate logic by using predicates such as '— is an abstract thing' or '— is a sensible object larger than 5 cm in diameter'.

So, as before, '$\exists$' means 'exists' in only one sense, which can be qualified, but doesn't also cover 'not exists'. Hence, (7) should come out false if the predicate 'winged horse' is not true of any thing, even if (1) is classified as true for whatever reason. So the evaluation of the existential quantifier will be as before.

Evaluation of the existential quantifier

$\sigma \vDash \exists x\, A(x)$ iff for some τ that differs from σ at most in what it assigns x, $\tau \vDash A(x)$

Thus, existential generalization, $A(c) \rightarrow \exists x\, A(x)$, is no longer valid. And the classical relation between $\forall$ and $\exists$ no longer holds, for the following can be false:

(8) $\neg \forall x \neg$ (x is a winged horse) $\rightarrow \exists x$ (x is a winged horse)

In a model with universe all living creatures, the antecedent can be true if (1) is classified as true, while the consequent would be false. We'll see in Section E, though, that we can model within these semantics the view that (8) is true.

3. Summary of the semantics for languages without equality

Since we need not alter our evaluations of the propositional connectives, the semantics for classical predicate logic with non-referring names amounts to making only the following changes to the semantics we previously gave for classical predicate logic:

---

[3] This view is usually modeled by assuming a second universe of non-existent objects. See, for example, *Bencivenga, 1986* and *Lambert, 1991*.

We allow for non-referring names in assignments of references (2).
We modify the condition on consistency and extensionality of predications (5).
We modify the evaluation of the universal quantifier (6).

We took atomic predications as primitive in our previous work, and (2), (5), and (6) could have been used for classical predicate logic with only referring names, since we're only drawing distinctions we previously ignored. So these semantics when used in a model in which all names refer give the same evaluation of wffs as before. Except for allowing non-referring names in our language, *there is nothing new in our semantics for classical predicate logic with non-referring names.*[4]

4. Equality

The equality predicate '=' is syncategorematic, a single interpretation for every model, relative to the universe of the model. So we should give an interpretation for it when we use non-referring names that is compatible with the interpretation when all names refer. We shall require that if $\sigma(t)$ and $\sigma(u)$ are defined, then $\sigma \vDash t = u$ iff $\sigma(t)$ is the same object as $\sigma(u)$.

Now consider:

Pegasus = Marilyn Monroe

This can't be true, because 'Pegasus' doesn't refer to any thing, while 'Marilyn Monroe' does. When one side of an equality refers and the other doesn't, the proposition is false. Even taking other semantic values into consideration, two names cannot refer in the same way if one refers to something and the other doesn't. Since for every assignment of references, $\sigma(x)$ does refer to something, we'll also have that $\sigma \nvDash c = x$ if c does not refer.

Now consider:

Pegasus = the horse beloved by Bellerophon

Both sides of the equality symbol are non-referring names. Any reason we have for saying this equality is true must be beyond the scope of classical predicate logic, for we have no other semantic value than reference to ascribe to names. (In a predicate logic in which subject matter is considered, for example, we might argue that this identity is true because the two sides have overlapping subject matter, regardless of their not having reference.) So just as with other atomic predications, we take as primitive whether $c = d$ is true when both c and d do not refer. But some restrictions are needed.

First, consider: Pegasus = Pegasus. Whatever identity means, this must be true: That's just how we use identity. This view can be better defended when other

[4] So far as I can tell, these semantics are not equivalent to any others proposed as a free logic (see *Bencivenga, 1986* and *Lambert, 1991*).

semantic values are ascribed to names, saying, perhaps, that both sides of the equality symbol pick out in the same way, though they pick out nothing. Further, whatever value we give for $c = d$ we must give to $d = c$. And equalities should be transitive, too. We stray as little as possible from classical predicate logic.

There are some like Russell, though, who hold that 'Pegasus = Pegasus' should be false, since 'Pegasus' does not refer. We'll see in Section E how we can model that view within these semantics.

To summarize, letting t, u, v stand for any terms, c, d stand for any names, and σ, τ stand for any assignments of references, we have the following.

Restrictions on the evaluation of the equality predicate

(9) i. If both $\sigma(t)$ and $\sigma(u)$ are defined, then $\sigma \vDash t = u$ iff $\sigma(t)$ is the same object as $\sigma(u)$.

ii. If only one of $\sigma(t)$ and $\sigma(u)$ is defined, then $\sigma \nvDash t = u$.

iii. If both c and d do not refer, then for every σ and τ, $\sigma \vDash c = d$ iff $\tau \vDash c = d$.

iv. For all t, $\sigma \vDash t = t$.

v. For all t and u, $\sigma \vDash t = u$ iff $\sigma \vDash u = t$.

vi. For all t, u, and v, if $\sigma \vDash t = u$ and $\sigma \vDash u = v$, then $\sigma \vDash t = v$.

The condition on consistency and extensionality of predications must now take into account the interpretation of the equality predicate.

Condition on consistency and extensionality of predications with equality

(10) For all atomic wffs $Q(t_1, \ldots, t_n)$ and $Q(u_1, \ldots, u_n)$ and for any assignments σ and τ, if for each $i \leq n$, $\sigma \vDash t_i = u_i$, then $\sigma \vDash (Q(t_1, \ldots, t_n))$ iff $\tau \vDash (Q(u_1, \ldots, u_n))$.

Thus, the semantics for languages with the equality predicate require only (9) and (10) in addition to the previous semantics. Note that (10) is equivalent to the condition on consistency and extensionality of predications for models in which all names refer (p. 85) since we've only replaced the clause '$\sigma(t_i) = \sigma(u_i)$' there with '$\sigma \vDash t_i = u_i$'.

This completes the description of the semantics for classical predicate logic with identity and non-referring names. In the last section of this chapter we'll look at how to give a mathematical abstraction of these semantics.

Exercises for Sections A–C

1. What is free logic?
2. Describe the evaluation of wffs of a semi-formal language when the universe is empty.
3. Why does it make more sense to formalize reasoning with non-referring names in a logic that takes account of semantic values other than truth-value and reference?
4. Why not allow for two different assignments of reference σ and τ in a model such that σ(Pegasus) is defined and τ(Pegasus) is not defined?
5. Show that (6) gives the same evaluation of the universal quantifier as we previously used for models in which all names refer.
6. Why don't we change the evaluation of '$\exists$' in models with non-referring names?
7. a. What conditions are put on the assignment of truth-values to atomic equalities?
   b. Why do we take 'Pegasus = Pegasus' to be true?
8. Why do we take predicates to be extensional in the semantics above?
9. State the condition on consistency and extensionality of predications for languages with equality and non-referring names.
10. Explain: Except for allowing non-referring names in our language, there is nothing new in our semantics for classical predicate logic with non-referring names.
11. Show that the following can be false in a model:
    $\neg \exists x$ (x is a mythological creature) $\rightarrow \forall x \neg$ (x is a mythological creature)
12. Consider the interpretation:

    $L(\neg, \rightarrow, \wedge, \vee, \forall, \exists, =; P_0, P_1, \ldots, c_0, c_1, \ldots)$
    $\downarrow$
    $L(\neg, \rightarrow, \wedge, \vee, \forall, \exists, =;$ is a dog, is a cat, is a horse, eats grass, has wings, is bigger than, is the father of, loves; Buddy, Juney, Richard L. Epstein, Pegasus, Garfield, Bellerophon$)$

    universe: all animals that have ever lived

    Here 'Buddy' names a Mongolian Shepherd dog I now have (roughly the size of a German Shepherd), 'Juney' is a dog I loved who died in a tragic accident, and 'Pegasus', 'Garfield', and 'Bellerophon' do not refer. Predications are interpreted as in ordinary speech for things that exist, and for non-referring names they are interpreted according to the stories we know that use them. Evaluate the following, justifying your evaluations.

    a. Buddy = Juney
    b. Buddy = Pegasus
    c. Bellerophon = Garfield
    d. $\exists x$ (x = Garfield)
    e. $\neg \forall x \neg$ (x = Garfield)
    f. Richard L. Epstein loves Juney.
    g. Bellerophon loves Pegasus.
    h. Garfield is the father of Pegasus.
    i. $\exists x$ (x loves Pegasus)
    j. $\neg \forall x \neg$ (x loves Pegasus)
    k. Garfield is a cat.
    l. Pegasus is bigger than Garfield.
    m. Garfield is bigger than Buddy.
    n. $\forall x \neg$(x is a cat $\wedge$ x is a horse)
    o. $\forall x$ (x is a horse $\rightarrow \neg$(x has wings))
    p. $\neg \exists x$ (x is a horse $\wedge$ x has wings)
    q. $\forall x \exists y$ (y loves x)
    r. $\exists y \forall x \neg$(x loves y)

13. Show that for any model M and closed wff A in which x is not free,

$$\mathsf{M} \vDash \forall x\, \mathrm{A} \text{ iff } \mathsf{M} \vDash \mathrm{A} \text{ iff } \mathsf{M} \vDash \exists x\, \mathrm{A}.$$

14. We can also lift the restriction that names refer in the logic of quantifying over names (Chapter XV.F). The only changes to those semantics is to take assignments of references as defined above and place the restrictions (9) on equality along with the conditions on consistency and extensionality of predications (10). In particular, the universal and existential quantifiers for variables are evaluated as they were before:

$\sigma \vDash \forall x\, \mathrm{A}(x)$ iff for every τ that differs from σ at most in what it assigns x, $\tau \vDash \mathrm{A}(x)$

$\sigma \vDash \exists x\, \mathrm{A}(x)$ iff for some τ that differs from σ at most in what it assigns x, $\tau \vDash \mathrm{A}(x)$

The universal and existential quantifiers over names are evaluated as they were before in models of the logic of quantifying over names:

$\sigma \vDash \forall a_i\, \mathrm{A}(a_i)$ iff $\sigma \vDash \mathrm{A}(b/a_i)$ for every name b

$\sigma \vDash \exists a_i\, \mathrm{A}(a_i)$ iff $\sigma \vDash \mathrm{A}(b/a_i)$ for some name b

(a) Define a new quantifier in this logic:

$\underline{\forall} x_i\, \mathrm{A}(x_i) \equiv_{\mathrm{Def}} \forall x_i\, \mathrm{A}(x_i) \wedge \forall a_i\, \mathrm{A}(a_i/x_i)$

a_i is the least name variable that does not appear in $\mathrm{A}(x_i)$

Show that for every assignment of references σ:

$\sigma \vDash \underline{\forall} x\, \mathrm{A}(x)$ iff σ validates $\forall x\, \mathrm{A}(x)$ according to (6) above.

(b) Define a new quantifier:

$\underline{\exists} x_i\, \mathrm{A}(x_i) \equiv_{\mathrm{Def}} \exists x_i\, \mathrm{A}(x_i) \wedge \exists a_i\, \mathrm{A}(a_i/x_i)$

a_i is the least name variable that does not appear in $\mathrm{A}(x_i)$

Show that (8) is true when formalized using this quantifer instead of $\exists x$.

(c) Explain why the use of the quantifier $\underline{\exists} x$ allows us to more fully model the view that there are non-existent things such as Pegasus and the present King of France.

D. An Axiomatization

I'll present here an axiomatization of classical predicate logic with equality and non-referring names for the language $L = L(\neg, \rightarrow, \forall, \exists, =; P_0, P_1, \ldots, c_0, c_1, \ldots)$.

Our axiomatization must yield a collection of theorems of classical predicate logic, since every model of classical predicate logic with only referring names is a model here, too. Hence, every theorem (consequence) of classical predicate logic with non-referring names must also be a theorem (consequence) of classical predicate logic.

We take the axioms of classical predicate logic without equality from Chapter X (p. 168) except for axiom schema 5.b, which we saw was invalid for non-referring names (example (8) above). For equality we add the axiom of identity as before (p. 178), but replace axiom scheme 8 for the extensionality of predications with a new axiom scheme 8′ below that entails scheme 8 relative to the other axioms. We also need to add a new group of axioms governing non-referring names, Group VI. (Axiom scheme 9 of Group V will be for languages containing function symbols.)

Classical predicate logic with equality and non-referring names
in $L(\neg, \rightarrow, \forall, \exists, =; P_0, P_1, \ldots, c_0, c_1, \ldots)$

I. Propositional axioms
The axiom schema of **PC** in $L(\neg, \rightarrow)$, where A, B, C are replaced by predicate logic wffs and the universal closure is taken:

$\forall \ldots \neg A \rightarrow (A \rightarrow B)$

$\forall \ldots B \rightarrow (A \rightarrow B)$

$\forall \ldots (A \rightarrow B) \rightarrow ((\neg A \rightarrow B) \rightarrow B)$

$\forall \ldots (A \rightarrow (B \rightarrow C)) \rightarrow ((A \rightarrow B) \rightarrow (A \rightarrow C))$

II. Axioms governing $\forall$

1. $\forall \ldots (\forall x\, (A \rightarrow B) \rightarrow (\forall x\, A \rightarrow \forall x\, B))$ *distribution of* $\forall$
2. When x is not free in A,
   a. $\forall \ldots (\forall x\, A \rightarrow A)$ *superfluous quantification*
   b. $\forall \ldots (A \rightarrow \forall x\, A)$
3. $\forall \ldots (\forall x \forall y\, A \rightarrow \forall y \forall x\, A)$ *commutativity of* $\forall$
4. When term t is free for x in A,
   $\forall \ldots (\forall x\, A(x) \rightarrow A(t/x))$ *universal instantiation*

III. Axioms governing the relation between $\forall$ and $\exists$

5.a. $\forall \ldots (\exists x\, A \rightarrow \neg \forall x \neg A)$

6. $8\forall \ldots (\forall x\, A(x) \rightarrow \exists x\, A(x))$ *all implies exists*

IV. Axioms for equality

7. $\forall x\, (x = x)$

8′. For every n-ary atomic predicate P, *extensionality of predications*
$\forall \ldots (\bigwedge_i (t_i = u_i) \rightarrow (P(t_1, \ldots, t_n) \rightarrow P(u_1, \ldots, u_n))$

VI. Axioms for non-referring names

10. $\forall \ldots (\forall y\, (\exists x\, (x = y) \rightarrow \neg B(y/x)) \rightarrow \neg \exists x\, B(x))$

11. When x is not free in A,
   a. $\forall \ldots (\exists x\, A \rightarrow A)$ *superfluous quantification*
   b. $\forall \ldots (A \rightarrow \exists x\, A)$

12. $\forall \ldots ((A(t/x) \wedge \exists x\, (x = t)) \rightarrow \exists x\, A(x))$ *existential generalization for referring names*

Rule $\dfrac{A, A \rightarrow B}{B}$ where A and B are closed formulas *modus ponens*

To show that this is a complete axiomatization I will modify the proof in Chapter X for classical predicate logic. Lemma X.1 on the relation of consistency, completeness, and negation still holds. I'll leave to you to show that the axiomatization is sound, that is, if $\Gamma \vdash A$, then $\Gamma \models A$. Lemmas X.3–X.8 and Lemma X.10 also hold, since none of those use Axiom 5.b. Note that Lemma X.10 is:

For any formula $B(x)$ with one free variable x,
if $\Gamma \vdash B(c/x)$ and c does not appear in any wff in Γ, then $\Gamma \vdash \forall x\, B(x)$.

I'll modify the constructions in Theorems X.11 and X.12, using the same notation as there, writing 'Axiom' for 'axiom scheme'.

Theorem 1 Let Γ be a consistent set of closed wffs of L. Then there is a collection of closed wffs Σ in $L(v_0, v_1, \ldots)$ such that:

a. $\Gamma \subseteq \Sigma$.

b. Σ is a complete and consistent theory.

c. If $\exists x\, B \in \Sigma$ and x is free in B, then for some m, $B(v_m/x) \in \Sigma$ and $\exists x\, (x = v_m) \in \Sigma$.

d. If $\neg\forall x\, B \in \Sigma$ and x is free in B, then for some m, $\neg B(v_m/x) \in \Sigma$.

e. For every wff $B(x)$ in $L(v_0, v_1, \ldots)$ with one free variable, if for each i, $B(v_i/x) \in \Sigma$, then $\forall x\, B(x) \in \Sigma$.

f. For every v_i, $\exists x\, (x = v_i) \in \Sigma$.

Proof Let $A_0, A_1, \ldots$ be a numbering of all the closed wffs of the expanded language $L(v_0, v_1, \ldots)$. Let '$\vdash$' refer to derivations in this language.

Define Σ by stages:

$$\Sigma_0 = \Gamma$$

$$\Sigma_{n+1} = \begin{cases} \Sigma_n \cup \{\neg A_n\} & \text{if } \Sigma \vdash \neg A_n. \\ \Sigma_n \cup \{A_n\} & \text{if } \Sigma \nvdash \neg A_n,\ A_n \text{ is not } \exists x\, B \text{ or } \neg\forall x\, B, \text{ and} \\ & x \text{ is free in B.} \\ \Sigma_n \cup \{A_n, B(v_m/x), \exists x\,(x = v_m)\} & \\ & \text{if } \Sigma \nvdash \neg A_n \text{ and } A_n \text{ is } \exists x\, B,\ x \text{ is free in B, and } v_m \\ & \text{is the least of } v_0, v_1, \ldots \text{ that does not appear in } \Sigma_n. \\ \Sigma_n \cup \{A_n, \neg B(v_m/x)\} & \\ & \text{if } \Sigma \nvdash \neg A_n \text{ and } A_n \text{ is } \neg\forall x\, B,\ x \text{ is free in B, and } v_m \\ & \text{is the least of } v_0, v_1, \ldots \text{ that does not appear in } \Sigma_n. \end{cases}$$

$$\Sigma = \bigcup_n \Sigma_n$$

I'll show that Σ satisfies (a)–(f). Part (a) follows by construction.

(b) We first show by induction that for each n, Σ_n is consistent. For $n = 0$ it's true by hypothesis. If it's true for n and Σ_{n+1} is defined by the first case, it's immediate. If Σ_{n+1} is defined by the second case it follows by induction and Lemma X.1.

Suppose that Σ_{n+1} is defined by the third case. Then $\Delta = \Sigma_n \cup \{\exists x\, B(x)\}$ is consistent by induction and Lemma X.1. Suppose Σ_{n+1} is not consistent. Then $\Delta \vdash \neg(\exists x\,(x = v_m) \wedge B(v_m/x))$. So by **PC**, $\Delta \vdash \exists x\,(x = v_m) \rightarrow \neg B(v_m/x)$. Hence, by Lemma X.10, as only finitely many formulas from Δ are used in that proof, there is a y not appearing in B such that $\Delta \vdash \forall y\,(\exists x\,(x = y) \rightarrow \neg B(y/x))$. But then by Axiom 10, $\Delta \vdash \neg\exists x\, B(x)$, which is a contradiction.

Suppose now that Σ_{n+1} is defined by the fourth case. Then by Lemma X.1 and **PC**, $\Delta = \Sigma_n \cup \{\neg\forall x\, B\}$ is consistent. Suppose that Σ_{n+1} is not consistent. Then $\Delta \vdash B(v_m/x)$. But then by Lemma X.10, for some y, $\Delta \vdash \forall y\, B(y)$. Hence by Axiom 2.b and Axiom 4, $\Delta \vdash \forall x\, B(x)$, which contradicts the consistency of Δ.

It then follows that Σ is consistent, for if it were not then some finite subset of it would be inconsistent, and hence some Σ_n would be inconsistent.

By construction, for every A, either $A \in \Sigma$ or $\neg A \in \Sigma$. So Σ is complete, and hence by Lemma X.1, Σ is also a theory.

Parts (c) and (d) follow by construction, since if $A \in \Sigma$, then for some n, A is A_n and the appropriate formulas are put into Σ at stage $n + 1$.

(e) I'll show the contrapositive. Suppose $\forall x\, B(x) \notin \Sigma$. Then by (b), $\neg\forall x\, B(x) \in \Sigma$. Hence by (d), for some m, $\neg B(v_m/x) \in \Sigma$. So by consistency, $B(v_m/x) \notin \Sigma$.

(f) There are infinitely many distinct formulas of the form $\exists x\, B(x)$ with x free that are in Σ, for example (using Axioms 7 and 6), $\exists x_i\,(x_i = x_i)$. So for every j, there is some stage n where $\exists x\,(x = v_j)$ is put into Σ. ■

Theorem 2 Löwenheim's theorem Every consistent collection of closed wffs in L has a countable model.

Proof Let Γ be a consistent collection of closed wffs of L. In $L(v_0, v_1, \ldots)$, let $\Sigma \supseteq \Gamma$ be as in Theorem 1. We define an interpretation M for $L(v_0, v_1, \ldots)$.

$U = \{v_i : i \geq 0\}$

v_i is interpreted by v_i

c_i is interpreted by v_j iff $(v_j = c_i) \in \Sigma$

$\sigma \models t = u$ iff $\sigma(t)$ is v_i and $\sigma(u)$ is v_j and $(v_i = v_j) \in \Sigma$, or $(t = u) \in \Sigma$

$\sigma \models P_i^n(t_1, \ldots, t_n)$ iff $P_i^n(u_1, \ldots, u_n) \in \Sigma$
where u_i is $\sigma(t_i)$ if $\sigma(t_i)$ is defined, and u_i is t_i if $\sigma(t_i)$ is not defined.

Define a relation $\simeq$ on U by: $v_j \simeq v_k$ iff $(v_j = v_k) \in \Sigma$.

As in Chapter X.B, you can show that this is an equivalence relation. The model for Σ will be M/≃, the definition of which is justified by Axiom 8.a, since scheme 8 (p. 178) is a syntactic consequence of Axiom 8.a and the other axioms, as you can show.

To show that M/≃ is a model, we begin by showing that the evaluation of '=' satisfies the conditions at (9).

(i) If both $\sigma(t)$ and $\sigma(u)$ are defined, say $\sigma(t) = [v_j]$ and $\sigma(t) = [v_k]$, then $\sigma \vDash t = u$ iff $(v_j = v_k) \in \Sigma$ iff $[v_j] = [v_k]$.

(ii) If $\sigma(t) = [v_j]$ and $\sigma(u)$ is undefined, then u is c, a name symbol in L. Suppose that $\sigma \vDash t = c$. Then, using that ≃ is an equivalence relation, $(v_j = c) \in \Sigma$. But then $\sigma(c) = v_j$, which is a contradiction. So $\sigma \nvDash t = c$. When $\sigma(u) = [v_j]$ and $\sigma(t)$ is undefined, proceed similarly.

(iii) The only case in which both $\sigma(t)$ and $\sigma(u)$ are both undefined is when t is c and u is d for some name symbols in the language. Then $\sigma \vDash c = d$ iff $(c = d) \in \Sigma$ iff for every τ, $\tau \vDash c = d$.

(iv) By Axiom 7 and Axiom 4, for all t, $\sigma \vDash t = t$.

(v) and (vi) The derivations in Theorem X.18 (pp. 178–179) hold for this axiom system, too.

I'll let you show that M/≃ satisfies the condition on consistency and extensionality of predications using Axiom 8′. Hence, M/≃ is a model, which I'll call **N**. It remains to show that **N** is a model of Σ.

I'll let you show by induction on the length of wffs that for every wff A in $L(v_0, v_1, \ldots)$, for any σ, if $\sigma(x) = [v_m]$, then $\sigma \vDash A(x)$ iff $\sigma \vDash A(v_m/x)$.

Now I'll establish by induction on the length of wffs that for every closed wff A in that language,

(‡) $\quad$ **N**$\vDash$A iff A$\in \Sigma$

For atomic wffs, I'll leave to you the case when A is $t = u$. If A is $P_i^n(t_1, \ldots, t_n)$, then (‡) follows by definition and Axiom 8′.

Suppose (‡) is true for all wffs shorter than A. I'll leave all the cases to you except when A is $\forall x$ B or $\exists x$ B.

Suppose that A is $\forall x$ B. If x is not free in B, then by Axiom 2, $\forall x$ B$\in \Sigma$ iff B$\in \Sigma$. So by induction, $\forall x$ B$\in \Sigma$ iff **N**$\vDash$B. I'll let you show **N**$\vDash$B iff **N**$\vDash \forall x$ B when x is not free in B (Exercise 14 of the last section). So $\forall x$ B$\in \Sigma$ iff **N**$\vDash \forall x$ B.

Suppose A is $\forall x$ B and x is free in B. Suppose $\forall x$ B$(x) \in \Sigma$. Since Σ is a theory, by Axiom 4, for all c, B$(c/x) \in \Sigma$, and hence for all c, **N**$\vDash$B(c/x). So also for every σ, if $\sigma(x) = [v_m]$, then $\sigma \vDash$B(x). Hence, **N**$\vDash \forall x$ B.

In the other direction, suppose **N**$\vDash \forall x$ B(x). Then for every σ, for any i, if $\sigma(x) = [v_i]$, then $\sigma \vDash$B(v_i/x), and so **N**$\vDash$B(v_i/x), and by induction B$(v_i/x) \in \Sigma$. Hence by part (e) of the conditions on Σ from Theorem 1, $\forall x$ B$(x) \in \Sigma$.

Now suppose A is $\exists x$ B. If x is not free in B, proceed as for $\forall x$ B using Axiom 11. So suppose A is $\exists x$ B and x is free in B.

If $\exists x\, B(x) \in \Sigma$, then x must be the only variable free in $B(x)$. Since Σ satisfies condition (c) of Theorem 1, for some m, $B(v_m/x) \in \Sigma$. Hence by induction, $N \vDash B(v_m/x)$, and so for any σ, if $\sigma(x) = [v_m]$, then $\sigma \vDash B(x)$. So $N \vDash \exists x\, B(x)$.

Now suppose $N \vDash \exists x\, B(x)$. Then for some σ, $\sigma \vDash B(x)$. Take one such σ, where $\sigma(x) = [v_i]$. Then $\sigma \vDash B(v_i/x)$. Since $B(v_i/x)$ is closed, $N \vDash B(v_i/x)$, so $B(v_i/x) \in \Sigma$. Since Σ satisfies part (f) of Theorem 1, $\exists x\, (x = v_i) \in \Sigma$. Hence, by Axiom 12, $\exists x\, B(x) \in \Sigma$.

To complete the proof of the theorem, define a structure N' for L by taking N and deleting the interpretations of the v_i's (the universes are the same). Note that N is countable. For any closed wff A in L, $N \vDash A$ iff $N' \vDash A$ by the partial interpretation theorem (Theorem V.15), which you can show holds here, too. So $N \vDash \Gamma$. ■

Our main theorem now follows from Theorem 2 (compare Chapter X.A).

Theorem 3 In $L(\neg, \rightarrow, \forall, \exists, =; P_0, P_1, \ldots, c_0, c_1, \ldots)$ with non-referring names,

a. For any collection of closed wffs Σ in L, Σ is a complete and consistent theory iff there is a countable model M such that $\Sigma = Th(M)$.

b. For any model M of L, there is a countable model M^* such that $Th(M) = Th(M^*)$.

c. ***Strong completeness*** $\Gamma \vdash A$ iff $\Gamma \vDash A$.

d. ***Compactness*** Γ has a model iff every finite subset of Γ has a model.

Exercises for Section D

1. Show that the axiomatization is sound, that is, if $\Gamma \vdash A$, then $\Gamma \vDash A$.
2. Show that if $\Gamma \vdash A$ in this axiomatization, then $\Gamma \vdash A$ in classical predicate logic (for models in which all names refer).
3. Give a syntactic justification that every instance of axiom scheme 8 (p. 178) is a consequence of the other axioms.
4. Why can't we replace Axiom 10 by $\forall x\, \exists y\, (x = y)$?
5. Show that $M/\simeq$ in the proof of Theorem 2 satisfies the condition on consistency and extensionality of predications.
6. Show in the proof of Theorem 2 for $M/\simeq$ by induction on the length of wffs that for every wff A in $L(v_0, v_1, \ldots)$, for any σ, if $\sigma(x) = [v_m]$, then $\sigma \vDash A(x)$ iff $\sigma \vDash A(v_m/x)$.
7. Show that the partial interpretation theorem (Theorem V.15) holds for models with non-referring names.

E. Examples of Formalization

1. **Pegasus is a winged horse.**
   ***Therefore*, something is a winged horse.**

Analysis On our interpretation of the ordinary English in our models, the formalization of the example is:

$$\frac{\text{Pegasus is a winged horse.}}{\exists x\ (x \text{ is a winged horse})}$$

The inference is invalid: In any model in which 'Pegasus' does not refer, the conclusion is false even if the antecedent is true. Existential generalization fails.

There are some, though, who say Example 1 is valid. This seems to me a remnant of the idea that the use of any name entails existence of a reference, which is what we set out to deny. There simply are no winged horses.

That view, though, becomes more respectable if we argue that there is a difference between 'there exists' and 'there is' (or 'something'). The former requires existence, and it is that we have modeled with '$\exists$' in our system. But the latter does not. That is, the following is invalid:

Pegasus is a winged horse.
Therefore, there exists a winged horse.

But Example 1 is valid, as is the following:

Pegasus is a winged horse.
Therefore, there is a winged horse.

We can define within our system a quantifier to model that view of 'there is'.

The generous existence quantifier $\exists_G x\ A(x) \equiv_{Def} \neg \forall x \neg A(x)$

Then the following example of existential generalization is valid:

$$\frac{\text{Pegasus is a winged horse.}}{\exists_G x\ (x \text{ is a winged horse})}$$

And when x is the single variable free in A and c is a name, the following is valid:

$A(c/x) \vDash \exists_G x\ A(x)$

2. **Everything that is a horse is a mammal.**
   Pegasus is a horse.
   ***Therefore*, Pegasus is a mammal.**

Analysis On our interpretation of the ordinary English in our models (Example 1, Chapter VII), the formalization of Example 2 is:

$\forall x$ (x is a horse $\rightarrow$ x is a mammal)
Pegasus is a horse.
____
Pegasus is a mammal.

This is valid: It follows from universal instantiation and *modus ponens*.

Some say the example is not valid. We could have a model in which the universe is all animals that have ever lived, for example, and the premises are true even though 'Pegasus' does not refer, while the conclusion is false. On this view, 'everything' is interpreted as meaning 'Every existing thing': 'all' includes 'and there exists'. That is the interpretation of 'all' which is rejected in standard classical logic formalizations: 'All cats that are dogs are loyal' is counted as true. Mathematicians use 'all' without including 'and there exists', and it is the formalization of mathematics we set out to model in classical predicate logic. Nonetheless, we can model the view that 'all' includes 'and there exists' by defining a new quantifier.

The restricted universal quantifier $\forall_R x\, A(x) \equiv_{Def} \neg \exists x \neg A(x)$

Then we could formalize Example 2 as:

$\forall_R x$ (x is a horse $\rightarrow$ x is a mammal)
Pegasus is a horse.
____
Pegasus is a mammal.

This is equivalent to:

$\neg \exists x$ (x is a horse $\wedge$ $\neg$(x is a mammal))
Pegasus is a horse.
____
Pegasus is a mammal.

And that is invalid: the premises are true in the model described above, but the conclusion is false. Instantiation fails for the restricted universal quantifier.

3. **Everything that is loved by Bellerophon is a winged horse.**
   Pegasus is loved by Bellerophon.
   ***Therefore*, something that is loved by Bellerophon is a winged horse.**

Analysis On our interpretation of the ordinary English in our models, the formalization of Example 3 is:

$\forall x$ (x is loved by Bellerophon $\rightarrow$ x is a winged horse)
Pegasus is loved by Bellerophon.
____
$\exists x$ (x is loved by Bellerophon $\wedge$ x is a winged horse)

This is invalid: the conclusion could be false in a model if there are no winged horses, yet the premises could be true.

However, we can use the other ways to interpret the quantifiers described in the last two examples to formalize this example:

$\forall x$ (x is loved by Bellerophon $\rightarrow$ x is a winged horse) *valid*
Pegasus is loved by Bellerophon.

$\exists_G x$ (x is loved by Bellerophon $\wedge$ x is a winged horse)

$\forall_R x$ (x is loved by Bellerophon $\rightarrow$ x is a winged horse) *valid*
Pegasus is loved by Bellerophon.

$\exists x$ (x is loved by Bellerophon $\wedge$ x is a winged horse)

$\forall_R x$ (x is loved by Bellerophon $\rightarrow$ x is a winged horse) *valid*
Pegasus is loved by Bellerophon.

$\exists_G x$ (x is loved by Bellerophon $\wedge$ x is a winged horse)

4. Pegasus is Pegasus.

Analysis On our interpretation of the ordinary English in our models, the formalization of Example 4 is:

Pegasus = Pegasus

In all our models this is true: It's an instance of identity, $\forall x\ (x = x)$, which is valid.

Some say the example is false because Pegasus does not exist.[5] We can model that view by using the following defined predicate

---

Restricted equality $x =_R x \equiv_{\text{Def}} \exists y\ (y = x \wedge x = x)$

where y is the least variable different from x

Then $\forall x\ (x =_R x)$ will fail in any model in which there is a non-referring name. We can have 'Pegasus = Pegasus' is true, but 'Pegasus $=_R$ Pegasus' is false.

Similarly, given any predicate $P(x)$ with the single variable x free, we can make the following definition.

The restriction of the predicate P

$P_R(x) \equiv_{\text{Def}} \exists y\ (y = x \wedge \text{A}(x))$

where y is the least variable that does not appear in $P(x)$

Then $P_R(c/x)$ is false in a model when c is a name that does not refer.

[5] See, for example, the discussion in Chapter VIII of *Predicate Logic* of methods of how to eliminate descriptive names.

Exercises for Section E

1. a. Show that the following is invalid:
   $\forall_R x\, A(x) \leftrightarrow \forall x\, (A(x) \wedge \exists x\, A(x))$
   b. Formalize and discuss the following, using $\forall_R\, x$ to formalize 'all'.
      i. All dogs bark
      ii. All dogs that are cats cough up hairballs.
      iii. All real numbers whose square root is –1 are positive.
2. Show that the following are valid:
   a. $\forall x\, A(x) \rightarrow \exists_G x\, A(x)$
   b. $\forall_R x\, A(x) \rightarrow \exists x\, A(x)$
   c. $\forall_R x\, A(x) \rightarrow \exists_G x\, A(x)$
   d. $\exists_G x\, A_R(x) \leftrightarrow \exists x\, A(x)$
   e. $\forall x\, A_R(x) \leftrightarrow \exists x\, A(x)$
3. Given a predicate $A(y_1, \ldots, y_n)$ where $n > 1$, define a restricted predicate $A_R(y_1, \ldots, y_n)$ as in Example 4.
4. Formalize the following. Either choose one formalization that you believe is best and argue for that, or compare the various ways to formalize the example.
   a. There is at most one winged horse and Pegasus is a winged horse. Therefore, something is a winged horse.
   b. Every dog barks. Therefore, Cerebus, if he exists, barks.
   c. No one has ever seen a winged horse. Therefore, there are no winged horses.
   d. Pegasus is a winged horse, so there is a winged horse, though none exist.
   e. Bellerophon owned a flying horse, though of course no one ever anywhere really owned a flying horse.
   f. Pegasus is not bigger than Fred Kroon, since Pegasus doesn't exist.
   g. If Pegasus exists, then he is bigger than Fred Kroon.
   h. Pegasus is the horse that was loved by Bellerophon.
   i. Everything that is a horse does not fly. Pegasus is a horse. Therefore, Pegasus does not fly.

F. Classical Predicate Logic with Names for Partial Functions

1. Partial functions in mathematics

Mathematicians regularly use function names to create terms that name nothing. Such function names are meant to stand for partial functions, or as mathematicians say, functions whose domain is not the entire universe. For example, the name 'tan' stands for the tangent function in studies of the real numbers, and '$\tan(\pi/2)$' has no reference, nor does '$\tan(x)$' when x stands for $-3\pi/2$. The name '–' is used for the subtraction function on natural numbers, and '$5 - 7$' stands for no natural number.

Mathematicians try to avoid using compound names that do not refer by saying a function such as tangent is defined only for numbers other than $m\pi + n\pi/2$ for m any integer and n an odd integer; an expression such as '$\tan(\pi/2)$' is not a legitimate term. But to follow that line in our formalizations creates a serious problem. It's not trivial to determine for what values $\cot(\tan(\sqrt{x}+\sin(y+\pi/3))+\cot(z))$ is defined. If to decide whether a concatenation of symbols of the formal language is a term we have to be able to decide an existence question, then the semantics become thoroughly enmeshed with the formation rules of the language. We would not be able to give an inductive definition of the formal language.

In this section I'll present a formal system for reasoning about partial functions in which compound terms need not refer, extending what we have already done.[6]

2. Semantics for partial functions

The semantics for languages with names for partial functions is a modification of the semantics for languages with non-referring singular names. We have the following conditions on assignments of references.

Assignments of references with function names

(11) i. For every variable x, $\sigma(x)$ is defined.

ii. For every closed term u, if $\sigma(u)$ is defined, then for every τ, $\tau(u)$ is defined and $\tau(u)$ is the same object as $\sigma(u)$.

iii. For every term t and every function name f in the language, if $\sigma(t)$ is not defined, then for every sequence of terms $t_1, \ldots, t, \ldots, t_n$, $\sigma(f(t_1, \ldots, t, \ldots, t_n))$ is not defined.

Condition (iii) reflects the usual practice that, for example, $\sin(\tan(\pi/2))$ is undefined because $\tan(\pi/2)$ is undefined.

The condition on the consistency and extensionality of predications (10) remains the same, since it was previously framed for any non-referring terms. But now we add a similar requirement for functions.

Extensionality of functions

(12) For all terms $f(t_1, \ldots, t_n)$ and $f(u_1, \ldots, u_n)$ and for any assignment σ, if for each $i \leq n$, $\sigma \vDash t_i = u_i$, then $\sigma \vDash f(t_1, \ldots, t_n) = f(u_1, \ldots, u_n)$.

However, we need something more:

(13) Assignments of references that agree on all the variables in two terms agree on those terms.

This seems an essential part of what we mean by saying that applications of functions are extensional. Yet the restrictions (9) on the equality predicate plus (12)

[6] See Chapter IX.E of *Predicate Logic* for a survey of ways that have been proposed to reason with descriptive phrases such as 'the wife of —' as partial functions.

do not give us (13), for we could have a model satisfying both (9) and (12) in which c and d do not refer, $\sigma(x) = \tau(x)$, yet $\sigma \models f(x, c) = d$ while $\tau \not\models f(x, c) = d$.
So we modify the restriction on the evaluation of the equality predicate (9).

Restrictions on the evaluation of the equality predicate with functions

(14) (9.i–vi) except that (iii) is replaced by:

$\text{iii}_{\text{functions}}$. If for every variable x that appears in t or u, $\sigma(x) = \tau(x)$, then $\sigma \models t = u$ iff $\tau \models t = u$.

I'll let you show that (12) plus (14) is equivalent to the condition on extensionality of functions for models in which there are no partial functions (Chapter VIII.C, p. 144).

The only other modification is to the evaluation (6) of universal quantifiers.

(15) $\sigma \models \forall x\, A$ iff i. For every assignment of references τ that differs from σ at most in what it assigns as reference to x, $\tau \models A(x)$.
and
ii. For every τ and every term t free for x in A, $\tau \models A(t/x)$.

Clause (ii) requires that we survey not only all elements of the universe as we did when names refer, but all terms, too, along with all assignments to those. Note that the first clause is now covered by the second clause, so we can simplify.

Evaluation of the universal quantifier

(16) $\sigma \models \forall x\, A$ iff for every assignment of references τ and every term t free for x in A, $\tau \models A(t/x)$

I'll let you show that (16) is equivalent to the evaluation of universal quantifiers in models in which the only non-referring names are atomic as well as in models in which all names refer. We don't use (16) for models in which all names refer because the much simpler evaluation using only (i) works. All we're doing here is drawing distinctions that we previously ignored. Except for allowing non-referring names and names for partial functions in our language, *there is nothing new in our semantics for classical predicate logic with non-referring names and partial functions.*

3. Examples

a. The extended real number system consists of the real number system to which two symbols, $+\infty$ and $-\infty$ have been adjoined, with the following properties:

(a) If x is real, then $-\infty < x < +\infty$, and

$x + \infty = +\infty, \ x - \infty = -\infty, \ \ldots$

Rudin, 1964

We can consider a *complex number* as having the form $a + b\,i$ where a and b are real numbers and i, which is called the *imaginary unit*, has the property that $i^2 = -1$. . . .

In performing operations with complex numbers we can proceed as in the algebra of real numbers, replacing i^2 by -1 when it occurs.

1. *Addition* $(a + b\,i) + (c + d\,i) = a + b\,i + c + d\,i = (a + c) + (b + d)\,i$

Spiegel, 1964

Mathematicians often aren't clear whether they're using symbols such as '∞' and 'i' as names of things that exist or as nonreferring names, simply giving the rules for which atomic predications using them are true and which are false.

The main reason to introduce the symbols '∞' and '$-\infty$', however, is to have values for functions that would otherwise be undefined. For example, $\tan(\pi/2) = \infty$ and $\tan(3\pi/2) = -\infty$. If we read the new symbols as non-referring names, we can model such reasoning with the semantics above. We can model reasoning with expressions such as '$\tan(\pi/2) = \tan(\pi + \pi/2)$' and '$\tan(\pi/2) \neq \tan(3\pi/2)$'.

b. In recursive function theory, mathematicians use a notion of equality for non-referring terms:

$f(x) \simeq g(x) \equiv f(x)$ and $g(x)$ are both defined and equal, or both are undefined

All undefined terms are taken as equal, since undefined terms all arise in the same way: a calculation does not halt. We can model such reasoning with the semantics given above.

c. Here is a simple model that shows how complicated these semantics can become:

$\mathsf{U} = \{1, 2, 3, 4, 5\}$

referring names: none.

non-referring names: *a*, *b*, *c*, *d*.

function symbol f Interpreted as the partial function f, where $\mathsf{f}(1) = 2$ and $\mathsf{f}(3) = 3$.

equality

The evaluation is the least collection of pairs of assignments of references and equality wffs given the conditions of (14) plus, for all σ:

$\sigma \vDash a = b$	$\sigma \vDash f(a) = f(d)$
$\sigma \nvDash a = c$	$\sigma \vDash f(b) = f(c)$
(so by 9.vi, $\sigma \nvDash b = c$)	$\sigma \vDash f(c) = f(x)$ when $\sigma(x) = 2$
$\sigma \nvDash a = d$	$\sigma \vDash f(f(c)) = f(x)$ when $\sigma(x) = 4$
$\sigma \nvDash b = d$	
$\sigma \nvDash c = d$	

binary predicate symbol P P is interpreted as the universal function on U, that is, for every σ, x, and y, $\sigma \vDash P(x, y)$, and for all σ:

$\sigma \nvDash P(a, y)$ when $\sigma(y) = 4$

$\sigma \nvDash P(x, c)$ when $\sigma(x) = 2$

$\sigma \nvDash P(f(x), c)$ when $\sigma(x) = 4$

d. Modify the previous model by first taking A, an undecidable set of natural numbers. Writing 'f^n' for the iteration n times of f, set:

$\sigma \vDash f^n(c) = f(x)$ iff $\sigma(x) = 4$ for $n > 2$ and $n \in$ A

Even though the resulting model M has a finite universe and only finitely many names, functions, and predicates, the set of wffs true in M is not decidable. In particular, $\{n\colon \mathsf{M} \vDash \exists x\, (f^n(c) = f(x))\}$ is not decidable.

In contrast, for any model in which there are no partial functions and which has a finite universe and only finitely many names, functions, and predicates, the set of wffs valid in the model is decidable, as you can show. Hence, we have shown the following, in contrast to classical predicate logic (Chapter VIII.E).

Theorem 4 Functions cannot in general be translated into predicates in classical predicate logic with non-referring names and partial functions.

We could ensure the translation of functions into predicates by allowing for partial predicates, saying, for example, that 'Pegasus is bigger than Juney' has no truth-value. But that is a major departure from classical logic which does not seem justified by the enjoinder to make logic free of existential assumptions.

To my knowledge, no one proposes models as complicated as (d). Stipulations are usually made, such as that for all σ, for all non-referring terms t, u, $\sigma \nvDash t = u$, or as in (a) above, certain symbolic elements are added to the universe that determine the collection of equalities in a constructive manner. The evaluations of predicates are also usually simplified, for example, requiring $\sigma \nvDash \mathrm{P}_i^n(t_1, \ldots, t_n)$ if $\sigma(t_i)$ is not defined for some i.

4. An axiomatization

Classical predicate logic with non-referring names and partial functions
We add just one axiom scheme to our previous list.

V. Axioms for functions

9′. For every n-ary function symbol f, *extensionality of functions*
$\forall \ldots (\bigwedge_i (t_i = u_i) \rightarrow (\mathrm{f}(t_1, \ldots, t_n) = \mathrm{f}(u_1, \ldots, u_n))$.

Since this axiom is true in all models in which all names refer and all functions are total, if $\Gamma \vdash A$ in this axiomatization, then A is a consequence of Γ in classical predicate logic. I'll let you show that the original scheme 9 (p. 180) is a syntactic consequence of scheme 9.a.

The proof of Theorem 3 for this logic is a simple modification of what we did previously. Axiom 9.a allows us to define $M/\simeq$ in the proof of Theorem 2. We only need to change the evaluation of the equality predicate to:

$\sigma \vDash t = u$ iff one of the following holds

i. $\sigma(t) = \sigma(u) \in \Sigma$

ii. $(t = u) \in \Sigma$

iii. The variables appearing in t are $y_1, \ldots, y_n$, the terms appearing in u are $z_1, \ldots, z_m$, and $(t(\sigma(y_1), \ldots, \sigma(y_n)) = u(\sigma(z_1), \ldots, \sigma(z_m))) \in \Sigma$

Exercises for Section F

1. Why is it a bad idea to allow names for partial functions in the formal language yet require that every term refer by stipulating that, for example, 'tan($\pi/2$)' is not a well-formed term?
2. Argue either for or against the convention: If $\sigma(t)$ is not defined, then for every sequence of terms $t_1, \ldots, t, \ldots, t_n$, $\sigma(f(t_1, \ldots, t, \ldots, t_n))$ is not defined.
3. Show that (12) is equivalent to the condition on extensionality of functions for models in which there are no partial functions (p. 144).
4. Show that (15) is equivalent to the evaluation of universal quantifiers in models in which the only non-referring names are atomic and also in models in which all names refer.
5. Show that if $\Gamma \vdash A$ in this axiomatization, then $\Gamma \vdash A$ in classical predicate logic with only referring names and total functions.
6. Give a syntactic derivation to show that every instance of axiom scheme 9 (p. 180) is a consequence of the other axioms.
7. Give an example of an area of mathematics in which partial functions are used and say whether all undefined terms are set equal.
8. Show that the set of wffs valid in a model with a finite universe, finite number of names and predicates, and no partial functions is decidable.
9. Show that when function symbols are allowed in the language of quantifying over names, the definition of $\underline{\forall} x\, A(x)$ given in Exercise 14 for Sections A–C need not give a wff that is evaluated as $\forall x\, A(x)$ according to (16).

G. A Mathematical Abstraction of the Semantics

A mathematical model is defined as in Chapter IX.C with the following changes:

1. Assignments of reference are partial functions σ from the set of terms of the language to the universe such that:
   i. For every x, $\sigma(x)$ is defined.
   ii. For every closed term t, one of the following holds:
      a. If $\sigma(t)$ is defined, then for every assignment of references τ, $\tau(t)$ is defined and $\sigma(t) = \tau(t)$.
      b. If $\sigma(t)$ is undefined, then for every τ, $\tau(t)$ is undefined.
   iii. If $\sigma(t)$ is not defined, then for every sequence of terms $t_1, \ldots, t, \ldots, t_n$, $\sigma(f(t_1, \ldots, t, \ldots, t_n))$ is not defined.

   If for all σ, $\sigma(t)$ is not defined, we say t *does not refer*.

2. A set $\mathsf{E} \subseteq \{((\sigma, t), (\sigma, u))$: both $\sigma(t)$ and $\sigma(u)$ are undefined$\}$ is given such that:
   i. For every term t, $((\sigma, t), (\sigma, t)) \in \mathsf{E}$.
   ii. For all terms t and u, $((\sigma, t), (\sigma, u)) \in \mathsf{E}$ iff $((\sigma, u), (\sigma, t)) \in \mathsf{E}$.
   iii. For all terms t, u, and v, if $((\sigma, t), (\sigma, u)) \in \mathsf{E}$ and $((\sigma, u), (\sigma, v)) \in \mathsf{E}$, then $((\sigma, t), (\sigma, v)) \in \mathsf{E}$.
   iv. If for every variable x that appears in either t or u, $\sigma(x) = \tau(x)$, then $((\sigma, t), (\sigma, u)) \in \mathsf{E}$ iff $((\tau, t), (\tau, u)) \in \mathsf{E}$.

3. For every σ and terms t and u, $\sigma \vDash t = u$ iff $\sigma(t)$ and $\sigma(u)$ are both defined and $\sigma(t) = \sigma(u)$ or $((\sigma, t), (\sigma, u)) \in \mathsf{E}$.

4. For each P_i^n there is a set:

   $\mathsf{P}_i^n \subseteq \{(\sigma, P_i^n(t_1, \ldots, t_n))$: $t_1, \ldots, t_n$ is a sequence of terms and σ is an assignment of references$\}$

   such that

   If $(\sigma, P_i^n(t_1, \ldots, t_n)) \in \mathsf{P}_i^n$, and for each j, $\sigma \vDash t_j = u_j$, then $(\sigma, P_i^n(u_1, \ldots, u_n)) \in \mathsf{P}_i^n$.

5. $\sigma \vDash P_i^n(t_1, \ldots, t_n)$ iff $(\sigma, P_i^n(t_1, \ldots, t_n)) \in \mathsf{P}_i^n$.

6. The universal quantifier is evaluated as at (16).

It will never be the case in these semantics that two distinct predicate symbols have the same interpretation. Rather, we can have that *two predicate symbols are interpreted equivalently*: for all terms $u_1, \ldots, u_n$,

$$(\sigma, P_i^n(u_1, \ldots, u_n)) \in \mathsf{P}_i^n \text{ iff } (\sigma, Q_i^n(u_1, \ldots, u_n)) \in \mathsf{Q}_i^n$$

I will leave to you to show that mathematical models satisfy all the conditions for a model given in Section C.3, C.4, and F.2.

XXII The Liar Paradox

A. The Self-Reference Exclusion Principle

The liar paradox in its simplest form is the assertion:

> This sentence is false.

On the face of it, if it is true then it is false, and if it is false then it is true.

In giving semantics for classical predicate logic, we chose to avoid dealing with the liar paradox by adopting the self-reference exclusion principle (Chapter IV.D):

> We exclude from consideration in our logic sentences that contain words or predicates that refer to the syntax or semantics of the language for which we wish to give a formal analysis of truth.

At the time I presented this as a simplifying restriction.

But Alfred Tarski, *1933*, held that any language that contains predicates and words that can be used to formulate a liar paradox will necessarily be inconsistent. In particular, any ordinary language such as English, he argued, is inherently inconsistent, for it contains the predicate '— is true' and the means to name its own sentences.

He held that the self-reference exclusion principle is not too limiting a restriction for our reasoning: Our semi-formal languages built in accord with it are adequate for mathematics and science, in which we can assert a proposition, but not assert that it is true. We can reason about the world, but reasoning about our reasoning must take place in another language. The only consistent resolution of the liar paradox, Tarski held, is to expunge 'true' and other semantic and syntactic predicates from our language: The liar paradox is not a proposition.

Tarski's resolution yields a split between logic and metalogic. We reason about the world in our formal language, L_1, but to reason about that language and its semantics, we need a *meta-language* L_2 that contains a copy of L_1 and predicates and names for the semantic analysis of L_1. In particular, in L_2 we have the predicate '— is a true wff of L_1', but to avoid the liar paradox in our meta-language we cannot have the predicate '— is a true wff of L_2'. We need a meta-meta-language, L_3, to reason about the wffs true in L_2. Continuing on, we can see that Tarski's solution to the liar paradox is not just a split between language and meta-language, but requires an infinite sequence of languages, each using the same inductive definition of truth in a model. There is no formalization of '— is true' in any of the languages, but only of '— is true in a lower language'. We cannot formalize the assumptions we used in developing the semantics for classical predicate logic, such as 'Every proposition is true or false.'

According to Tarski, then, without regimenting the reasoning in this text very carefully according to levels of language, all the material presented here is inconsistent, not meaningless, but certain to lead to contradictions—as is the material in Tarski's 1941 textbook, *Introduction to Logic*.[1]

Saul Kripke, *1975*, argued that Tarski's semantic analysis fails to model important uses of the predicate '— is true' that he says are intuitive and correct. There is nothing paradoxical or wrong in asserting '2 + 2 = 4 and what I have just said is true.' He also points out that Tarski's notion of levels of languages is counterintuitive. Suppose that Dick and Fred say simultaneously:

Dick: What Fred is saying is not true.
Fred: What Dick is saying is true.

These must be assigned different levels of Tarski's hierarchy of languages or we have a paradox. Yet there is no reason why one should be in a lower level than the other.

Kripke offers a resolution of the liar paradox in which the predicate '— is true' is in the formal language. He abandons Tarski's hierarchy of languages, instead using an infinite hierarchy of models. But he abandons more: the liar paradox, according to Kripke, is a proposition but it has no truth-value. The classical laws of logic no longer hold. In particular, the formalization of 'Every proposition is true or false' is not true.

Others have proposed formal systems and semantics that incorporate the predicate '— is true' in the formal language. But each of those in some way denies some law or basic principle of classical predicate logic. You can read a nontechnical analysis of many of those in *Kirkham, 1992*; for technical summaries and comparisons you can consult *Yablo, 1985*, *Hellman, 1985*, or *Burgess, 1986*.

[1] For example, on p. 20 there he says of sentences in our ordinary language, 'When we utter the negation of a sentence, we intend to express the idea that the negation is false; if the sentence is false, its negation is true, while otherwise its negation is false.'

B. Buridan's Resolution of the Liar Paradox

The 14th Century logician Jean Buridan presented in his *Sophismata* a very different resolution of the liar paradox in which the liar paradox is taken to be a proposition, yet all the laws and principles of classical logic hold. In this section I'll describe informally how Buridan's ideas, as presented in translation in *Hughes, 1982*, make sense within the framework of the semantics we have for classical predicate logic. In the next section I'll use these ideas to give a formal theory.[2]

To begin, we must give up some of the simplifications we made in formalizing reasoning. The very first definition we made in this book was:

> A *proposition* is a written or uttered declarative sentence used in such a way that it is true or false, but not both.

In Chapter I.B we simplified our conception of propositions to allow for words to be treated as types. To do this, we had to agree that we would not use indexicals: words such as 'I' or 'now' whose meaning depends on context.

Then we made the much more substantial assumption that propositions can be treated as types:

> In any discussion in which we use logic we will consider a sentence to be a proposition only if any other sentence or phrase that is composed of the same words in the same order can be assumed to have the same properties of concern to logic during that discussion. We therefore identify equiform sentences or phrases and treat them as the same sentence.

But consider:

box **R**→	The sentence in box **R** contains forty-seven letters.	The sentence in box **R** contains forty-seven letters.

The two sentences are equiform: they contain the same words in the same order, with the same punctuation. But semantically they are very different: One refers to itself, one does not. To test the truth of the left-hand one we need to consider whether it itself satisfies the predicate '— contains forty-seven letters', but that's not the case with the right-hand one. There is a clear and perceptible difference that is obscured by identifying them. There is no reason to assume ahead of time that the conditions under which they are true are the same.

Note that these sentences contain no indexical words. Just having the means to create self-reference in the language allows for equiform sentences to have different properties of concern to logic. We abandon the assumption that propositions are types.

---

2 See *Epstein, 1992* for a description of how this work differs from the work of Buridan and others such as Kripke and *Gupta, 1982*.

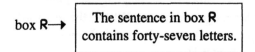

Propositions A proposition is a sentence that is uttered or written at a specific time used in such a way that it is true or false, but not both. Propositions come into existence and cease to exist just as any other objects in the world. Until they come into existence they are not among the objects under discussion.

In what follows I'll use 'uttered' for 'uttered or written'. In Example 2 below I'll discuss how this definition can be used by a platonist.

In Buridan's solution we recognize distinctions we previously chose to ignore. But we also need to make two assumption that, if not implicit in our previous semantic analyses, are certainly compatible with them.

The principle of truth-entailment If A is true, then any subsequent utterance of 'A is true' is true. And conversely, if 'A is true' is true, then A is true.

Truth is not arbitrary If all the facts are in, then there are no arbitrary choices to be made in determining the truth-value of a proposition. In particular, a proposition cannot be true simply because we choose to call it true.

The latter assumption rules out sentences such as 'That's strike three' when uttered by an umpire at a baseball game, sentences that linguists call *performatives*.

I'll describe informally now how truth-values can be assigned to wffs that contain the predicate '— is true'. To make this easier to understand, I'll talk about how we come to know truth-values, but that's just a convenient expository device.

First, 'all the facts are in' can be understood in terms of our previous semantic analysis as meaning that the predicates and names of the language are interpreted in a model. For a wff A of any first-order language and model, I'll call the *material conditions for* or *of* A the result of applying to A the inductive definition of truth in a model, as in the informal analyses in Chapter IV.

Some propositions are uttered. We'd like to know which are true and which are false. We know how to analyze the material conditions of each proposition. And because of the principle of truth-entailment, we know that we must be consistent: If in analyzing A we conclude that A is true, we shouldn't later have to retract and say that A is false.

We survey all the ways we could assign truth-values to the propositions uttered simultaneously with A. For each of those assignments we see if the facts, that is, the material conditions of A, force us after sufficient analysis to assert that A is true rather than false on pain of inconsistency. If we are always forced to assert that A is true, then A must be true and there is nothing arbitrary about that. On the other hand, if for even one of these possible assignments: (a) We would be forced to assert that A is false on pain of inconsistency, or (b) the material conditions of A force us to alternately assert A and to retract our assertion, then A is false. It would be an arbitrary choice to avoid that case, and truth is not arbitrary.

Suppose we've concluded that A is false, yet the material conditions of A would force us to assert that A is true if we begin with that assumption. Then it is not inconsistent to assert that A is false, for the material conditions of A are only part of its truth-conditions. Since A is false, it must be because it is either arbitrary or inconsistent with the facts to assume that A is true.

How does this resolve the liar paradox? Consider the following version:

box W →	The sentence in box W is false.

I'll use the name '*c*' for the sentence in box W. To assume that *c* is true leads to a contradiction, namely, *c* is false. Hence, *c* is false. This doesn't entail in turn that *c* is true, for that would be inconsistent, even though the material conditions of *c* hold. But then what's wrong with the following reasoning?

(1) *c* is false.

That is,

(2) The sentence in box W is false.

So by the principle of truth-entailment:

(3) 'The sentence in box W is false' is true.

Hence,

(4) *c* is true.

This reasoning depends on an equivocation. The step from (2) to (3) is justified, but only if what is named by the quoted words at (3) is the sentence at (2). The step from (3) to (4) is justified only if what is named by the quoted words at (3) is *c*. But (2) and *c* are not the same: (2) is true and *c* is false.

The use of quotation names is a convention designed for sentence types and is inherently ambiguous with respect to equiform utterances. I'll use quotation names in this chapter only when there's little chance of misunderstanding and the examples are more easily understood in that form. We cannot use quotation names in our formal system.

The truth-conditions for propositions can now be summarized.

The truth conditions for a wff A is true iff (i) the material conditions for A hold, and (ii) it is consistent and not arbitrary that the material conditions for A hold.

This definition is compatible with our earlier semantic analyses, for it only recognizes distinctions we previously ignored. For example, consider the two propositions:

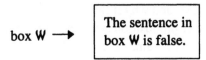

Ralph is a dog.   Ralph is a dog.

These must have the same truth-value in any model. That's because the truth-value of each does not depend on it itself satisfying some predicate. So if we do the usual first-order analysis of the truth-value of one of these, or of any other proposition of classical first-order logic, we are not led back to it again, but rather are led only to truth-conditions corresponding to its parts and eventually to atomic predications. That's why the inductive definition of truth works, and why it fails to deal adequately with the liar paradox. Since any other proposition equiform with these two will be analyzed identically with reference to the same objects and predicates, it will have the same truth-value as they have. Hence, we may identify these two propositions and any other proposition equiform with them; we may harmlessly speak of them as being the same proposition. Or from a platonist perspective, we may say that they pick out the same proposition.

When, then, is 'Ralph is a dog' true? If you were to say that it's exactly when Ralph is a dog, you'd be right. But you wouldn't be giving the truth-conditions for the proposition. Those are: Ralph is a dog, and it is consistent that Ralph is a dog, and there is nothing arbitrary about that. It's just that for this proposition, as for any other that is not self-referential, the latter clauses add nothing. If Ralph is a dog, then it is consistent and not arbitrary that the material conditions of these two propositions hold. If you doubt that, then the inductive definition of truth in a usual first-order classical predicate logic model should be questionable.

Exercises for Sections A and B

1. a. What reasons do we have for wanting to lift the restriction of the self-reference exclusion principle?
   b. What reasons does Tarski give for why we should adopt the self-reference exclusion principle?
2. Show that the following dialogue is paradoxical if both utterances are at the same time:
   Dick: What Fred is saying is not true.
   Fred: What Dick is saying is true.
3. What is comparable to Tarski's hierarchy of languages in the use of second-order logic as an extension of first-order logic?
4. Argue either for or against the principle of truth-entailment.
5. a. Argue either for or against the principle that truth is not arbitary.
   b. Give two examples of performative sentences.
6. Explain how the principle that truth is not arbitrary contributes to the evaluation of the truth-value of wffs.
7. Explain why part (ii) of the truth-conditions for a wff is redundant for wffs that do not contain any parts that can create self-reference.

8. Analyze: 'This sentence is true.'
9. *Kripke, 1975* resolves the liar paradox by saying that we can reason with such self-referential sentences, but they have no truth-value. That is, a wff can be true, false, or have undefined truth-value.
   a. How could you modify the definition of 'proposition' to take account of this view?
   b. Show that on this view the *strengthened liar paradox* is paradoxical: 'This sentence is false or has undefined truth-value.'

C. A Formal Theory

In this section I present semantics for a formal language of classical predicate logic that incorporates the predicate '— is true'.

To begin, a language is not a completed infinite collection of sentences, though it may be harmless to assume so when we invoke the self-reference exclusion principle. Rather, a language is a vocabulary and a way to generate sentences from the vocabulary. For our formal languages, the method of generating sentences is given by the inductive definition of 'wff' and 'closed wff'.

Let L be the usual language of first-order logic:

$$\mathrm{L} = \mathrm{L}(\neg, \rightarrow, \forall, \exists, =, \mathrm{P}_0, \mathrm{P}_1, \ldots, \mathrm{f}_0, \mathrm{f}_1, \ldots, \mathrm{c}_0, \mathrm{c}_1, \ldots)$$

Assume we have a realization of the predicate symbols, function symbols, and names, along with a universe and a model M for that realization. These will stay fixed throughout the following discussion.[3]

Add to the language L a unary predicate symbol 'T', with the usual formation rules for wffs. This is meant to be interpreted as '— is true' in all models of the language, and hence is syncategorematic.

In order to formalize 'Every proposition . . .' and 'Some proposition . . .' we could add another unary predicate 'P' to be interpreted as '— is a proposition', expanding the universe of M to include closed wffs of our expanded language. But that complicates the definition of truth in a model too much, for '$\forall x$ (x is a dog)' could be true in M but not true in the expanded universe, since x could take a proposition as reference. It is much easier to expand the language of L to include a second sort of variables, as discussed in Chapter XIV.H. The usual variables x_0, $x_1, \ldots$ will continue to be used to range over the universe of M, and the variables

[3] It was suggested to me that it's necessary to give a theory of strings for tokens—that is, specific instances of types—prior to discussing the formal language, as *Tarski ,1933* does for types. But that assumes informal mathematics is prior to and justifies formal logic, which I do not think is right (see Chapter II.F of *Propositional Logics* for a discussion). In any case, *A. A. Markov, 1954*, gives a mathematical analysis of algorithms on concrete letters, alphabets, and words that is adequate for the formal theory I present here. See Chapter 2.5 of *Kirkham, 1992* for answers to other objections to the use of tokens as truth-bearers.

$x^{\mathrm{p}}_0, x^{\mathrm{p}}_1, \ldots$ are meant to range over the propositions, that is, closed wffs of this language. Let L_T be this expanded language with 'T' and new variables.

We also have to allow for names of wffs or parts of wffs. But to do that we need to consider the order in which the propositions of the semi-formal language are uttered. That order is given; it is part of the world. Given our capacities for generating propositions, we can assume the order in which they are uttered is discrete, so we can speak of the propositions at stage *n*. Any sentences of L_T can be uttered as propositions at stage *n*, and we call the collection of them $\mathbf{S}_n$. But we make some restrictions on names and naming.

Naming wffs At stage *n* or any later stage we may name any of the sentences so far uttered and use those names in forming new propositions. In particular, those names may be used to form propositions in $\mathbf{S}_n$. Thus, we can expand at this stage the list of names in L_T.

All names refer. In particular, if $A \in \mathbf{S}_n$ and '*b*' is a name of a proposition and '*b*' appears in A, then *b* has already been uttered or is uttered at this stage.

The next assumption allows us to treat those cases where the sentences uttered at a particular time contain interlocking references and hence must be analyzed simultaneously. Though not essential, it will resolve some otherwise puzzling situations (Example 2 below).

Interlocking references We impose a logical ordering on the sentences (closed wffs) uttered at stage *n* by dividing them into two parts. In the second part goes any proposition B such that there is another proposition A and:

a. The method of analysis of the truth-value of A described below gives the same truth-value regardless of whether B had been uttered or not.

b. The truth-value of B in the analysis below (or even whether B is a proposition) depends on whether A has been uttered and analyzed before B.

As a matter of convenience, rather than talking about the first part of stage *n* and the second part of stage *n*, I'll assume that by renumbering the stages these are two successive stages.

Now I'll describe the method of evaluating the truth-values of wffs in $\mathbf{S}_n$.

The truth-values of the propositions in $\mathbf{S}_n$

Each proposition on being uttered is true or false, and its truth-value never changes. But how we come to see what its truth-value is can best be described in substages by assigning hypothetical truth-values according to the rules below.

Let $\mathsf{M}_0 = \mathsf{M}$. For $n \geq 1$, let M_{n-1} be the model for L plus all wffs of $\bigcup_{m<n} \mathsf{S}_m$. That is, M_{n-1} is M plus the interpretation T_{n-1} of 'T', the collection of true propositions of $\bigcup_{m<n} \mathsf{S}_m$. Denote by F_{n-1} the collection of false propositions in $\bigcup_{m<n} \mathsf{S}_m$. The phrase the *given facts* now refers to M_{n-1}.

The variables $x^{\mathrm{P}}_0, x^{\mathrm{P}}_1, \ldots$ will now range over $\bigcup_{m<n} \mathsf{S}_n$.

Informally, we first make an hypothesis about which wffs of S_n are true. We collect those and the wffs in T_{n-1} and label them as $\mathsf{T}_n(0)$. Then we consider the model M augmented by interpreting 'T' as $\mathsf{T}_n(0)$ according to the usual semantics for first-order models. The set of sentences true in that model is $\mathsf{T}_n(1)$. Now we use $\mathsf{T}_n(1)$ to interpret 'T' to obtain another first-order model. This process of taking the semantic consequences of the hypothetical truth-values $\mathsf{T}_n(k)$ can be continued indefinitely. If the facts are such that A really is true, then eventually we will be forced to recognize this, for A will stabilize as being in $\mathsf{T}_n(k)$ for all large k. On the other hand, we may have one of the following:

For all large k, A is not in $\mathsf{T}_n(k)$.

For arbitrarily large j and k, A is in $\mathsf{T}_n(k)$ and not in $\mathsf{T}_n(j)$.

If either of these should be the case on even one assumption about which propositions of S_n are true, that is, for even one choice of $\mathsf{T}_n(0)$, then A is false. That's because it is not necessary, relative to the facts, that A be true. A is true if and only if for each such analysis A will be in $\mathsf{T}_n(k)$ from some point onward.

Formally, for $n \geq 1$, suppose $\mathsf{S}_n = \{A_1, \ldots, A_m\}$. Let $\alpha_1, \ldots, \alpha_{2^m}$ be a list of all subsets of S_n. For *each* α_i we have the following *substage analysis*.

Substage 0:

If $n = 0$, then $\mathsf{T}_n(\alpha_i, 0) = \alpha_i$

$$\mathsf{F}_n(\alpha_i, 0) = (\mathsf{S}_n - \alpha_i).$$

If $n > 0$, then $\mathsf{T}_n(\alpha_i, 0) = \mathsf{T}_{n-1} \cup \alpha_i$

$$\mathsf{F}_n(\alpha_i, 0) = \mathsf{F}_{n-1} \cup (\mathsf{S}_n - \alpha_i).$$

Substage $k + 1$:

$\mathsf{T}_n(\alpha_i, k+1) = \mathsf{T}_{n-1} \cup \{A \in \mathsf{S}_n$: using the usual inductive definition of truth in a first-order model, A is true in the model M where the variables $x^{\mathrm{P}}_0, x^{\mathrm{P}}_1, \ldots$ range over all propositions of $\bigcup_{m \leq n} \mathsf{S}_m$, interpreting T as $\mathsf{T}_n(\alpha_i, k)\}$.

$$\mathsf{F}_n(\alpha_i, k+1) = \mathsf{F}_{n-1} \cup \{A \in \mathsf{S}_n\colon A \notin \mathsf{T}_n(\alpha_i, k+1)\}.$$

We define the set of propositions *true under the hypothesis* α_i, and those *false under the hypothesis* α_i:

$\mathsf{T}_n(\alpha_i) = \mathsf{T}_{n-1} \cup \{A \in \mathsf{S}_n\text{: there is some } m \text{ such that for all } k \geq m,\ A \in \mathsf{T}_n(\alpha_i,\ k)\}$

$\mathsf{F}_n(\alpha_i) = \mathsf{F}_{n-1} \cup \{A \in \mathsf{S}_n\text{: } A \notin \mathsf{T}_n(\alpha_i)\}$

The propositions in $\bigcup_{m \leq n} \mathsf{S}_m$ that are *true* are those that are true under any assumption about the truth-values of the propositions in S_n. That is, they are the propositions that are in $\mathsf{T}_n(\alpha_i)$ for every α_i.

$$\mathsf{T}_n \ = \ \bigcap_i \mathsf{T}_n(\alpha_i)$$

The *false* propositions are those that are not true:

$$\mathsf{F}_n \ = \ \{A \in \bigcup_{m \leq n} \mathsf{S}_m\text{: } A \notin \mathsf{T}_n\}$$

This completes the definition of M_n for $n \geq 1$.

We can view our language as the collection of wffs whenever we terminate our analysis, that is, $\bigcup_{m \leq n} \mathsf{S}_m$ for some n. Or we may think of L_T as what we would get if "we went on forever". With respect to the given facts, the collection of *true propositions* is $\mathsf{T} = \bigcup_n \mathsf{T}_n$, and the *false propositions* are those in $\mathsf{F} = \bigcup_n \mathsf{F}_n$.

Since we're assuming a fixed model M, I'll say simply 'true' or 'false' now. Sometimes when the context makes clear that I'm talking about a substage analysis I'll say that A is true (false) when I mean it is hypothetically true (false).

Note that for every substage analysis, at every substage k, $\mathsf{T}_{n-1} \subseteq \mathsf{T}_n(\alpha_i,\ k)$ and $\mathsf{F}_{n-1} \subseteq \mathsf{F}_n(\alpha_i,\ k)$. The truth-values of the propositions in $\bigcup_{m < n} \mathsf{S}_m$ are not affected by any newly uttered propositions.

Theorem 1

a. ***Tertium non datur*** Every proposition is true or false but not both.

b. If A is an instance of a classical tautology, then A is true.

c. If A is an instance of a classical contradiction, then A is false.

Proof: Part (a) I'll leave to you. For (b), let $A \in \mathsf{S}_n$. If A is an instance of a classical tautology, then for any choice of $\mathsf{T}_n(0)$, $A \in \mathsf{T}_n(k)$ for all $k \geq 1$, so A is true. Part (c) then follows similarly. ■

Theorem 2 ***Truth-entailment*** If A is true and '*a*' is a name of A, and B is a proposition of the form T(*a*), then B is true. Conversely, if B is true then A is true.

Proof: I'll leave to you the case when $A \in \mathsf{T}_n$ and $B \in \mathsf{S}_k$ for some $k > n$. The assumption on interlocking references rules out the case where A and B are both in S_n unless B is A itself. In that case A is false, as I'll show in Example 3 below. ■

Theorem 2 would hold even were we not to make the assumption on interlocking references: If A and B are both in S_n, then for some m, all $k \geq m$, $A \in \mathsf{T}_n(k)$, hence all for all $k \geq m + 1$, $B \in \mathsf{T}_n(k)$. So $B \in \mathsf{T}_n$, too.

Theorem 3 If there are exactly m propositions in S_n, the truth-value of every wff in S_n can be determined by making at most $2^m(2^m + 1)$ substage analyses.

Proof: Let $\mathsf{S}_n = \{A_1, \ldots, A_m\}$. The truth-value of A at any substage of any α_i-analysis depends solely on which predicates apply to which objects in M and which of $A_1, \ldots, A_m$ are true at the previous substage. So the truth-values of $A_1, \ldots, A_m$ at any substage are completely determined by the truth-values assigned at the preceding substage. Thus, if any combination of truth-values for $A_1, \ldots, A_m$ appears at a substage k and again at substage $k + r$, then substages $k, k+1, \ldots, k+r-1$ are repeated forever in that sequence, as nothing additional can enter into the calculations. There are 2^m different possible combinations of truth-values for $A_1, \ldots, A_m$, so by substage $2^m - 1$ (recall we start at substage 0) every combination that can appear in a particular analysis will have appeared, and at substage 2^m some combination of values must be repeated.

Thus, to determine the truth-value of any $A \in \mathsf{S}_n$ relative to any particular α_i, let $k < s \le 2^m$ be the first substages such that the values assigned to $A_1, \ldots, A_m$ at substage k are the same as at substage s. If for all substages $k, \ldots, s$, A is evaluated as true, then A is true under the hypothesis α_i. If at some substage between k and s A is evaluated as false, then A is false under the hypothesis α_i. There are 2^m possible choices for α_i, and for each we need consider at most $2^m + 1$ substages. Hence, to determine the truth-value of A, we need consider at most $2^m(2^m + 1)$ substage analyses. ∎

Theorem 3 shows that relative to the model M there is a constructive procedure for determining the truth-values of all atomic predications . No infinite procedure must be accomplished before the truth-value of an atomic proposition can be decided.

Exercises for Section C

1. What restrictions do we place on the naming of the wffs in our (semi-) formal language?
2. Describe in your own words how to determine the truth-value of a proposition that is in L_T but not in L.
3. Show Theorem 2 when $A \in \mathsf{T}_n$ and $B \in \mathsf{S}_k$ for some $k \ge n$.
4. Show that the bound in Theorem 3 can be improved. That is, if m is the number of wffs uttered in S_n, find some function $t(m) < 2^m(2^m + 1)$ such that the truth-value of every wff in S_n can be determined by making at most $t(m)$ substage analyses.
(Hint: If in the substage analysis of $\mathsf{T}_n(\alpha_i)$, $\mathsf{T}_n(\alpha_i, k) = \alpha_j$ for some $j \ge i$, then $\mathsf{T}(\alpha_i)$ need not be calculated separately.)

D. Examples

The examples I'll present here are meant to test the aptness of the formal theory to model the informal resolution of the liar paradox and our conception of truth. In most of them if a proposition is false I'll exhibit only one substage analysis to show that. To indicate that '*b*' is the name of a proposition B, I'll write $\underset{b}{\overset{\mathrm{B}}{|}}$.

1. This sentence is false.

$$\underset{a}{\overset{\neg \mathrm{T}(a)}{|}}$$

Analysis The example is false. Suppose it occurs at stage $n \geq 0$. Suppose that it's false, that is, in $\mathsf{F}_n(0)$ (where here and in succeeding examples I'll suppress the index 'α_i'). Then at substage 1, $a \in \mathsf{T}_n(1)$, as *a* asserts that *a* does not satisfy the predicate that interprets T, and we are interpreting T as $\mathsf{T}_n(0)$. That is, if we assume *a* to be false, then *a* is true. In the same way, $a \in \mathsf{F}_n(2)$; if we assume *a* to be true, then it is false. This vacillation continues. So it is inconsistent to take *a* to be true, and hence *a* is false.

Since *a* is false, its material conditions hold. They do not hold "after the fact", that is, after our analysis, but rather the moment that *a* is uttered. At that time, perhaps, we didn't know that and felt we had recourse to our substage analysis because there is no "fact of the matter" as to *a*'s truth to begin with. But that confuses the way we come to know that *a* is false, and explicate it, with whether *a* is false.

Our substage method allows us to conclude that *a* can't be true. It does not, however, lead to a contradiction to take *a* to be false. The substage method only shows whether it is consistent for the material conditions of *a* to hold. For *a* to be true its material conditions must hold, but also it must be consistent and not arbitrary that they hold. For *a* to be false its material conditions may hold, yet one of the other clauses of the truth-conditions for *a* fails. In contrast to the principle of truth-entailment:

If 'A is false' is not true, it need not be the case that A is true.

If 'A is false' is inconsistent, it need not be the case that A is true.

2. This sentence is false. 'This sentence is false' is false.

$$\underset{a}{\overset{\neg \mathrm{T}(a)}{|}} \qquad\qquad \underset{b}{\overset{\neg \mathrm{T}(a)}{|}}$$

Analysis In the previous example we saw that *a* is false. Hence *b* is true, as it asserts that *a* is false. Formally, *b* is analyzed at a later stage than *a* by the assumption on interlocking references, as it is possible to analyze *a* without analyzing *b*, but not vice versa. Since *a* is false it will be false at every stage and substage at which *b* is analyzed, hence *b* will be true. Here is a case where the assumption on interlocking references matters, for if these propositions were to be analyzed simultaneously, we'd have that *b* is false, as it would vacillate between being in $\mathsf{T}_n(k)$ and in $\mathsf{F}_n(k)$ in alternation with *a*. Yet *a* is false: No alternation of *b*'s truth-value in the substages arises once we recognize that, that is, incorporate it into our model.

Here we have equiform closed formal wffs that, even if uttered at the same time (though analyzed sequentially), have opposite truth-values. No indexicals appear in them.

This example also illustrates how to form the contradictory of a sentence. Given a sentence *c* we know that if we form *another* proposition $\neg\mathrm{T}(c)$ it will have the opposite truth-value to that of *c*, even if it is equiform with *c*. On the other hand, simply negating a false proposition may yield one that is also false: the proposition *a*, which is $\neg\,\mathrm{T}(a)$, is false, yet $\mathrm{T}(a)$ and $\neg\neg\,\mathrm{T}(a)$ are both false, as you can show.

So if you want to say that the proposition over there is wrong, not true, you'll have to say that. Don't say 'not A', at least if A has 'true' in it or has names that via some chain of references lead to a proposition with 'true' in it, for only if the material conditions suffice for the consistency conditions does the use of 'not' have the same effect as 'it's not true that', which negates all the truth-conditions.

For a platonist who believes that abstract propositions are what are true or false, the two sentences of our example represent or point to different abstract propositions. Just because two sentence utterances or inscriptions are equiform, even in a formal language, does not entail that they represent the same abstract proposition. If abstract propositions are to be defended as anything more than fictions, then their nature and the nature of our language must determine when two sentences refer to the same abstract proposition. Since the relationship between abstract propositions and our language is so unclear, I've chosen to deal with what we have: sentences, as utterances or inscriptions.[4]

It may be possible to recast much of this theory for sentence types, but each sentence type would have to be fully indexed. Any aspect of sentence utterances that can differ for distinct uterances within a sentence type could, I suspect, be used to reformulate the paradoxes. The only way I see to ensure that there is no (logically) discernible difference among sentences in a type is if there is only one in each.

Platonists, however, argue that we should not take utterances as the bearers of truth because we cannot give unambiguous answers to the questions such as: What is a sentence? What constitutes a use of a sentence? When has one been used as an

---

[4] *Gupta, 1982*, p.4, argues that we need abstract propositions because there are unexpressed truths just as there are unnamed objects. But I know there are unnamed objects because I can see and feel them; I do not know any unexpressed truths until they are expressed, barring mysticism from the scope of logic.

assertion or even put forward for discussion? They avoid these questions by taking things inflexible, rigid, timeless as propositions. But then in order to reason, we have to answer the questions: What is the relation of these formal theories of mathematical symbols to our arguments, discussions, and search for truth? How can we tell if this utterance is an instance of that abstract proposition? How do we use logic? The emphasis on formalism and mathematics in logic has gone too far if we cannot relate our work to its intended use, its original motivation. Logic is not a formal mathematical theory; it is, at its best, formal theories used to explain difficult problems about how to reason correctly.

How we characterize the use of a sentence has a simple, though not rigid answer. We must have agreement to do logic, we must speak the same language, be able and willing to follow each other's arguments when uttered or written. And in any particular instance, we must agree whether a sentence that has been put forward has a truth-value. If we cannot agree on a certain case then we cannot reason together with it. That does not mean that we then do different logics, or that logic is psychological; it means only that we differ on borderline cases. I take as a primitive notion the utterances of sentences that may be assigned truth-values. That notion may be further explicated, and has been done so by others. But agreement by cases, or in general as with a first-order language, is all that we need to reason together.

3. This sentence is true.

$$\begin{array}{c} T(a) \\ | \\ a \end{array}$$

Analysis The example is called the *truth-teller*, and it is false. At whatever stage a is analyzed we may begin by assuming it false. Then at all further substages it will be false. So it is false.

We could have begun by assuming that a is true, and then it would have been true at all subsequent substages. But that does not show that a is true, for it would have been arbitrary to restrict ourselves to that assumption in our substage analysis. We can't make a proposition true just by calling it so. Relative to our linguistic conventions and the facts of the matter, there is nothing arbitrary about a proposition being true.

The example is not false by an arbitrary choice, though. It is false because it fails to be true. The false propositions are those that are not true, those whose material conditions fail, or which are contradictory, or which are arbitrary. The odd, anomalous, paradoxical sentences, ones that we cannot use as a basis for good reasoning, are all false: *tertium non datur*. In *Propositional Logics* I show that both mathematically and philosophically almost all the propositional logics that have been proposed as models of reasoning can be understood as taking falsity as the default truth-value. The only logics I know that can be understood as taking truth as the default truth-value are paraconsistent logics, in which a non-trivial theory may contain both a proposition and its negation (see Chapter IX of *Propositional Logics*). According to the motivations of those systems, we should set up a dual to the formal system of Section C by requiring that it not be arbitrary that A is false, calling a proposition true if it is not false: Truth would be the default truth-value. We would have:

$$\mathsf{F}_n(\alpha_i) = \mathsf{F}_{n-1} \cup \{A \in \mathsf{S}_n : \text{there is some } m \text{ such that for all } k \geq m,\ A \in \mathsf{F}_n(\alpha_i, k)\}$$

$$\mathsf{T}_n(\alpha_i) = \mathsf{T}_{n-1} \cup \{A \in \mathsf{S}_n : A \notin \mathsf{F}_n(\alpha_i)\}$$

4. Dick says, 'What Fred is saying is false.' Fred says, 'What Dick is saying is false.' And that is all that either of them says.

$$\underset{a}{\neg T(b)} \qquad \underset{b}{\neg T(a)}$$

Analysis Both propositions are false. If they are analyzed at stage n, begin by assuming them both false. Then at the first substage they are both (hypothetically) true, and hence at the next substage false, etc.

5. Dick says, 'What I am saying is true, and what Fred is saying is true if and only if what Fred is saying is true.' Fred says, 'What Dick is saying is false.' And that is all that either says.

$$\underset{a}{T(a) \wedge (T(b) \leftrightarrow T(b))} \qquad \underset{b}{\neg T(a)}$$

Analysis The proposition a is the truth-teller in a new guise, but it must be analyzed simultaneously with b.

If we begin by assuming that a and b are both false, then at substage 1, a is false and b is true, and at every subsequent substage a is false and b is true. So a is false. But we may not therefore conclude that b is true, for the analysis of b is not independent of that of a. If we begin an analysis by assuming both a and b are true, then at all subsequent substages a is true and b is false. So b is not necessarily true. Therefore, b is false.[5]

---

[5] In the first presentation of this formal system I used only one substage analysis for S_n in which every proposition in S_n was initially assumed to be false, reasoning that for A to be true its truth must be forced on us even if we assume the contrary. The condition for the truth of A then is 'A is not true' is inconsistent, rather than 'A is consistent'. With that analysis every example here comes out the same except for the present one, in which we would have that a is false and b is true. This may seem more intuitive, but only at the cost of ignoring that a and b must be analyzed simultaneously. To use only the one substage analysis now seems to me to involve an arbitrary assumption about the possible truth-values of the propositions under discussion.

6. **Dick says, 'What Fred is saying is false.' Fred says, 'What Walter is saying is false.' Walter says, 'What Dick is saying is false or Ralph is a dog.' And that is all that any of them says.**

$$\begin{array}{ccc} \neg T(b) & \neg T(c) & \neg T(a) \vee D(r) \\ | & | & | \\ a & b & c \end{array}$$

Analysis Suppose we analyze these at stage n. If at substage 0 we begin by assuming them false, then at substage 1 each is true. At this point we have two cases depending on whether Ralph is a dog. Suppose Ralph is not a dog. Then since the material condition of the second disjunct is never satisfied, c will be validated at a substage if and only if a is not true. So at any substage all of a, b, and c will have the same hypothetical truth-value, which will alternate from substage to substage. So they are all false.

The analysis is more interesting if Ralph is a dog. Regardless of what we take for $\mathsf{T}_n(0)$, at all substages $k \geq 1$, $c \in \mathsf{T}_n(k)$. So $b \in \mathsf{F}_n(k)$ at all substages $k \geq 2$, and so at all substages $k \geq 3$, $a \in \mathsf{T}_n(k)$, though $a \in \mathsf{F}_n(2)$. So a and c are true while b is false. Note that b is false regardless of whether Ralph is a dog, so b is necessarily false, though a and c are contingent.

Here we have an example of a proposition, a, that is true but does not settle down to appear so until the third substage (compare Exercise 9 below).

7. **Every proposition is true or false.**

$$\forall x^{P}_{1}\ T(x^{P}_{1}) \vee \neg T(x^{P}_{1})$$

Analysis The example is true, as we showed in Theorem 1. But if it is uttered at stage n, then the propositions it refers to are those in $\bigcup_{m \leq n} \mathsf{S}_m$. If an equiform proposition is uttered at a later stage, it will refer to a larger class of propositions, and that will be true, too. The proposition is a necessary truth, it is true no matter when uttered.

Doesn't this example show the need for abstract propositions? It seems there is no sentence that we can use to assert that every proposition, no matter when uttered, is true or false, that is, no sentence that timelessly captures the principle of *tertium non datur*. But our discussion has already pointed to such a sentence:

'Every proposition is true or false' is a necessary truth.

8. **This sentence is true or false.**

$$\begin{array}{c} T(a) \vee \neg T(a) \\ | \\ a \end{array}$$

Analysis As a substitution instance of the previous example (a notion that applies to sentence utterances as well as sentence types, since we're assuming that words are types), this must be true. And it is: At whatever stage it's uttered, whether it is first assumed to be false or true, at all further substages it is evaluated as true.

9. The first disjunct of this sentence is true or the first disjunct of this sentence is false.

$$\mathrm{T}(a) \vee \neg \mathrm{T}(a)$$
$$|$$
$$a$$

Analysis A part of a proposition is not a proposition, though it may be equiform with a proposition. So as only propositions are true or false, no part of a proposition can be true or false. Thus, the material conditions for each disjunct fail, and the example is false. This does not contradict that every instance of Example 7 is true, since we can substitute only names of propositions for the variables x^{P}_{i}.

But if a part of a proposition is not a proposition, how can we use the inductive definition of truth in the model M? Consider:

(5) 'Ralph is a dog $\vee$ Howie is a cat' is true
iff 'Ralph is a dog' is true or 'Howie is a cat' is true
iff Ralph is a dog or Howie is a cat.

We don't need to assume in (5) that 'Ralph is a dog' and 'Howie is a cat' are propositions and have truth-values, where the quoted words are meant to name the parts of the original proposition. Rather, (5) is a (usually) harmless way of abbreviating the correct analysis. Let '*d*' name: Ralph is a dog $\vee$ Howie is a cat.

d is true iff the material conditions for *d* hold and
(*) it is consistent and not arbitrary that they hold
iff (the material conditions for the first disjunct of *d* hold) or
(the material conditions for the second disjunct of *d* hold)
and (*)
iff Ralph is a dog or Howie is a cat and (*)

If we ignore (*), which is redundant for non-self-referential propositions, then we have the usual inductive definition of truth. But the way we have come to it is by the more cautious route, forced on us by our work with self-referential sentences, that talks of the material conditions of the parts of a proposition. If we wish to deal only with non-self-referential sentences, as when we invoke the self-reference exclusion principle, then it is perfectly harmless to identify the material conditions of a proposition with its truth-conditions and to say that parts of a proposition are true or false. But when we reason with self-referential sentences, the question of their existence is sometimes crucial to the analysis of their truth-value, and parts of a proposition do not exist as completed utterances in their own right the way a proposition does.

If parts of propositions were to be considered as propositions, we would get an anomaly. Call the second disjunct b, and the entire proposition c.

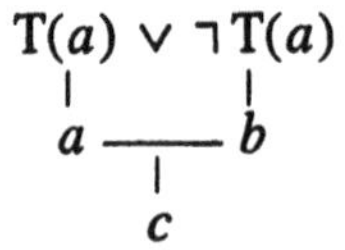

Suppose c is analyzed at stage n. If we assume that a is true, then at all subsequent substages it will be true and b will be false, and c will be true. If we assume that a is false, then at all subsequent substages it will be false and b will be true, and c will be true. Thus we would have that a is false, b is false, and c is true. That has to be wrong, since it means that c is true even though its material conditions fail. Alternatively, if we were to analyze the truth-value of c after a and b, we would have that c is false, and the laws of classical logic would no longer hold.

10. No proposition has the word 'true' in it.

$$\neg \exists x^{P_1}\, W(x^{P_1})$$

Analysis *Gupta, 1982*, has shown how we may include predicates such as '— has the word 'true' in it' in our usual models of first-order logic; the self-reference they engender is apparently harmless. So suppose that '— has the word 'true' in it' is in our model M and our example is in language L. Suppose also that Example 10 is uttered at stage 0. Then it is true. Suppose that at the next stage the following proposition is uttered: T(b), where b names Example 10. And then a proposition c equiform with Example 10 is uttered at stage 2. Then c is false. It is not that Example 10 has changed its truth-value; it is still false, for it refers to the time of its utterance, namely, stage 0. If you like, all propositions are indexed by their time of utterance. At stage 2, c is false because the world has changed.

Example 10 is true or false depending on its time of utterance. Its truth-value depends on what sentences have already been uttered (at an earlier stage or the same stage). Sentences are part of the world, and the order in which they are uttered matters, just as it matters whether Plato died after Socrates for the truth-value of 'Plato wrote a description of the death of Socrates'.

Self-reference is never really harmless if utterances are taken as the bearers of truth-values. For that reason, if we wish to add predicates referring to the syntax or semantics of the language, we should add them as we do the predicate '— is true'.

It might seem that by considering propositions only in the order in which they are uttered we have introduced a hierarchy into this analysis. But there is only one language here, which can be extended by ostensive naming, and one set of rules for forming sentences. English is just one language, yet new sentences appear all the time. Nor is the substage analysis a hierarchy of models. Propositions are true or false when uttered, and the finite number of hypothetical models used to analyze that

(Theorem 3) are no more a hierarchy than are the rows of a truth-table used to analyze a proposition in classical propositional logic.

11. No proposition is negative.

$$\neg \exists x^{p}_{1}\ \mathrm{N}(x^{p}_{1})$$

Analysis Let's assume that, as discussed in Example 10, we've added the predicate symbol 'N' to our formal language to be interpreted as '— is negative'. Then Example 11 is false no matter when uttered, as it itself must satisfy the predicate '— is negative'. Hence it's impossible that it be true; it is necessarily false. This should not be confused with the claim that things may be as the proposition asserts them to be, that is, the material conditions for the proposition could hold: It could have been that no language-using creatures evolved. But any way in which the material conditions of the example hold, the example itself would not exist.

We make a distinction. A proposition is *possible* means that there is a way its material truth-conditions could be satisfied. Example 11 is possible. A proposition is *possibly true* means that it could be true, and for a proposition to be true, it must exist. Example 11 is not possibly true.

Still, when we want to analyze some proposition of Aristotle for the first time, don't we say that it's true or false even though it doesn't now exist? We analyze propositions that we believe have the same truth-conditions as Aristotle's: those that are equiform with his purported utterances or inscriptions, or translations of those. So long as they involve no self-reference it is harmless to make the claim that they are Aristotle's, understanding by this that we are making an identification.

12. Plato was Athenian entails that 'Plato was Athenian' is true.

Analysis The ambiguity of quotation names requires us to make this example more precise. Let's take a formal version:

$$\underset{\substack{| \\ a}}{\mathrm{A}(p)} \quad \text{therefore} \quad \mathrm{T}(a)$$

The inference is not valid. Plato was Athenian, yet it is possible for a not to exist. For a sentence to be true, it must exist; for the material conditions for $\mathrm{T}(a)$ to hold, a must exist. Something has to have been uttered.

Example 12 is not an instance of the principle of truth-entailment, which says that if A is true, then 'A is true' is true. For A to be true, it must exist.

In showing that this inference was not valid, I described a way the world could be in which the material conditions of the premise hold, but that of the conclusion fail. That is,

(6) An inference is valid iff it is not possible for the material conditions of the premises to hold and the conclusion fail.

Compare that to the usual definition of validity that we use for non-self-referential propositions:

(7) An inference is valid iff it is not possible for the premises to be true and the conclusion false.

According to (7), Example 12 would be valid, since for the premise to be true, it must exist. But (7) would invalidate:

(8) Every proposition is affirmative.
Therefore, no proposition is negative.

Yet (8) should be valid, since if the world is as the premise describes it, the world is certainly as the conclusion describes it. But if the world is as the premise describes, that is, the material conditions of the premise hold, then the conclusion is not true, for it does not exist. It is (6) that we want for validity.

13. No proposition has the word 'Juney' in it.
If no proposition has the word 'Juney' in it, then Juney is not Juney.
***Therefore*, Juney is not Juney.**

$\neg \exists x^{P}_{1}\, J(x^{P}_{1})$

$\neg \exists x^{P}_{1}\, J(x^{P}_{1}) \rightarrow j \neq j$

$j \neq j$

Analysis Suppose we reason: The material conditions for the first premise could hold. If the second premise is then uttered, its material conditions would hold, since its antecedent would fail. Yet the material conditions for the conclusion never hold. So the inference is invalid.

That reasoning is wrong. The material conditions for the second premise can hold only if its antecedent fails. In that case, the material conditions of the first premise fail; the example is valid. The correct definition of validity requires a phrase that was always implicit in our applications of it.

Valid inferences An inference is valid iff it is not possible for the material conditions of the premises to hold and the conclusion to fail *in the same way and at the same time.*

There are, then, two different versions of *modus ponens*:

A, A→B, therefore B.

If A is true and A→B is true, then B is true.

The first is valid according to our definition, as I'll let you show. It is not clear that the second is valid, though I have not been able to formulate a counterexample.

Exercises for Section D

1. Give an example of two formal wffs that are equiform yet have opposite truth-values. Does any indexical part of speech (formal or informal) appear in either of them?
2. Show that for any proposition c, $\neg T(c)$ has the opposite truth-value from c.
3. Show that for Example 2, $T(a)$ and $\neg\neg T(a)$ are both false.
4. a. Explain why the assumption that propositions are types is a simplification rather than a necessary part of the development of the syntax and semantics of classical predicate logic.
   b. Why is it appropriate to make the assumption that propositions are types for the standard language of classical predicate logic?
   c. Why does taking the bearers of truth-values to be abstract propositions not avoid the difficulties in answering questions such as: What is a sentence? What constitutes a use of a sentence?
5. Why is the truth-teller false?
6. a. Why is $\{S_n : n \geq 0\}$ not a hierarchy of languages?
   b. Why are the substage analyses not a hierarchy of models?
7. a. Show that the following is a valid inference: A, A→B, therefore B.
   b. Show that every inference schema that is classified as valid in first-order classical predicate logic is still valid in the formal system of Section C according to the definition of validity given in Example 13.
8. Formalize and determine the truth-values of the following:
   a. This sentence is false or God exists.
   b. If this sentence is true, then Ralph is a dog.
   c. If a proposition is true, its negation is false.
   d. No proposition is both true and false.
   e. Every proposition equiform with this one is false.
9. Show that for any natural number $m \geq 1$ we can find a proposition that is true but in some substage analysis does not settle down until at least stage m. (Hint: Modify Example 6 to have m propositions.)
10. Argue either for or against: Saying that a proposition that Aristotle uttered is true or false is like saying that 'Socrates is a man' is true even though Socrates does not now exist.
11. What examples in this section would have a different truth-value if we took truth to be the default truth-value in our formal system, as discussed in Example 3?

E. One Language for Logic?

In the writings of Buridan there is no distinction between logic and metalogic. One language and system of reasoning is to suffice. Nor did Frege or Russell and Whitehead employ a distinction between logic and metalogic. Their goal was to create one precise perspicuous language for reasoning in which logical relations among propositions would be evident.

The self-referential paradoxes seemed to make that goal unattainable. Frege's system was inconsistent; Russell and Whitehead had recourse to a stratified language that is in essence a hierarchy. Since Tarski's analysis of truth, most modern logicians have taken metalogic to be distinct from logic.

A satisfactory analysis of self-referential paradoxes in a language with its own truth-predicate suggests that a re-incorporation of logic and metalogic is possible. I present some examples here that set out problems and directions for that project.

14. No proposition equiform with this proposition is true.

$$\neg\exists x^{p}_{1}\ (E(x^{p}_{1}, a) \underset{\overset{|}{a}}{\wedge} T(x^{p}_{1}))$$

Analysis The example is false, for to assume it is true leads to a contradiction. By the same reasoning, any wff equiform with this formal wff is also false.

But then how are we to say that all those wffs are false? We use a proposition that is not equiform with the example. For example, the following is true:

$$\neg\exists x^{p}_{1}\ (T(x^{p}_{1}) \underset{\overset{|}{c}}{\wedge} E(x^{p}_{1}, a))$$

But in classical logic a and c are equivalent. Adding the predicate '— is equiform with —' to our language will complicate our analysis of what we mean by two propositions being equivalent.

15. The material conditions of this proposition do not hold.

Analysis Informally, the proposition is false, since to assume it true would lead to an inconsistency.

But then do the material conditions of the example hold? That depends in part on how we extend the formalization of our principles when more than one new semantic predicate is added to a usual first-order language and model. Technical difficulties arise in the substage analysis, and in this particular case we need to be clear about exactly what part of the analysis is to be indentified with the material conditions of the proposition being considered. This proposition does not illustrate a flaw or strengthened liar paradox, but only a challenge to extend our methods to incorporate more semantic predicates in our language.

16. The set of all sets each of which is not a member of itself is a member of itself.

Let $d = \{x: x \notin x\}$. The formalization is: $d \in d$.

Analysis When we make the mathematical abstraction of the semantics of classical predicate logic, we say that all objects satisfying a predicate P form a set P, and identify the condition that P applies to an object o with the condition that $\mathsf{o} \in \mathsf{P}$. So '— is an element of —' is a semantic predicate as much as '— applies to —'. We should expect to find variations of the liar paradox within any system of set theory. In particular, $d \in d$ iff $d \notin d$.

The solution to this problem given in axiomatic set theories based on classical predicate logic is to restrict the ways that objects can be collected into sets. Then it can be proved that no set described as d exists, assuming the formal theory is consistent.

Those restrictions require that no set x satisfies $x \in x$. So if there were a set d as in this example, it would be the set of all sets, which we already know does not exist (Theorem IX.1). To reason about such collections, the formal theory of sets is often expanded to include a new kind of variable that ranges over collections such as these, called *classes*. Those restrictions on what objects can be collected together as a set look very much like the stratification into hierarchies of languages or models that have been used to deal with the liar paradox.

We could, however, formalize set theory in the formal system of Section C without using those restrictions on what counts as a set. Then $d \in d$ is just false, for to assume it true would lead to a contradiction. Similarly, $d \notin d$ is false. To assert that the example is false, we need again to form the contradictory, $\neg$T(Example 16).

17. The name of this sentence is 'Rudolfo'.

Analysis The role of naming in reasoning needs to be further clarified. In this chapter the naming of propositions, though essential to logic, is not part of the theory itself. Consider:

Ralph is a dog.
    |
    a

The part above the vertical line is the proposition. The vertical line and the letter below it represent the act of pointing to the proposition and uttering the name '*a*' (a type) or sticking a label onto the proposition. The entire complex of signs can't be a proposition that says '*a* is the name of 'Ralph is a dog'', for then where would the proposition be that '*a*' names? It can't be part of the latter, for a part of a proposition is not a proposition (Example 9).

To reason about such a complex we could view labelling as an incomplete process: We implicitly assume a description of what it is we are pointing to when we name. By making such descriptions explicit we could restrict ourselves to using only descriptive phrases as names, for example, ¬T(the single sentence that is Example 16). Such phrases can be incorporated into the logic. But as I discussed in Chapter VIII.D of *Predicate Logic*, this approach miscontrues the role of names in reasoning, which is why we rejected it when trying to incorporate nonreferring names into classical predicate logic (compare Chapter XXI.B).

It seems to me that naming is metalogical and can't be incorporated within our logic. Naming is pointing; pointing is prior to but not part of language. We can reason about naming, but the naming we do to connect our reasoning to the world is not part of our language. Even if one language were sufficient for logic, naming is an act essential to logic but not part of logic.

XXIII On Mathematical Logic and Mathematics

Concluding Remarks

One of the goals of this book has been to try to clarify the relation of classical mathematical logic to mathematics. By 'mathematics' I do not mean a body of knowledge, but the practice of mathematics, particularly reasoning in mathematics.

Mathematical logic is a formal model for reasoning in mathematics. Though it may be motivated by reflecting on mathematical practice, it is not considered wrong when we find that some or many mathematicians do not reason in accord with it. No surveys of mathematicians, no inspection of papers of mathematicians could establish the incorrectness of mathematical logic. So, it seems, classical mathematical logic is a prescriptive model: This is the correct way to reason, given these assumptions about the nature of reasoning, language, and the world.

But mathematicians are not criticized for not writing their papers in formal logic. Some, like *Fallis, 1998*, argue that mathematical proofs appearing in journals could nonetheless be formalized in classical mathematical logic. But that "could be" is a matter of faith, not demonstration. I have not seen any journal article written in formal logic since Peano and his group tried that, nor have I see even one article appearing in a mathematical journal later converted to a formal proof in classical mathematical logic as part of a project to establish that "could be".

The reason is that it is extraordinarily difficult to write a paper in a formal system for a significant new result in algebra, real analysis, differential equations, or . . . . I know no one who can. No steps can be left to the reader. A formal proof is akin to a computer program, and one cannot read a computer program and quickly assess the intent of the author when one finds a gap—or a mistake. So that "could be" is left hanging. Even in the chapters on the formalization of geometry in this text, the formal theorems are listed in their order, but the proofs are for the greater part done semantically, using informal reasoning.

Some argue that we can do all of mathematics in classical first-order mathematical logic by formalizing our set theory within that. But first-order set theory is nothing more than a truncation of second-order logic to have tractable models, and the debates about the nature of the assumptions of second-order logic are badly served by that deformation. To do mathematics in the tradition of reasoning about only objects and relations among objects, we need to quantify over all possible subsets of a set, and indeed over all sets—however one understands that. And first-order mathematical logic does not allow that. First-order set theory is useful in elucidating consequences of assumptions about sets, but it is not a theory in which to formalize all of mathematics.

So what is the raison d'être of classical mathematical logic? What is the justification, and what is the utility?

The history of the subject can be our guide. Mathematical logic was developed as part of the investigation of the foundations of mathematics. It was meant to reveal the assumptions underlying mathematics as it was done in the latter half of the 19th and early 20th centuries, leading, it was hoped, to a firm basis on which inconsistencies and paradoxes in the study of the real numbers could be resolved. It was also engendered as part of a long hoped-for formalization of all of Euclidean and non-Euclidean geometry, serving as the culmination of Euclid's methods: a clear, formal method meant to be the epitome of reasoning.

The utility of classical mathematical logic is in revealing the assumptions about reasoning and the world on which much of mathematics is based. The justification is in making clear what we need to assume in order to get the mathematics we accept as standard: what follows from what in mathematics. Classical mathematical logic is the framework in which to study consequences of assumptions in mathematics.

To think that classical mathematical logic is the study of truth is a view most mathematicians whom I know would not accept. It matters little to them whether numbers exist in one sense rather than another, nor whether there is something in the world that validates the theorems of Euclid. They are, it seems, interested in what follows from what. Many understand mathematics as modeling or as extensions of models of ordinary experience (see Chapter XII.A). To that extent they may be interested in applications of the mathematics they do, as in any science (compare the discussion of models and modeling in *Epstein, 2004*). But for the most part, mathematics is the study of consequences of certain assumptions, and classical mathematical logic clarifies that.

But classical mathematical logic has also transformed—or as some might say, deformed—the practice of mathematics. Or perhaps mathematical logic is only a reflection of the changes in mathematics engendered by the attempts to clarify the foundations of the real number system in the 19th century. What has been lost since then is any sense in ordinary mathematics of functions as processes. It has become commonplace to claim that confusions in understanding the nature of the real numbers and functions led people to believe that there could not be a nowhere differentiable everywhere continuous function on the real line. Yet whatever notion of function that was held at that time was not simply a relation between sets. Functions were understood as processes, and processes cannot, on the face of it, be that way. If we have a formalization of the nature of the real numbers that cannot model the intuition behind a function being what is drawn by a curved line on a piece of paper, then perhaps the formalization is wrong. What has been jettisoned from mathematics since the late 19th century is any reasoning about process as a distinct metaphysical category that cannot be reduced to objects and relations among objects. What has also been lost is any reasoning about substances, such as gold or mud, as a distinct metaphysical category, which is what Aristotelian logic could easily accommodate (compare the discussion in *Epstein, 1994*).

In this text I hope to have shown that classical mathematical logic can be useful in setting out clearly the assumptions on which much of mathematics can be based. But as we have seen in this text, there are serious limitations to the scope of classical mathematical logic, even setting aside its unsuitability for formalizing reasoning about processes. Classical first-order mathematical logic is highly unsuitable for formalizing all but a very small portion of mathematics. And there are very serious and rightly contentious debates about the assumptions necessary to do mathematics in second-order logic. What classical mathematical logic can do is provide a framework for studying those different views.

Appendix The Completeness of Classical Predicate Logic Proved by Gödel's Method

---

A. Description of the Method

Really it's the Löwenheim-Skolem-Herbrand-Gödel method, the history of which is detailed in *Dreben and van Heijenoort, 1986.*

Every wff is equivalent to one in prenex normal form (pp. 107–108), and if in prenex normal form then its matrix, for any assignment of references to variables, is evaluated as a **PC**-wff. Kurt Gödel exploited this to use the completeness theorem for **PC** when a model is needed to illustrate that if a wff is valid, then it is a theorem.

As this proof is tied to normal forms of classical predicate logic, it will be tied much more to derivations in classical logic than the Henkin-style proof of Chapter X. We will need a number of metatheorems about the syntax of classical logic; they are all consequences of the completeness theorems in Chapter X, but they must be derived syntactically here. Gödel, *1930*, p. 105, says of most of them, 'The proofs are not given, since they are in part well known, in part easy to supply.' This is almost true for the axiom system he uses because it has many rules that in our system must be established as derived rules. In particular, he takes as primitive rules (i) that in any theorem any wff can be uniformly substituted for any predicate symbol and a theorem results, and (ii) that any alphabetic variant of a theorem is a theorem. In establishing these as metatheorems I will not supply all the details, as I would were this the only completeness theorem in the text.

I'll take just $\neg$, $\rightarrow$, $\forall$ as primitive, treating $\wedge$, $\vee$, and $\exists$ as defined symbols. Because the proof relies on normal forms, the extension of the proof to the full language would require more work than the few observations noted in Chapter X.

Other proofs of completeness that use rather different methods are those of *Abraham Robinson, 1951* (see also *Abraham Robinson, 1974*) and ones using ultraproducts, as in *Bell and Slomson, 1969*, though they are not so very different from the two methods I present in this volume. See *Zygmunt, 1973* for a survey of those and other methods. *Surma, 1973A* is an important source for information about the history of completeness proofs for both propositional and predicate logic.

B. Syntactic Derivations

We take the language L(¬, →, ∀, =, P_0, P_1, . . . , c_0, c_1, . . .), where $\exists x$ is an abbreviation of $\neg\forall x\neg$, and each of ∧, ∨, ↔ has its usual definition in terms of ¬ and → (pp. 21 and 32). The axioms are those of the schema of Groups I and II (pp. 168) and axiom scheme 6 of Group III ('all implies exists', p. 176). I'll write 'Axiom' for 'axiom schema'. We may use everything proved in Chapter X through Lemma 9, since those proofs do not depend on the completeness theorem.

In what follows I'll adopt some abbreviations. For any sequence of variables $z_1, \ldots, z_n$,

$\forall\vec{z}$ means $\forall z_1 \ldots \forall z_n$

$\exists\vec{z}$ means $\exists z_1 \ldots \exists z_n$

$Q\vec{z}$ means $Q_1 z_1 \ldots Q_n z_n$ where for each i, Q_i is ∀ or ∃.

I'll first establish a very general theorem about substitution in wffs.

Theorem 1 Given any wffs A, B, C(A), and C(B) such that:

1. $z_1, \ldots, z_n$ is a list of all the variables free in A or B.
2. C(A) contains as a part either A or any formula obtained from A by replacing $z_1, \ldots, z_n$ with variables other than those that are bound in A or B.
3. C(B) differs from C(A) only in that some but not necessarily all places in C(A) where A or a formula derived from A as above appears, B or a formula derived from B by the same substitutions of variables appears.
4. $y_1, \ldots, y_m$ are all the variables free in either C(A) or C(B).

Then $\vdash \forall\vec{y}(\forall\vec{z}(A \leftrightarrow B) \rightarrow (C(A) \leftrightarrow C(B)))$.

Proof The proof is by induction on the number of appearances of ∀, ¬, → in C(A) that do not appear in A. Suppose there are none. Then the theorem asserts:

$$\vdash \forall\vec{y}(\forall\vec{z}(A \leftrightarrow B) \rightarrow (A(y_i/z_j) \leftrightarrow B(y_i/z_j)))$$

Here the substitutions are for some collection of the z_j's. This follows by repeated use of Axiom 4 and **PC**.

Now suppose it is true for less complex wffs and C(A) is D(A)→E(A). By induction: $\vdash \forall\vec{y}(\forall\vec{z}(A \leftrightarrow B) \rightarrow (D(A) \leftrightarrow D(B)))$ and $\vdash \forall\vec{y}(\forall\vec{z}(A \leftrightarrow B) \rightarrow (E(A) \leftrightarrow E(B)))$. So by **PC**, $\vdash \forall\vec{y}(\forall\vec{z}(A \leftrightarrow B) \rightarrow ((D(A) \rightarrow E(A)) \leftrightarrow (D(B) \rightarrow E(B))))$. If C(A) is ¬D, the proof is similar.

If C(A) is $\forall w$D(A), we have by induction:

$$\vdash \forall \ldots \forall w \ldots (\forall\vec{z}(A \leftrightarrow B) \rightarrow (D(A) \leftrightarrow D(B)))$$

So by permuting the quantifiers in the closure (an extension of Lemma X.9 that I will leave to you) and distributing (Axiom 1) we have in succession:

$$\vdash \forall\vec{y}(\forall w\, \forall\vec{z}\,(A \leftrightarrow B) \rightarrow \forall w\,(D(A) \leftrightarrow D(B)))$$

$$\vdash \forall\vec{y}(\forall w\, \forall\vec{z}(A \leftrightarrow B) \rightarrow (\forall w\, D(A) \leftrightarrow \forall w\, D(B)))$$

But w is not one of the y_i's. So $\vdash \forall\vec{y}(\forall\vec{z}(A \leftrightarrow B) \rightarrow \forall w\, \forall\vec{z}(A \leftrightarrow B))$. Hence by the transitivity of →, $\vdash \forall\vec{y}(\forall\vec{z}(A \leftrightarrow B) \rightarrow (\forall w\, D(A) \leftrightarrow \forall w\, D(B)))$ ■

Theorem 2 The Tarski-Kuratowski algorithm **(*syntactic*)** If x is not free in B, then:

a. $\vdash\forall\ldots(\forall x\,A \leftrightarrow \neg\exists x \neg A)$

b. $\vdash\forall\ldots(\exists x\,A \leftrightarrow \neg\forall x \neg A)$

c. $\vdash\forall\ldots(\neg\forall x\,A \leftrightarrow \exists x \neg A)$

d. $\vdash\forall\ldots(\neg\exists x\,A \leftrightarrow \forall x \neg A)$

e. $\vdash\forall\ldots((\exists x\,A) \wedge B \leftrightarrow \exists x\,(A \wedge B))$

f. $\vdash\forall\ldots(B \wedge (\exists x\,A) \leftrightarrow \exists x\,(B \wedge A))$

g. $\vdash\forall\ldots((\forall x\,A) \wedge B \leftrightarrow \forall x\,(A \wedge B))$

h. $\vdash\forall\ldots(B \wedge (\forall x\,A) \leftrightarrow \forall x\,(B \wedge A))$

i. $\vdash\forall\ldots((\exists x\,A) \vee B \leftrightarrow \exists x\,(A \vee B))$

j. $\vdash\forall\ldots(B \vee (\exists x\,A) \leftrightarrow \exists x\,(B \vee A))$

k. $\vdash\forall\ldots((\forall x\,A) \vee B \leftrightarrow \forall x\,(A \vee B))$

l. $\vdash\forall\ldots(B \vee (\forall x\,A) \leftrightarrow \forall x\,(B \vee A))$

m. $\vdash\forall\ldots((B \rightarrow \exists x\,A) \leftrightarrow \exists x\,(B \rightarrow A))$

n. $\vdash\forall\ldots((B \rightarrow \forall x\,A) \leftrightarrow \forall x\,(B \rightarrow A))$

o. $\vdash\forall\ldots((\exists x\,A \rightarrow B) \leftrightarrow \forall x\,(A \rightarrow B))$

p. $\vdash\forall\ldots((\forall x\,A \rightarrow B) \leftrightarrow \exists x\,(A \rightarrow B))$

q. $\vdash\forall\ldots((\exists x\,A \vee \exists x\,B) \leftrightarrow \exists x\,(A \vee B))$

Proof Left to the reader (remember Gödel's remark?). Note that once (m)–(p) are established, the rest follow by **PC** using the definitions of $\wedge$, $\vee$, and $\exists$. ∎

Recall from Chapter V (pp. 107–108) that a formula is in *prenex normal form* if it has a variable-free matrix, every variable that appears in the prefix of B appears also in the matrix of B, and no variable appears twice in the prefix of B. We saw there that for every wff there is a wff in prenex normal form which is semantically equivalent to it. Mimicking that proof and using Theorems 1 and 2, you can show the following.

Lemma 3 For any closed wff A, there is a closed wff B in prenex normal form with $\vdash A \leftrightarrow B$.

Call B an *immediate alphabetic variant* of A if it arises by replacing in A a variable x everywhere by some variable z that does not appear in A. Call B an *alphabetic variant* of A if there is a sequence of wffs $A = A_1, \ldots, A_n = B$ each of which is an immediate alphabetic variant of the preceding wff in the sequence. For example, let A be $\forall x\,P(x) \vee \exists y\,\forall x\,R(x, y)$; then

$\forall w\,P(w) \vee \exists y\,\forall w\,R(w, y)$ is an immediate alphabetic variant of A.

$\forall w\,P(w) \vee \exists z\,\forall w\,R(w, z)$ is an alphabetic variant of A (two steps).

$\forall y\,P(y) \vee \exists x\,\forall y\,R(y, x)$ is an alphabetic variant of A (four steps).

Note that if A* is an alphabetic variant of A, then A is an alphabetic variant of A*.

Lemma 4 If A is closed and in prenex normal form, and A* is an alphabetic variant of A, then $\vdash A \leftrightarrow A^*$.

Proof By the transitivity of $\rightarrow$ (p. 171) we may assume that A* is an immediate alphabetic variant of A.

Suppose A* is A with every occurrence of x replaced by z. Since x appears in A and there is no superfluous quantification, by Axiom 3 we can assume that A is of the form $Q\vec{y}\,Q_1 x\,B$, where B is in prenex normal form and x does not appear in $Q\vec{y}$. I'll let you show that for any A, if w does not appear free in A, both $\vdash\forall\ldots\forall x\,A \rightarrow \forall w\,A(w/x)$ and $\vdash\forall\ldots\exists x\,A \rightarrow \exists w\,A(w/x)$.

So we have $\vdash Q\vec{y}(Q_1x\, B \leftrightarrow Q_1z\, B(z/x))$. Applying Theorem 2, we have $\vdash Q\vec{y}Q_1x\, B \leftrightarrow Q\vec{y}Q_1z\, B(z/x)$. ∎

Theorem 5 ***Substitution of alphabetic variants*** If A is closed and A* is an alphabetic variant of A, then $\vdash A \leftrightarrow A^*$.

Proof By our previous work we may assume that we have a sequence of wffs $A = A_1, \ldots, A_n = B$ where B is in prenex normal form and for each $i \geq 1$, $\vdash A_i \leftrightarrow A_{i+1}$, and either A_{i+1} is a Tarski-Kuratowski equivalent of A_i, or is an alphabetic variant of A_i, or is A_i with a superfluous quantification deleted.

Further, we can assume that if A* arises from A by using the new variables $z_1, \ldots, z_m$, then none of those appear in any of $A_1, \ldots, A_n$.

Let $A^* = A_1^*, \ldots, A_n^* = B^*$ be a sequence where each A_i^* is A_i with the same replacements that yield A* from A. Then this sequence yields an equivalent prenex normal form B* for A*.

Since B* is an alphabetic variant of B, by Lemma 4, $\vdash B \leftrightarrow B^*$. And $\vdash A \leftrightarrow B$ as well as $\vdash A^* \leftrightarrow B^*$. Hence, $\vdash A \leftrightarrow A^*$. ∎

Theorem 6 ***Substitution of a formula for a predicate*** Let A be a closed wff and P an n-ary predicate symbol that appears in A. Let B be any wff such that:

i. B has exactly n variables free.

ii. No variable y bound in B appears as $P(z_1, \ldots, y, \ldots, z_n)$ in A.

Let A(B/P) be the result of replacing every occurrence of P in A by B, retaining the same variables at each occurrence, that is, for $P(z_1, \ldots, z_n)$ take $B(z_1, \ldots, z_n)$. In that case, if $\vdash A$, then $\vdash A(B/P)$.

Proof The proof proceeds by induction on the length of a proof of A.

Suppose A has a proof of length 1. Then A is an axiom. The axioms are instances of schema, and A(B/P) is an instance of the same schema as A(P) (condition (ii) in the statement of the theorem is required to ensure that this is correct for Axiom 4).

Suppose now that the theorem is true for all A with proofs of length $\leq m$, and A has a proof of length $m + 1$. If A is an axiom we are done as before. So assume A has a proof $C_1, \ldots, C_{m+1} = A$, where for some $j, k \leq m$, C_j is $C_k \rightarrow A$. Let $C_1^*, \ldots, C_m^* = A^*$ be alphabetic variants of $C_1, \ldots, C_m$, using the same replacements in each case, such that no variable bound in B appears in them. Then this is also a proof, since if C_i is an instance of an axiom scheme, so is C_i^*. Hence, by induction $\vdash C_k^*(B/P) \rightarrow A^*(B/P)$ and $\vdash C_k^*(B/P)$. Hence $\vdash A^*(B/P)$. But A*(B/P) is an alphabetic variant of A(B/P). Hence by Theorem 5, $\vdash A(B/P)$. ∎

C. The Completeness Theorem

First I'll consider $L(\neg, \rightarrow, \forall, P_0, P_1, \ldots)$. Then I'll show how to extend the work to the language containing names and '='.

Lemma X For any closed wff A in prenex normal form, there is a closed wff B in prenex normal form whose prefix begins with $\forall$ such that $\vdash A \leftrightarrow B$.

Proof If A is not already of that form, it has the form $\exists x\, C$. Let y be the first variable not appearing in A. Let H(y) be $P_1^1(y) \vee \neg P_1^1(y)$. By Lemma X.7, $\vdash \forall y\, H(y)$. Hence by **PC**, $\vdash [\exists x\, C \wedge \forall y\, H(y)] \leftrightarrow \exists x\, C$. By the Tarski-Kuratowski algorithm, $\vdash \exists x\, C \wedge \forall y\, H(y) \leftrightarrow \forall y\, \exists x\, (C \wedge H(y))$. Hence, $\vdash \forall y\, \exists x\, (C \wedge H(y)) \leftrightarrow \exists x\, C$. Now either we are done, or C has the form $Q\vec{w}\, D$, where D is quantifier-free. In that case, apply the Tarski-Kuratowski algorithm again to get, $\vdash \exists x\, C \leftrightarrow \forall y\, \exists x\, Q\vec{w}\, (D \wedge H(y))$. ■

The *degree* of a formula in prenex normal form that begins with $\forall$ is $n + 1$ if there are n alternations $\exists\forall$ in its prefix for $n \geq 0$. We will show two lemmas:

Lemma Y For formulas in prenex normal form beginning with $\forall$:
If for every formula B of degree n, B has a model or $\vdash \neg B$, then
for every formula B of degree $n + 1$, B has a model or $\vdash \neg B$.

Lemma Z For every formula B in prenex normal form that begins with $\forall$ that is of degree 1, B has a model or $\vdash \neg B$.

Assuming Lemmas Y and Z for the moment, we can show the completeness theorem.

Theorem 7 Completeness of classical predicate logic In $L(\neg, \rightarrow, \forall, P_0, P_1, \ldots)$, for every closed wff B, $\vdash B$ iff $\models B$.

Proof We already have by Theorem X.2 that if $\vdash B$, then $\models B$. By Lemma X, we may assume that B is in prenex normal form beginning with $\forall$. So suppose $\models B$. Then $\neg B$ has no model. So by Lemmas Y and Z, $\vdash \neg\neg B$. So by **PC**, $\vdash B$. ■

I'll now establish Lemmas Y and Z.

Proof of Lemma Y Suppose that for every formula C of degree n, $\vdash \neg C$ or C has a model. Let A be a formula of degree $n + 1$:

(A) $\quad \forall \vec{y}\, \exists \vec{z}\, Q\vec{w}\; B(\vec{y}, \vec{z})$

where Q begins with a universal quantifer. Let P be any predicate symbol of the appropriate arity that does not occur in A, and consider:

(β) $\quad \forall \vec{y}\, (\forall \vec{z}\, [P(\vec{y}, \vec{z}) \rightarrow Q\vec{w}\; B(\vec{y}, \vec{z})] \wedge \exists \vec{z}\, P(\vec{y}, \vec{z}))$

Let $\vec{x}$ be a sequence of as many variables as there are in $\vec{z}$, each of which does not appear in β. Applying the Tarski-Kuratowski algorithm and Lemma 4, we can transform β into:

(γ) $\quad \forall \vec{y}\, \forall \vec{z}\, Q\vec{w}\, \exists \vec{x}\, ([P(\vec{y}, \vec{z}) \rightarrow Q\vec{w}\; B(\vec{y}, \vec{z})] \wedge P(\vec{y}, \vec{x}))$

That is, $\vdash \beta \leftrightarrow \gamma$. And so $\models \beta$ iff $\models \gamma$.

Since γ has degree n, $\vdash \neg\gamma$ or γ has a model. So $\vdash \neg\beta$ or β has a model.

I'll let you show that $\models \beta \rightarrow A$. So if β has a model, then A has a model. It remains to show that if $\vdash \neg\beta$, then $\vdash \neg A$.

Suppose we have $\vdash \neg \forall \vec{y}\, (\forall \vec{z}\, [P(\vec{y}, \vec{z}) \rightarrow Q\vec{w}\; B(\vec{y}, \vec{z})] \wedge \exists \vec{z}\, P(\vec{y}, \vec{z}))$.
Then by Theorem 6, we have:

$\vdash \neg \forall \vec{y}\ (\forall \vec{z}\ [\mathrm{Q}\vec{w}\ \mathrm{B}(\vec{y}, \vec{z}) \to \mathrm{Q}\vec{w}\ \mathrm{B}(\vec{y}, \vec{z})] \wedge \exists \vec{z}\ \mathrm{Q}\vec{w}\ \mathrm{B}(\vec{y}, \vec{z}))$

By **PC**, we have $\vdash \forall \vec{y}\ \forall \vec{z}\,(\mathrm{Q}\vec{w}\ \mathrm{B}(\vec{y}, \vec{z}) \to \mathrm{Q}\vec{w}\ \mathrm{B}(\vec{y}, \vec{z}))$, and so

$$\vdash \forall \vec{y}\ (\forall \vec{z}\,[\mathrm{Q}\vec{w}\ \mathrm{B}(\vec{y}, \vec{z}) \to \mathrm{Q}\vec{w}\ \mathrm{B}(\vec{y}, \vec{z})] \wedge \exists \vec{z}\ \mathrm{Q}\vec{w}\ \mathrm{B}(\vec{y}, \vec{z})) \leftrightarrow \forall \vec{y}\ (\exists \vec{z}\ \mathrm{Q}\vec{w}\ \mathrm{B}(\vec{y}, \vec{z})).$$

Hence by Theorem 1, $\vdash \neg \mathrm{A}$. ∎

Proof of Lemma Z Let B be a formula in prenex normal form that is of degree 1:

$$\forall y_1 \ldots y_k \exists z_1 \ldots z_m\, \mathrm{A}(\vec{y}, \vec{z}\,)$$

where A is quantifier-free.

We first need an ordering of k-tuples of the variables of our language. Let $(x_{i_1}, \ldots, x_{i_k})$ precede $(x_{j_1}, \ldots, x_{j_k})$ iff $i_1 + \cdots + i_k < j_1 + \cdots + j_k$ or $i_1 + \cdots + i_k = j_1 + \cdots + j_k$ and for the first m such that $i_m \neq j_m$, $i_m < j_m$.

Let the nth k-tuple be $(x_{n_1}, \ldots, x_{n_k})$. We define:

$$\mathrm{B}_n \equiv_{\mathrm{Def}} \mathrm{A}(x_{n_1}/y_1, \ldots, x_{n_k}/y_k, x_{(n-1)m+1}/z_1, \ldots, x_{nm}/z_m)$$

Note (and this depends on the ordering above) if $n > 1$, then all of $x_{n_1}, \ldots, x_{n_k}$ appear already in some formulas B_p, $p < n$. Define, associating to the left,

$$\mathrm{C}_n \equiv_{\mathrm{Def}} \mathrm{B}_1 \wedge \cdots \wedge \mathrm{B}_n$$

$$\mathrm{D}_n \equiv_{\mathrm{Def}} \exists \ldots \mathrm{C}_n$$

Now we assign to each atomic wff a propositional variable such that distinct wffs are assigned distinct variables. Explicitly, letting q_n designate the nth prime (2 is the 0th):

$$\mathrm{P}_j^k(x_{i_1}, \ldots, x_{i_k}) \text{ is assigned } p_{2^j 3^{i_1} 5^{i_2} \ldots q_k^{i_k}}.$$

Now for each n we define a propositional formula:

$\mathrm{E}_n \equiv_{\mathrm{Def}} \mathrm{C}_n$ with each atomic wff replaced by its propositional variable.

We have two possibilities:

(1) For some n, E_n has no model.

(2) For all n, E_n has a model.

First we'll consider (1). In this case we have $\vDash_{\mathbf{PC}} \neg \mathrm{E}_n$. So by the completeness theorem for **PC**, $\vdash_{\mathbf{PC}} \neg \mathrm{E}_n$. Hence by **PC**, $\vdash \forall \ldots \neg \mathrm{C}_n$. So by **PC**, we have $\vdash \neg \exists \ldots \mathrm{C}_n$, that is, $\vdash \neg \mathrm{D}_n$. We need now that for every n, $\vdash \mathrm{B} \to \mathrm{D}_n$, and then we will have $\vdash \neg \mathrm{B}$.

Sublemma For every n, $\vdash \mathrm{B} \to \mathrm{D}_n$.

Proof We proceed by induction on n. For $n = 1$, we want

$$\vdash \forall \vec{y}\ \exists \vec{z}\ \mathrm{A}(\vec{y}, \vec{z}) \to \exists x_0\, \exists x_1 \ldots \exists x_m\, \mathrm{A}(x_0, \ldots, x_0, x_1, \ldots, x_m).$$

We have by Theorem 5,

$$\vdash \forall \vec{y}\ \exists \vec{z}\ \mathrm{A}(\vec{y}, \vec{z}) \to \forall \vec{y}\ \exists x_1 \ldots \exists x_m\, \mathrm{A}(\vec{y}, x_0, \ldots, x_0, x_1, \ldots, x_m)$$

Then via Axioms 4, 1, and 2a, $\vdash \forall \vec{y}\, \exists x_1 \ldots \exists x_m\, \mathrm{A}(\vec{y}, x_0, \ldots, x_0, x_1, \ldots, x_m) \to \forall x_0\, \exists x_1 \ldots \exists x_m\, \mathrm{A}(x_0, \ldots, x_0, x_1, \ldots, x_m)$. And we then have by Lemma X.8.b,

$$\forall x_0\, \exists x_1 \ldots \exists x_m\, A(x_0, \ldots, x_0, x_1, \ldots, x_m) \rightarrow$$
$$\exists x_0\, \exists x_1 \ldots \exists x_m\, A(x_0, \ldots, x_0, x_1, \ldots, x_m).$$

So we are done by the transitivity of $\rightarrow$.

Suppose we have the sublemma for $1 \leq m \leq n$. We wish to show:

$$\vdash B \rightarrow \exists x_0\, \exists x_1 \ldots \exists x_{nm}\, ((B_1 \wedge \cdots \wedge B_{n-1}) \wedge B_n).$$

Since (and this depends on our ordering) $x_{(n-1)m+1}, \ldots, x_{nm}$ do not appear in $B_1, \ldots, B_{n-1}$, we have by the Tarski-Kuratowski algorithm:

$$\vdash \exists x_0\, \exists x_1 \ldots \exists x_{nm}\, ((B_1 \wedge \cdots \wedge B_{n-1}) \wedge B_n) \leftrightarrow$$
$$(\exists x_0\, \exists x_1 \ldots \exists x_{(n-1)m}\, (B_1 \wedge \cdots \wedge B_{n-1}) \wedge \exists x_{(n-1)m+1} \ldots \exists x_{nm}\, B_n).$$

We then have just as in the case when $n = 1$, $\vdash B \rightarrow \exists x_{(n-1)m+1} \ldots \exists x_{nm}\, B_n$. Hence by induction, $\vdash B \rightarrow D_n$. And we are done with the sublemma and case (1).

Now we turn to case (2) where we will show that B has a model. Suppose that each E_n has a model. We can (constructively) find an assignment of truth-values to the propositional variables appearing in E_n that yields T as the truth-value of E_n. For each n there is at least one such validating assignment and at most finitely many. Note that an assignment that validates E_n also validates E_j for $j < n$.

We now assign truth-values to each of the propositional variables in E_n for all n by induction. The procedure will satisfy:

(3) The assignment at stage n validates $E_1, \ldots, E_n$, and for infinitely many $j > n$ there is at least one validating assignment for E_j that uses the assignment given at stage n for the propositional variables in $E_1, \ldots, E_n$.

Stage n: The assignment of truth-values to the propositional variables in E_i, $i < n$, is retained. Let $q_0, \ldots, q_m$ be the list of propositional variables in alphabetic order that appear in E_n and that do not appear in E_i for any $i < n$.

Substage 0: If in infinitely many of the validating assignments for E_j for $j \geq n$ that use the assignment for stage $n - 1$ the value for q_0 is T, then assign q_0 value T now; otherwise assign it F.

Substage $j + 1$: If in infinitely many of the validating assignments for E_j for $j \geq n$ that use the assignment of stage $n - 1$ and substages 1 to j, q_{j+1} is assigned T, then assign q_{j+1} value T now; otherwise assign it F.

Note that the choice of assignments at stage n is not necessarily constructive.

To show that (3) holds, note that at each substage and at the end of the stage, the values for the q_i's have been chosen so that there are validating assignments for infinitely many E_j, $j > n$ that include those values for the variables in E_k, $k < n$ and the q_i's. And a validating assignment for E_n also validates E_k for $k < n$.

Now define a truth-value assignment υ for all propositional variables by:

If q appears in some E_n, $\upsilon(q)$ = the value assigned to q at the least such stage n.

If q does not appear in E_n for any n, $\upsilon(q) = F$.

Then υ is a **PC**-model, and for all n, $\upsilon(E_n) = T$.

Now we construct an abstract mathematical predicate logic model that validates B. The universe of the model will be the collection of natural numbers. Interpret each n-ary predicate symbol that does not appear in B as the collection of all n-tuples of natural numbers. If P_i^n appears in B, set:

$(k_1, \ldots, k_n) \in$ the interpretation of P_i^n iff $v(p_r) = T$
where p_r is the propositional variable assigned to $P_i^n(x_{k_1}, \ldots, x_{k_n})$.

It remains to show that in this model B, that is, $\forall \vec{y}\, \exists \vec{z}\, A(\vec{y}, \vec{z})$, is true. Given any σ, we have $\sigma(y_j) = i_j$ for $j < k$. Then $(i_1, \ldots, i_k)$ is the nth k-tuple of natural numbers for some n. Recall that:

$$B_n \text{ is } A(x_{i_1}/y_1, \ldots, x_{i_k}/y_k, x_{(n-1)m+1}/z_1, \ldots, x_{nm}/z_m)$$

Now $v(E_n) = T$. So letting τ agree with σ on $\vec{y}$ and $\tau(z_j) = (n-1)m + j$ for $1 \le j \le n$, we have $\tau \vDash A(\vec{y}, \vec{z})$. So $\sigma \vDash \forall \vec{y}\, \exists \vec{z}\, A(\vec{y}, \vec{z})$, and so $\forall \vec{y}\, \exists \vec{z}\, A(\vec{y}, \vec{z})$ is true in this model. ■

We have shown the completeness theorem for classical predicate logic. Gödel did not prove the strong completeness theorem in *Gödel, 1930*, but he did prove the compactness theorem, from which it is easy to show strong completeness, as I will do below. The next lemma is from *Gödel, 1929*.

Lemma 8 For any collection of wffs Σ in $L(\neg, \to, \forall, P_0, P_1, \ldots)$, either Σ has a model, or for some $A_1, \ldots, A_n$ in Σ, $\vdash \neg(A_1 \wedge \cdots \wedge A_n)$.

Proof By Lemma X, we can assume that every wff in Σ is in prenex normal form and begins with $\forall$. We now reduce the problem to showing it for the case when Σ has only formulas of degree 1.

If Σ has formulas that are not of degree 1, for each A in Σ by successive reductions as in the proof of Lemma Y we can show that there is a formula β_A of degree 1 such that $\vDash \beta_A \to A$ and if $\vdash \neg\beta_A$ then $\vdash \neg A$. So let $\Gamma = \{\beta_A : A \text{ is in } \Sigma\}$. If Γ has a model, then so, too, does Σ. If there are $\beta_{A_1}, \ldots, \beta_{A_n}$ such that $\vdash \neg(\beta_{A_1} \wedge \cdots \wedge \beta_{A_n})$, then by **PC**, $\vdash \neg(A_1 \wedge \cdots \wedge A_n)$. So we may assume that the formulas in Σ are:

$$\forall \vec{y}_1\, \exists \vec{z}_1\, A_1(\vec{y}_1, \vec{z}_1), \ldots, \forall \vec{y}_n\, \exists \vec{z}_n\, A_n(\vec{y}_n, \vec{z}_n), \ldots$$

Let y_j^i, z_j^i denote k-tuples of $x_0, x_1, \ldots$, where k is the number of variables in $\vec{y}_i$ or $\vec{z}_i$ respectively. In the proof of Lemma Z, instead of B_n take (associating to the left):

$$\begin{aligned} H_n \equiv_{\text{Def}} \; & [A_1(y_1^1, z_1^1) \vee A_1(y_2^1, z_2^1) \vee \cdots \vee A_1(y_n^1, z_n^1)] \\ & \wedge [A_2(y_1^2, z_1^2) \vee \cdots \vee A_2(y_{n-1}^2, z_{n-1}^2)] \\ & \wedge \cdots \wedge [A_{n-1}(y_1^{n-1}, z_1^{n-1}) \vee A_{n-1}(y_2^{n-1}, z_2^{n-1})] \\ & \wedge A_n(y_1^n, z_1^n) \end{aligned}$$

where:

1. y_j^i is the jth k-tuple (k the number of variables in $\vec{y}_i$).
2. The variables in $z_i^1, z_{i-1}^2, \ldots, z_1^i$ are all different from one another and from all variables occurring to the left and above the diagonal on which they lie, the least such k-tuple being chosen in each instance.

Then for each n, for each $i \leq n$,

$$\vdash \exists \vec{y}_1 \forall \vec{z}_1 A_1 \wedge \cdots \wedge \exists \vec{y}_n \forall \vec{z}_n A_n \rightarrow \exists \ldots H_n$$

Using the same assignment of variables to atomic wffs as in the proof of Lemma Z, for each n associate to H_n a propositional formula J_n.

Suppose that for some n, J_n has no model. Then $\models_{PC} \neg J_n$. Hence $\vdash \forall \ldots \neg H_n$. So $\vdash \neg \exists \ldots H_n$. So $\vdash \neg(A_1 \wedge \cdots \wedge A_n)$.

If on the other hand each J_n has a model, then we proceed as in the proof of Lemma Z to establish that Σ has a model. ■

Corollary 9 For any collection of wffs Σ in $L(\neg, \rightarrow, \forall, P_0, P_1, \ldots)$, if Σ has a model, then Σ has a model whose universe is the natural numbers.

Theorem 10 ***Compactness*** For any collection of wffs Σ in $L(\neg, \rightarrow, \forall, P_0, P_1, \ldots)$, Σ has a model iff every finite subset of Σ has a model.

Proof If Σ has a model, then every finite subset of Σ has a model. In the other direction, suppose Σ has no model. Then by Lemma 8, for some $A_1, \ldots, A_n$ in Σ, $\vdash \neg(A_1 \wedge \cdots \wedge A_n)$. Hence $\{A_1, \ldots, A_n\} \subseteq \Sigma$ has no model. ■

Theorem 11 ***Strong completeness*** In $L(\neg, \rightarrow, \forall, P_0, P_1, \ldots)$, $\Gamma \vdash A$ iff $\Gamma \models A$.

Proof We already have that if $\Gamma \vdash A$, then $\Gamma \models A$. In the other direction, suppose $\Gamma \nvdash A$. So for any finite subset of Γ, say $B_1, \ldots, B_n$, $\nvdash B_1 \wedge \cdots \wedge B_n \rightarrow A$. Hence by the completeness theorem, $\neg(B_1 \wedge \cdots \wedge B_n \rightarrow A)$ has a model, which is a model of $\{B_1, \ldots, B_n, \neg A\}$. So every finite subset of $\Gamma \cup \{\neg A\}$ has a model. Therefore by the compactness theorem, $\Gamma \cup \{\neg A\}$ has a model. So $\Gamma \nvDash A$. ■

For the language with '=', Gödel proceeds just as we did in Chapter X.B.

Theorem 12 In $L(\neg, \rightarrow, \forall, =, P_0, P_1, \ldots)$, $\vdash A$ iff $\models A$.

Gödel did not consider in his paper the question of completeness for a system with name symbols. But the extension is not difficult.

Theorem 13 In $L(\neg, \rightarrow, \forall, P_0, P_1, \ldots, c_0, c_1, \ldots)$, $\Gamma \vdash A$ iff $\Gamma \models A$.

Proof I'll do the case when there is only one name symbol, c, in the language and leave the generalization to you. As usual, if $\Gamma \vdash A$, then $\Gamma \models A$.

For the other direction, let z be a new variable that does not appear in the language, and expand the language to include it. For each $D \in \Gamma$, let $D(z)$ be D with all occurrences of c in D (if any) replaced by z. Let $\Gamma(z) = \{\forall z\, D(z) : D \in \Gamma\}$. Though c may not actually appear in A, I will nonetheless write A as $B(c)$.

Examining the proof of Lemma X.9, you can establish that $\Gamma \vdash B(c)$ iff $\Gamma(z) \vdash \forall z\, B(z)$. Now let us assume that $\Gamma \nvdash B(c)$ in the language with names. Then $\Gamma(z) \nvdash \forall z\, B(z)$ in the language with names. So even more so, there is no proof of $\forall z\, B(z)$ from $\Gamma(z)$ in the language without names. Hence by Theorem 11, there is a model M of the language without names such that for every $D \in \Gamma$, $M \models \forall z\, D(z)$ and $M \not\models \forall z\, B(z)$. So $M \models \exists z \neg B(z)$. Let σ be an assignment such that $\sigma \models \neg B(z)$. Then we can take M to be a model of the language with c, where $\sigma(c) = \sigma(z)$. For that language, $M \models D$ for every $D \in \Gamma$, and $M \not\models B(c)$. So $\Gamma \not\models A$. ■

A good exercise to test your understanding of the methods of this appendix is to prove Lemmas X, Y, Z and Theorem 7 using $\exists\forall$ forms instead of $\forall\exists$ forms.

Summary of Formal Systems

Propositional Logic

PC *in* L(¬, →) axiomatized with schema §II.D.1

1. ¬A → (A → B)
2. B → (A → B)
3. (A → B) → ((¬A → B) → B)
4. (A → (B → C)) → ((A → B) → (A → C))

rule $\dfrac{\text{A, A} \rightarrow \text{B}}{\text{B}}$ *modus ponens*

PC *in* L(¬, →, ∧, ∨) axiomatized with schema §II.D.5

PC *in* L(¬, →) plus

5. A → (B → (A ∧ B))
6. (A ∧ B) → A
7. (A ∧ B) → B
8. A → (A ∨ B)
9. B → (A ∨ B)
10. (A → C) → ((B → C) → ((A ∨ B) → C))

rule $\dfrac{\text{A, A} \rightarrow \text{B}}{\text{B}}$

Classical Predicate Logic

In $L(\neg, \rightarrow, \forall, \exists; P_0, P_1, \ldots, c_0, c_1, \ldots)$ axiomatized with schema §X.A.1

I. Propositional axioms

The axiom schema of **PC** in $L(\neg, \rightarrow)$, where A, B, C are replaced by wffs of L and the universal closure is taken.

II. Axioms governing ∀

1. $\forall \ldots (\forall x\,(A \rightarrow B) \rightarrow (\forall x\, A \rightarrow \forall x\, B))$ *distribution of* ∀
2. When x is not free in A,
   a. $\forall \ldots (\forall x\, A \rightarrow A)$ *superfluous quantification*
   b. $\forall \ldots (A \rightarrow \forall x\, A)$
3. $\forall \ldots (\forall x \forall y\, A \rightarrow \forall y \forall x\, A)$ *commutativity of* ∀
4. When t is free for x in A, $\forall \ldots (\forall x\, A(x) \rightarrow A(t/x))$ *universal instantiation*

III. Axioms governing the relation between ∀ and ∃

5. a. $\forall \ldots (\exists x\, A \rightarrow \neg\forall x \neg A)$
   b. $\forall \ldots (\neg\forall x \neg A \rightarrow \exists x\, A)$

rule $\dfrac{A,\ A \rightarrow B}{B}$ where A and B are closed formulas

In languages without name symbols

Add to Group III axioms: §X.A.4

6. $\forall \ldots (\forall x A(x) \rightarrow \exists x A(x))$ *all implies exists*

In languages without the existential quantifier §X.A.4

Define $\exists x\, A \equiv_{\text{Def}} \neg\forall x \neg A$. Delete Schema 5.a and 5.b of Group III.

In languages with equality §X.B.1

IV. Axioms for equality

7. $\forall x\,(x = x)$ *identity*
8. For every n-ary atomic predicate P, *extensionality of atomic predications*

   $\forall \ldots \forall x\, \forall y\, (x = y \rightarrow$

   $(P(z_1, \ldots, z_k, x, z_{k+1}, \ldots, z_n) \rightarrow P(z_1, \ldots, z_k, y, z_{k+1}, \ldots, z_n))$

In languages with function symbols §X.B.2

V. Axioms for functions

9′. For every n-ary function symbol f, *extensionality of functions*

$\forall \ldots \forall x\, \forall y\, (x = y \rightarrow$

$(f(z_1, \ldots, z_k, x, z_{k+1}, \ldots, z_n) = f(z_1, \ldots, z_k, y, z_{k+1}, \ldots, z_n))$

Arithmetic

S, the *theory of successor* in L(= ; S, o) §XV.A

S1 $\forall x \forall y (Sx = Sy \rightarrow x = y)$

S2 $\forall x (o \neq Sx)$

S3 $\forall y (y \neq o \rightarrow \exists x (y = Sx))$

T_n $\forall x (S^n x \neq x)$ for $n \geq 1$ (n is a metalogical abbreviation)

Q in L(= ; S, ⊕, ∗, o) §XV.B.1

Q1 $\forall x \forall y (Sx = Sy \rightarrow x = y)$

Q2 $\forall x (0 \neq Sx)$

Q3 $\forall y (y \neq o \rightarrow \exists x (y = Sx))$

Q4 $\forall x (x \oplus o = x)$

Q5 $\forall x \forall y (x \oplus Sy = S(x \oplus y))$

Q6 $\forall x (x * o = o)$

Q7 $\forall x \forall y (x * Sy = (x * y) \oplus x)$

$Q \equiv_{\text{Def}} Q1 \wedge Q2 \wedge Q3 \wedge Q4 \wedge Q5 \wedge Q6 \wedge Q7$

Induction$_1$, *the principle of first-order induction* §XV.C.1

$\forall \ldots ([A(o) \wedge \forall x (A(x) \rightarrow A(Sx))] \rightarrow \forall x A(x))$

for every wff A in which x appears free.

***Peano Arithmetic*, PA,** in L(= ; S, ⊕, ∗, o) §XV.C.1

Q plus all instances of *Induction*$_1$

Arithmetic *in* L(= ; S, ⊕, ∗, o) §XV.C.1

Th(<$\mathbb{N}$; +1, +, ·, 0>)

Induction, the second-order axiom of induction §XV.E

$\forall X ((X(o) \wedge \forall y (X(y) \rightarrow X(Sy))) \rightarrow \forall x X(x))$

PA$_2$, *second-order Peano arithmetic* in L_2(=, S, ⊕, ∗, o) §XV.E

The semantic consequences of $Q \wedge$ *Induction*.

Z-Arithmetic, the *theory of the integers* in L(= ; <, ⊕, ∗) §XVI.A.2

Th(**Z**), where **Z** = <$\mathbb{Z}$; less than, +, ·>

Q-Arithmetic, the *theory of the rationals* in L(= ; <, ⊕, ∗) §XVI.A.2

Th(**Q**), where **Q** = <$\mathbb{Q}$; less than, +, ·>

Linear Orderings

Ord, the *theory of orderings* in L(= ; <) §XIII.A

O1 $\forall x \neg(x < x)$ *anti-reflexivity*

O2 $\forall x \forall y (x < y \rightarrow \neg(y < x))$ *anti-symmetry*

O3 $\forall x \forall y \forall z ((x < y \wedge y < z) \rightarrow x < z)$ *transitivity*

LO, the *theory of linear orderings* in L(= ; <) §XIII.A

Ord

O4 $\forall x \forall y (x < y \vee y < x \vee x = y)$ *trichotomy*

DLO, the *theory of dense linear orderings without endpoints*, in L(= ; <) §XIII.A

LO

O5 $\forall x \forall y (x < y \rightarrow \exists z (x < z \wedge z < y))$ *density*

O6 $\forall x \exists y (x < y)$ *no last endpoint*

O7 $\forall x \exists y (y < x)$ *no first endpoint*

Dense linear orderings with no first element in L(= ; <) Exercise 8, §XIII.A

Th(*O1–O6*, ¬*O7*)

Dense linear orderings with no last element in L(= ; <) Exercise 8, §XIII.A

Th(*O1–O5*, ¬*O6*, *O7*)

Dense linear orderings with endpoints element in L(= ; <) Exercise 8, §XIII.A

Th(*O1–O5*, ¬*O6*, ¬*O7*)

Dedekind's axiom (*the continuity axiom*) §XVII.C

$$\forall \ldots [\forall X \forall Y (\exists x \exists y (X(x) \wedge Y(y)) \wedge \forall x \forall y ((X(x) \wedge Y(y)) \rightarrow x \leq y)$$
$$\rightarrow \exists z \forall x \forall y ((X(x) \wedge Y(y)) \rightarrow x \leq z \leq y))]$$

Countable density

$$\exists X (\mathrm{Count}(X) \wedge \forall x \forall y (x < y \rightarrow \exists z [X(z) \wedge (x < z < y)]))$$ §XVII.C

The first-order scheme of Dedekind's axiom (*Dedekind's axiom*$_1$) §XVII.D

$$\forall \ldots ([\exists x \exists y (A(x) \wedge B(y)) \wedge \forall x \forall y (A(x) \wedge B(y) \rightarrow x \leq y)]$$
$$\rightarrow \exists z \forall x \forall y (A(x) \wedge B(y) \rightarrow x \leq z \leq y))$$

where A is any wff in which x is free, B is any wff in which y is free, and neither y nor z appear in A, and neither x nor z appear in B

RLO$_2$, the second-order theory of the *linear order of the real line* in L$_2$(= ; <) §XVII.C

DLO

countable density

Dedekind's axiom

Groups

Groups

G, the *theory of groups* in $L(= ; \circ, {}^{-1}, e)$ §XII.A

G1 $\forall x_1 (e \circ x_1 = x_1 \wedge x_1 \circ e = x_1)$

G2 $\forall x_1 (x_1 \circ x_1^{-1} = e \wedge x_1^{-1} \circ x_1 = e)$

G3 $\forall x_1 \forall x_2 \forall x_3 ((x_1 \circ x_2) \circ x_3 = x_1 \circ (x_2 \circ x_3))$

GK, *Kleene's theory of groups* in $L(= ; \circ, {}^{-1}, e)$ §XII.A

GK1 $\forall x_1 (x_1 \circ e = x_1)$

GK2 $\forall x_1 (x_1 \circ x_1^{-1} = e)$

G3

Note: **G** = **GK**.

G* in $L(= ; \circ, {}^{-1})$ §XII.B.1

*G*1* $\exists ! x_2\, E(x_2)$

*G*2* $\forall x_2 (E(x_2) \rightarrow \forall x_1 (x_1 \circ x_1^{-1} = x_2))$
$\wedge \forall x_2 (E(x_2) \rightarrow \forall x_1 (x_1^{-1} \circ x_1 = x_2))$

G3

where $E(x_2) \equiv_{\text{Def}} \forall x_1 (x_2 \circ x_1 = x_1 \wedge x_1 \circ x_2 = x_1)$

G# in $L(= ; \circ, e)$ §XII.B.2

G1

G3

$\forall x_1 \exists ! x_2\, I(x_1, x_2)$

where $I(x_1, x_2) \equiv_{\text{Def}} (x_1 \circ x_2 = e \wedge x_2 \circ x_1 = e)$

G† in $L(= ; \circ)$ §XII.B.2

G† 1 $\exists ! x_2\, E(x_2)$

G† 2 $\forall x_1 \exists ! x_2 (\forall x_3\, E(x_3) \rightarrow (x_1 \circ x_2 = x_3 \wedge x_2 \circ x_1 = x_3))$

G3

Abelian G, the *theory of abelian groups* §XII.A

In any of the languages of groups add:

Abelian $\equiv_{\text{Def}} \forall x_1 \forall x_2 (x_1 \circ x_2 = x_2 \circ x_1)$

Inf G, the *theory of infinite groups* §XII.A

In any of the languages of groups add:

$\{E_{\geq n} : n \geq 1\}$ $E_{\geq n}$ is defined on p. 183.

Divisible Groups

The division axioms $D_n \equiv_{\text{Def}} \forall x \, \exists y \, (x = n\,y)$ for $n \geq 1$

$n\,y$ is a metalogical abbreviation

GD*n* for any $n \geq$, in L(= ; ⊕, o) §XVII.B

AbelianG

$\exists x \, \exists y \, (x \neq y)$

D_n

GD$_{\mathsf{U}}$ for any $\mathsf{U} \subseteq \mathbb{N}$, in L(= ; ⊕, o) §XVII.B

AbelianG

$\exists x \, \exists y \, (x \neq y)$

$\{D_n : n \in \mathsf{U}\}$

GD, the *theory of divisible groups* in L(= ; ⊕, o) §XVII.B

AbelianG

$\exists x \, \exists y \, (x \neq y)$

$\{D_n : n \geq 1\}$

Ordered Divisible Groups

ODG, the theory of *ordered divisible groups* in L(=; <, ⊕, o) §XVII.B

DLO

GD

OA $\forall x \, \forall y \, \forall z \, (y < z \rightarrow x \oplus y < x \oplus z)$ *the order-addition axiom*

OGDed$_1$ in L(= ; <, ⊕, o) §XVII.D

DLO

Abelian G

D_2

OA

Dedekind's axiom$_1$ *scheme*

Note: **ODG** = **OGDed**$_1$

RODG$_2$, the second-order theory of *real ordered divisible groups* in L_2(=; <, ⊕, o)

DLO §XVII.D

AbelianG

D_2

OA

Dedekind's axiom

Fields

F, the *theory of fields* in $L(=; \oplus, *, \mathrm{o}, \text{ł})$ §XVII.E.1

AbelianG

$\forall x\, (x * \text{ł} = x)$

$\forall x\, \forall y\, \forall z\, (x * (y * z) = (x * y) * z)$

$\forall x\, \forall y\, (x * y = y * x)$

$\forall x\, (x \neq \mathrm{o} \rightarrow \exists y\, (x * y = \text{ł}))$

$\forall x\, \forall y\, \forall z\, (x * (y \oplus z) = (x * y) \oplus (x * z))$

$\text{ł} \neq \mathrm{o}$

OF, the *theory of ordered fields* in $L(=; <, \oplus, *, \mathrm{o}, \text{ł})$ §XVII.E.2

F, **LO**, *OA*

$\forall x\, \forall y\, ((\mathrm{o} < x \wedge \mathrm{o} < y) \rightarrow \mathrm{o} < x * y)$ *the order-multiplication axiom*

RCF, the *theory of real closed fields* in $L(=; <, \oplus, *, \mathrm{o}, \text{ł})$ §XVII.E.3

OF

$\forall x\, \exists y\, (x = y^2 \vee x = {\sim}y^2)$

$\forall x_0\, \forall x_1 \ldots \forall x_{2n}\, \exists y\, (x_0 \oplus (x_1 * y) \oplus \cdots \oplus (x_{2n} * y^{2n}) \oplus y^{2n+1} = \mathrm{o})$

(a scheme of axioms, one for each natural number n, writing y for x_{2n+1})

TRCF, *Tarski's theory of real closed fields* in $L(=; <, \oplus, *, \mathrm{o}, \text{ł})$ §XVII.E.3

OF plus the scheme:

$\forall \ldots [\, (\exists x\, A(x) \wedge \exists w\, \forall x\, (A(x) \rightarrow w < x)) \rightarrow$
$\exists z\, \forall y\, (\forall x\, (A(x) \rightarrow y < x) \rightarrow y \leq z)]$

where A is any wff in which x is free and in which neither y nor z appear

DRCF, the *Dedekind theory of real closed fields* in $L(=; <, \oplus, *, \mathrm{o}, \text{ł})$ §XVII.E.3

OF

Dedekind's axiom$_1$ (scheme)

Note: **RCF** = **TRCF** = **DRCF**.

RCF$_2$, the *second-order theory of real closed fields* in $L_2(=; <, \oplus, *, \mathrm{o}, \text{ł})$ §XVII.E.3

OF

Dedekind's axiom

The least upper bound principle $\forall X\, ([\exists x\, X(x) \wedge \exists y\, \forall x\, (X(x) \rightarrow x \leq y)] \rightarrow$ §XVII.E.3
$\exists z\, (\forall x\, (X(x) \rightarrow x \leq z) \wedge \forall y (\forall x\, (X(x) \rightarrow x \leq y) \rightarrow z \leq y)))$

Note: **RCF$_2$** = Th_2(**OF** $\wedge$ *the least upper bound principle*) §XVII.Appendix

Geometry

One-dimensional geometry

B1, the *theory of one-dimensional betweenness* in L(=; B) §XVIII.B.1

B1-1 $\exists x \exists y \; x \neq y$

B1-2 $\forall x \forall y \; B(x, y, x) \rightarrow x = y$ *the one-point set is convex*

B1-3 $\forall x \forall y \forall z \; B(x, y, z) \rightarrow B(z, y, x)$ *direction is irrelevant*

B1-4 $\forall x \forall y \forall z \forall w \; B(x, y, z) \wedge B(x, z, w) \rightarrow B(y, z, w)$ *transitivity*

B1-5 $\forall x \forall y \exists z \; y \neq z \wedge B(x, y, z)$ *extension*

B1-6 $\forall x \forall z \; x \neq z \rightarrow \exists y (x \neq y \wedge y \neq z \wedge B(x, y, z))$ *density*

B1-7 $\forall x \forall y \forall z \; B(x, y, z) \vee B(y, z, x) \vee B(z, x, y)$ *dimension*

B1$_{01}$, *one-dimensional betweenness with parameters* in L(=; B, o, ł) §XVIII.B.4

B1
$o \neq ł$

Countable density-B $\exists X\, (\mathrm{Count}(X) \wedge \forall x \forall y \; x \neq y \rightarrow \exists z\, (X(z) \wedge B(x, z, y)))$ §XVIII.B.6

Dedekind's axiom-B (*continuity axiom for the line*) §XVIII.B.6

$\forall \ldots \forall X \forall Y ([\exists x \exists y\, (X(x) \wedge Y(y)) \wedge \exists w \forall x \forall y\, (X(x) \wedge Y(y)$
$\rightarrow B(w, x, y))] \rightarrow \exists z \forall x \forall y\, (X(x) \wedge Y(y) \rightarrow B(x, z, y)))$

B1*, the *second-order theory of one-dimensional betweenness* in L_2(=; B) §XVIII.B.6

B1
countable density-B
Dedekind's axiom-B

C1, a *theory of one-dimensional congruence* in L(=; ≡) §XVIII.C.1

C1-1 $\exists x \exists y \; (x \neq y)$

C1-2 $\forall x \forall y \; (xy \equiv yx)$ *reflexivity*

C1-3 $\forall x \forall y \; (xy \equiv zz \rightarrow x = y)$

C1-4 $\forall x \forall y \forall z \forall w \forall r \forall s \; (xy \equiv rs \wedge zw \equiv rs \rightarrow xy \equiv zw)$ *transitivity*

C1-5 $\forall x \forall z \exists y \; (xy \equiv yz)$ *midpoint*

C1-6 $\forall x \forall y \; (x \neq y \rightarrow \exists z\, (x \neq z \wedge xy \equiv yz))$ *extension*

C1-7 $\forall x \forall y \forall z \forall w \forall r \; (x \neq z \wedge y \neq w \wedge rx \equiv rz \wedge ry \equiv rw \rightarrow xy \equiv zw)$
dimension

Division axioms $C1\text{-}5_n \equiv_{\text{Def}} \forall x_0\ \forall x_n\ (x_0 \neq x_n \rightarrow A_n)$ for $n \geq 2$

$$A_2 \equiv_{\text{Def}} \exists x_1\ (x_0 x_1 \equiv x_1 x_2 \wedge x_0 \neq x_2)$$

$$A_{n+1} \equiv_{\text{Def}} \exists x_n\ (A_n \wedge x_{n-1} x_n \equiv x_n x_{n+1} \wedge x_{n-1} \neq x_{n+1}) \text{ for } n \geq 2$$

$C1_{Div}$, the *complete theory of one-dimensional congruence* in L(= ; ≡) §XVIII.C.6

C1

C1-5$_p$ for every prime $p > 2$

E1, *one-dimensional geometry* in L(= ; B, ≡) §XVIII.D.1

B1

$C1_{Div}$

BC1 $\forall x\, \forall y\, \forall z\, \forall r\, \forall s\, \forall t\ ((xy \equiv rs \wedge yz \equiv st \wedge zx \equiv tr) \wedge B(x, y, z))$
$\rightarrow B(r, s, t)$ *isometries preserve betweenness*

$E1_{01}$ in L(= ; B, ≡, o, ł) §XVIII.D.1

E1

o ≠ ł

Dedekind's axiom$_1$*-B*

$$\forall \ldots ([\exists x\, \exists y\, (C(x) \wedge D(y)) \wedge \exists w\, \forall x\, \forall y\, (C(x) \wedge D(y) \rightarrow B(w, x, y))]$$
$$\rightarrow \exists z\, \forall x\, \forall y\, (C(x) \wedge D(y)) \rightarrow B(x, z, y))$$

where C is any wff in which x is free, D is any wff in which y is free, and neither y nor z appear in C, and neither x nor z appear in D

E1-D, the *Dedekind axiomatization of one-dimensional geometry* in L(= ; B, ≡) §XVIII.D.3

B1

C1

BC1

Dedekind's axiom$_1$*-B*

Note: **E1-D = E1** .

Axioms *B-5, B-6, C-1, C-3* can be deleted (Exercise XVIII.D.4).

E1*, the second-order theory of *one-dimensional geometry* in L_2(= ; B, ≡) §XVIII.D.4

B1

C1

BC1

Dedekind's axiom-B

Two-dimensional Euclidean geometry

E2, *basic two-dimensional Euclidean geometry* in $\mathrm{L}(=;\mathrm{B},\equiv)$ §XIX.A

E1 $\exists x\, \exists y\ x \neq y$

E2 $\forall x\, \forall y\, \exists r\, \exists s\ \mathrm{B}(x, y, r) \wedge yx \equiv ys \wedge yr \equiv ys \wedge sx \equiv sr$

E3 $\forall x\, \forall z\, \exists y\ \mathrm{B}(x, y, z) \wedge yx \equiv yz$

E4 $\forall x\, \forall y\ \mathrm{B}(x, x, y)$

E5 $\forall x\, \forall y\ \mathrm{B}(x, y, x) \rightarrow x = y$

E6 $\forall x\, \forall y\, \forall z\, \forall r\, \forall s\ \mathrm{B}(x, y, z) \wedge \mathrm{B}(r, z, s) \rightarrow$
$\exists p\, (\mathrm{B}(r, y, p) \wedge \mathrm{B}(x, p, s))$

E7 $\forall x\, \forall y\, \forall z\, \forall w\ \mathrm{B}(x, y, z) \wedge \mathrm{B}(x, z, w) \rightarrow \mathrm{B}(y, z, w)$

E8 $\forall x\, \forall y\, \forall z\, \forall w\, \forall r\, \forall s\ (xy \equiv rs \wedge zw \equiv rs) \rightarrow xy \equiv zw$

E9 $\forall x\, \forall y\, \forall z\, \forall r\, \forall s\, \forall t\ ((xy \equiv rs \wedge yz \equiv st \wedge zx \equiv tr) \wedge \mathrm{B}(x, y, z))$
$\rightarrow \mathrm{B}(r, s, t)$

E10 $\forall x\, \forall y\, \forall z\, \forall r\, \forall s\, \forall t\ (\mathrm{B}(x, y, z) \wedge \mathrm{B}(r, s, t) \wedge$
$yx \equiv yz \wedge sx \equiv sz \wedge yr \equiv yt \wedge sr \equiv st) \rightarrow xr \equiv zt$

E11 $\forall x\, \forall y\, \forall z\, \exists r\, \exists s\ \mathrm{B}(x, y, r) \wedge \mathrm{B}(y, x, s) \wedge zr \equiv zs$

E12 $\forall x\, \forall y\, \forall z\, \forall r\, \forall s\ (x \neq y \wedge \mathrm{B}(x, y, z) \wedge xr \equiv xs \wedge yr \equiv ys) \rightarrow zr \equiv zs$

E13 $\forall x\, \forall y\, \forall z\, \forall r\, \forall s\ (r \neq s \wedge xr \equiv xs \wedge yr \equiv ys \wedge zr \equiv zs) \rightarrow$
$\mathrm{B}(x, y, z) \vee \mathrm{B}(y, z, x) \vee \mathrm{B}(z, x, y)$

E14 $\forall x\, \forall y\, \forall z\ \neg \mathrm{B}(x, y, z) \wedge \neg \mathrm{B}(y, z, x) \wedge \neg \mathrm{B}(z, x, y) \rightarrow$
$\exists r\, (rx \equiv ry \wedge ry \equiv rz)))$

$\mathbf{E2}_{\mathbf{01\underline{1}}}$ in $\mathrm{L}(=;\mathrm{B},\equiv,\mathrm{o},\text{ł},\underline{\text{ł}})$ §XIX.B.9

E2

$\mathrm{o} \neq \text{ł} \wedge \mathrm{o}\text{ł} \perp \mathrm{o}\underline{\text{ł}} \wedge \mathrm{o}\text{ł} \equiv \mathrm{o}\underline{\text{ł}}$

E2+, *elementary two-dimensional Euclidean geometry* in $\mathrm{L}(=;\mathrm{B},\equiv)$ §XIX.E

E2

Dedekind's axiom$_1$*-B*

Note: Betweenness (B) is definable (Exercise XIX.E.4).

E2*, *second-order theory of two-dimensional Euclidean geometry* in $\mathrm{L}_2(=;\mathrm{B},\equiv)$ §XIX.E

E2

Dedekind's axiom-B

Classical Predicate Logic with Non-Referring Names

Classical predicate logic with equality and non-referring names §XIX.E
in $L(\neg, \rightarrow, \forall, \exists, =; P_0, P_1, \ldots, f_0, f_1, \ldots, c_0, c_1, \ldots)$

I. Propositional axioms

As for classical predicate logic with referring names (p. 476 above).

II. Axioms governing ∀

As for classical predicate logic with referring names (p. 476 above).

III. Axioms governing the relation between ∀ and ∃

5.a. $\forall \ldots (\exists x\, A \rightarrow \neg \forall x \neg A)$

6. $\forall \ldots (\forall x\, A(x) \rightarrow \exists x\, A(x))$

IV. Axioms for equality

7. $\forall x\, (x = x)$

8′. For every n-ary atomic predicate P,
$\forall \ldots (\bigwedge_i (t_i = u_i) \rightarrow (P(t_1, \ldots, t_n) \rightarrow P(u_1, \ldots, u_n))$

V. Axioms for functions §XIX.F.4

9′. For every n-ary function symbol f,
$\forall \ldots (\bigwedge_i (t_i = u_i) \rightarrow (f(t_1, \ldots, t_n) = f(u_1, \ldots, u_n))$.

VI. Axioms for non-referring names

10. $\forall \ldots (\forall y\, (\exists x\, (x = y) \rightarrow \neg B(y/x)) \rightarrow \neg \exists x\, B(x))$

11. When x is not free in A,

a. $\forall \ldots (\exists x\, A \rightarrow A)$ *superfluous quantification*

b. $\forall \ldots (A \rightarrow \exists x\, A)$

12. $\forall \ldots ((A(t/x) \wedge \exists x\, (x = t)) \rightarrow \exists x\, A(x))$ *existential generalization for referring names*

Rule $\dfrac{A,\ A \rightarrow B}{B}$ where A and B are closed formulas

Classical Predicate Logic with Name Quantification

As for second-order logic, all systems are given in terms of semantic consequence.

Add to the language of first-order classical predicate logic:

name variables a_i for $i \geq 0$

quantifiers $\forall a_i$, $\exists a_i$

L_{qn}, the *first-order logic of quantifying over names* §XV.F

QN, the theory of *named arithmetic* in $L_{qn}(= ; S, \oplus, *, o)$ §XV.F

Q

AN $\forall x_1 \exists a_0 (x_1 = a_0)$

OF-n, the theory of *ordered fields* in $L_{qn}(= ; <, \oplus, *, \sim, ^{-1}, o, \mathfrak{l})$ p. 325

OF

$\forall x (x \oplus (\sim x) = o)$

$\forall x (x \neq o \rightarrow x * x^{-1} = \mathfrak{l}) \wedge (o^{-1} = o)$

Rational N-Arithmetic in $L_{qn}(= ; <, \oplus, *, \sim, ^{-1}, o, \mathfrak{l})$ p. 325

OF-n

AN

Real N-Arithmetic, the *theory of real closed subfields of the reals* in $L_{qn}(= ; <, \oplus, *, \sim, ^{-1}, o, \mathfrak{l})$ p. 325

RCF (with **OF/n** replacing **OF**)

AN

$\forall x \forall y (P(y) \rightarrow \exists q(o < (q \sim x) \wedge (q \sim x) < y^{-1}))$ *approximation*

Bibliography

- Only works cited in the text or elsewhere in the bibliography are listed.
- Citations are to the most recent English reference listed, unless otherwise noted.

ARNAULD, Antoine, and Pierre NICOLE
1662 *La Logique ou l'art de penser*
Edition of 1683 translated as *Logic or the Art of Thinking*
by Jill Vance Buroker, Cambridge University Press, 1996.

ARTIN, Emil, and Otto SCHREIER
1927 Algebraische Konstruktion reeler Körper
Abh. Math. Sem. Univ. Hamburg, vol. 5, pp. 85–99.

BELL, J., and A. B. SLOMSON
1969 *Models and Ultraproducts*
North-Holland.

BENCIVENGA, Ermanno
1986 Free logics
Chapter 6 of Volume III of GABBAY and GUENTHER, pp. 373–426.

BERNAYS, Paul
1926 Axiomatische Untersuchung des Aussagen-Kalküls der *Principia mathematica*
Mathematische Zeitschrift, vol. 25, pp. 305–320.

BEZBORUAH, A., and J. C. SHEPHERDSON
1976 Gödel's second incompleteness for **Q**
Journal of Symbolic Logic, vol. 41, pp. 503–512.

BOOLE, George
1854 *The Laws of Thought*
Macmillan. Reprinted 1954, Dover.

BORSUK, Karol, and Wanda SZMIELEW
1960 *Foundations of Geometry*
North-Holland. Translation of *Podstawy Geometrii*,
Panstwowe Wydawnictwo Naukowe (PWN), Warsaw, 1955.

BOYER, Carl B.
1968 *A History of Mathematics*
Wiley.

BURGESS, John P.
1986 The truth is never simple
The Journal of Symbolic Logic, vol. 51, pp. 663–681.

CANTOR, Georg
1895–1897 *Contributions to the Founding of the Theory of Transfinite Numbers*
Translated from the German with an introduction by Phillip E. B. Jourdain,
The Open Court Co., 1915. Reprinted 1955, Dover.

COPI, Irving M.
1971 *The Theory of Logical Types*
Routledge & Kegan Paul.

DEDEKIND, Richard
1872 *Stetigkeit und irrationale Zahlen*
Translated as 'Continuity and irrational numbers' in Dedekind, *Essays on the Theory of Numbers*, The Open Court Co., 1901. Reprinted 1963, Dover.
DE MORGAN, Augustus
1847 *Formal Logic*
Reprinted 1926, The Open Court Co.
DREBEN, Burton and Jean VAN HEIJENOORT
1986 Note to Gödel's dissertation
In GÖDEL, 1986, pp. 44–59.
ENDERTON, Herbert B.
1972 *A Mathematical Introduction to Logic*
Academic Press.
EPSTEIN, Richard L.
1990 *Propositional Logics* (The Semantic Foundations of Logic)
Kluwer. 2nd ed. with corrections, Wadsworth, 2001.
1992 A theory of truth based on a medieval solution to the liar paradox
History and Philosophy of Logic, vol. 13, pp. 149–177.
1994 *Predicate Logic* (The Semantic Foundations of Logic)
Oxford University Press. Reprinted 2001, Wadsworth.
2001 *Five Ways of Saying "Therefore"*
Wadsworth.
2004 On theories and models with applications to economics
Bulletin of Advanced Reasoning and Knowledge, vol. 2, pp. 77–98.
(Also online at <www.AdvancedReasoningForum.org>.)
EPSTEIN, Richard L., and Walter A. CARNIELLI
1989 *Computability*
Wadsworth & Brooks/Cole. 2nd ed., 2000, Wadsworth.
EUCLID. *See* HEATH.
FALLIS, Don
1998 Review of Hersh, 'Proving is convincing and explaining', 1993
The Journal of Symbolic Logic, vol. 63, pp. 1196–1200.
FEFERMAN, S.
1960 Arithmetization of metamathematics in a general setting
Fundamenta Mathematicae, vol. 49, pp. 35–92.
FLAGG, Robert
1978 On the independence of the Bigos-Kalmár axioms for sentential calculus
Notre Dame Journal of Formal Logic, vol. XIX, pp. 285–288.
FRAENKEL, Abraham A., Yehoshua BAR-HILLEL, and Azriel LEVY
1973 *Foundations of Set Theory*
North-Holland.
FREGE, Gottlob
1879 *Begriffschrift*
L. Nebert, Halle. Translated by S. Bauer-Mengelberg as *Begriffschrift, a formula language* in VAN HEIJENOORT, 1967.
GABBAY, D., and F. GUENTHNER (eds.)
1983–1989 *Handbook of Philosophical Logic, Volumes I–IV*
D. Reidel (Kluwer).

GÖDEL, Kurt
1929 *Über die Vollständigkeit des Logikkalküls*
Doctoral dissertation, University of Vienna.
Reprinted with an English translation as 'On the completeness of the calculus of logic' in GÖDEL, 1986, pp. 60–101.
1930 Die Vollständigkeit der Axiome des logischen Funktionenkalküls
Monatshefte für Mathematik und Physik, vol. 37, pp. 349–360.
Reprinted with an English translation as 'The completeness of the axioms of the functional calculus of logic' in GÖDEL, 1986, pp. 102–123.
1986 *Collected Works, Volume I*
ed. S. Feferman et al., Oxford University Press.

GOLDFARB, Warren D.
1979 Logic in the twenties: the nature of the quantifier
The Journal of Symbolic Logic, vol. 44, pp. 351–369.

GRZEGORCZYK, Andrzej
1974 *An Outline of Mathematical Logic*
D. Reidel. Translation of *Zarys Logiki Matematycznej*, PWN, Warsaw, 1969.

GUPTA, Anil
1982 Truth and paradox
Journal of Philosophical Logic, vol. 11, pp. 1–60.

HAUSDORFF, Felix
1957 *Set Theory*
Chelsea. Translation of *Mengelehre*, 3rd ed., J. R. Autman, de Gruyter, 1937.

HAZEN, Allen
1983 Predicative logics
Chapter 5 of Volume I of GABBAY and GUENTHER, pp. 1–131.

HEATH, Thomas L., translator and commentator
1956 *The Thirteen Books of Euclid*
Dover, 1956.

HELLMAN, Geoffrey
1985 Review of papers by Martin and Woodruff, Kripke, Gupta, and Herzberger
The Journal of Symbolic Logic, vol. 50, pp. 1068–1071.

HENKIN, Leon
1954 Boolean representation through propositional calculus
Fundamenta Mathematicae, vol. 41, pp. 89–96.

HERSTEIN, I. N.
1964 *Topics in Algebra*
Blaisdell/Ginn and Co.

HILBERT, David
1899 *Grundlagen der Geometrie*
Translated as *The Foundations of Geometry* by E. J. Townshend, The Open Court Co., 1902.

HUGHES, G.E.
1982 *John Buridan on Self-Reference*
Cambridge University Press.

ISAACSON, Daniel
1996 Arithmetical truth and hidden higher-order concepts
In *The Philosophy of Mathematics*, ed. W. Hart, Oxford U. Press, pp. 203–224.

KALMÁR, László
1935 Über die Axiomatisierbarkeit des Aussagenkalküls
Acta Scientiarum Mathematicarum, vol. 7, pp. 222–243.
KIRKHAM, Richard L.
1992 *Theories of Truth*
MIT Press.
KLEENE, Stephen Cole
1967 *Mathematical Logic*
Wiley.
KORDOS, Marek, and Lesław SZCZERBA
1976 *Geometria dla nauczycieli*
Panstwowe Wydawnictwo Naukowe, Warsaw.
KREISEL, Georg
1967 Informal rigour and completeness proofs
In *Problems in Philosophy of Mathematics*, ed. I. Lakatos, North-Holland.
KREISEL, Georg, and J. L. KRIVINE
1967 *Elements of Mathematical Logic*
North-Holland.
KRIPKE, Saul
1975 Outline of a theory of truth
Journal of Philosophy, vol. 82 (19), pp. 690–716. Reprinted in MARTIN.
1976 Is there a problem about substitutional quantification?
In *Truth and Meaning*, eds. G. Evans and J. McDowell, Clarendon, Oxford.
LAMBERT, Karel
1991 *Philosophical Applications of Free Logic*
Oxford University Press.
LEWANDOWSKI, Grzegorz M., and Lesław SZCZERBA
1980 An axiom system for the one-dimensional theory of the betweenness relation
Warsaw University, Białystok Division, Institute of Mathematics, Internal Report No. 5.
LIPSCHUTZ, Seymour
1964 *Set Theory*
Schaum's Outline Series in Mathematics, McGraw-Hill.
ŁOS, Jerzy
1951 An algebraic proof of completeness for the two-valued propositional calculus
Colloquium Mathematicum, vol. 2, pp. 236–240.
MARKOV, A. A.
1954 *The Theory of Algorithms*
Tr. Mat. Inst. Steklov., XLII. English translation, Israel Program for Scientific Translations Ltd., Jerusalem, 1971.
MARTIN, Robert L.
1984 *Recent Essays on Truth and the Liar Paradox*
Oxford University Press.
MATES, Benson
1965 *Elementary Logic*
Oxford University Press. 2nd ed., 1972.

MENDELSON, Elliott
1973 *Number Systems and the Foundations of Analysis*
Academic Press.
1987 *Introduction to Mathematical Logic*
Wadsworth & Brooks/Cole, 3rd ed. (First ed., 1964, D. Van Nostrand.)
MILL, John Stuart
1874 *A System of Logic*
Harper & Brothers, 8th ed.
MOORE, Gregory H.
1982 *Zermelo's Axiom of Choice*
Springer-Verlag.
MORSCHER, Edgar, and Peter SIMON
2001 Free logic: A fifty-year history and an open future
In *New Essays in Free Logic*, eds. E. Morscher and A Hieke, Kluwer.
MOSZYNSKA, Maria
1986 Review of Schwabhäuser, Tarski, and Szmielew.
The Journal of Symbolic Logic, vol. 51, pp. 1073–1075.
PARIS, J., and L. HARRINGTON
1977 A mathematical incompleteness in Peano arithmetic
In *Handbook of Mathematical Logic*, ed. J. Barwise, North-Holland, pp. 1133–1142.
PEANO, Giuseppe
1889 *Arithmetices principia, nova methodo exposita*
Bocca, Turin. Translated as 'The principles of arithmetic, presented by a new method', in VAN HEIJENOORT, 1967, pp. 85–97.
POST, Emil L.
1921 Introduction to a general theory of elementary propositions
American Journal of Mathematics, vol. 43, pp. 163–185.
Reprinted in VAN HEIJENOORT, 1967, pp. 264–283.
ROBINSON, Abraham
1951 *On the Metamathematics of Algebra*
North-Holland.
1956 *Complete Theories*
North-Holland.
1974 *Introduction to Model Theory and to the Metamathematics of Algebra*
North-Holland.
ROBINSON, Julia
1949 Definability and decision problems in arithmetic
The Journal of Symbolic Logic, vol. 14, pp. 98–114.
1965 The decision problem for fields
In *Theory of Models*, ed. J. W. Addison, North-Holland, pp. 299–311.
RUDIN, Walter
1964 *Principles of Real Analysis*
McGraw-Hill, 2nd ed.
RUSSELL, Bertrand. *See* WHITEHEAD and RUSSELL
SHAPIRO, Stewart
1991 *Foundations without Foundationalism*
Clarendon Press, Oxford.

SHAPIRO, Stewart—*contd.*
1997 *Philosophy of Mathematics: Structure and Ontology*
Oxford University Press.
SCHWABHÄUSER, Wolfram, Alfred TARSKI, and Wanda SZMIELEW
1983 *Metamathematische Methoden in der Geometrie*
Springer-Verlag.
SHOENFIELD, Joseph
1967 *Mathematical Logic*
Addison-Wesley.
SIMPSON, Stephen G.
1999 *Subsystems of Second-Order Arithmetic*
Springer-Verlag.
SPIEGEL, Murray R.
1964 *Complex Variables*
Schaum's Outline Series in Mathematics, McGraw-Hill.
SURMA, Stanisław J.
1973 A (ed.) *Studies in the History of Mathematical Logic*
Polish Academy of Sciences, Warsaw.
1973 B A history of the significant methods of proving Post's theorem about the completeness of the classical propositional calculus
In SURMA, 1973 A, pp. 19–32.
SZCZERBA, Lesław
1977 Interpretability of elementary theories
In *Logic, Foundations of Mathematics and Computability Theory*, eds. R. I. Butts and J. Hintikka, D. Reidel, pp. 129–145.
1979 Elementary means of creating entities
In *Conference on Technology of Science*: *Mathematics in Social Sciences and Semiotics*, ed. R. Matuszewski, (Towarzystwo Naukowe Płockie), pp. 13–27.
1980 Interpretations with parameters
Zeitschrift für mathematische Logik und Grundlagen der Mathematik, vol. 26, pp. 35–39.
1986 Tarski and geometry
The Journal of Symbolic Logic, vol. 51, pp. 907–941.
See also KORDOS and SZCZERBA; LEWANDOWSKI and SZCZERBA.
SZMIELEW, Wanda
1948 Decision problem in group theory.
Proceedings of the Tenth International Congress of Philosophy, vol. 2, pp. 763–766.
See also BORSUK and SZMIELEW; SCHWABHÄUSER, TARSKI and SZMIELEW.
TAKEUTI, Gaisi
1975 *Proof Theory*
North-Holland. 2nd ed., 1987.
TARSKI, Alfred
1930 Über einige fundamentale Begriffe der Metamathematik
Comptes Rendus des séances de la Société des Sciences et des Lettres de Varsovie, cl. iii, vol. 23, pp. 22–29. Translated as 'On some fundamental concepts of metamathematics' in TARSKI, 1956, pp. 30–37.

TARSKI, Alfred—*contd.*
1933 The concept of truth in formalized languages
Reprinted in TARSKI, 1956, pp. 152–278, where a detailed publication history of it is given.
1940 *The completeness of elementary algebra and geometry*
Page proofs only, Les Editions Hermann, Paris. Published 1967 by Institut Blaise Pascal, Paris, 100 copies only. Reprinted in TARSKI, 1986, vol. 4, pp. 289–346.
1941 *Introduction to Logic and to the Methodology of Deductive Sciences*
Oxford University Press. 2nd ed., revised 1946.
1951 *A Decision Method for Elementary Algebra and Geometry*
University of California Press. Also in TARSKI, 1986, vol. 3, pp. 297–367.
1953 Undecidability of the elementary theory of groups.
In TARSKI, MOSTOWSKI, AND ROBINSON, pp. 77–87.
1956 *Logic, Semantics, Metamathematics*
2nd ed. with corrections (1983), ed. J. Corcoran, Hackett.
1959 What is elementary geometry?
The Axiomatic Method with Special Reference to Geometry and Physics, eds. L. Henkin, P. Suppes, and A. Tarski, North-Holland, pp. 16–29.
1986 *Collected Papers*
In four volumes, eds. Steven R. Givant and Ralph N. McKenzie, Birkhäuser.
See also SCHWABHÄUSER, TARSKI and SZMIELEW.

TARSKI, Alfred, and Steven GIVANT
1999 Tarski's system of geometry
The Bulletin of Symbolic Logic, vol. 5, pp. 175–214.

TARSKI, Alfred, Andrzej MOSTOWSKI, and Raphael M. ROBINSON
1953 *Undecidable Theories*
North-Holland.

VAN DER WAERDEN, B. L.
1953 *Modern Algebra*
Translated by Fred Blum, Frederick Ungar, New York.

VAN HEIJENOORT, Jean, ed.
1967 *From Frege to Gödel: A Source Book in Mathematical Logic, 1879–1931*
Harvard University Press.

WHATELY, Richard
1827 *Elements of Logic*
J. Mawman, London.

WHITEHEAD, Alfred North, and Bertrand RUSSELL
1910–13 *Principia Mathematica*
Cambridge University Press.

YABLO, Stephen
1985 Truth and reflection
Journal of Philosophical Logic, vol. 14, pp. 297–349.

ZYGMUNT, Jan
1973 A survey of the methods of proof of the Gödel-Malcev's completeness theorem
In SURMA, 1973A, pp. 165–237.

Index of Notation

Metavariables

Logical Semantics

Second-Order Predicate Logic

Set Theory

Sets of Numbers

Syntactic Symbols and Abbreviations in Formal Systems

Semantic Symbols in Geometry

Index

- *Italicized* page numbers indicate a definition, statement, or quote; 'n' indicates a footnote.
- Exercises, noted with an 'e', are indexed only for material not found elsewhere.
- See *Index of Notation* for all notation.
- See *Summary of Formal Systems* for list of all formal systems.
- See Chapter XX.B for a list of all translations in the text.

Ralph